Plant Cell Vacuoles

AN INTRODUCTION

PLANT CELL VACUOLES

AN INTRODUCTION

Deepesh N. De

CSIRO
PUBLISHING

National Library of Australia Cataloguing-in-Publication entry

De, Deepesh Narayan.
   Plant cell vacuoles : an introduction

   Bibliography.
   Includes index.
   ISBN 0 643 06254 8.

   1. Plant vacuoles. I. Title.

   571.655

© CSIRO 2000

This book is available from:
CSIRO PUBLISHING
PO Box 1139 (150 Oxford Street)
Collingwood VIC 3066
Australia

Tel: (03) 9662 7666   Int: +(613) 9662 7666
Fax: (03) 9662 7555   Int: + (613) 9662 7555
Email: sales@publish.csiro.au
http://www.publish.csiro.au

Printed in Australia by Brown Prior Anderson

*Dedicated
to
The Directors and the Faculty,
past and present,
of the
Indian Institute of Technology
Kharagpur*

PREFACE

Every author has excuses, and not necessarily lame ones, for writing a book and inflicting it on unsuspecting readers who are already burdened with the exploding amount of information on their topic of study. However, this volume is expected to lessen the burden of readers to some extent, if their interest lies, even remotely, with plant cell structure and function. I may be exonerated from guilt even though many reviews are available, including those in *Advances in Botanical Research* (Vol. 25) and the post-Gutenberg era of electronic communication, only due to the fact that a comprehensive treatise or a *vade mecum* is not available on this ubiquitous, multifaceted and indispensable organelle of the plant cell.

For more than four decades, eat DNA–drink DNA–sleep DNA has mesmerised biologists and many other scientists from crystallographers to geographers to forensic experts. No wonder the vacuole was almost thrown into the dustbin of history. The neglect meted out to the vacuole in cell biology can be gauged by its definition as a 'large fluid-filled organelle' in a textbook on molecular cell biology (1986), the authorship of which includes a Nobel Laureate (Darnell, Lodish and Baltimore). Out of 1100 pages, only half a page is devoted to the vacuole and the word 'tonoplast' is not even mentioned. Another textbook on Molecular Biology of the Cell (1994), whose authors include the Nobel Laureate J.D. Watson) defines a vacuole as a single-membrane-bound vesicle functioning in 'space-filling and also intracellular digestion'. Out of 1300 pages, barely one page is devoted to vacuoles. This unfortunate state of affairs has motivated me to write this treatise on an organelle without which a plant cell cannot long survive. Whereas certain plant cells are known to contain no mitochondrion, plastid nor even a nucleus, no cell exists without a vacuole or at least a provacuole. The benign neglect of vacuoles has been harmful to students and perhaps caused the potential of vacuoles in important cells of economic plants to be under-exploited.

Thanks to a few die-hard souls and editors of a few journals—especially *Plant Physiology*, *The Plant Cell* and *Planta*—research on vacuoles is very much alive. It is contributing greatly to our knowledge of cell function and will perhaps show that it can be ignored at the expense of maximising plant productivity for human welfare.

The renewed interest and rate of growth of research on vacuoles may be estimated from the total number of publications in the frontier areas presented in Chapter IV (Tonoplast composition), Chapter VI (Biogenesis), Chapter VII (Functions) and Chapter VIII (Retrospect and Prospect) for the last two decades. During the 10 years from 1978 to 1987 some 217 papers were published, compared to 383 papers in the following 10 years (1988 to 1997). Moreover, the quality and density of information in recent research publications has been of a very high standard. An enormous amount of information has been obtained by the assiduous work of a number of laboratories around the globe and for this reason the organelle needs to be discussed and respected from every aspect of cell biology and cell physiology. The landmarks in this period have been (a) the discovery of aquaporin in the tonoplast, (b) the elucidation of protein targeting to the vacuoles, (c) the establishment of the role of Golgi vesicles in vacuolar development, (d) the use of the patch-clamp technique for discernment of ion channels in the tonoplast, (e) comprehension of the role of the vacuole in detoxification of xenobiotics and in phytoremediation of heavy metals and (f) the use of tools of molecular genetics for engineering of the tonoplast proteins.

My intellectual foundation was laid in the early 1950s at the Cytogenetics Laboratory of the Botany Department of Calcutta University where Professor Arun Kumar Sharma taught me to look at chromosomes in preparations of meristematic cells in which vacuoles could not to be seen. My tryst with the vacuole was perhaps fixed in the mid-1950s when electron microscopy of *Tradescantia* staminal hair cells in Dr. Berwind P. Kaufmann's laboratory in the Carnegie Institution of Washington at Cold Spring Harbor revealed that the vacuoles were not separated from the rest of the cytoplasm by a phase boundary, but by a distinct membrane — the tonoplast. But what struck me most was that occasionally certain living cells of the staminal hair lacked the nucleus which had migrated to the next cell making it binucleate and the anucleate cell remained alive with active cyclosis around the large vacuole. The flame of this inexplicable observation was secretly nurtured in my subconscious for a decade until it was kindled by observation of the dramatic vacuolisation of the callus culture cells at J. Straub's laboratory in the Max-Planck-Institut für Züchtungsforschung in Cologne. I began writing this treatise two decades later, while attempting to comprehend the chromosome cycle of the highly vacuolate unicellular alga *Acetabularia* during a sabbatical leave at the Max-Planck-Institut für Zellbiologie in Ladenburg, Heidelberg. At that time, the late 1980s, it was barely possible to write a full-length text on this subject.

In view of the tremendous surge in the genetic engineering of plants for commercial products undertaken by biotechology industries and seed and agricultural companies, a comprehension of the functions and possibilities of vacuole manipulation has become imperative, since most of the targets of improvement—like overproduction and storage of fructans, starch, pharmaceuticals and resistance to disease, herbicide, drought and soil salinity—directly involve vacuoles. Thus, we can no longer afford to ignore the vacuole.

To make it a useful text all the available information, from the earliest to the most recent (up to June 1999), on all aspects, including methodologies, occurrence and diversity, tonoplast structure and biochemistry, molecular biology of biogenesis and diverse functions, is presented in a concise way. No course in cell biology, cell physiology or plant physiology can be complete without dealing adequately with the structure and functions of vacuoles.

Chapter I depicts how early workers with primitive microscopes were able to develop remarkable insights and raise fundamental questions on various aspects of the apparently not-so-important aqueous sac. Chapter II outlines various *Methodologies* which are used especially for vacuoles. The simple microscopic methods are presented in detail so that they are useful for beginners as well as researchers who may check a point quickly in different systems. The biophysical methods which are evolving fast and meant for specialists are also briefly described. Chapter III makes a complete survey of the *Occurrence and Distribution* of vacuoles in all the tissues of all groups of the plant kingdom. This survey, demonstrating the vacuole's ommipresence and diversity, would help researchers looking into an alternate system of vacuoles. The *Ultrastructure and Chemical Composition* of the vacuolar membrane in Chapter IV provides hints about the possible function of the vacuole and differentiates the tonoplast from the other membranes of the cell. The survey of *Vacuolar Contents* in Chapter V is an account of the diversity of substances, from the simplest to the most complex, exhibiting the plurality in storage as well as function. The master table on the distribution of the enzymes in the tonoplast and sap of various plants is expected to be a search tool. Chapter VI on *Biogenesis and Development* of vacuoles discusses the diversity of origin, as well as programmed ultrastructural and biochemical steps in the gradual development of the vacuole. Chapter VII on *Functions* of vacuoles reveals how the cell uses a simple compartment to satisfy its manifold needs. Finally, the overview of the organelle in Chapter VIII on *Retrospect and Prospect* brings out its uniqueness as well as its homology with other cellular organelles, its evolution and the possibilities of genetic engineering of tonoplast. Exploration into the immense possibilities of manipulating the vacuole for enhancing plant productivity has just commenced.

I did not dare to define the vacuole at the beginning of the book but did so only in the last chapter, since some biologists may have had reservations about calling it a distinct organelle. I sincerely believe that this treatise will indicate that the vacuole is not an abstruse structure fit only for the attention of specialists, but a prominent entity for plant scientists in general. Indeed, the vacuole has come of age to become a discipline or field of study which may be termed **vacuology**.

This book on vacuology may help obviate the necessity of searching or poring over thousands of pages scattered among hundreds of research publications of pre-CD-ROM–WWW literature. Adequate references are provided at the end of each chapter which may enable the more inquisitive reader to gain rapid access to the original research works. I have limited the illustrations to an appropriate level to validate and/or clarify the text wherever necessary. The original diagrams, which are gratefully acknowledged, have been provided in order to project the views of the workers.

In preparing the book, I have assumed that readers have taken courses in introductory plant sciences or biology with elementary knowledge of cell structure and function and general biochemistry.

I am aware that this book is not likely to be a textbook for a whole course, but I expect it to contribute substantially to any course on cell biology or plant cell physiology. I will feel rewarded if a section of teachers and students finds it useful and also if the book arouses more interest in vacuoles amongst biologists or at least overcomes some indifference to the topic.

In the expanding field of vacuology it may not always have been possible for me to keep pace with each step of advancement of knowledge and I may also have missed a point or two. I admit that errors may have crept in and some results misinterpreted. However, it has not been possible for me to have first-hand experience in diverse fields where the latest research has been undertaken with very modern techniques. If any important work has been omitted I beg the indulgence of the authors concerned.

I am dedicating this book to the Directors and Faculty of the Indian Institute of Technology, Kharagpur which nurtured me for more than three and a half decades through its academic atmosphere and tranquil life without the worry of daily bread or shelter.

I am grateful to numerous friends and fellow scientists who provided me with encouragement, suggestions and support with the literature search. I am thankful to all my students and colleagues in our Department of Agricultural and Food Engineering for discussion and material help. I am especially thankful to Prof. Rabindra Krishna Mukherjee of this Department and Prof. Hrishikesh Chatterjee of Vivekananda College, Calcutta for reading some of the chapters. I am also indebted to Prof. Pradyot Bhanja of Burdwan University who has shown keen interest in vacuole research. I am grateful to a number of vacuologists including Dr. Natasha Raikhel, Dr. C.A. Ryan, Dr. Y. Ohsumi, Dr. I.K. Vasil, Prof. I. Tsekos, Prof. M. Yamada, Prof. A.K. Bal and Dr. Rajeev Arora for providing me with original micrographs, and to a host of authors and publishers who have kindly permitted me to use the very revealing figures reproduced in the book.

My gratitude knows no bounds for the first and only lady in my life, my wife Mira, who is all-in-one to me, tolerating my negligence of family life even to the point of correcting the typescript. I am enormously indebted to the two angels in the form of my daughters, Babon (Srijata) and Chhoton (Sudeshna) and a seraphim, the latter's husband Zhenja (Dr. Evgenij Barsukov) who sang the tune, '— shall overcome someday —' from the beginning of the book until its final stage, and who searched the Web for the latest writings on vacuoles, although it is completely alien to his field of research. The long-hand writing would not have seen the light of the typescript were it not for the PC expertise and patience of Mr. Ashutosh Biswas. I am also thankful to Mr. Rajesh Kumar Mallah for helping me with the internet.

Mr Kevin Jeans, publisher, and Ms Meredith Lewis, editorial and production manager, CSIRO Publishing, deserve my special thanks for sound advice and constant support in publishing the book.

D.N. De
Kharagpur
July, 1999

Contents

Preface — vii

I Early studies — 1
1.1 Discovery of vacuoles — 1
1.2 Vacuolar contents — 1
1.3 Physiology and biogenesis — 2
1.4 Nature of vacuoles — 3
1.5 Summary — 5
1.6 References — 5

II Methodologies — 8
2.1 *In situ* methods — 8
    2.1.1 Light microscopy — 8
    2.1.2 Fluorescence microscopy — 9
        2.1.2.1 Autofluorescence — 10
        2.1.2.2 Secondary fluorescence — 10
    2.1.3 Electron microscopy — 12
    2.1.4 Cytochemical localisation — 13
        2.1.4.1 Protocol for selected enzymes — 14
2.2 *In vitro* methods — 17
    2.2.1 Isolation of intact vacuoles — 17
        2.2.1.1 Early methods — 17
        2.2.1.2 Compact tissues — 18
        2.2.1.3 Suspension cells — 19
        2.2.1.4 Leaf tissues — 19
        2.2.1.5 Stomatal guard cells — 20
        2.2.1.6 Yeast cells — 22
    2.2.2 Isolation of vacuolar sap and tonoplast — 22
        2.2.2.1 Extraction of sap — 22
        2.2.2.2 Tonoplast from intact vacuole — 23
    2.2.3 Direct extraction of tonoplast fraction — 23
        2.2.3.1 Discontinuous gradient — 24
        2.2.3.2 Continuous gradient — 24
    2.2.4 Tonoplast markers — 25
2.3 Physiological methods — 25
    2.3.1 Determination of pH — 25
        2.3.1.1 Weak acid/base distribution — 25
        2.3.1.2 Fluorescence probes — 26
        2.3.1.3 Microelectrodes — 26
        2.3.1.4 Phosphorus NMR spectroscopy — 27

	2.3.2 Determination of osmotic values	27
	2.3.3 Special methods	28
	2.3.3.1 Patch-clamp technique	29

2.4 Summary — 29
- 2.4.1 *In situ* methods — 29
- 2.4.2 *In vitro* methods — 30
- 2.4.3 Physiological methods — 30

2.5 References — 31

III Occurrence and distribution of vacuoles — 38

3.1 Procaryotes — 38
3.2 Algae — 39
3.3 Fungi — 42
3.4 Bryophyta — 45
3.5 Pteridophyta — 46
3.6 Gymnosperms — 47
3.7 Angiosperms — 47
- 3.7.1 Meristem — 47
- 3.7.2 Epidermis — 48
- 3.7.3 Cortex — 48
- 3.7.4 Endodermis — 48
- 3.7.5 Vascular tissues — 48
- 3.7.6 Reproductive tissues — 50
- 3.7.7 Specialised cells — 54

3.8 Shape, number and volume — 56
3.9 Summary — 59
3.10 References — 60

IV Ultrastructure and chemical composition of tonoplast — 64

4.1 Ultrastructure — 64
4.2 Chemical analysis of tonoplast — 67
- 4.2.1 Lipids — 67
- 4.2.2 Proteins — 67
- 4.2.3 Tonoplast enzymes — 70
  - 4.2.3.1 Proton ATPase — 70
  - 4.2.3.2 Proton-PPase: — 72
  - 4.2.3.3 Ca^{2+}-ATPase — 72

4.3 Summary — 73
- 4.3.1 Ultrastructure — 73
- 4.3.2 Chemical composition — 74

4.4 References — 74

V Vacuolar contents — 79

5.1 Solids and particulate inclusions — 79
- 5.1.1 Crystals — 79
- 5.1.2 Aleurone grains — 80

		5.1.3 Lipid bodies	83
		5.1.4 Isoprenoids, polyterpenes, rubber, resin, and essential oil	84
		5.1.5 Volutin and polyphosphate	84
	5.2	Colloidal or dissolved substances	85
		5.2.1 Phenolics	85
		5.2.2 Anthocyanin and anthoxanthin pigments	86
		5.2.3 Alkaloids and steroids	88
		5.2.4 Sugars	88
		5.2.5 Organic acids	89
		5.2.6 Amino acids	89
		5.2.7 Hormones	90
		5.2.8 Inorganic ions	90
		5.2.9 Miscellaneous substances	93
	5.3	Enzymes	93
	5.4	Summary	103
		5.4.1 Solids and particulate inclusions	103
		5.4.2 Colloidal and dissolved substances	103
		5.4.3 Enzymes	104
	5.5	References	104

VI Biogenesis and development of vacuoles 115

6.1	Early views	115
6.2	*De novo* origin	115
	6.2.1 *De novo* liposome	115
	6.2.2 *De novo* Golgi bodies	116
	6.2.3 Growth of tonoplast	116
	6.2.3.1 Regeneration of tonoplast	117
	6.2.3.2 Synthesis and direct transfer of lipids to tonoplast	117
	6.2.3.3 Synthesis and direct transfer of proteins to tonoplast	119
6.3	Origin through lipid utilisation	119
6.4	Origin of aleurone vacuoles	121
6.5	Origin through lytic processes	122
6.6	Origin from membranous components	124
	6.6.1 Origin from endoplasmic reticulum	124
	6.6.2 Origin from plasma membrane	125
	6.6.3 Origin from Golgi bodies	125
	6.6.3.1 Golgi-vesicle-vacuole	126
	6.6.3.2 ER-Golgi-vacuole	126
6.7	Inheritance of vacuoles in yeast	128
6.8	Biochemical aspects of biogenesis	128
	6.8.1 Vesicular transport	128
	6.8.2 SNARE hypothesis on vesicular transport	131
	6.8.3 Vesicular transport in yeast: SNARE mechanism	132
	6.8.4 SNARE mechanism in plants	133
	6.8.5 Mechanics of fusion	135
	6.8.6 Prevacuoles as intermediate structures	137
	6.8.7 Membrane trafficking	139
	6.8.7.1 Independent sorting of membrane and lumen protein	139

                6.8.7.2 Membrane protein sorting   139
                6.8.7.3 Lumen protein sorting   139
        6.8.8 Transport pathways   141
                6.8.8.1 VPS-dependent or CPY pathway   142
                6.8.8.2 Alternative/ALP pathway   142
                6.8.8.3 Cytoplasm-to-vacuole targeting (Cvt) pathway   142
    6.9 Sorting signals for proteins   143
        6.9.1 Positive signalling for plant proteins   143
        6.9.2 Direct pathway   147
        6.9.3 Default mechanism   147
                6.9.3.1 Absence of sorting signal   147
                6.9.3.2 Garbage pathway   148
                6.9.3.3 Molecular chaperons   148
        6.9.4 Autophagy   149
                6.9.4.1 Concluding remarks   149
    6.10 Summary   149
    6.11 References   151

VII Functions of vacuoles   163
    7.1 Transport of molecules across membranes   163
    7.2 Uptake of water   164
        7.2.1 Aquaporin   164
                7.2.1.1 Histological considerations   165
    7.3 Uptake of ions   167
        7.3.1 General considerations   167
        7.3.2 Accumulation of ions   168
                7.3.2.1 Primary active transport   169
                7.3.2.2 Secondary active transport   170
                7.3.2.3 Uptake of anions   172
                7.3.2.4 Role of ion channels   173
    7.4 pH regulation by vacuoles   173
        7.4.1 Role of oxido-reductases   176
    7.5 Salt tolerance   176
    7.6 Osmoregulation   177
        7.6.1 Turgor for growth   178
        7.6.2 Tonoplast permeability   178
        7.6.3 The mechanism of osmoregulation   178
                7.6.3.1 Osmoregulation by sugars   178
                7.6.3.2 Turgor movements   180
    7.7 Role of vacuole in calcium homeostasis   183
        7.7.1 Ca^{2+} transporters and channels   184
                7.7.1.1 Ca^{2+} as second messenger   185
                7.7.1.2 Ca^{2+}- binding proteins   185
    7.8 Metabolic functions   186
        7.8.1 Crassulacean acid metabolism   186
        7.8.2 Homeostasis of amino acid levels   187
        7.8.3 Sucrose accumulation and transport   188

	7.8.4 Biosynthesis inside vacuoles	191
	7.8.4.1 Processing of sugars	191
	7.8.4.2 Biosynthesis of glycosides	191
	7.8.4.3 Anthocyanins	191
	7.8.4.4 Ethylene metabolism	192
	7.8.4.5 Processing of storage proteins	193
	7.8.4.6 Biosynthesis of alkaloids	194
7.9	Lytic functions	196
	7.9.1 Hydrolytic activity	196
	7.9.2 Proteolytic activity	196
	7.9.2.1 Degradation of extraneous proteins	197
	7.9.2.2 Proteolysis of mutant proteins	197
	7.9.3 Nucleolytic activity	197
	7.9.4 Autophagy	198
	7.9.4.1 Organelle breakdown inside vacuoles	198
	7.9.4.2 Membranous enclosure of cytoplasm	198
	7.9.4.3 Invagination of tonoplast	198
	7.9.5 Heterophagy	201
	7.9.5.1 Types of heterophagy	201
	7.9.5.2 Plasmalemmasomes	201
	7.9.5.3 Accumulation of breakdown products	202
	7.9.5.4 Host–pathogen interactions	202
	7.9.6 Lytic activity in differentiation	208
	7.9.6.1 Autophagy in vascular differentiation	208
	7.9.6.2 Autophagy in laticifer development	211
	7.9.6.3 Autophagy in cotyledonary development	211
	7.9.6.4 Autophagy in gametogenesis	211
	7.9.6.5 Autophagy in senescence	212
7.10	Detoxification of SO_2	213
7.11	Detoxification of xenobiotics	214
7.12	Phytoremediation of heavy metals	216
7.13	Scavenging of active oxygen species	219
7.14	Summary	220
7.15	References	225

VIII Retrospect and prospect — 249

8.1	Definitions	249
8.2	Significance of vacuoles	249
8.3	Uniqueness of vacuoles	251
8.4	Homology	252
	8.4.1 Homology of tonoplast with plasma membrane	252
	8.4.1.1 Evolution of the vacuole	255
	8.4.2 Homology with other membranes	255
	8.4.3 Homology with Golgi apparatus	256
	8.4.4 Homology with single-membraned organelles	256
	8.4.5 Homology with lysosomes	256
	8.4.5.1 Vacuoles are not lysosomes	259

8.5	Prospect of genetic engineering of tonoplast	259
	8.5.1 Engineering against water stress	260
	8.5.2 Engineering of proton pumps	260
	8.5.3 Engineering for storage and processing	262
	8.5.4 Engineering for phytoremediation and detoxification	263
	8.5.5 Engineering for protein targeting	264
	8.5.6 Engineering for disease resistance	264
	8.5.7 Other possibilites	264
8.6	Summary	265
8.7	References	267

Index 273

1
Early studies

The history of cell biology begins in 1665 with Robert Hooke who observed cells in cork as well as in many other tissues. However, Hooke did not recognise the vacuolar compartment in any definite way. More than a century later, in 1776, Spallanzani made a detailed study of a structure which he erroneously considered to be involved in breathing. It was actually a contractile vacuole in a protozoon.

1.1 Discovery of vacuoles

In the same year, Corti (1776) and Fontana (1776) discovered cytoplasmic streaming, but without any appreciation of the vacuole. Sixty years passed before Dujardin (1835, 1841) recognised the formation of aqueous optically empty spaces in unicellular animals as peculiar to living cytoplasm. He coined the term **vacuole** for any of the large customarily transparent vesicles in mature cells which are limited by a membrane. Meyer (1835), in his illustrations on cytoplasmic streaming, indicated the presence of a large central vacuole characteristic of a mature cell. Studying the living cells of algae and other aquatic plants, Schleiden (1842) could differentiate the streaming cytoplasm from the vacuolar sap which he called 'transparent cell juice'. Then, in 1844, Nägelli asserted that vacuoles were normal morphological components of plant protoplasm.

With the aid of primitive microscopes, workers observed and described vacuoles in many organisms including desmids where they contained gypsum crystals (Cohn 1854). In his classic book, Unger (1855) documented his first microscopic observation on the uptake of stains (coloured fruit sap from *Phytolacca decandra*) by individual cells. He also recorded the phenomenon of plasmolysis and deplasmolysis of cells.

1.2 Vacuolar contents

That vacuoles may contain dissolved matter also became clear to Nägeli (1844). According to him, the watery fluid of the single-celled giant alga *Valonia* tasted saltier than its surrounding medium, the sea water. This was corroborated by Famintzin (1860). Much later in 1904 Meyer, through chemical analysis, was the first to state that indeed the cells contained various salts including sodium, potassium, calcium, magnesium, chloride, sulfate, nitrate and phosphate. Moreover, the sap contained more potassium and less sodium than sea water.

Chemists have long known that even small amounts of tannin can be detected by simple methods like precipitating with iron salts or alkali treatments. Iron salts produce blue, green or black colouration, and alkali treatments produce yellow to orange with tannin. On the basis of this principle, Wigand (1862) and Wiesner (1862) were the first to detect tannin in the cell sap. That other types of organic substances besides tannins could be present in vacuoles was realised when Hartig (1855, 1856), Holle (1858) and Maschke (1858) studied aleurone grains. It became apparent that in storage tissues like cotyledons, the vacuoles became specialised to accumulate colloidal substances and even proteins. Thus aleurone grains could be located in vacuoles (Maschke 1858). Later, Gris (1864) and Wakker (1888) showed that the grains themselves were partially dehydrated vacuoles.

1.3 Physiology and biogenesis

Parallel to these studies, physiological aspects, especially on water relations of plant cells, attracted the attention of microscopists in the middle of the nineteenth century. Nägeli (1855) and Pringsheim (1854) described the process of plasmolysis, by which water moves out of the vacuole if the cell is placed in a hypertonic solution leading to a decrease in vacuolar volume. Finally the whole protoplasm shrinks producing a space between the protoplast and the cell wall.

This was the state of knowledge during the 1850s and 1860s, all limited by the primitive stage of the microscope. However, microscopists received a great boost when the superior devices built on the basis of Ernst Abbe's research on light diffraction, numerical aperture, aberrations, etc., became available around 1872.

The microscopical observations of Charles Darwin in 1875 first pointed out the curious phenomenon of the pigmented bodies, later recognised as vacuoles, in the cells of the tentacles of the sundew *Drosera*, an insectivorous plant. When the tentacles were stimulated by an insect, the large pigmented vacuoles broke up into smaller rods and vacuoles. Later, these small vacuoles fused and the cells returned to their initial stage. This reversible phenomenon of breakdown and formation of vacuoles was later confirmed by subsequent workers. The great Dutch botanist, Hugo de Vries, who is best known for his Mutation Theory, made a significant contribution to vacuolor research. As early as 1877, de Vries recognised the importance of vacuoles in the study of osmosis and suggested that the elongation of cells is based upon two processes: the osmotic uptake of water by the vacuole and the stretching of the cell wall.

Although in 1828 Dutrochet had realised the role of turgor in maintaining the form and rigidity of plant parts, it was de Vries (1884) who articulated the concept of turgor pressure, semipermeability of the membrane and the role of vacuole. De Vries realised the importance of vacuoles in the investigation of osmosis and pointed out that the cytoplasmic layer in large plant cells apparently acted as a perfectly semi-permeable membrane. He regarded vacuoles as permanent and differentiated cytoplasmic components. De Vries should also be credited for the development of the most widely used, though not very precise, technique for determining the osmotic value of the cell sap. The technique of incipient plasmolysis consists of placing specimens in fluids of known osmotic strength and recording the osmotic value of the solution in which about one-half of the cells are not plasmolysed while the other half are barely plasmolysed. The average osmotic pressure in the cells of the tissue under investigation is considered to be equal to the osmotic pressure of the solution in which this condition obtains. It must be admitted that some of his methods were unique and led to far-reaching concepts.

This is evident from his study on the sap composition and osmotic pressure of the cell. De Vries extracted sap from the leaves and stems of many different plants. He then determined the relative ability of these cell saps to plasmolyse epidermal cells of *Rhoeo*, in comparison with the effect of various concentrations of aqueous solutions of KNO_3 on *Rhoeo* cells. He analysed the cell sap quantitatively for its chief constituents. Then he determined the relative ability of these individual components to plasmolyse *Rhoeo* cells. Thus he calculated the portion of the total osmotic pressure contributed by the various sap constituents. He concluded that the major osmotically active solutes in the vacuoles are organic acids, chiefly malic acid and inorganic salts of Ca, Mg and K, as well as sugars. More critical investigations indeed showed that de Vries was not far off the mark and the sap constituents listed by him account for two-thirds or more of the osmotic values reported.

One of de Vries' (1885) most interesting observations was the demonstration that a method of 'Golgi-stain' could be used to stain the plant vacuole. He reduced osmium in plant vacuoles from 1% solution of osmium tetroxide in 10% potassium nitrate. He rightly considered the reduction of osmium to be due to the presence of tannin and not to the presence of protein, as widely believed. He also discovered that vacuoles can be fixed with salts of other heavy metals like $HgCl_2$,

$AgNO_3$, $CuSO_4$, etc. In his famous 1885 essay, de Vries challenged the prevalent concept that vacuoles arise *de novo*. He proposed that vacuoles originate from minute cytoplasmic particles which he termed 'tonoplast'. These tonoplasts were supposedly small plastid-like bodies which occurred in meristematic cells and multiplied only by division of the pre-existing tonoplasts. As they absorbed water they enlarged and became vacuoles, the tonoplast itself developing into the vacuolar membrane (tonoplasma). De Vries (1886) confirmed the fragmentation of the coloured vacuoles in the tentacles of *Drosera rotundifolia* and the changes which accompanied the movements of the tentacles.

Along with the cytomorphic aspects of the vacuole, its physiological features were also investigated before the end of the nineteenth century. Pfeffer (1906), a contemporary of de Vries, was essentially a plant physiologist and contributed to the various physiological aspects of the cell. In a detailed investigation he established that the plasma membrane, as well as the vacuole membrane, is semi-permeable. He also explained the process of plasmolysis. Pfeffer (1886) developed a vital staining procedure with various dyes and also observed the accumulation of dye in vacuole sap. He was the first to conclude that tannin of the cell sap is primarily responsible for dye accumulation. He worked on vacuoles of diverse organisms, including a myxomycete, *Chondrioderma difforme*, in which he could experimentally induce *de novo* vacuole formation by granules of asparagin. He observed that the newly formed vacuoles may fuse with each other or with other plasmodial vacuoles (Pfeffer 1890). Janse (1887) studied passive permeability and explained the slow deplasmolysis observed in experiments with certain non-electrolytes as external hypertonic solutions.

Went (1888, 1890), a student of de Vries, contributed to the knowledge of vacuoles by establishing that vacuoles are present in all types of cells including the reproductive cells of certain algae. He established that even in meristematic cells minute vacuoles can be observed. Went did not endorse the view that vacuoles originate from massive plastid-like 'tonoplast', but thought they arise from pre-existing vacuoles by division or strangulation. Thus vacuoles form a continuous lineage. Strasburger (1898) considered the protoplasm as mostly alveolar in structure and suggested that vacuoles are formed by the enlargement and coalescence of alveoli.

The twentieth century ushered in a new era in genetics with the rediscovery of Mendelism. Hand-in-hand was the development of nuclear cytology and appreciation of chromosome behaviour during mitosis and meiosis. Although the growth of science does not follow the Gregorian calendar, it is useful for recording the history of science. Studies on vacuoles continued and more and more information on their diversity and function was accumulated.

1.4 NATURE OF VACUOLES

A series of studies by P.A. Dangeard (1916) and his son P. Dangeard (1920) greatly boosted our understanding of the vacuole. They suggested that the vacuolar system is a permanent constituent of the cell and coined the word 'vacuome' — which is a collective term for the vacuolar system of a single cell consisting of one or more vacuoles. Pensa (1917) recorded the presence of minute vacuoles in the meristematic cells and traced their subsequent development as canaliculi to form the large central vacuole (Fig. 1.1) in conformity with the earlier observation of Bensley (1910). The study of development of aleurone grains in the vacuoles of *Ricinus* and other plants is another major contribution by P. Dangeard (1923). Numerous observations of Guilliermond (1923), whose researches on mycology began early this century, contributed to the knowledge of vacuoles in fungi and recognised the autonomous nature of the vacuome, favouring the view that the vacuolar material itself, rather than the cavity containing it, is of significance.

As the twentieth century progressed, a number of other researchers, including Küster (1939) and Strugger (1940), continued to work on various aspects of vacuole structure and function.

4 Plant cell vacuoles: an introduction

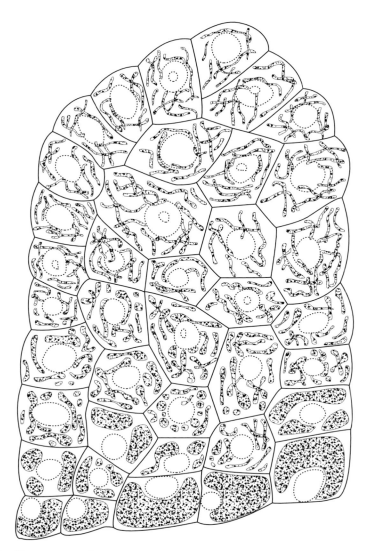

Figure 1.1 The filamentous anthocyanin-filled vacuoles (represented by stippling) swell, anastomose and fuse to form one large vacuole per cell in the teeth of young living rose leaflets (after A. Pensa 1917).

Topics like vital staining, vacuolar uptake of dyes, continuity of vacuoles, and homology of vacuoles with animal Golgi bodies were studied and debated. On the other hand, plant physiologists like Hoagland (1929) and Osterhout (1936) studied water relations of plant cells. Along with other researchers they demonstrated that the vacuolar sap of marine algae like *Valonia* and *Nitella* may contain more salts than sea water, thus revealing that vacuoles can accumulate salts against an osmotic gradient. Hoagland demonstrated that the pH of the cell sap may remain constant even when the pH of the surrounding medium is changed considerably.

In addition to the prevalent microscopic and physiological techniques, new methods were developed. Plowe (1931) refined and gainfully employed the methods of micromanipulation. He had been able to draw out the total cellular content of a partially plasmolysed cell to show the structural stability of the tonoplast, along with other physical features of the onion epidermal protoplasm. By means of a microneedle he isolated the vacuoles from the rest of the cytoplasm.

At the same time, Chambers and Höfler (1931) demonstrated that the isolated vacuoles retained their characteristic properties and also their red anthocyanin pigments. Techniques of centrifugation of living cells were developed and yielded a considerable amount of data on the specific gravity and osmotic value of vacuoles. Mottier (1899) centrifuged a wide variety of living cells and reported that the colourless fully hydrated vacuoles were lighter than the cytoplasm and were displaced toward the centripetal end of the cell. Milovidov (1930) centrifuged barley roots vitally stained with neutral red. He demonstrated that the meristematic cells contained denser vacuoles which settled at the bottom of the cells, whereas those of the elongated mature cells were lighter than the rest of the cytoplasm. By means of an air-driven ultracentrifuge, Beams and King (1935) could obtain a force as high as 400 000 g and when bean root-tips were subjected to it, the cell organelles were placed in separate layers. They observed that the vacuoles in the root-tip cells were lighter than the rest of the cytoplasm.

A number of publications, especially those by Zirkle (1937), Guilliermond (1941), Küster (1951) and Dangeard (1956), record many of the contributions and discoveries of the early vacuologists.

1.5 SUMMARY

It is a marvel that during the last couple of centuries early scientists working with primitive instruments were able to develop deep insight and discover basic truths in all fields of science. However, the rate of progress has been very uneven. Although the history of cell biology began with Robert Hooke in 1665, it took two long centuries before Nägelli in 1855 established that vacuoles were normal morphological components of plant cells, that they could contain dissolved substances even in higher concentration than the surrounding medium, and were capable of plasmolysis. Unger (1855) articulated the real significance of the vacuole by documenting the uptake of plant pigments as well as the role of the vacuole in plasmolysis and deplasmolysis. Over the next few decades of the nineteenth century the major discoveries about vacuoles centred around the presence of dissolved substances like salts and tannin and undissolved substances like protein crystalloids and oxalate crystals. Due to the availability of Ernst Abbe's superior microscopes, cytological observations became more precise. Darwin (1875) noted reversible breakdown and formation of pigmented vacuoles. A major conceptual advancement on turgor pressure and semi-permeability of membranes, brought about by De Vries (1884), showed that the osmotically active solutes in the vacuoles were organic acids and sugar, as well as inorganic salts of Ca, Mg and K. De Vries also suggested that vacuoles may originate from some pre-existing structure within the cell. Soon it became clear that vacuoles could fuse with each other as well as with other membranous structures. Early in the twentieth century understanding of vacuolar structure and function increased and the identity and autonomy of vacuoles, together with their participation in diverse cellular activities, became established.

1.6 REFERENCES

Beams, H.W. and King, R.L., The effects of ultracentrifuging on the cells of the root tip of the bean (*Phaseolus vulgaris*), *Proc. Royal Soc.* (London), **118**, 264, 1935.

Bensley, R.R., On the nature of the canalicular apparatus of animal cells, *Biol. Bull.*, **19**, 179, 1910.

Chambers, R. and Höfler, K., Micrurgical studies on the tonoplast of *Allium cepa*, *Protoplasma*, **12**, 338, 1931.

Cohn, F., Untersuchungen über die Entwicklungsgeschichte der mikroskopischen Algen und Pilze, *Acta Kaiser Leopold Acad.*, **24**, 101, 1854.

Corti, B., Sur la circulation d'un fluide, découverte en diverse plantes, *Obs. Phys. Hist. Nat.*, **8**, 232, 1776.
Dangeard, P.A., Sur les corpuscles metáchromatiques des Levures, *Bull. soc. Myc. France*, **32**, 27, 1916.
Dangeard, P., Sur la metachromatin et les composés tanniques des vacuoles, *Compt. Rend. Acad. Sci., Paris*, **171**, 1016, 1920.
Dangeard, P., Recherches de biologic cellulaire. Evolution du systeme vacuolaire chez les vegetaux, *Le Botaniste*, **15**, 1, 1923.
Dangeard, P., Le vacuome de la cellule végétale, *Protoplasmatologia*, III. D., 1., 1956, Springer-Verlag, Berlin.
Darwin, C., Insectivorous Plants, 1875
De Vries, H., Untersuchungen über die mechanischen Ursachen der Zellstreckung, ausgehend von der Einwirkung von Salzlösungen auf den Turgor wachsender Pflanzenzellen, Engelmann, Leipzig, 1877.
De Vries, H., Eine Methode zur Analyse der Turgorkraft, *Jahrb. Wiss. Bot.*, **14**, 427, 1884.
De Vries, H., Plasmalytische Studien über die Wand der Vakuolen, *Jahrb. Wiss. Bot.*, **16**, 465, 1885.
De Vries, H., Über die aggregation im Protoplasma von *Drosera rotundifolia.*, *Bot. Zeit*, **44**, 17, 1886.
Dujardin, F., Recherches sur les organismes inférieurs, *Ann. Sci. Nat. Zool.*, 2e serie, **4**, 343, 1835.
Dujardin, F., Histoire naturelle des Zoophytes: Infusoires, Roret, Paris, 1841.
Dutrochet, H.E., Nouvelles recherches sur léndosmose et l'exosmoses, 1828.
Famintzin, A., Beitrag zur Kenntnis der *Valonia utricularis*, *Bot. Zeit*, **18**, 341, 1860.
Fontana, F., Sur les movements plantes, *Obs. Phys. Hist. Nat.*, **7**, 285, 1776.
Gris, A., Recherches anatomiques et physiologiques sur la germination, *Ann. Sci. Nat. Bot.*, **2**, 5, 1864.
Guilliermond, A., Sur les constituants morphologiques du cytoplasme de la cellule végétal, *Arch. Anat. Mier.*, **20**, 1, 1923.
Guilliermond, A., The cytoplasm of the plant cell, Chronica Botanica, Waltham, Mass., 1941.
Hartig, T., Über des Klebermehl, *Bot. Zeit*, **13**, 881, 1855.
Hartig, T., Weitere Mitteilungen, das Klebermehl (Aleurone betreffend), *Bot. Zeit*, **14**, 257, 1856.
Hoagland, D.R. and Davis, A.R., The intake and accumulation of electrolytes by plant cells, *Protoplasma*, **6**, 610, 1929.
Holle, G. von, Über die Formen und die Bau der vegetablischen Proteinkörper, *Heidelb. Jahrb. Lit.*, **349**, 1858.
Hooke, R., Of the schematisme or texture of cork, and of cells and pores of some other such frothy bodies. *Micrographia*, London, 1665.
Janse, J.M., Versl. Mededel. Koninkl. Ned. Akad. Wetenschap. Amsterdam, Afd. Natuurk, 3e Reeks, **4**, 332, 1887.
Küster, E., Vital-staining of plant cells, *Bot. Rev.*, **5**, 351, 1939.
Küster, E., Die Pflanzenzelle, 2nd Ed., Fischer, Jena, 1951, 473.
Maschke, O., Über die Bau und die Bestandtheile der Kleberbläschen, *Bot. Zeit*, **17**, 409, 1858.
Meyer, A., Orientierende Untersuchungen über Verbreitung, Morphologie und Chemie des Volutins, *Bot. Zeitg.*, **62**, 113, 1904.
Meyer, F. J. F., Nouvelles observations sur la circulation du sue cellulaire dans les plantes, *Amer. Sci. Nat. Bot.*, **4**, 257, 1835.
Milovidov, P.E., Einfluß der Zentrifugierung aus das Vacuom, *Protoplasm*, **10**, 452, 1930.
Mottier, D. M., The effect of centrifugal force upon cells, *Amer. Bot.*, **13**, 325, 1899.
Nägeli, C., Pflanzenphysiologische Untersuchungen, Nägeli, C. and Cramer, C. Eds. Schulthess, Zürich, 1855.
Nägeli, C., Zellkern, Zellbildung und Zellenwachstum bei den Pflanzen, *Z. Wiss. Bot.*, **1**, 34, 1844.

Osterhout, W.J.V., The absorption of electrolytes in large plant cells, *Bot. Rev.*, **2**, 283, 1936.
Pensa, A. Fatti e considerazioni a proposito di aluine formazioni nelle cellule vegetale. *Monitore Zool. Ital.* **28**, 9, 1917.
Pfeffer, W., Über Aufnahme von Anilinfarben in lebende Zellen. Untersuch. *Bot. Inst., Tübingen*, **2**, 179, 1886.
Pfeffer, W., The Physiology of Plants, 2nd Ed., Translated and edited by A.J. Ewartt, Vol. III., Clarendon Press, Oxford, 1906.
Pfeffer, W., Zur Kenntnis der Plasmahaut und der Vakuolen, nebst Bemerkungen über den Aggregatzustand des Protoplasmas und über osmotische Vorgänge. *Abb. Math. Phys. Kgl. Sachs, Ges. Wiss.*, **16**, 185, 1890.
Plowe, J., Membranes in the plant cells, I. and II. *Protoplasma*, **12**, 196, 1931.
Pringsheim, N., Untersuchungen über den Bau und die Bildung der Pflanzenzelle, Abt. 1: Grundlinien einer Theorie der Pflanzenzelle., Hirschwald, Berlin, 1854.
Schleiden, M. J., Grundzüge der wissenschaftlichen Botanik, Berlin, 1842.
Spallanzani, L., Opuscoli di fisica animale e vegetabile, Societá Tipografica, Modena, Italy, 1776.
Strasburger, E., Die pflanzlichen Zellhäute, *Jahrb. Bot.*, **31**, 511, 1898.
Strugger, S., Fluoreszenzmikroskopische Untersuchungen über die Aufnahme und Speicherung des Acridin-orange durch lebende und tote Pflanzenzellen, Zeit. *Naturwissenschaften*, **73**, 97, 1940.
Unger, F., Anatomie und Physiologie der Pflanzen, C.A. Hartleben, Leipzig, 1855.
Wakker, T., Studien über die Inhaltskörper der Pflanzenzellen. *Jahrb. Wiss. Bot.*, **19**, 423, 1888.
Went, F.A.F.C., Die Vermehrung der normalen Vakuolen durch Teilung. *Jahrb. Wiss. Bot.*, **19**, 295, 1888.
Went, F.A.F.C., Die Entstehung der Vakuolen in den Fortpflanzungszellen, *Jahrb. Wiss. Bot.*, **21**, 299, 1890.
Wiesner, J., Einige Beobachtungen über Gerbe- und Farbstoffe der Blumenblätter, *Bot. Zeit*, **20**, 389, 1862.
Wigand, A., Einige Sätze über die physiologische Bedeutung des Gerbestoffes und der Pflanzenfarbe, *Bot. Zeit*, **20**, 121, 1862.
Zirkle, C., The Plant Vacuole, *Bot. Rev.*, **3**, 1, 1937.

II
Methodologies

2.1 IN SITU METHODS

2.1.1 Light microscopy

The pioneer workers were able to observe and collect a good deal of worthwhile information on the structure, development and functions of vacuoles with primitive microscopes before apochromats, Köhler's illumination or coated lenses were known. Over the course of two centuries, innumerable studies on vacuoles in both living and fixed cells were conducted by various types of treatments, colouring with dyes and experimental finesse. Vacuoles generally appear as hyaline, transparent or vacant areas in the cytoplasm, often occupying a considerable amount of space inside a cell. They can be made to appear distinct with fixation or with some mild treatment in living condition. One early technique for rendering the vacuoles distinct was to place cells in a 10–15% aqueous solution of potassium nitrate (de Vries 1885; Went 1888). It was also realised early on that vacuoles containing tannins, chloride, etc., could be blackened by salts of heavy metals like silver and osmium. Oil droplets and fatty acids also produced such blackening (Dangeard 1923; Zirkle 1932).

The pioneer cytologists noticed that vacuoles stained very easily with various dyes in both living and dead condition. Vacuoles do not stain in the same sense that a chromosome or cell wall stains, since most vacuoles contain very little of stainable matter, but as they accumulate various dyes the sap becomes coloured. Many basic dyes accumulate liberally in vacuoles (Kuster 1921). Each dye has a characteristic pH range at which it will permeate. Neutral red, brilliant cresyl blue, methylene blue, pyronin, etc., penetrate rapidly from basic solutions, slower from slightly acid solutions and not at all from buffers in the more acid ranges. On the other hand, methyl red, ethyl red, briliant green, etc., penetrate more rapidly from acid solutions and slowly, if at all, from alkaline buffers. Rhodamine accumulation in vacuoles, however, takes place uniformly over a wide range of pH (Bailey 1931). Experimentation with dyes on living plant cells and tissues is rather easy and a number of staining techniques are available (Stadelman and Kinzel 1972). However, even for such simple experiments, the novice may come out with an inexplicable staining pattern or totally misleading results. One of the early methods of identifying a vacuole was to stain with neutral red, which has a pH less than 6.0. A very reproducible protocol suggested by Peterson (1979) is as follows:

   i) Stain cells in 0.01% solution of neutral red at pH 7.2 in 0.08 M phosphate buffer.
   ii) Rinse with the same buffer.
   iii) Mount in buffer.

Neutral red accumulates in the vacuole of living plant cells, but not in those of dead cells or other organelles. Gupta et al. (1982) suggested the following method for vacuoles which may contain a number of hydrolytic enzymes:

   i) Treat tissue with 25–100 ppm of neutral red dissolved in 0.9% sodium chloride in darkness for 1–4 h.

ii) Mount in a drop of water.

Lysosome-like vacuoles take a deep red stain.

On the basis of critical studies on dye uptake, early workers realised (Dangeard 1916; Bailey and Zirkle 1931) that two types of vacuoles exist: those which contain tannin and other phenolics stain differently from those which are without tannin. Höfler (1947, 1949) articulated the difference with the aid of uptake of acridine orange for fluorescence microscopy. The cells with low pH and dissolved phenolics were designated as 'full cell sap' and those which accumulated the dye by an ion trap mechanism were designated as 'empty cell sap'.

Stadelman and Kinzel (1972) gave a detailed staining protocol for 'full' and 'empty cell sap in the scale leaf cells of onion (*Allium cepa*).

i) Stain tissue section in phosphate buffered neutral red solution (1:10 000) at pH 8.0 to 8.3 for 10–20 min.
ii) Rinse and mount in the same buffer.

The epidermal cells of the abaxial outer side of the scale leaf have full cell sap and stain violet, and those of the adaxial (inner) epidermis with empty cell sap exhibit a dull brick red colour.

Similarly, toluidine blue can be used:

i) Stain tissue sections in phosphate buffered toluidine blue solution (1:10 000) at pH 11 for 20 min.
ii) Rinse and mount in the same buffer.

The full cell sap of the outer epidermis stains green/blue, while the empty cell sap of the parenchyma stains violet. In some of the epidermal cells there will be a precipitation of the stain in the form of droplets with intense blue colour.

Many dyes change to become colourless or leuco compounds by the action of reducing agents such as sodium thiosulphite *in vitro*. They become non-ionic and show a high penetrating rate compared to their ionic form. After easy penetration of the membranes they regain their colour by oxidation (Baker 1958). The vacuoles of enzymatically isolated protoplasts from flower petals and mesophyll can be successfully stained by the leucobases of methylene blue and toluidine blue by the following method developed in the author's laboratory (unpublished):

i) Prepare leucomethylene blue and leucotoluidine blue as follows (Gatenby and Beams 1950): Add 25 mL of 0.1 N sodium thiosulphite and 4 mL of 0.1 N HCl to 100 mL of 0.01% aqueous solution of the dye. Stir well and store in darkness for about 24 h. The colorless leucobase is formed within 24 h. Store the solution in dark bottles in a refrigerator till use.
ii) Stain protoplasts in 0.7 M mannitol in a drop of the dye on a microscope slide for 5 min.
iii) Rinse and mount in 0.7 M mannitol.

Only vacuoles of living protoplasts accumulate the dye and no other cell compartment is stained. Vacuoles of the dead cells fail to stain.

2.1.2 Fluorescence microscopy

Certain substances absorb incident radiation at various wavelengths and convert some of the absorbed energy into light of longer wavelengths in the visible range. If this emission occurs with negligible delay (10^{-9} sec) the phenomenon is known as fluorescence. In fluorescence microscopy the preparation becomes self-luminous, while the radiation exciting the luminosity does not contribute to the image formation. The fluorescing part of the preparation appears bright, usually coloured, against a dark background. The light emitted by a high intensity mercury vapour lamp filtered through an exciter filter is of shorter wavelength and is used for excitation. Just as the non-absorbed components of light falling upon a substance give its inherent colour, the primary fluorescence where it occurs is mainly a function of its chemical constitution. On the other hand, secondary fluorescence is produced by a substance only when another chemical substance, e.g. a fluorochrome, is conjugated to it.

2.1.2.1   Autofluorescence

Primary fluorescence, or autofluorescence, includes all luminescence phenomena which are produced by irradiating untreated biological materials with blue and violet light (480 nm to 400 nm) or ultraviolet light (<400 nm). Tissues are also known to exhibit ultraviolet fluorescence, that is fluorescence in the UV range, when excited by far UV. A good deal of work on autofluorescence has been conducted with animal tissues (Hemperl 1934; Sjostrand 1945). The fluorescence of organic compounds of biological significance has been reviewed by Udenfriend (1962), and Konev (1967) reviewed the information on proteins and nucleic acids. A wide range of substances which are autofluorescent are known in plants. The chief among them are photosynthetic pigments. The indole acids, gibberellins, coumarines, caffeic and chlorogenic acids, flavones, anthocyanins, alkaloids and various toxic substances are also autofluorescent (Goodwin 1953). Most of these substances are, however, difficult to detect by fluorescence microscopy because of their high dilution in the cell sap. However, nucleoproteins, due to their density in nucleus, are more easily detectable by autofluorescence. In spite of this, capable researchers can use autofluorescence to their advantage. Larcher (1953) could differentiate dead cells from living cells on the basis of vacuolar autofluorescence. In a number of animal tissues, Koenig (1963) found that particles which contain acid phosphatase fluoresce when irradiated with UV light. It was not clear whether the fluorescence was associated with the lysosomal membrane or with the lysosomal content. Autofluorescence is weakest in fresh tissue and is intensifield by dehydration, fixation and other procedures. Acetic ethanol fixation or gentle hydrolysis (Feulgen reaction type) increases the intensity. Because of its low intensity and quick decay, the study of autofluorescence has never been popular with the microscopists. However, the availability of highly sophisticated microspectrophotometers permitting very precise fluorescence analysis, should make study on autofluorescence of vacuolar contents very rewarding. Palevitz *et al.* (1981 a, b) have successfully used vacuolar autofluorescence to study vacuolar movement and development.

2.1.2.2   Secondary fluorescence

As far as secondary fluorescence studies with the aid of fluorescent dyes are concerned, the staining protocol is similar to that of standard non-fluorescent dyes. However, the great advantage is that a very low concentration of the dye, as low as 20 ppm (Gupta and De 1983), can be used to detect the vacuole. The major disadvantage is that the high intensity excitation radiation can induce damage or death to the cells. Moreover, the fluorochrome may undergo photooxidation and photodegradation. Many methods of vital fluorochroming of plant cells were developed by early workers. The following one, for example, is simple and reproducible.

1. Fluorochroming with rhodamine B:
    i) Stain cells for 5–10 min in 0.1% aqueous solution of rhodamine B.
    ii) Wash in distilled water to bring out differentiation.

The full cell sap fluoresces golden brown and the empty cell sap does not fluoresce.

In the staining protocol recommended by Stadelman and Kinzel (1972), more critical results may be obtained in relation to other cell compartments. Destaining with $CaCl_2$ solution, as developed by Schümmelfeder (1956), may be helpful. The treatment with $CaCl_2$ is based on the fact that the colourless solutions compete with dye ions for the available sites. Thus the cell is destained and more dye is transferred to the cytoplasm and the vacuole, producing a sharp contrast in the fluorescence of the cells. The method is as follows:

i) Stain cells or tissue sections in phosphate buffered acridine orange solution (1:5000 or 1:10,000) at pH 3.8 and pH 8.5, for 10–20 min.
ii) Wash cells in respective buffers in three consecutive dishes, agitating well.
iii) Mount in the respective buffers.
iv) Alternatively, after (ii) the tissue stained at pH 3.8 may be destained in 0.2–0.5 M $CaCl_2$ in distilled water for 20 min and observed.

Table 2.1 Fluorescence pattern of *Rhoeo* staminal hair cells treated with aqueous acridine orange solutions for various periods in darkness.

Concentration	Period	6 hr	12 hr	24 hr	48 hr
1 ppm	Nuclei	NF	Perceptible light green	Very faint green	Faint green
	Cytoplasm	NF	NF	NF	NF
	Vacuoles	NF	NF	NF	NF
10 ppm	Nuclei	Perceptible light green	Faint green	Light green	Green
	Cytoplasm	NF	NF	Pale green	Light green
	Vacuoles	NF	NF	NF	NF
20 ppm	Nuclei	Green	Green	Dark green	Yellow
	Cytoplasm	Very light green	Light green	Green	Green
	Vacuoles	NF	Pale yellow	Brilliantly fluorescing	Size of vacuoles increase
50 ppm	Nuclei	Light orange	Orange	Orange	Orange
	Cytoplasm	Light green	Green	Greenish yellow	Yellow
	Vacuoles	Bright orange	Size increases	Colourless vacuoles showing orange red particles	Fluorescent particles form 'aggregate'
100 ppm	Nuclei	Yellow orange	Orange	Deep orange	Whole filament deep red. Nuclei and cytoplasm not distinguishable
	Cytoplasm	Yellow	Yellow	Yellow	
		Yellow orange	Orange red particles	Number of particles increases	

NF, No fluorescence

At pH 3.8, the cell walls stain red and the cytoplasm and vacuole of the empty cells do not stain. Treatment with $CaCl_2$ solution induces green fluorescence. At pH 8.5 the empty vacuoles accumulate the dye and fluoresce red, the cell wall remains unstained and the cytoplasm and nucleus fluorese green. In contrast, the full vacuoles show green fluorescence.

The fluorescence pattern of the cellular compartments of *Rhoeo* staminal hair cells vitally stained with various concentrations of aqueous solution of acridine orange for different periods supports the reproducibility and validity of the vital staining procedure (Gupta and De 1983) (Table 2.1). One fluorescent dye which has recently found wide application is 6-carboxy fluorescein (CF) which was used earlier as a tracer in plants. Goodwin *et al.* (1990) reported that CF is accumulated by the vacuoles of the epidermal cells of *Egeria densa* leaves. CF as such is membrane impermeable in its dissociated state and can be loaded into the cells as CF-diacetate. Shepherd *et al.* (1993) reported that vacuoles are loaded with CF if the living hyphae of the fungus *Pisolithus tinctorius* are exposed to an unbuffered 20 µg/mL solution of CFDA for 10 min followed by washing with the culture medium. Another derivative of CF, 2'7-bis-(2-carboxy ethyl)-5-(and-6) carboxy fluorescein (BCECF), has been used in a remarkable way not only for loading of plant vacuoles but also for precise monitoring of vacuolar pH and effect of inhibitors (Brauer *et al.* 1995, 1997). The process of loading the vacuole with the dye essentially consists of submerging the cells in 0.1 mM $CaCl_2$ containing 3 µM BCECF-AM, (the membrane-permeable acetoxy- methyl ester of BCECF) for 30 min and then rinsing in 0.1 mM $CaCl_2$ for 30 min. In aleurone protoplasts, 10 µM of BCECF-AM for 40 min exhibited complete sequestration of the dye in the vacuoles (Swanson and Jones 1996). Monochlorobimane (MCB) is a compound that fluoresces only when conjugated to glutathione or other sulfhydryl-containing molecules. In living cells uptake of fluorescent MCB requires ATP. In recent years a number of lipophilic dyes like CDCFDA (5-(and-6)-carboxy-2',7'-dichlorofluorescein diacetate), DASPMi (2-4-dimethyl-aminostyryl)-N-methylpyridiniumiodide), FM4-64 (N-(3-triethylammoniumpropyl)-4-(p-diethylamminophenyl-hexatrienyl)pyridinium dibromide) have been used as fluorescent vital stains in diverse tissues. CDCFDA appears to be a good vital fluorescent dye for the vacuolar lumen of yeast cells (Pringle et al. 1989). The protocol for vacuolar lumen staining with CDCFDA is as follows:

i) Yeast cultures are re-suspended in YPD (yeast extract-bactopeptine-dextrose) medium containing 0.1 M citrate-KOH at 30°C and adjusted to pH 4.0.
ii) CDCFDA is added (10 µM final concentration).
iii) Cells are incubated at the same temperature for 15 min.
iv) Cells are viewed under a fluorescence microscope with a barrier filter of 450-490 nm.

Only the vacuolar lumen appears fluorescent.

FM4-64 appears to be specific for the tonoplast of yeast (Vida and Emr 1995). The protocol for a specific ring staining pattern of the tonoplast of yeast by FM4-64 is as follows:

i) Yeast cells are cultured in YPD medium at 20–30°C, harvested and re-suspended in the same medium.
ii) FM4-64 is added to 20–40 µM from a stock solution of 16 mM in DMSO followed by incubation with shaking for 10–15 min at 30°C.
iii) Cells are re-suspended in fresh YPD medium and are viewed under a fluorescence microscope with a barrier filter of 546 nm.

Only the tonoplast fluoresces as a ring.

Finally, a fluorescent probe should be mentioned. Lucifer Yellow CH (LYCH), which is intrinsically a membrane-impermeable dye, has been found to enter the central vacuole of a wide range of plant cells by a process presumably of fluid-phase endocytosis. The dye does not stain the cytoplasm but only the vacuole (Oparka 1991). Wright and Oparka (1989) reported that the uptake of LYCH was biphasic with respect to substrate concentration and sensitive to pH, low temperature and a range of metabolic inhibitors. Wright et al. (1992) showed that the LYCH uptake by vacuoles of the protoplasts was dependent on the stage of development of the vacuole and only relatively mature vacuoles showed LYCH fluorescence. Thus, the sequestration of LYCH proceeds along with the discrete development of the vacuole. Their data negated simple diffusion and fluid phase endocytosis of the dye and suggested that LYCH uptake takes place by the highly co-ordinated membrane transport systems on both plasmalemma and tonoplast. Hence LYCH may be advantageously used for study of vacuoles in a living cell.

2.1.3 Electron microscopy

Both types of electron microscopy, i.e. transmission and scanning, aided with proper electron staining have been widely used to study the tonoplast as well as the vacuolar contents. However, specific staining reactions for each of the different intracellular membranes can only be obtained by either cytochemical or immunochemical reactions of the specific enzymes or proteins. A phosphotungstic acid-chromic acid staining procedure has been developed for plasma membranes by Lembi et al. (1971). Hall and Flowers (1976) have shown that ruthenium red can be employed for in vitro staining of both plasma membrane and tonoplast with the aid of glutaraldehyde. Haschke (1991) has modified the technique by simultaneous application of ruthenium red and glutaraldehyde to obtain selective staining of the tonoplast. The protocol, as detailed below, shows that specific binding of ruthenium red to tonoplast takes place in in vivo conditions in intact tissue. The living tissue pieces are incubated with 0.1% (w/v) ruthenium red in 50 mM K_2HPO_4/KH_2PO_4 (at pH 7.2 and 2.5% (v/v) glutaraldehyde for 1 h at 25°C. The tissue is washed in the same buffer six times, for 5 min each. Post-fixation in 1% OsO_4/uranylacetate and processing are carried out as usual. Only the tonoplast shows significant staining.

2.1.4 Cytochemical localisation

One of the attributes of the various types of vacuoles has been that they contain a number of enzymes, especially the hydrolases. Certain enzymes, like acid phosphatases and esterases, may be present in all vacuoles. In spite of recent advances in the development of techniques of isolation of intact vacuoles leading to specific determination of various vacuolar components there are distinct advantages of *in situ* detection of the enzymes in different cellular organelles. Cytochemical methods are highly useful when different types of vacuoles co-exist in a cell, when a given tissue contains a heterogenous assemblage of diverse cells or when it is necessary to know the precise localisation of the enzyme with the tonoplast or vacuole sap. The sophistication of quantitative microspectrophotometry has opened a wider door to cytochemical studies and the techniques of X-ray microanalysis can be used for quantification of metal deposits in tissue sections.

The technique of sub-cellular localisation of enzymes was first developed by Gomori (1939, 1952). The technique is based on the principle that an enzyme can be detected *in situ* by its action when incubated with an appropriate substrate leading to the formation of a product. Since the product is generally soluble, cytochemical localisation requires immediate precipitation in a detectable form by what is called a 'simultaneous capture mechanism'. Gomori (1939, 1956) developed the technique of detection of phosphatase by incubating the tissue with suitable phosphate ester as substrate and lead nitrate. Inorganic phosphate released by enzymic activity is immediatly precipitated as insoluble lead phosphate. To make it more visible, the phosphate can be converted to a black sulphide by exposing the tissue to a weak solution of hydrogen sulphide or ammonium sulphide (metal salt technique). Alternatively, the incubation mixture may include substrates that are altered by the corresponding enzymes to a form that can then bind covalently to dyes. Thus, acid phosphatase digests napthylphosphates to produce naphthols which react readily with dyes of the diazonium-salt type to form coloured precipitates (azo-dye technique).

Many of the cytochemical methods developed for light microscopic detection have been or could be adopted for electron microscopic localisation. For EM cytochemistry it is essential that the reaction product is electron-dense so that it may directly be observed in ultrathin sections with the aid of transmission electron microscopy. In view of the greater resolution and higher magnification offered by EM, ultrastructure of the cell should be preserved with appropriate fixation. Artefacts due to diffusion of reaction product and non-specific retention of electron-dense substances must be carefully avoided. The techniques of enzyme cytochemistry involve three major stages:

1. *Specimen preparation*. This involves fixation in an appropriate fixative which should ideally (i) preserve cell morphology and fine structure, (ii) preserve and immobilise enzyme activity, and (iii) induce favourable membrane permeability. Although coagulant fixatives like ethanol or acetone are excellent for retaining enzyme activity, they are unacceptable as far as the preservation of cellular structure is concerned. Non-coagulating fixatives inducing covalent linkages, especially the aldehydes, viz. formaldehyde, glutaraldehyde and acrolein, are generally favoured. Heavy metal fixatives, like osmium tetroxide and potassium permanganate, destroy enzyme activity. In general, 1–3% glutaraldehyde or 4% formaldehyde is acceptable for most cases, with the pH maintained near neutral by an appropriate buffer. The standard method of embedding in paraffin may adversely affect most enzyme activity. Freeze-sectioning with the aid of a cryo-microtome is recommended for light microscopy. For electron microscopy, ultrathin sectioning of resin embedded tissue block is carried out after incubation for cytochemical staining.

2. *Incubation for enzyme action and staining*. This is the most crucial step. It involves the action of the enzyme on the substrate administered as the incubation medium at appropriate pH, buffer and temperature for the optimum time. The reaction product is captured immediately at the site of the enzyme to produce metal deposits as coloured or electron-dense product. Appropriate controls should be maintained to check the validity of the enzyme action. The incubation may be made highly specific for a given enzyme and should be critically designed to produce high resolution enzyme localisation in the cell structure.

3. *Post-incubation processing.* For light microscopy this may involve washing and mounting in a proper medium. For electron microscopy, this step includes osmication, standard dehydration, embedding in resin and ultrathin sectioning, with or without electron staining.

2.1.4.1 Protocol for selected enzymes

The principles outlined above show that cytochemical localisation of enzymes requires a number of steps to be correctly handled. Several books and reviews are available on various aspects of the different enzymes (Sexton and Hall 1978; Gahan 1984; Jensen 1962). Because the enzymes appear in a greatly diluted form in aqueous vacuoles their localisation has often been frustrating. However, it has been possible to indicate the location of H^+-ATPase on the cytoplasmic surface of tonoplast with the use of very precise methodology (Franceschi and Lucus 1982; Hall et al. 1980; Balsamo and Uribe 1988). The detailed techniques for some common enzymes are given below:

Acid phosphatase: a. Lead salt procedure

Light microscopy (Gomori 1956; Jensen 1956)
- i) Use fresh freeze-dried, freeze-substituted or cold neutral formalin(4%)-fixed sections or paraffin sections directly on slides with Haupt's adhesive or subbed in 1% gelatin.
- ii) Incubate at 37°C for 0.5 h–18 h with freshly prepared medium containing:
    10 mM sodium β-glycerophosphate (pH 5.0) adjusted with acetic acid
    50 mM acetate buffer (pH 5.0)
    3.6 mM Pb $(NO_3)_2$
- iii) Wash in several changes of distilled water.
- iv) Treat with a dilute yellow ammonium sulphide solution for 5 min.
- v) Rinse in distilled water, place in 95% ethanol and in absolute ethanol. (Deparaffinize if paraffin sections used.)
- vi) Mount in glycerine jelly or euparol.

The sites of acid phosphatase activity appear as black deposits.

Electron microscopy: (Holt and Hicks 1961; Marty 1978)
- i) Fix tissue blocks in 3% glutaraldehyde in 0.1 M cacodylate buffer, pH 7.4 at 0–4°C for 2 h.
- ii) Rinse in same buffer.
- iii) Cut thick sections (40–50 μm) in a tissue chopper or cryo-microtome and collect in cold buffer containing 0.22 M sucrose.
- iv) Wash with quick changes in graded buffer from pH 7.4 to 5.0 and finally in 0.1 M acetate buffer pH 5.0 (2 min).
- v) Incubate at 37°C in β-glycerophosphate medium as described for light microscopy.
- vi) Wash in several changes of water (omit ammonium sulphide reaction).
- vii) Osmicate with 1% osmium tetroxide solution for 1 h at 0–4°C.
- viii) Dehydrate, and process for usual electron microscopy. Electron dense deposits of lead phosphate indicate sites of enzyme activity.

Control: An appropriate control is absolutely essential. The control consists of:
- tissue incubated in medium without sodium β-glycerophosphate;
- tissue incubated in medium with 0.01 M NaF which inhibits phosphatase activity; and
- tissue placed in ammonium sulfide solution without being placed in the substrate (for LM only).

For the demonstration of acid phosphatases, *viz.* β-glycerophosphatase, CMPase (cytidine 5-monophosphatase) and NADPase (β-nicotinamide adenine diphosphatase) in the cells of green alga, *Gleomonas kupfferi*, Domozych (1989) successfully used the following method. After proper fixation and washing, the cells were treated with 0.1 M sodium acetate buffer at pH 5 twice for 10 min each. Cells were then incubated in the following phosphatase-labelling media for 60 min at

37°C in dark : 5.0 mL 0.2 M NaOAC buffer (pH 5.0), 1 mL 50 M CaCl$_2$, 1.0 mL 50 mM MgCl$_2$, 1.2 mL Pb(NO$_3$)$_2$ 1.8 mL D-H$_2$O and 10 mg either β-glycerophosphate, CMP or NADP. After incubation, cells were washed in 0.1M NaOAC buffer (pH 5.0) twice for 10 min each followed by washing in cacodylate buffer and processing for electron microscopy.

Acid phosphatase: b. Azo dye method
Light microscopy (Rutenberg and Seligman 1955; Zalokar 1960; Pearse 1960)
  i) Use fresh freeze-substituted, frozen, cold formalin-fixed sections or deparaffinized paraffin-embedded sections in water.
  ii) Incubate at 25°C for 1.5–2 h in the medium prepared by dissolving 25 mg of sodium 6-benzoyl-2-naphthyl phosphate in 80 mL of water and adding 20 mL of 0.5 acetate buffer at pH 5.0 and 2.0 g of NaCl.
  iii) Wash sections in three changes of cold saline (0.85% NaCl).
  iv) Stain with cold aqueous (1 mg/mL) solution of a diazonium salt or tetraazotised diorthoanisidine, made alkaline with sodium bicarbonate, for 3–5 min with gentle agitation.
  v) Wash in three changes in cold saline.
  vi) Fix (the unfixed sections) in cold formalin.
  vii) Mount in glycerine jelly.

Controls may be maintained as for PbS procedure, including a direct staining without the incubation (Step ii).

Blue colouration due to di-coupling of the dye indicates high acid phosphatase activity, and low activity is detected by mono-coupling leading to purple or red colouration.

Electron microscopy: (Bowen 1974; Esau and Charvat 1975)
  i) Prepare tissue blocks or thick sections after usual fixation with glutaraldehyde.
  ii) Prepare incubation medium as follows:
      a. Dissolve 5 mg of naphthol AS-BI phosphate in 0.25 mL dimethyl formamide.
      b. Dilute with 25 mL of water.
      c. Add 25 mL of 0.2 M acetate buffer (pH 5.2).
      d. Dissolve 30 mg p-nitrobenzene diazonium tetrafluoroborate.
      e. Add 0.1 mL of 10% MnCl as activator.
  iii) Incubate at room temperature for 1.5–3 h.
  iv) Wash with cold buffer.
  v) Osmicate and process for electron microscopy.

Fine highly electron dense areas indicate acid phosphatase activity.

Alkaline phosphatase: a. Adenosine triphosphatase
Electron microscopy (Balsamo and Uribe 1988; Hall *et al.* 1980; Sanchez-Aguayo *et al.* 1991)
  i) Fix tissue blocks in 0.5% glutaraldehyde-2% formaldehyde in 100 mM cacodylate buffer (pH 7.2) for 1 h at 4°C.
  ii) Wash in same buffer supplemented with sucrose for 2 h.
  iii) Wash in 50 mM tris-maleate buffer (pH 7.0) for 30 min.
  iv) Incubate for 2 h at room temperature in the medium containing: 2 mM ATP, 2 mM Ca(NO$_3$)$_2$ or Mg(NO$_3$)$_2$, 3.6 mM Pb(NO$_3$)$_2$ and 50 mM tris-maleate buffer.
  v) Wash in tris-maleate, followed by cacodylate buffer for 30 min.
  vi) Post-fix in 1% OsO$_4$ in cacodylate byffer for 1 h at 4°C.
  vii) Process for electron microscopy.

Result: Electron dense deposits are noted on various membranes.
Controls:
  a. Incubate without any ATP.

b. Incubate with addition of 0.1 mM sodium vanadate to inhibit plasma membrane ATPase.
c. Incubate with addition of 50 mM KNO_3 to inhibit tonoplast ATPases.
d. Incubate with both vanadate and potassium nitrate.

Fine electron dense areas indicate ATPase activity.

b. Other alkaline phosphatases

Domozych (1989) demonstrated four alkaline phosphatases by the following protocol: For IDPase (inosine 5-diphosphatase), ITPase (inosine 5-triphosphatase), ATPase (adenosine 5-triphosphatase) and TTPase (thiamine pyrophosphatase), after proper fixation and washing, cells were treated with 0.1 M tris-maleate (pH 7.2) buffer twice for 10 min each. The cells were incubated for 60 min at 37°C in the dark in the following phosphatase-labelling solution: 4 mL of IDP, ITP, ATP or TPP substrate solution (4.5 mg/mL in $D-H_2O$), 1.6 mL $D-H_2O$, 9.0 mL 0.2M Tris-maleate buffer (pH 7.2), 1 mL 90 mM $CaCl_2$, 1 mL 90 mM $MgCl_2$ and 2.4 mL freshly prepared $Pb(NO_3)_2$. After incubation, the cells were washed in 0.1 M Tris-maleate buffer (pH 7.2) twice for 10 min each, followed by washing in cacodylate buffer, and processed for electron microscopy.

Aryl sulfatase

Light microscopy (Rutenberg *et al.* 1952; Avers 1961)

The similarity between the sulfatase and phosphatase reactions has led to the development of cytochemical procedures using the same principle.
i) Use tissue sections after proper fixation (as stated earlier) and bring to water.
ii) Wash with cold saline.
iii) Incubate for 5–30 min at 37°C with the medium prepared as follows:
  a. Dissolve 25 mg of potassium-6-bromo-2 naphthyl sulfate in 80 mL of hot 0.85% NaCl solution.
  b. Add 20 mL of 0.5 M acetate buffer at pH 6.1.
iv) Wash with two changes for 10 min each in cold saline solution.
v) Treat with a fresh cold solution of naphthanil diazo-blue B (1 mg/mL) in 0.05 M phosphate buffer at pH 7.6 for 3–5 min, agitating gently.
vi) Wash in two changes of cold saline and two changes of distilled water.
vii) Mount in glycerin-jelly or water.

Appropriate controls should be run (1) without any substrate in the reaction medium, (2) omitting the incubation medium, or (3) using a heat-killed tissue.

A blue–purple or blue–red–pink colouration indicates decreasing intensity of aryl sulfatase activity.

Electron microscopy (Poux 1966)

The procedure for EM of acid phosphatase can be followed for fixation and tissue preparation. The incubation medium consists of: 15–30 mg p-nitrocatechol sulfate; 0.16 mL 24% lead nitrate; and 5 mL veronal acetate buffer pH 5.5.

Dense areas indicate aryl sulfatase activity.

Non-specific esterases

Light microscopy (Holt 1958)
i) Use tissue sections after usual fixation and bring to water.
ii) Incubate at 37°C for 5 min–1 h in the medium containing: 0.1 mL absolute ethanol containing 1.5 mg of 5-bromo-4-chloroindoxyl acetate; 2.0 mL 0.1 M tris buffer at pH 8.5; 0.5 mL 0.05 M potassium ferricyanide; 0.5 mL 0.05 M ferrocyanide; 0.1 mL 1 M calcium chloride; 5.0 mL 2 M sodium chloride; 1.8 mL $D-H_2O$.

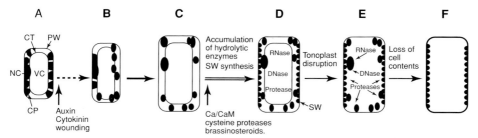

Figure 7.15 Cellular events that occur during transdifferentiation of *Zinnia* Mesophyll cells into TEs. **a** isolated mesophyll cell; **b** de-differentiated cell; **c** TE precursor cell; **d** immature TE; **e** maturing TE; **f** mature TE. Transdifferentiation is induced by wounding and a combination of auxin and cytokinin. Broken, single and double arrows show the progress of stages I, II, and III, respectively. The transition from stage II to stage III appears to be regulated by calcium (Ca)/CaM, cysteine proteases, and endogenous brassinosteroids. At the start of stage III, genes that are involved in both secondary wall synthesis and autolysis are expressed. Hydrolytic enzymes, such as DNases, RNases, and proteases, may accumulate in the vacuole. The disruption of the tonoplast causes these enzymes to invade the cytoplasm and attack various organelles, resulting in the formation of a mature TE that has lost its cell contents. CP, chloroplast; CT, cytoplasm; NC, nucleus; PW, primary wall; SW, secondary wall; VC, vacuole (from Fukuda 1997). With permission of the American Society of Plant Physiologists.

1969). Apart from the particulate occurrence of the enzymes, various hydrolytic enzymes in the sieve elements have been detected using histochemical and cytochemical methods (Wardlaw 1974). The presence of acid phosphatase containing digestive vacuoles with membranous contents in the companion cells has been recorded by Esau and Charvat (1975).

It has long been known that meristematic cells destined to be primary xylem cells lose their cell contents by general breakdown of the cytoplasm (Wodzicki and Brown 1973). During the last decade, a number of laboratories have been engaged in research on the differentiation of tracheary elements (TE) in suspension cultures of *Zinnia* cells and a lot of information on the ultrastructure, biochemistry and molecular genetics is available (vide review by Fukuda 1996). Whereas *in planta* the procambial cells differentiate into a primary xylem, *in vitro* the mesophyll cells of *Zinnia* in suspension can be induced directly to form TE (Fukuda and Komamine 1980) (Fig. 7.15). On the basis of cytological and biochemical processes and identification of numerous genetic markers, the trans-differentiation is divided into three stages (Fukuda 1997). Stage I involves the functional 'de-differentiation' process including disarray of chloroplast, cessation of photosynthesis and initiation of calmodulin. In stage II, along with preparation for secondary wall (SW) synthesis, the process of programmed cell death (PCD) is initiated with the synthesis of hydrolytic enzymes like cysteine protease (Minami and Fukuda 1995), serine protease (Beers and Freeman 1997) and a number of endonucleases. Several Zn^{2+}-activated DNases and RNases are also closely associated with TE differentiation. The signal peptides of the N-termini of the proteases and the endonucleases suggest that their precursors are probably transported into the vacuole where they are processed to the activated form. It is interesting that the optimal pH of both cysteine protease and DNase is 5.5 which corresponds to the pH in the vacuole. Groover et al. (1997) noted the presence of small vacuoles which cause partial loss of density of the cytoplasm at this stage. Recently, Groover and Jones (1999) reported a 40 kD serine protease secreted during TE differentiation and concomitant SW synthesis, and specific proteolysis of the extracellular matrix is necessary and sufficient to trigger Ca^{2+} influx.

Stage III involves autolysis and secondary wall formation. The influx of Ca^{2+} is followed by rapid collapse of the large hydrolytic vacuole and cessation of cytoplasmic streaming. After the disruption of the tonoplast, organelles with single membranes such as Golgi bodies and ER

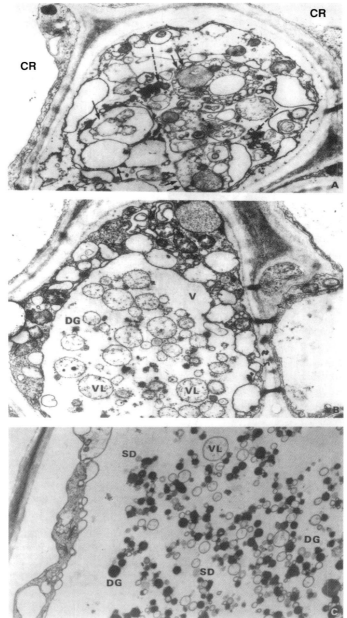

Figure 7.16 Stages in the development of laticifer in *Calotropis gigantea*. **a** A portion of a laticifer at its middle stage of development showing the process of extensive degeneration of cellular organelles and vacuolation. Autophagy is evident in a few membrane-bound cytoplasmic masses (double arrow) in which numerous membrane-bound vesicles are undergoing degeneration. Other vacuoles or vesicles (single arrow) show an advanced state of degeneration. Electron-dense particulate matter (broken arrow) starts to develop (× 8600). **b** An advanced state of development of a laticifer in which a large central vacuole (V) contains numerous vesicles (VL) containing some granular matter. Electron-dense globules (DG) are all more or less spherical discrete structures (× 8600). **c** The central vacuole of a mature laticifer shows three types of structures: membrane-bound vesicles (VI), electron-dense globules (DG) and semidense particles (SD), (×12500) (A.T. Roy and D.N. De, *Ann. Bot.*, **70**, 443, 1992). With permission of Academic Press.

become swollen and then rupture. Subsequently, organelles with double membranes are degraded (as well as the nucleus). The TE dramatically lose most of their organelles within a few hours after disruption of the tonoplast and the entire contents of the cells disappear within 6 hrs after the first visible evidence of SW thickening.

7.9.6.2 Autophagy in laticifer development

Autophagic activity has also been recorded in the formation of laticifers of many plants. In cells of the shoot apex of *Euphorbia characias*, Marty (1970, 1971) demonstrated extensive autophagic activity which led to the degradation of a large amount of the cytoplasm in the central region of the cell, leaving a thin layer of cytoplasm along the plasma membrane. Esau and Kosaki (1975) reported partial autolysis of cytoplasmic components in vacuoles of articulated laticifers of *Nelumbo nucifera*. The laticifer formation in *Calotropis gigantea* has been studied by Roy and De (1992). The transformation of the parenchymatous cells starts with the enlargement of the small vacuoles and formation of new vacuoles, concomitant with the degeneration of cell organelles. This leads to formation of various cytoplasmic islands with diverse membranous components. At a later stage, the central region is occupied by a large vacuole housing vesicles of various dimensions and small spheres of slightly electron dense latex particles (Fig. 7.16).

As well as general and random autolysis of the cytoplasm, selective autophagy has been noted in various cases of cellular differentiation. Gifford and Stewart (1968) noted in the shoot-apical cells of *Bryophyllum* and *Kalanchoe* that proplastids containing certain inclusions are transferred to vacuoles by tonoplast invagination. Villiers (1971) observed that, in the cells of *Fraxinus* embryos subjected to prolonged dormancy, vacuoles engulfed specifically proplastids and mitochondria, which were presumably damaged.

7.9.6.3 Autophagy in cotyledonary development

One of the best examples of lytic activity is shown in the protein bodies in the cells of seed cotyledons. For example, in the 17-day-old developing cotyledonary cells of carrot, a few lipid bodies invaginated in the vacuole and distinct amoeboid-like lipid bodies crossing the tonoplast were observed (Dutta *et al.* 1991). In leguminous seeds, storage protein together with substances like phytin and lectin is accumulated in structures which are equivalent to vacuoles. During seedling growth the reserve proteins are mobilised, the protein bodies appear empty and their outlining single membranes fuse to form the central vacuole (vide Chapter VI). Van der Wilden *et al.* (1980) have demonstrated the presence of hydrolases α-mannosidase, N-acetyl-β-glucosaminidase, ribonuclease, acid phosphatase, phosphodiesterase and phospholipase D. Their ultrastructural studies indicate that invagination of the protein body membrane causes portions of the cytoplasm to be internalised within the protein bodies, resulting in the formation of autophagic vesicles.

7.9.6.4 Autophagy in gametogenesis

The process of gametogenesis is the most important cell differentiation in the life cycle of a plant. Sangwan *et al.* (1989) have shown that during microsporogenesis of *Datura*, autophagic vacuoles are formed in the cytoplasm of the early meiotic cells and during pollen development cytoplasmic organelles pass into the large central vacuole and undergo degradation. They suggest that these two processes are related to the overall turnover of cytoplasmic constituents and are probably related to the sporophytic to gametophytic transition. Finally, autophagy is also involved in the release of sperm. The sperm cell is a very precisely programmed cell destined for fusion with the egg nucleus. The cytoplasm of the sperm cell is not only expendable, but its dissolution is essential for liberating the sperm nucleus prior to fusion. Ultrastructure study of isolated sperm cells of *Zea mays* by Wagner and Dumas (1989) has revealed that the cytoplasm of the sperm cell undergoes dissolution by both types of autophagy, i.e. by concentric stacking of ER and by direct degradation of the cytoplasmic mass in the pre-existing vacuole.

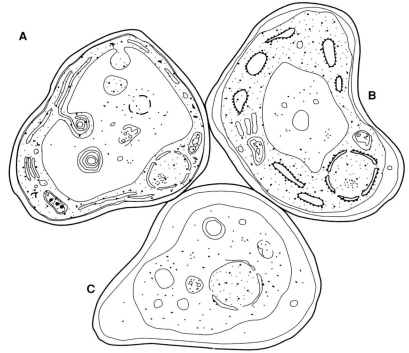

Figure 7.17 Schematic diagrams of the changes in the ultrastructure of mesophyll cells during senescence of the corolla of *Ipomoea purpurea* (Morning Glory). **a** Autophagic activity of the vacuole; invaginations of the tonoplast result in sequestration of cytoplasmic material into the lysosomal compartment (i.e. the central vacuole according to the interpretation of Matile). **b** Shrinkage of the vacuole, dilution of the cytoplasm and inflation of cytoplasmic membrane systems. **c** Autolysis is initiated by ultimate breakdown of the tonoplast (from Matile and Winkenbach 1971). With permission of Oxford University Press.

7.9.6.5 Autophagy in senescence

Senescence is the ultimate state of differentiation leading to programmed cell death. The petals of mature flowers provide a striking example of rapid senescence. By means of biochemical and ultrastructural techniques Matile and Winkenbach (1971) followed the changes in the activity of the acid hydrolases and breakdown of cytoplasmic organisation of the wilting corolla of Morning Glory (*Ipomoea purpurea*). As the corolla undergoes senescence the hydrolytic enzymes, viz. DNase, RNase, protease and β-glucosidase, increase markedly, and the various organelles undergo progressive degeneration with increasing vacuolar activity. At first the invagination of the tonoplast draws the cytoplasmic constituents within the lumen of the vacuole which acts as the lytic compartment for the cytoplasm and gradually the organelles are degraded. Further activity of the hydrolases leads to the bursting of the tonoplast and digestion of the nucleus (Fig. 7.17). The senescing petals of carnation (*Dianthus caryophyllus*) exhibited ultrastructural changes in mesophyll vacuoles from the pre-senescent stage where occasional rupture of the tonoplast was noticed. In the pre-climacteric petals, the vacuoles of some mesophyll cells were seen to contain numerous small vesicles and some mitochondria and in some other cells tonoplast appeared to burst, resulting in a general mixing of the cell contents. As the petals passed from the climacteric to the post-climacteric stage, rupture of the tonoplast led to large-scale ultrastructural disorganization (Smith *et al.* 1992). Thus, in programmed cell death autophagy of cytosolic constituents by the vacuole and/or rupture of tonoplast causing release of hydrolases leads to cellular disintegration. Smart (1994) recorded that the activity of several hydrolytic enzymes, most notably cysteine proteases localised in the vacuole, increases dramatically during leaf senescence. This catabolism frees carbons and nitrogen for mobilisation to the shoots and roots. The senescing cell and its organelles are disassembled in an ordered fashion: chloroplasts diassemble before other organelles by autolysis rather than by autophogy.

Matile (1992) stated that mesophyll chloroplasts are converted into 'gerontoplast' with degenerating internal ultrastrure during leaf senescence and yellowing, accompanied by the disassembly of thylakoid pigment-protein complexes and breakdown of chlorophyll into phytol and water-soluble porphyrin derivatives. The water-soluble porphyrins are exported from the gerontoplast and accumulated ultimately in the vacuoles of senescing cells where they remain until the end of the senescence period (Matile *et al.* 1988). This intracellular excretion storage of linear tetrapyrroles is conducted by ABC transporters in the tonoplast and serves to protect the cytosol from photo-oxidative damage which accompanies production of free radicals (Hinder *et al.* 1996). Finally however, the tonoplast bursts and autolysis is complete.

7.10 DETOXIFICATION OF SO_2

As sulfur is an essential element for plant nutrition and sulfates are the principal form of uptake of the element, detoxification of SO_2 is only an issue when there is excess sulfate in the soil or SO_2 is present in the air. As has been discussed earlier, an excess of a substance in the cytoplasm may be sequestered in the vacuole. In some form or other, excess sulfur is finally detoxified by the vacuole. When barley seedlings are grown hydroponically in the presence of high concentrations of potassium sulfate, the sulfate concentration in the mesophyll vacuoles increases by a factor of about 10 or more. Similar results were obtained with experiments on isolated vacuoles (Kaiser *et al.* 1989). Although uptake of sulfate ions cannot be considered in isolation, sulfate tolerance of barley mesophyll cells depends on the capacity of the sulfate transporter of the tonoplast.

On the other hand, gaseous uptake of SO_2 into the leaves occurs by passive diffusion. Inside the cytoplasm SO_2 is hydrated, and the hydration products HSO_3^-, SO_3^{2-} and H^+ are trapped in the slightly alkaline cytoplasm, and not in the acidic vacuole (Pfanz *et al.* 1987).

In most plants, excess SO_4 is stored in vacuoles, where it remains metabolically inactive. Accumulation of $BaSO_4$ in *Chara* and sulfuric acid in *Desmarestia* are examples of extreme cases . Irrespective of the form of available sulfur, i.e. whether it is sulfur dioxide, hydrogen sulfide or cysteine, it is stored as SO_4. When there is excess sulfur in the reduced form it may accumulate as glutathione, which appears to be the storage form of reduced sulfur in higher plants; but the storage glutathione in the vacuole is not yet clear. (Only a part of glutathione is found in the vacuole.)

In *Lemna minor*, as well as in carrot cells, sulfate is actively transported at both plasmalemma and tonoplast (Thoiron *et al.* 1981; Cram 1983a, b). However when excess sulfate is made available, the tonoplast regulates the influx by feedback inhibition (Cram 1983a). Rennenberg (1984) outlined a model for the influx of sulfate into cellular compartments. With increasing sulfate concentration in the cytoplasm, vacuolar sulfate concentration will increase. This will cause a decrease in the active influx of sulfate and the efflux by facilitated diffusion will increase until both fluxes are equal. Negative feedback control by sulfate may prevent the vacuolar sulfate concentration from increasing indefinitely with increasing external supplies of sulfate. Kaiser *et al.* (1989) have shown that sulfate accumulation is powered by a tonoplast ATPase. The main cation accompanying sulfate during net uptake into vacuoles is K^+ by the K^+/H^+ antiport mechanism. The role of vacuoles in the absorption of atmospheric SO_2 has also been emphasised by the work of Kaiser *et al.* The sulfate-pumping capacity of tonoplast of barley mesophyll cells is more than sufficient to cope with transport requirements for sulfate when leaves are exposed to SO_2 pollution within permissible limits (140 µg SO_2 m^{-3}). Moreover, vacuolar detoxification capacity appears to be limited by the vacuole's capacity for proton storage, not by that for sulfate storage. The phytotoxic sulfite and bisulfite anions may be detoxified by either reduction or oxidation. The primary product of reduction of H_2S may be emitted and the main secondary products are cysteine and organic compounds. However, it appears that the major detoxification mechanism is oxidation to sulfate, which is actively sequestered in the vacuole by a symport mechanism in which Ca^{2+}, K^+ and Mg^{2+} serve as dominating cations for sulfate in coniferous trees (Hüve et al. 1995).

7.11 Detoxification of xenobiotics

Non-nutrient, foreign soluble or particulate substances which may enter the plant cell have been considered in the section on heterophagy. But, xenobiotic chemicals which are cytotoxic must be detoxified to the process of detoxification by plant cells. There is no sharp line of demarcation between heterophagy and detoxification, except that detoxification deals with xenobiotics and herbicides. Xenobiotics are diverse synthetic chemicals and herbicides are lipophilic electrophiles that penetrate the barriers surrounding the symplast, i.e. the extracelluar matrix and plasma membrane as well as the other endomembrane systems. Plant cells are able to metabolise and detoxify many xenobiotics by a variety of enzymatic reactions. Ishikawa (1992) described the metabolism and detoxification of xenobiotics in three phases. In phase I (activation), the compound is oxidized, reduced, or hydrolysed to expose or introduce a functional group of the appropriate reactivity for phase II enzymes. In phase II (conjugation), the activated derivative is conjugated with hydrophilic substances, such as glutathione, glucuronic acid or glucose. In phase III (elimination), the conjugate is either excreted to the extracellular medium or sequestered in an intracellular compartment. Kreuz et al. (1996) and Rea et al. (1998) suggest phase IV in which the transported conjugates are further substituted and degraded to yield transport inactive derivatives. After phase II the resulting conjugates are i) generally inactive toward the initial target site; ii) more hydrophilic and less mobile in the plant than the parent substance; or iii) susceptible to further processing which may include secondary conjugation, degradation and compartmentation (Kreuz et al. 1996). The phase I process of oxidation, reduction or hydroxylation may be carried out by a number of enzymes, of which cytP450-dependent mono-oxygenases play a pivotal role. There are a number of CytP450 forms which may mediate hydroxylation of aromatic rings or alkyl groups, or hetero-atom release (Barret 1995). After oxidation or hydrolysis, which usually introduces a hydroxy, amino or carboxylic acid function, the herbicide is subjected to rapid glycosylation by a glucosyltransferase utilising UDP-GLc as the usual sugar donor. The glucosyltransferases which are needed for biosynthesis of secondary metabolites are present in the plant cytosol either as soluble or membrane-bound forms. On the other hand, numerous studies have established the existence of glutathione-S-transferase (GST) families comprising isozymes possessing varying degrees of substrate specificity for particular herbicides (Kreuz 1993). GST conjugates glutathione (GSH) to the electrophilic substrate (X–Z) with concomitant displacement of a nucleophil (Z, eg. halogen, phenolate or alkyl sulfoxide):

$$X - Z + GSH \xrightarrow{GST} X - GS + HZ$$

The GSH conjugates of herbicides in plants may undergo extensive processing to Cys or N-malonylcysteine and other conjugates. Since the two classes of herbicide complexes, viz. glucosylated and glulathione-conjugated, may be harmful to the cytosol, they are either processed further or sequestered in a cellular compartment. No information is available on the excretion of the detoxified xenobiotic to the extracellular matrix or apoplast across the plasma membrane. However, a number of workers have investigated the mechanism of vacuolar transport. Vacuoles are known to accumulate a wide range of unrelated compounds varying in molecular weights, steric configurations and charges, many of which may be due to the presence of a unidirectional transporter in the tonoplast. Oparka et al. (1996) suggested that a probenecid-sensitive uptake mechanism on the tonoplast may function as a part of a detoxification system for the removal of xenobiotics from the cytosol. Martinoia et al. (1993) studied the uptake of a number of glutath-

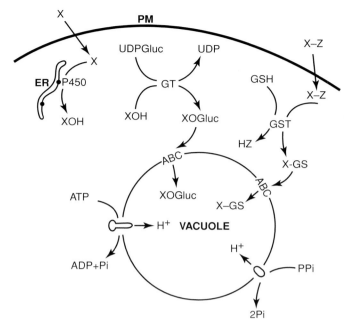

Figure 7.18 Schematic representation of the role of ATP-binding cassette (ABC) transporters in detoxification of xenobiotic X and X–Z. Z, the nucleophile moiety displaced in the GST-catalysed reaction. PM, plasma membrane; P450, CytP450 monooxygenase; GT, glucosyl transferase, GSH, glutathione; GST, glutathione S-transferase

ione conjugates by isolated barley vacuoles and concluded that they accumulated in the vacuole even when the vacuolar H$^+$-ATPase was inhibited and the pH gradient between the cytosol and the vacuole was abolished, indicating the presence of a specific transport ATPase. Moreover, it is possible that only a single transporter recognises all the different forms of the glutathione conjugates (Tommasini et al. 1993).

Li et al. (1995a, b) have shown that, unlike most other characterised organic solute transport in plants, uptake of the model compound S-2,4-dinitrophenyl glutathione (DNP-GS) and glutathione-S-conjugates by *Vigna* tonoplast vesicles is directly energised by MgATP. The detection of glutathione-S-conjugate transporter satisfied the minimum requirement of an element involved in the detoxification of xenobiotics, viz. enhancement of its activity in response to exposure of the intact organism to the target compound. The degradation of GSH conjugates to the Cys conjugate has often been observed for herbicides and may represent a further detoxification step. Uptake of herbicide glucosides increases in the presence of Mg-ATP and unlike glutathione conjugates, their uptake is inhibited by other glucosides (Gaillard et al. 1994). The ATP-binding-cassette (ABC) family of transporters present in the tonoplast may function to sequester xenobiotics, as well as endogenous compounds into the vacuole (Fig. 7.18). Xenobiotics, like other compounds in the cytoplasm, will require glutathione canjugation to be transported across the tonoplast via the ABC transporter (Swanson and Jones 1996). Recently, Lu et al. (1998) identified a gene, *AtMRP2* from *Arabidopsis*, that encodes a multispecific ABC-transporter competent in the transport of both GS-conjugates as well as chlorophyll catabolites. It is intriguing that the same gene is responsible for both and it is now accepted (vide review by Rea et al. 1998) that the members of the ATP-binding cassette (ABC) transport superfamily, the largest protein family known, are capable of a multitude of transport functions. They are directly energised by Mg^{2+}-ATP rather than by the electrogenic pumps, V-ATPase and PPase. A major subclass of the superfamily, MRPs (or multidrug resistance associated proteins as named by their animal prototypes), are represented by GS-X pumps (glutathione-conjugate or multispecific organic anion Mg^{2+}-ATPase), which participate in the transport of exogenous and endogenous amphipathic anions and glutathionated compounds from the cytosol into the vacuole. Several other such pumps have been detected and implicated in the transport of i) large amphipathic organic ions like chlorophyll catabolites (Alfenito et al. 1998), bile acid-like substances e.g. digitonin, sterol or lipid derivatives and glucuronate conjugates; ii) glucose conjugates; and iii) phytochelatins. Thus they are pivotal in herbicide detoxification, cell pigmentation, alleviation of oxidative damage, and storage of antimicrobial compounds. They may also be involved in channel activity and/or heavy metal chelates (discussed in the next section).

For Phase IV, or transformation, some information is available on storage and excretion of the end product. Wolf et al. (1996) recorded that exposure of barley leaves to alachlor results in massive vacuolar accumulation of alachlor-GS. Colemen et al. (1997) demonstrated in situ glutathionation of monochlorobimane to bimane-GS and accumulation of the latter in vacuoles of cultured cells. Very high accumulative capacity vacuolar GS-conjugate pumps have been demonstrated in studies by Li et al. (1995) and Wolf et al. (1996). Measurements of DNP-GS(S-(2,4-dinitrophenyl)-glutathione) by vacuolar membrane vesicles and of the distribution of the glutathionated herbicide, alachlor-GS, between the vacuolar and non-vacuolar compartments of barley leaves yield and accumulation ratio above 50. The detoxified substance may not always be stored in the vacuole. The transport of bound residues to the extracellular matrix has been noted in certain cases (Sandermann 1992).

7.12 PHYTOREMEDIATION OF HEAVY METALS

The removal of pollutants from the environment by green plants, a process known as phytoremediation (Cunningham and Berti 1993; Raskin et al. 1994) is well known. Plants have a unique sorption potential for a large array of metallic cations from the environment. The uptake of toxic metals could be either through adsorption and/or absorption and is possible through some physiological adaptiveness and homoeostasis. The metals are transported to various compartments, especially to apparent intercellular or intracellular free space, vacuoles and other such bodies, to sequester the metals away from the main metabolic pathways. Formation of immobilised heavy metal containing crystals have been described for Cd, Co, Fe, Pb, Sr and an increase in Ca-oxalate crystals observed under heavy exposure to the metal. In general, the metals may form metal-organic acid, metal-sulfide, metal-phytate, or metal-peptide complexes in plants.

Zinc

When exposed to Zn salts, intravacuolar bodies in root meristematic cells of the grass *Festuca rubra* were found to contain Zn (Davies et al. 1991). Exposure to zinc salts actually induces vacuolation to various degrees. An exposure to a subtoxic concentration of Zn can induce a 2.93 fold and 6.78 fold increase in total vacuolar volume fraction in rice and wheat root meristematic cells, whereas rye roots exhibit no increase in vacuolation (Davies et al. 1992). In the case of Zn-tolerance Mathys (1977) suggested that Zn was chelated in the cytosol by malate, and this complex was delivered to the vacuole where it was sequestered as Zn-oxalate. A high malate content can be correlated with high Zn tolerance. However, this model has been criticised for a lack of critical experimental evidence. Thurman and Rankin (1982) have shown that the metal may indeed be bound to organic acids in the vacuole. Godbold et al. (1984) have demonstrated a high correlation between citrate content and Zn-accumulation in the root cortical cells of *Deschampsia caespitosa* and argued that citrate may have a role in Zn-tolerance. In the same species, Van Steveninck et al. (1987) showed that Zn is complexed with phytate in small vacuoles in root cortical cells. This observation is especially interesting in view of the fact that Zn^{2+} can bind tightly at the multiple phosphate groups of phytate (inositol hexophosphate).

X-ray microanalysis of Zn accumulation in *Thlaspi caerulescens*, a known hyperaccumulator of metals, by Vázquez et al. (1994), showed that Zn was compartmentalised in vacuoles of roots as well as leaves. When exposed to high concentrations of Zn, the plant exhibited a higher concentration of the metal in leaves than in roots. Zn storage is achieved by the formation of Zn-rich crystals in vacuoles of epidermal and subepidermal cells of leaves. Both the Zn/P element ratios found in the crystals and the absence of Mg indicate that, in contrast to other plant species, phytate is not the main storage form for Zn in *I. caerulescens*. On the other hand, vacuolar citrate may be the central mechanism for sequestration of Zn in plants exposed to either low or high levels of the metal (Wang *et al.* 1992). They used computer modeling of chemical equilibria to predict the metal-ligand species in the vacuoles of suspension culture cells of tobacco.

Cadmium

Like Zn, cadmium can also make organic acid complexes and be retained in the vacuole. Computer simulation also showed that in moderate concentrations the metal may form stable complexes as Cd-citrate despite an abundance of malate (Wang *et al.* 1991). On the other hand, a number of reports showed that Cd-peptides are chiefly located in vacuoles of leaf protoplasts of plants or cultured cells grown under Cd-pollution conditions. (Vögeli-Lange and Wagner, 1990; Krotz *et al.* 1989). Electron probe microanalyses by Heuillet *et al.* (1986) and Van Stevenick *et al.* (1990) provide conclusive evidence that vacuoles are the site of deposition of Cd. A class of cysteine-rich, metal-binding proteins, known as phytochelatins, may have an important role in cellular tolerance to toxic metals (Tomsett and Thurman 1988; Rauser 1990). Phytochelatins consist of repeating units of γ-glutamylcysteine followed by a C-terminal glycine [poly-(γ-Glu-Cys)$_n$-Gly peptides]. Although certain studies have shown that Zn is not generally bound to phytochelatin in vacuoles (Wang *et al.* 1992), on exposure to Cd a significant proportion of intracellular Cd is bound as a heterogenous Cd-phytochelatin sulfide complex, production of which is necessary for Cd-detoxification. The complex is localised in the vacuole (Steffens 1990; Wagner 1994). Salt and Wagner (1993) identified a Cd^{2+}/H^+ antiport on the tonoplast, which may be responsible for Cd-accumulation in the vacuole. Salt and Rauser (1995) have reported that when the oat roots were exposed to moderate to high levels of Cd, the metal was found to be primarily associated with the phytochelatin. The authors have shown that Cd-complexes of the phytochelatin are transported across the tonoplast by a MgATP-dependent, vanadate-sensitive, possibly ATP-binding cassette type (ABC) transporter which utilises an ATPase-proximate energy source for active organic solute transport. On the basis of studies on Cd-sensitive mutants of yeast *Schizosaccharomyces pombe*, Oritz *et al.* (1994) have shown that an ABC-transporter associated with the tonoplast is responsible for transport of both phytochelatin and Cd-phytochelatin complexes. It may be concluded that the vacuole is intimately associated with the process of detoxification of metals. Whereas a moderate amount of the metal can be complexed with organic acid inside the vacuole, high doses of the metal may cause, at least in a number of cases, the induction of phytochelatin synthesis and subsequent transport of the metal-phytochelatin complex to the vacuole. On the other hand, a novel gene *YCF1* or cadmium factor (a close homologue of *MRP1* of the ABC family), which confers resistance to cadmium salts, was isolated from yeast cells. Recently, Li *et al.* (1997) studied the substrate requirements, kinetics and Cd^{2+}/glutathione stoichiometry of cadmium uptake and the molecular weight of the transport-active complex which demonstrated that YCF1 selectively catalyses the transport of bis(glutathionato) cadmium. Recently, Ghosh *et al.* (1999) have shown that in yeast cells the same gene is responsible for removal of arsenite (As III) from the cytosol by sequestering it in the vacuole as the glutathione conjugate. Hence, the cadmium-factor gene encodes a MgATP-energised glutathione-S-conjugate transporter responsible for vacuolar sequestration of organic compounds after their S-conjugation with glutathione.

Nickel

Although the vacuole has long been suspected of being a site for accumulation of nickel in plant roots, no direct evidence is yet available. Lately Gries and Wagner (1998) have made a critical attempt to study Ni transport along with Ca and Cd into the tonoplast vesicles of oat roots. Although

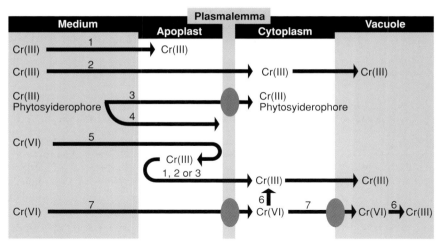

Figure 7.19 Chromium uptake model based on current hypothesis of micronutrient uptake. (1) retention of Cr(III) in the apoplast; (2) Diffusion of Cr(III) through plasmalemma and tonoplast; (3) Binding of Cr(III) by phytosiderophores and transport of the complex through plasmalemma (in monocots); (4) binding of the Cr(III) phytosiderophore complex to plasmalemma; (5) reduction of Cr(VI) to Cr(III) by root reductase; (6) reduction of Cr(VI) in cytoplasm or vacuole (7) transport of Cr(VI) by anion carriers (from Barceló and Poschenrieder 1997). Courtesy J. Barceló. With permission of Franco-Angeli.

Ni was associated with the vesicles, its relative rate of accumulation/association is very low compared to that of Ca and Cd. Protonophores and the potential Ni ligands citrate and histidine, nucleoside triphosphates or PPi did not stimulate Ni association with vesicles. Comparison of Ni versus Ca and Cd associated with vesicles, using various membrane perturbants, indicate that while Ca and Cd are rapidly and principally antiported to the vesicle sap, Ni is only associated with the membrane in a hardly dissociable condition. Hence, it has been concluded that no Ni^+/H^+ antiport exists in oat tonoplast and the vacuole is not a major compartment for Ni accumulation.

Chromium

Because chromium does not cause widespread environmental problems and phytotoxicity due to Cr under field conditions is unknown, Cr toxicity has not been widely investigated. Of the common oxidation states of Cr in inorganic compounds, +II, +III and +VI, the Cr(III) and Cr(VI) are biologically significant. On the basis of diverse mutagenic and carcinogenic effects of Cr(VI) on various organisms, Gebhart (1984) regarded it as a highly cytotoxic substance. Vázquez *et al.* (1987) studied the ultrastructural changes in the cells of *Phaseolus vulgaris* which were treated with Na_2CrO_4, that is, Cr in hexavalent form, in the nutrient solution. They noted that besides other ultrastructural damages, the membranes, especially of the vacuole in root and stem cells, were damaged and certain precipitates were also detected in the vacuoles. Membrane damage by Cr(VI) has also been described in the fronds of two fresh water plants, *Lemna minor* and *Pistia stratiotis* (Bassi et al. 1990). In view of i) the similar effective ionic radius of Cr(III) and Fe(III); ii) understanding of the mechanism of absorption of Fe as a micronutrient (Welch 1995); iii) binding of Cr(III) by phytosiderophores (plant homologues of animal non-heme Fe-transport protein); and iv) certain other similarities in uptake of the two metals, Barceló and Poschenrieder (1997) proposed a model of Cr uptake (Fig. 7.19). According to this model, what is finally accumulated in the vacuole is the trivalent form of Cr. However, the mechanism of transport of Cr(III) across the tonoplast is not known. Since Cr(III) tightly binds to carboxyl groups of free amino acids as well as to those in proteins, it may form metaloprotein complexes and also some low molecular weight complexes containing nicotinic acid, glycine, glutamic acid and cysteine (Toepfer *et al.* 1977). The formation of Cr-phytochelatin remains to be confirmed. As discussed earlier, it is possible that certain ABC transporters could be involved in sequestration of Cr-conjugates in the vacuole.

Aluminium

Unlike chromium, aluminium may cause toxicity problems in acid soils (pH <5.0) for crop productivity, which can be ameliorated by application of lime or gypsum. Al accumulates mainly in the apoplast and crosses the plasma membrane slowly, so toxicity occurs mainly in the apoplast (vide review by Kochian 1995). Aluminium was not detectable by specific staining in root tip vacuoles of either tolerant or sensitive wheat exposed for 48 hr to Al (Tice *et al.* 1992). Electron probe X-ray microanalysis by several authors (Marienfeld *et al.* 1995; Lazof *et al.* 1997) failed to detect any Al deposit inside the cytoplasm or vacuoles of roots of plants treated with Al. However, Barceló *et al.* (1996) detected Al in electron-dense deposits in the root tip vacuoles of maize plants treated with Al for 24 to 120 hr. Recently, Vázquez *et al.* (1999) have reported that, in the roots of a tolerant maize variety, no Al was detected in cell walls after 24 hr, but an increase in vacuolar Al was observed after 4 hr of exposure. After 24 hr, higher amounts of Al, P and Si were observed in the vacuolar deposits. Moreover, the mineral contents of P, Ca, Zn and Mg in these deposits are similar to those reported for phytate deposits from maize (Mikus *et al.* 1992). Thus formation of phytate inside the vacuole may be responsible for Al tolerance.

It appears that the vacuole plays a distinct role as the terminal point in phytoremediation of a number of metals. A toxic metal may be conjugated to low molecular organic acids, glutathione peptides, phytochelatin or phytate. Certain metals may be transported by proton antiporters and others may be conjugated in the cytoplasm to be sequestered in the vacuole by specific ABC transporters. Thus, when a cell is subjected to an undesirable metal or even an essential metal in high dosage, it uses certain effective metabolic pathways to conjugate it and finally store it in the safe custody of a vacuole in an inert form. The tolerance or resistance of a plant depends, at least partly, on this capacity for detoxification and sequestration in the vacuole.

7.13 SCAVENGING OF ACTIVE OXYGEN SPECIES

Various stress conditions like wounding, attack of pathogens, and exposure to xenobitics, heavy metals, UV and ionising radiation (reviewed by Inz and Van Montagu 1995), induce an oxidative system with formation of molecular species of active oxygen: superoxide (O_2), hydrogen peroxide (H_2O_2), hydroxyl radical (OH), singlet oxygen ($^1O_2$) or peroxyl radical. These active oxygen species (AOS) are normally involved in major aerobic biochemical reactions like respiratory or photosynthetic electron transport and are produced in copious quantities by several enzyme systems. While endogenous AOS are readily utilised by the cell metabolism, exogenous AOS have to be rapidly processed or scavenged to prevent oxidative damage. AOS are strongly nucleophilic and react with biomolecules to generate reactive electrophiles. In this respect glutathione is extremely important because it acts both as a reducing agent protecting the cell from AOS and as a nucleophile protecting the cell from the electrophilic products of AOS action. Although a direct connection between oxidative stress and the GS-X pump on the tonoplast has not been established, it is known that a glutathione S-conjugate pump on the tonoplast transports oxidised glutathione (GSSG) into the lumen (Li *et al.* 1995) and many of the oxidative stresses are known to induce specific glutathione S-transferases (vide review by Rea *et al.* 1998). In addition to glutathione, a number of efficient antioxidants like ascorbate, carotenoids and flavonoids are found in plant cells. Many studies have indicated that flavonoids, which are the most common secondary metabolite in vascular plants, can directly scavenge AOS. Flavonoids are largely localised in vacuoles and it is unlikely that various active radicals will pass through

220 Plant cell vacuoles: an introduction

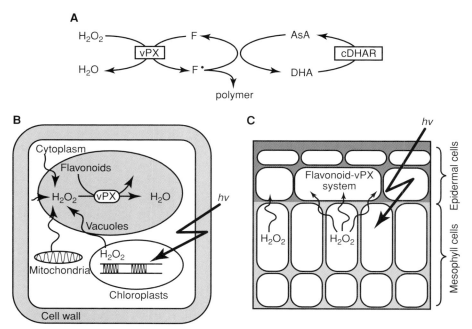

Figure 7.20 A proposed diagram for the protective function of flavonoids during stress and growth. **A** Scheme of the H_2O_2-scavenging mechanism by flavonoids. vPX, Vacuolar peroxidase; F, flavonoid; F; flavonoid radical; AsA, ascorbic acid; DHA, dehydroascorbic acid; hv, light energy; and cDHAR, cytosolic dehydroascorbic acid reductase. The diffusive nature of H_2O_2 enables vPX to scavenge it in vacuoles, even if the generating site is other than a vacuole (B). This concept can be expanded to the cell–cell interaction. The photoproduced H_2O_2 may leak out from mesophyll cells and be scavenged in epidermal cells that have a high flavonoid content (C) (from Yamasaki *et al.* 1997). With permission of the American Society of Plant Physiologists.

the tonoplast, since degradation of lipids of the tonoplst would be their first hit. However, unlike other AOS, H_2O_2 is stable and able to diffuse across membranes. On the other hand, flavonoids are electron donors for peroxidation and Takahama (1989) has shown that the flavonol-peroxidase reaction may act as an H_2O_2-scavenging system *in vivo* as well as *in vitro*. Yamasaki *et al.* (1997) reported that the flavonoids in the leaves of a tropical tree *Schefflera arboricola* have the potential to act as electron donors to peroxidase and that vacuolar peroxidase participates in H_2O_2-induced oxidation of flavonoids. Thus, the flavonoid-peroxidase reaction can function as a mechanism for H_2O_2-scavenging in plants (Fig. 7.20). Although the scheme is attractive, further investigations are necessary to determine its validity and efficiency in other cell systems.

7.14 SUMMARY

Gross function

The gross function of the vacuole is **storage of water** along with **dissolved** or **undissolved** substances in its lumen. The transport of molecules across the tonoplast takes place either by **passive transport** which is independent of energy, or by **active transport** which requires energy.
- Passive transport may take place by **simple diffusion** or **facilitated diffusion** aided by certain **transport proteins** which come in two classes: i) **carrier proteins** which bind transiently with solute, and include **transporters** and **group translocators**; and ii) **channel proteins** which consist of **porins**, **ion channels** and **ion transporters**.

- The ion channels are not coupled to energy, yet provide a very fast rate and ion specificity. They may be open or closed due to a stimulus of voltage across the membrane, mechanical stress or chemical ligands. They are indirectly aided by energy and conduct secondary active transport.
- Bulk flow of water across the tonoplast takes place through a porin, known as **aquaporin**, which is partly controlled by state of phosphorylation, salinity, etc.

Uptake of ions

The large accumulation of ions in the cell, especially in the vacuole, occurs chiefly through the two transport processes.

- **Primary active transport** is the process by which an ion is moved against a gradient of electro chemical potential energised by either of the two pumps, vacuolar type H^+-ATPase (V-ATPase) and H^+-pyrophosphatase (H^+-PPase). Both pumps on the tonoplast catalyse inward electrogenic H^+-tranlocation from cytosol to vacuolar lumen to establish inside positive trans-tonoplast H^+-electrochemical potential difference for the energisation of H^+-coupled solute transport. Both V-ATPase and H^+-PPase activities may coexist in the tonoplast but V-ATPase is the predominant proton pump and in many cases H^+-PPase takes over only when V-ATPase fails or is inhibited.
- **Secondary active transport** is the process in which the transporter is aided by pH gradient and/or electrical gradient. Three types of secondary active mechanisms are i) *Uniport*, which transports an ion or a small molecule against a concentration gradient, uncoupled to the movement of any other molecule; ii) *Antiport*, which transports an ion or small molecule by coupling its movement to that of another ion or molecule in the opposite direction, and iii) *Symport*, which co-transports an ion or molecule with another ion or molecule in the same direction.
- **Uptake of anions** generally takes place through symport or even uniports which are *anion channels*. The anions move through the anion channel in the tonoplast in response to positive membrane potential generated by V-ATPase. In addition, there may be specific transports in the tonoplast for various organic or inorganic anions.

Vacuolar pH regulation

The pH of the vacuole may vary considerably from the pH of the cytoplasm. In general, H^+-ATPase is chiefly responsible for maintaining the vacuolar pH. However, depending on the cell system and its metabolic activity, H^+-PPase may control vacuolar pH. Besides programmed pH changes, the vacuole may act as a buffer for additional or sudden changes in the pH of the cytoplasm. In many cases, the balance between the passive leakage of H^+ from vacuole to cytoplasm and active H^+ transport from cytoplasm to vacuole seems to play a key role in vacuolar pH regulation.

Salt tolerance

The salt tolerance of the cell or plant is derived from its capacity to tolerate sufficient ions to maintain growth while avoiding water deficit or excess of ions. Compared to the cytoplasm, vacuoles compartmentalise large amounts of ions and the tonoplast proton pumps and ion transporters are responsible for such accumulation. Membrane phospholipid may play a role in the process.

Osmoregulation

Osmoregulation has two components: turgor pressure and osmotic pressure. One of the vacuole's basic functions is the maintenance and modulation of turgor pressure by controlled uptake and loss of water. Adequate turgor pressure is essential to maintain the plant cell's rigidity as well as its growth. Due to the tonoplast's high water permeability, the vacuole is always in osmotic equilibrium with the cytoplasm. Salts and sugars are the main solutes which affect osmotic pressure and hence turgor pressure. Thus, osmoregulation is brought about by one or both the following processes: i) interconversion of organic solutes to soluble and insoluble components; and ii) ion transport across the tonoplast.

Turgor movements

Turgor movements in various plant cells, tissues and organs are due to changes in turgor pressure of the vacuoles of participating cells. The collapse and recovery of the vacuole responsible for guard cell movement is caused by operation of electrogenic pumps as well as K^+ channels, Ca^{++} channels of the tonoplast, in response to cytosolic changes in anions and Ca^{2+}. The turgor movement of leaf pulvinar motor cells also involves fluxes of water, K^+, Cl^- and Ca^{2+} from the vacuole. Tannin-bound Ca^{2+} and Ca^{2+}-oxalate play a secondary role.

Ca^{2+} Homeostasis

Ca^{2+} is the pivotal ion species which actively controls numerous physiological processes in the cytoplasm. Its concentration in the cytoplasm is maintained within a narrow range by influx or efflux of Ca^{2+} from the vacuole which may maintain the ion at a level of 3 to 4 orders of magnitude of the cytoplasm. The multiplicity of transporters and channels for Ca^{2+} suggests that the most prevalent divalent cation acts as an intracellular second messenger. A number of Ca^{2+}-binding proteins, including the most studied calmodulin, modulate many vacuolar fluxes of Ca^{2+} in response to diverse stimuli.

Metabolic functions

Vacuoles control a large number of metabolic functions of the cell.
- During **Crassulacean acid metabolism**, malic acid produced by dark assimilation of CO_2 is transported by a malate transporter in the tonoplast into the vacuole. During the day, malate in the vacuole is exported either by passive diffusion or mediated by a non-selective ion channel.
- In the **homeostasis of amino acids**, vacuoles play a significant role in yeast and other fungi where they maintain a high concentration of certain amino acids. In higher plants, specific transport systems for uptake of various amino acids by the vacuole have been demonstrated. However, the amino acids are maintained at an extremely low critical concentration and any excess is efficiently exported from the vacuole.
- Depending on the metabolic state, **sucrose accumulation** by the vacuoles of different cell types may take place by diffusion mediated by carrier protein or by sucrose/H^+ antiporter. In the storage cells of sugar cane and red beet, a multienzyme complex in the tonoplast may govern the formation and translocation of sucrose into vacuoles.
- **Biosynthesis inside vacuoles** may be conducted by certain enzymatic processes in the lumen. One simple example is that of conversion of sucrose to glucose and fructase by invertase in the vacuoles of ripening fruit cells. Again, in the vacuoles of Jerusalem artichoke, sucrose is converted to inulin by the action of a series of enzymes. A number of enzymes for uptake of anthocyanin and subsequent processing have been detected in the tonoplast and lumen. Anthocyanins may be trapped by conformational change or conjugated with protein for retention inside the vacuole. Similarly, ethylene is formed from its precursor.
- The best example of **vacuolar processing** is noted in the developing protein bodies of seeds. In the storage bodies, the vacuolar processing enzyme, cysteine protease, converts trimeric proglobulin to hexameric globulin, proglycinin to glycinin, or proricin to ricin in different plant cotyledons. Concanavaline A is first synthesised as Pro-Con A which undergoes combined action of endopeptidase and carboxypeptidase. Other vacuolar enzymes like proteinases undertake processing activity in different cell systems including yeast.
- Perhaps **biosynthesis of alkaloids** in the cells of the medicinal plant *Catharanthus roseus* is the best illustration of vacuolar compartmentalisation of specific metabolic activity. For example,

the precursor substances tryptamine and geraniol, produced in the cytosol, are transported to the vacuole and conjugated there to form strictosidine. This unstable gluco-alkaloid is readily passed out of the vacuole for removal of glucose moiety and subsequently converted to ajmalicine or cantharidine which may again enter the vacuole for peroxidation/dimerisation to form stable akalolids for storage.

Lytic functions
- The **lytic functions** of the vacuole are evident from the presence of both hydrolytic and proteolytic enzymes in its lumen. That the enzymes are present and are released by the alteration of the permeability or disruption of the tonoplast has been amply demonstrated. During intracellular turnover, programmed or induced differentiation, or death, vacuolar endopeptidases are effective in degradation of the cytoplasmic proteins and enzymes, but not the vacuolar hydrolases. Extraneous proteins, if introduced inside a vacuole, can be readily degraded. Mutant or defective proteins are also known to be taken up by the vacuole and digested. The presence of RNA-degrading enzymes has been reported in certain vacuoles.
- Another aspect of lytic functions is **autophagy**, which involves degradation of intracellular structures and substances after their ingestion by the vacuole. Besides organelle breakdown inside the vacuole, autophagy may be initiated by RER or some endomembrane enclosing and isolating a certain mass of the cytoplasm. The membrane-bound sac may produce lytic enzymes and gradually degrade its contents to form an autophagic vacuole. In other cases, cytoplasmic fragments and organelles invaginate the tonoplast. The intruding mass of the cytosol together with the organelles is pinched off from the tonoplast and finally digested. Thus, autophagy is a routine process in many differentiating and developmental systems. An example is provided by wheat storage proteins which are synthesised by ER to form protein bodies. These enter the vacuole by typical autophagy. On the other hand, perturbation in nutrient requirement may induce autophagy. In yeast, starvation may induce formation of autophagosome which fuses later with the vacuole. If the methanol-utilising yeast *Pichia* cells are shifted to ethanol or glucose as the carbon source, the constitutive peroxisomes become redundant and they are soon sequestered into autophagosomes for subsequent fusion with the vacuole.
- **Heterophagy**, which denotes ingestion of foreign non-nutrient substances, is not a normal process in plant cells. Heterophagy of dissolved substances may take place by various mechanisms. In certain cases, internalisation may involve fluid-phase endocytosis. Typical animal-like vesicle-mediated endocytosis is undertaken for ferritin-like substances.
- In plants, **host-pathogen** interaction does not involve typical animal-like phagocytosis. However, the well-studied *Rhizobium* infection of leguminous root cells exhibits some semblance of phagocytosis. The infection is initiated with the release of the bacteria close to the vacuole and the bacteria are protected against lytic enzymes by a peribacteroid membrane. These bacteria-housing vesicles may be considered as heterophagosome. The remnants of the infection thread can be noted in vacuoles and are readily digested.

    Certain substances produced by pathogens called **elicitors** may induce production of host-encoded proteins called **pathogenesis-related** (PR) proteins. Elicitors may induce transcription of genes for fungal wall-degrading enzymes like chitinase and glucanase which are stored in vacuoles. When the host cell is lysed, the vacuolar enzymes are released to destroy the pathogen.

    Hypersensitive reaction (HR) is indicated by necrosis of cells surrounding the point of infection. HR is caused by release of the vacuolar degrading enzymes which may be preceded by formation of active oxygen species (oxidative burst) in the elicited host cell. Another defence arsenal is a group of anti-microbial compounds called **phytoalexins** which are constituvely produced or elicited by pathogens. Phytoalexins are generally stored in vacuoles and are active when needed.

- As the first step towards **cellular differentiation**, autophagy in the form of ingestion of cytoplasmic organelles inside the vacuole or rupture of the vacuole is necessary.

  Cytochemical and biochemical studies or differentiation of sieve tubes and tracheary elements from the precursor cells have indicated that differentiation is invariably associated with the production and release of vacuolar enzymes. Extensive or partial autophagy in a series of cells is necessary for laticifer development in many plants.

  Autophagy has also been extensively recorded in the development of cotyledonary leaves and is a well-known process during embryo sac development. During microgametogenesis autophagy is evident in the development of pollen tubes from pollen grains as well as in the dissolution of the cytoplasm proximal to the sperm nuclei.

  Senescence of cells and tissues is always associated with uptake of cytoplasmic organelles or their breakdown substance by the vacuole. Ultimately the tonoplast is ruptured and the vacuolar enzymes dissolve most of the cell structure.

Detoxification of SO_2

Generally, the excess amount of sulfate which is absorbed by the roots is sequestered in the vacuole. When SO_2 is in excess in the atmosphere the sulfate-pumping activity of the tonoplast of the leaf mesophyll cells is generally sufficient to cope with excess sulfate formed in the cytoplasm.

Detoxification of xenobiotics

Xenobiotic or harmful compounds in the environment are detoxified by plant cells in four phases: i) activation, by which the xenobiotic is oxidised, reduced or hydrolysed; ii) conjugation by which the activated substance is conjugated with a hydrophilic substance; iii) elimination, in which the conjugate is secreted or sequestered in the vacuole; and iv) termination, in which the conjugate is converted to inert derivatives. A large number of xenobiotics are conjugated with glutathione in the cytosol and subsequently the glutathione-conjugate is transported into the vacuole by glutathione-conjugate pump, a member of the family of ATP-binding cassette transporters.

Phytoremediation of heavy metals

Phytoremediation involves uptake of heavy metals by plants from the environment. If absorbed in excess, Zn is generally conjugated with organic acids like oxalic acid or malic acid in the cytoplasm and later compartmentalised as Zn-deposits in the vacuole. However, Cd, in addition to conjugation with organic acids, is chiefly conjugated with a class of cysteine-rich metal-binding proteins, called phytochelatin and transported to the vacuole. Cd-glutathione conjugates may also be sequestered in the vacuole. Cr is usually compartmentalised in its trivalent (Cr (III)) form in the vacuole as a Cr-organic acid complex or even as a Cr-phytochelatin complex. In cases of excess Al, the metal is detected as Al-phytate in the vacuole.

Scavenging of active oxygen species

Scavenging of active oxygen species (AOS) is the latest in a long list of beneficial roles played by vacuoles in the life of a cell or a plant. Various stress conditions produce AOS, viz. superoxide, hydrogen peroxide, hydroxyl radical, singlet oxygen and peroxyl radical. Glutathione (GS) is the most important molecule which scavenges AOS in the cytoplasm and forms oxidised glutathione (GSSG) which is readily transported to the vacuole. When H_2O_2, a stable AOS, diffuses across the tonoplast, the flavonoids in the vacuole may act as a H_2O_2-scavenging system by flavonoid-peroxidase reaction.

7.15 REFERENCES

Abel, S. and Glund, K., Localization of RNA-degrading enzyme activity within vacuoles of cultured tomato cells, *Physiol. Plant.*, **66**, 79, 1986.

Abel, S., Blume, B. and Glund, K., Evidence for RNA-oligonucleotides in plant vacuoles isolated from cultured tomato cells, *Plant Physiol.*, **94**, 1163, 1990.

Abel, S., Krauss, G.J. and Glund, K., Ribonuclease in tomato vacuoles: high performance liquid chromatagraphic analysis of ribonucleolytic activities and base specificity, *Biochem. Biophys. Act.*, **998**, 145, 1989.

Agre, P., Sasaki, S. and Chrispeels, M.J., Aquaporins - a family of water channel proteins, *Am.J. Physiol.*, **265**, F461, 1993.

Akazawa, T. and Hara-Nishimura, I., Topographic aspects of biosynthesis, extracellular secretion, and intracellular storage of proteins in plant cells, *Ann. Rev. Plant Physiol*, **36**, 441, 1985.

Albersheim, P. and Valent, B.S., Host-pathogen interactions. VII. Plant pathogens secrete proteins which inhibit enzymes of the host capable of attacking the pathogen, *Plant Physiol.*, **53**, 684, 1974.

Albersheim, P., Jones, T.M. and English, P.D., Biochemistry of the cell wall in relation to infective processes, *Ann. Rev. Phytopathol.*, **7**, 171, 1969.

Alexandre, J., Lassalles, J.P., and Kado, R.T., Opening of Ca^{2+} channels in isolated red beet root vacuole membrane by inositol-1,4,5 triphosphate, *Nature*, **343**, 567, 1990.

Alfenito, M.R., Souer, E., Goodman, C.D., Buell, R., Mol, J., Koes, R., and Walbot, V., Functional complementation of anthocyanin sequestration in the vacuole by widely divergent glutathione S-transferases, *Plant Cell*, **10**, 1135, 1998.

Alibert, G., Boudet, A.M. and Rataboul, P., Transport of O-coumaric acid glucoside in isolated vacuoles of sweet clover, in Plasmalemma and Tonoplast: Their functions in the plant cell, Marme, D., Marre, E. and Hartel, R., Eds, Elsevier Biomedical Press, Amsterdam, 1982, 193.

Allen, G.J. and Sanders, D., Two voltage-gated cadmium release channels coreside in the vacuolar membrane of broad bean guard cells, *Plant Cell*, **6**. 685, 1994.

Allen, G.J., Muir, S.R. and Sanders, D., Release of Ca^{2+} from individual plant vacuoles by both InsP3 and cyclic ADP-ribose, *Science*, **268**, 735, 1995.

Amodeo, G., Escobar, A. and Zeiger, E., A cationic channel in the guard cell tonoplast of *Allium cepa*, *Plant Physiol.*, **105**, 999, 1994.

Anhalt, S. and Weissenböck, G, Subcellular localization of luteolin glucuronides and related enzymes in rye mesophyll, *Planta*, **187**, 83, 1992.

Aoki, K. and Nishida, K., ATPase activity associated with vacuoles and tonoplast vesicles isolated from the CAM plant, *Kalanchoe daigremontiana*. *Physiol. Plant*, **60**, 21, 1984.

Apostol, I., Heinstein, P. and Low, P., Rapid stimulation of an oxidative burst during elicitation of cultured plant cells: role in defense and signal transduction, *Plant Physiol.*, **90**, 109, 1989.

Askerlund, P., Calmodulin-stimulated Ca^{2+}-ATPases in the vacuolar and plasma membranes in cauliflower, *Plant Physiol.*, **114**, 999, 1997.

Baba, M., Ohsumi, M., Scott, S.V., Klionsky, D.J. and Ohsumi, Y. Two distinct pathways for targeting proteins from cytoplasm to vacuole/lysosome, *J. Cell. Biol.*, **139**, 1687, 1997.

Baba, M., Takashige, K., Baba, N. and Ohsumi, Y., Ultrastructural analysis of the autophagic process in yeast: Detection of autophagosomes and their characterization, *J. Cell Biol.*, **124**, 903, 1994.

Bacon, J.S.D., MacDonald, I.R., and Knight, A.H., The development of invertase activity in slices of root of *Beta vulgaris* L. washed under aseptic conditions, *Biochem. J.*, **94**, 175, 1965.

Bal, A.K., Vacuolation and infection thread in root nodules of soybean, *Cytobios*, **42**, 41, 1985.

Balsamo, R.A., and Uribe, E.G., Plasmalemma and tonoplast-ATPase activity in mesophyll protoplasts, vacuoles and microsomes of the Crassulacian-acid-metabolism plant *Kalanchoe daigremontiana*, *Planta*, **173**, 190, 1988.

Barceló, J., and Poschenrieder, Ch., Chromium in plants, in 'Chromium Environmental Issues', Canali, S., Tittarelli, F., and Sequi, P., Eds, Franco Angeli, Milan, 1997, 102.

Barceló, J., Poschenrieder, Ch., Vázquez, M.D. and Gunse, B., Aluminium phytotoxicity, *Fertilizer Res.*, **43**, 217, 1996.

Barrett, M., Metabolism of herbicides by cytochrome P450 in corn, *Drug Metab. Durg Interact.*, **12**, 299, 1995.

Bassi, M., Corradi, M.G., and Realini, M., Effects of chromium (VI) on two fresh water plants, *Lemna minor* and *Pistia stratiotes*. I. Morphological observations, *Cytobios*, **62**, 27, 1990.

Beers, E.P., and Freeman, T.B., Protease activity during tracheary element differentiation in *Zinnia* mesophyll cultures, *Plant Physiol.*, **113**, 873, 1997.

Berhane, K., Wiedersten, M., Engstron, A., Kozarich, J.W. and Mannervik, B., Detoxification of base propenals and other Alpha unsaturated aldehyde products of radical reactions and lipid peroxidation by human glutathione S-transferases, *Proc. Natl. Acad. Sci. USA*, **91**, 1480, 1994.

Bethke, P.C. and Jones, R.L., Ca^{2+}-calmodulin modulates ion channel activity in storage protein vacuoles of barley aleurone cells, *Plant Cell*, **6**, 277, 1994.

Bieleski, R.L., Phosphate pools, phosphate transport and phosphate availability, *Annu. Rev. Plant Physiol.*, **24**, 225, 1973.

Binzel, M.L., Hess, F.D., Bressan, R.A., and Hasegawa, P.M., Intracellular compartmentation of ions in salt adapted tobacco cells, *Plant Physiol.*, **86**, 607, 1988.

Bisson, M.A. and Kirst, G.O., *Lamprothamnium*, a euryhaline charophyte. I. Osmotic relations and membrane potential at steady state, *J. Exp. Bot.*, **31**, 1223, 1980.

Blackshear, P.J., Naim, A.C. and Kuo, J.F., Protein kinases 1988: a current perspective, *FASEB J.* **2**, 2957, 1988.

Blom, T.J.M., Sierra, M., Van Vilet, T.B., Franke-van Dijk, M.E.I., deKoning, P., van Iren, F., Verpoorte, R. and Libbenga, K.R., Uptake and accumulation of ajmalicine into isolated vacuoles of cultured cells of *Catharanthus roseus* (L.) G. Don. and its conversion into serpentine, *Planta*, **183**, 170, 1991.

Blumwald, E., and Poole, R.J., Na^+/H^+ antiport in isolated tonoplast vesicles from storage tissue of *Beta vulgaris*, *Plant Physiol.*, **78**, 163, 1985a.

Blumwald, E., and Poole, R.J., Nitrate storage and retrieval in *Beta vulgaris*: effect of nitrate and chloride on proton gradients in tonoplast vesicles, *Proc. Nat. Acad. Sci. US*, **82**, 3683, 1985b.

Blumwald, E., Cragoe, E.J., and Poole, R.J., Inhibition of Na^+/H^+ antiport activity in sugar beet tonoplasts by analogs of amiloride, *Plant Physiol.*, **85**, 30, 1987.

Boller, T., Die Arginin-Permease der Hefe vacuole, Ph.D.Thesis No. 5928, Swiss Fed. Inst. Technol., Zürich, 1977.

Boller, T., Induction of hydrolases as a defense reaction against pathogens, in 'Cellular and Molecular Biology of Plant Stress', Key, J.L. and Kosuge, T., Eds, Alan R. Liss, New York, 1985, 247.

Boller, T. and Alibert, G., Photosynthesis in protoplast from *Melilotus alba*: distribution of products beween vacuole and cytosol, *Z. Pflanzenphysiol.*, **110**, 231, 1983.

Boller, T., and Wiemken, A., Dynamics of vacuolar compartmentation, *Ann. Rev. Plant Physiol.*, **37**, 137, 1986.

Booth, J.W. and Guidotti, G., Phosphate transport in yeast vacuoles, *J.Biol. Chem.*, **272**, 20408, 1997.

Bowes, B.G., Electron microscopic observations on myelin like bodies and related membranous elements in *Glechoma hederacea*, L., *Z. Pflphysiol.*, **60**, 414, 1969.

Bowles, D.J., Defense related proteins in higher plants, *Annu. Rev. Biochem.*, **58**, 873, 1990.

Bowman, E.J. and Bowman, B.J., The H$^+$-translocating ATPase in vacuolar membranes of *Neurospora crassa*, in 'Biochemistry and Function of Vacuolar Adenosin-Triphosphatase in Fungi and Plants', Marin, B.P. Ed., Springer Verlag, Berlin, 1985, 131.

Boyer, I.S., Water Transport, *Ann. Rev. Plant Physiol.*, **36**, 473, 1985.

Bramm, J., Regulation of expression of calmodulin and calmodulin-related genes by environmental stimuli in plants, *Cell Calcium* **13**, 457, 1992.

Brauer, D., Conner, D. and Tu, S., Effects of pH on proton transport by vacuolar pumps from maize roots, *Physiol plant.*, **86**, 63, 1992.

Brauer, D., Otto, J. and Tu, S-I, Nucleotide binding is insufficient to induce cold inactivation of the vacuolar type-ATPase from maize roots, *Plant Physiol. Biochem.*, **33**, 555, 1995.

Brauer, D., Unkalis, J., Triana, R., Sachar-Hill, Y. and Tu, S-I., Effects of bafilomycin A1 and metabolic inhibitors on the maintenance of vacuolar acidity in maize root hair cells, *Plant Physiol.*, **113**, 809, 1997.

Brauer, M., Sanders, D. and Still, M., Regulation of photosynthetic sucrose synthesis: a role for calcium, *Planta*, **182**, 236, 1990.

Braun, Y., Hassidim, M., Lerner, H.R., and Reinhold, L., Evidence for a Na$^+$/H$^+$ antiporter in membrane vesicles isolated from roots of the halophyte *Atriplex nummularia*, *Plant Physiol*, **87**, 104, 1988.

Bredrode, F.F., Linthorst, H.J.M. and Bol, J.F., Differential induction of acquired resistance and PR gene expression in tobacco by virus infection, ethephon treatment, UV light and wounding, *Plant Mol. Biol.*, **17**, 1117, 1991.

Bremberger, C. and Lüttge, U., Dynamics of tonoplast proton pumps and other tonoplast proteins of *Mesembryanthemum crystallinum* L. during the induction of Crassulacean acid metabolism, *Planta*, **188**, 575, 1992.

Bremberger, C., Haschke, H., and Lüttge, U., Separation and purification of the tonoplast ATPase and pyrophosphatase from plants with constitutive and inducible crassulacean acid metabolism, *Planta*, **175**, 465, 1988.

Brisken, D.P., Thornby, W.R., and Wyse, R.E., Membrane transport in isolated vesicles from sugar beet taproot: Evidence for sucrose/H$^+$ antiport, *Plant Physiol*, **78**, 871, 1985.

Brown, S.G. and Coombe, B.G., Proposal for hexose group transport at the tonoplast of grape pericarp cell, *Physiol. Veg.*, **22**, 231, 1984.

Bryant, N.J. and Stevens, T.H., Vacuole biogenesis in *Saccharomyces cerevisiae*: Protein transport pathways to the yeast vacuole, *Microbiobiol. Mol. Biol. Rev.*, **62**, 230, 1998.

Burgess, J. and Lawrence, W., Studies of the recovery of tobacco mesophyll protoplasts from an evacuolation treatment, *Protoplasma*, **126**, 140, 1985.

Burgos, P.A. and Donaire, J.P., H$^+$-ATPase activities of tonoplast-enriched vesicles from nontreated and NaCl-treated jojoba roots, *Plant Sci.*, **118**, 167, 1996.

Buser-Suter, C., Wiemken, A., and Matile, P., A malic acid permease in isolated vacuoles of a crassulacean acid metabolism plant, *Plant Physiol.*, **69**, 456, 1982.

Bush, D.S., Regulation of cytosolic calcium in plants, *Plant Physiol.*, **103**, 7. 1993.

Buvat, R., Origin and continuity of cell vacuoles, in 'Origin and Continuity of Cell Organelles', Reinert, J. and Ursprung, H., Eds, Springer Verlag, Berlin, 1971, 127.

Caldwell, J.H., Brunt, J.V. and Harold, F.M., Calcium-dependent anion channel in the water mold *Blastocladiella emersonii*, *J. Membr. Biol.*, **86**, 85, 1986.

Campbell, N.A., and Thomson, W.W., Effects of lanthanum and ethylenediamine tetraacetate on leaf movements of *Mimosa*, *Plant Physiol.*, **60**, 635, 1977.

Canut, H., Alibert, G. and Boudet, M., Hydrolysis of intracellular proteins in vacuoles isolated from *Acer pseudoplatanus* L. cells, *Plant Physiol.*, **79**, 1090, 1985.

Canut, H., Alibert, G., Carrasco, A. and Boudet, A.M., Rapid degradation of abnormal proteins in vacuoles from *Acer pseudoplatanus* L. cells. *Plant Physiol.*, **81**, 460, 1986.

Canut, H., Carrasco, A., Rossignol, M., and Ranjeva, R., Is vacuole the richest store of IP_3-mobilizable calcium in plant cells? *Plant Sci.*, **90**, 135, 1993.

Canut, H., Dupre, M., Carrasco, A. and Boudet, A.M., Proteases of *Melilotus alba* mesophyll protoplasts. II General properties and effectiveness in degradation of cytosolic and vacuolar enzymes, *Planta*, **170**, 541, 1987.

Carpita, N., Sabularse, D., Montezinos, D. and Delmer, D.P., Determination of the pore size of cell walls of living plant cells, *Science*, **205**, 1144, 1979.

Cerana, R. Giromini, L. and Colombo, R., Malate-regulated channels permeable to anions in vacuoles of *Arabidopsis thaliana*, *Aust. J. Plant Physiol.*, **22**, 115, 1995.

Chang, A. and Fink, G.R., Targeting of the yeast plasma membrane [H^+] ATPase: A novel gene ASTI prevents mislocalisation of mutant ATPase to the vacuole, *J. Cell Biol.*, **128**, 39, 1995.

Chang, K. and Roberts, J.K.M., Observation of cytoplasmic and vacuolar malate in maize root tips by ^{13}CNMR spectroscopy, *Plant Physiol.*, **89**, 197, 1989.

Chanson, A., Fichmann, J., Spear, D., and Taiz, L., Pyrophosphate-driven proton transport by microsomal membranes of corn coleoptiles, *Plant Physiol.*, **79**, 159, 1985.

Chiou, T-J. and Bush, D.R., Molecular cloning, immunochemical localisation to the vacuole, and expression in transgenic yeast and tobacco of a putative sugar transporter from sugar beet, *Plant Physiol.* **110**, 511, 1996.

Chodera, A.J. and Briskin, D.P., Chlorate transport in isolated tonoplast vesicles from red beet (*Beta vulgaris* L.) storage tissue, *Plant Sci.*, **67**, 151, 1990.

Chrispeels, M.J. and Maurel, C., Aquaproins: The molecular basis of facilitated water movement through living plant cells, *Plant Physiol.*, **105**, 9, 1994.

Chvatchko, Y., Howald, I. and Riezman, H., Two yeast mutants defective in endocytosis are defective in pheromone response, *Cell*, **46**, 355, 1986.

Cole, L., Coleman, J., Kearns, A., Morgan, G. and Hawes, C., The organic anion transport inhibitor, probenecid, inhibits the transport of Lucifer Yellow at the plasma-membrane and the tonoplast in suspension-cultured plant cells, *J. Cell Sci.* **99**, 545, 1991.

Coleman, J.O.D., Randall, R., and Blake-Klaff, M.A.A., Detoxification of xenobiotics in plant cells by glutathione conjugation and vacuolar compartmentalization: a fluorescent assay using monochlorobimane, *Plant Cell Environ*, **20**, 449, 1997.

Cosgrove, D. J. and Hedrich, R., Stretch-activated chloride, potassium and calcium channels coexisting in plasma membranes of guard cells of *Vicia faba* L., *Planta*, **186**, 143, 1991.

Coulomb, C., and Buvat, R., Processus de degenerescence cytoplasmique partielle dans les cellules de jeunes racines de *Cucurbita pepo*, *Compt. Rend. Acad. Sci.*, Paris, **267**, 843, 1968.

Cram, W.J., Negative feedback regulation of transport in cells: the maintenance of turgor, volume and nutrient supply, in 'Encyclopedia of Plant Physiology', New Ser. 2A, Lüttge, U. and Pitman, M.G., Eds, Springer Verlag, Berlin, 1976, 284.

Cram, W.J., Characteristics of sulfate transport accross plasmalemma and tonoplast of carrot root cells, *Plant Physiol.*, **72**, 204, 1983a.

Cram, W.J., Sulfate accumulation is regulated at the tonoplast, *Plant Sci. Lett.*, **31**, 329, 1983b.

Cram, W.J., and Laties, G.G., The use of short term and quasi-steady influx in estimating plasmalemma tonoplast influx in barley root cells at various external and internal salt concentrations, *Aust. J. Biol. Sci.*, **24**, 633, 1971.

Cramer, C.L., Vaugh, L.E. and Davis, R.H., Basic amino acids and inorganic polyphosphates in *Neurospora crassa*: independent regulation of vacuolar pools, *J. Bactriol.*, **142**, 945, 1980.

Cunningham, S.D., and Berti, W.R., Remediation of contaminated soils with green plants: an overview. *In Vitro Cell. Dev. Biol.* **29**, 207, 1993.

d'Auzac, J., ATPase membranaire de vacuoles lysomales: les lutoids du latex d'*Hevea brasiliensis*, *Phytochemistry*, **16**, 1881, 1977.

d'Auzac, J., Crestin, H., Marin, B., and Lioret, C., A plant vacuolar system: the lutoids from *Hevea brasiliensis* latex. *Physiol. Veg.*, **20**, 311, 1982.

d'Auzac, J., Chrestin, H., and Marin, B., Biochemical and enzymatic components of a vacuolar membrane: Tonoplasts of lutoids from *Hevea latex, Methods Enzymol.*, **148**, 87, 1987.

Daman, S., Hewitt, J. Nieder, M. and Bennet, A.B., Sink metabolism in tomato fruit. II. Phloem unloading and sugar uptake, *Plant Physiol.*, **87**, 731, 1988.

Daniels, M.J., Mirkov, T.E., and Chrispeels, M.J., The plasma membrane of *Arabidopsis thaliana* contains a mercury-insensitive aquaporin that is a homolog of the tonoplast water channel protein TIP, *Plant Physiol.*, **106**, 1325, 1994.

Davies, C. and Robinson, S.P., Sugar accumulation in grape berries: Cloning of two putative vacuolar invertase cDNA and their expression in grapevine tissues, *Plant Physiol.*, **111**, 275, 1996.

Davies, J.M., Poole, R.J., Rea, P.A. and Sanders, D., Potassium transport into vacuoles energized directly by a proton-pumping inorganic pyrophosphatase, *Proc. Natl. Acad. Sci. USA*, **89**, 11701, 1992.

Davies, K.L., Davies, M.S. and Francis, D., Zinc-induced vacuolation in root meristematic cells of *Festuca rubra*, *Plant Cell Environ*, **14**, 399, 1991.

Davies, K.L., Davies, M.S. and Francis, D., Zinc-induced vacuolation in root meristematic cells of cereals, *Ann. Bot.*, **69**, 21, 1992.

De Boer, A.H., and Wegner, L.H., Regulatory mechanisms of ion channels in xylem parenchyma cells, *J. Exp. Bot.*, **48**, 411, 1997.

De Leon, J.L.D., Daie, J. and Wyse, R., Tonoplast stability and survival of isolated vacuoles in different buffers, *Plant Physiol.*, **88**, 251, 1988.

Delrot, S., Thom, M. and Maretzki, A., Evidence for a uridine-5'-di-phosphate-glucose protected p-chloromercuribenzene sulfonic acid-binding site in sugarcane vacuoles, *Planta*, **169**, 64, 1986.

Deus-Neuman, B. and Zenk, M.H., Accumulation of alkaloids in plant vacuoles does not involve an ion trap mechanism, *Planta*, **167**, 44, 1986.

Dietz, K-J., Jäger, R., Kaiser, G. and Martinoia, Amino acid transport across the tonoplast of vacuoles isolated from barley mesophyll protoplasts, *Plant Physiol.*, **92**, 123, 1990.

Douglas, T.J., NaCl effects on 4-desmethylsterol composition of plasma membrane-enriched preparations from citrus roots. *Plant Cell Environ.*, **8**, 687, 1985.

Drews, G., Ziser, K., Shrock-Vietor, U. and Golecki, J.R., Cellular responses of soybean to virulent and avirulent strains of *Pseudomonas syringae* pv glycinea, *Eur. J. Cell Biol.*, **46**, 369, 1988.

Düggelin, T., Schellenberg, M., Borttik, K. and Matile, P., Vacuolar location of lipofuscin- and proline-like compounds in senescent barley leaves, *J. Plant Physiol.*, **133**, 492, 1988.

Dürr, M., Urech, K., Boller, T., Wiemken, A., Schwenke, J. and Nagy, M., Sequestration of arginine by polyphosphate in vacuoles of yeast (*Saccharomyces cerevisiae*), *Arch. Microbiol.*, **121**, 169, 1979.

Dutta, P.C., Appleqvist, L.A., Gunnarson, S. and von Hofsten, A., Lipid bodies in tissue culture, somatic and zygotic embryo of *Daucus carota* L., *Plant Sci.*, **78**, 259, 1991.

Ebal, J. and Grisebach, H., Defense strategies of soybean against the fungus *Phytophthora megasperma* f. sp.glycinea: a molecular analysis. *Trend. Biochem. Sci.*, **13**, 23, 1988.

Echeverria, E. and Salvucci, M.S., Sucrose phosphate is not transported into vacuoles or tonoplast vesicles from red beet (*Beta vulgaris*) hypocotyl, *Plant Physiol.* **96**, 1014, 1991.

Edelman, J. and Jefford, T.G., The mechanism of fructan metabolism in higher plants as exemplified in *Helianthus tuberosus*, *New Phytol.*, **67**, 517, 1968.

Epstein, E., and Hagen, C.E., A kinetic study of the absorption of alkali cations by barley roots, *Plant Physiol.*, **27**, 457, 1952.

Ersek, T. and Kiraly, Z., Phytoalexins: warding off compounds in plants?, *Physiol. Planta*, **68**, 343, 1986.

Esau, K. and Charvat, I.D., An ultrastructural study of acid phosphatase localization in cells of *Phaseolus vulgaris* phloem by the use of azo dye method, *Tissue Cell*, **4**, 619, 1975.

Esau, K. and Cronshaw, J., Relation of tobacco mosaic virus to the host cells, *J. Cell Biol.*, **33**, 665, 1967.

Esau, K. and Kosaki, H., Laticifers in *Nelumbo nucifera* Gaertn.: distribution and structure. *Ann. Bot.* **39**, 713, 1975.

Evans, D.E., Regulation of cytoplasmic free calcium by plant cell membranes, *Cell Biol. Int. Report*, **12**, 383, 1988.

Faye, L. and Chrispeels, M.J., Transport and processing of the glycosylated precursor of Concanavalin A in jackbean, *Planta*, **170**, 217, 1987.

Faye, L., Greenwood, J.S., Herman, E.M., Stürm, A. and Chrispeels, M.J., Transport and posttransitional processing of the vacuolar enzyme alpha-mannosidase in jackbean cotyledons, *Planta*, **174**, 271, 1988.

Felle, H., Auxin oscillations of cytosolic free calcium and pH in *Zea mays* coleoptiles, *Planta*, **174**, 495, 1988.

Fineran, B.A., Organization of the tonoplast in frozen-etched root tips, *J. Ultrastruct. Res.*, **33**, 574, 1970.

Fineran, B.A., Ultrastructure of vacuolar inclusions in root tips, *Protoplasma*, **72**, 1, 1971.

Fisher, D.B., Hansen, D., and Hodges, T.K., Correlation between ion fluxes and ion-stimulated adenosine triphosphatase activity of plant roots, *Plant Physiol.*, **46**, 812, 1970.

Fleurat-Lessard, P. Ultrastructural features of the starch sheath cells of the primary pulvinus after gravistimulation of the sensitive plant (*Mimosa pudica* L.), *Protoplasma*, **105**, 177, 1981.

Fleurat-Lessard, P., and Millet, B.J., Ultrastructural features of cortical parenchyma cells, motor cells in stamen filaments of *Berberis canadensis* and tertiary pulvini of *Mimosa pudica*, *J. Exp. Bot.*, **35**, 1332, 1984.

Fleurat-Lessard, P. Frangne, N., Maeshima, M., Ratajczak, R., Bonnemain, J.L., and Martinoia, E., Increased expression of vacuolar aquaporin and H^+-ATPase related to motor all function in *Mimosa pudica* L., *Plant Physiol.* **114**, 827, 1997.

Flowers, T. J., Halophytes, in 'Ion Transport in Plant Cells and Tissues', Baker, D.A. and Hall, J.L. Eds, North-Holland, Amsterdam, 1975, Chap. 10.

Flowers, T.J., Chloride as a nutrient and as an osmoticum, Advances in Plant Nutrition, Tinker, P.B., Läuchli, A., Eds, Praeger, New York, 1988, Vol 3, 55.

Fowke, L.C. Investigations of cell structure using cultured plant cells and protoplasts, *Int. Congr. Plant Tissue Cell Cult.* **6** Mett, 19 (Abstract), 1986.

Fox, G.G. and Ratcliffe, R.G., ^{31}P-NMR observations on the effect of the external pH on the intracellular pH values in plant cell suspension cultures, *Plant Physiol.*, **93**, 512, 1990.

Frehner, M., Keller, F. and Wiemken, A., Localization of fructan metabolism in the vacuoles isolated from protoplasts of Jerusalem artichoke tubers (*Helianthus tuberosus* L.), *J. Plant Physiol.*, **116**, 197, 1984.

Fried, M., and Noggle, J.C., Multiple site uptake of individual ions by roots as affected by hydrogen ion, *Plant Physiol.*, **33**, 139, 1958.

Friemert, V., Heiniger, D., Kluge, M., and Ziegler, H., Temperature effects on malic acid efflux from the vacuoles and on the carboxylation pathways in Crassulacian-acid-metabolism plants, *Planta*, **174**, 453, 1988.

Fritsch, H. and Griesbach, H., Biosynthesis of cyanidin in cell cultures of *Haplopappus gracilis*, *Phytochemistry*, **14**, 2437, 1975.

Fukuda, H., Xylogenesis : Initiation, progression and cell death, *Ann Rev. Plant Physiol. Plant Mol. Biol.*, **47**, 299, 1996.

Fukuda, H., Tracheary element differentiation, *Plant Cell.*, **9**, 1147, 1997.

Fukuda, H., and Komamine, A., Establishment of an experimental system for the tracheary element differentiation from single cells isolated from the mesophyll of *Zinnia elegans*, *Plant Physiol.*, **65**, 57, 1980.

Fukuda, A., Yazaki, Y., Ishikawa, T., Koike, S. and Tanaka, Y., Na^+/H^+ antiporter in tonoplast vesicles from rice roots, *Plant Cell Physiol.*, **39**, 196, 1998.

Gahan, P.B. and Maple, A.J., The behaviour of lysosome-like particles during cell differentiation, *J. Exp. Botany*, **17**, 151, 1966.

Gahan, P.B. and McLean, J., Subcellular localization and possible functions of acid beta-glycerophosphatases and naphthal esterases in plant cells, *Planta*, **89**, 126, 1969.

Gaillard, C. Dufaud, A., Tommasini, R., Kreuz, K., Amrhein, N. and Martinoia, E., A herbicide antidote (safener) induces the activity of both the herbicide detoxifying enzyme and of a vacuolar transporter for the detoxified herbicide, *FEBS Lett.* **352**, 219, 1994.

Garbarino, J., and DuPont, F.M., NaCl induces a Na^+/H^+ antiport in tonoplast vesicles from barley roots, *Plant Physiol.*, **86**, 231, 1988.

Garbarino, J. and Du Pont, F.M., Rapid induction of Na^+/H^+ exchange activity in barley root tonoplast, *Plant Physiol.*, **89**, 1, 1989.

Gebhart, E., Mutagenität, Karzinogenität, Teratogenität, in 'Metalle in der Umwelt', Merian, E., Ed., Verlag Chemie, Weinheim, 1984, 237.

Gehring, C.A., Williams, D.A., Cody, S.H., and Parish, R.W., Phototropism and geotropism in maize coleoptile are spatially correlated with increases in cytosolic free calcium, *Nature*, **345**, 528, 1990.

Gelli, A. and Blumwald, E., Calcium retrieval from vacuolar pools: characterization of a vacuolar calcium channel, *Plant Physiol.*, **102**, 1139, 1993.

Getz, H.P., Accumulation of sucrose in vacuoles released from isolated beet root protoplasts by both direct sucrose uptake and UDP-glucose-dependent translocation, *Plant Physiol. Biochem.*, **25**, 573, 1987.

Getz, H.P. Sucrose transport in tonoplast vesicles of red beet roots is linked to ATP hydrolysis, *Planta*, **185**, 261, 1991.

Getz, H.P. and Klein, M., Characteristics of sucrose transport on the tonoplast of red beet (*Beta vulgaris* I.) storage tissue, *Plant Physiol.*, **107**, 459, 1995.

Getz, H.P., Grosclaude, J., Kurkdjian, A., Lelievre, F., Maretzki, A. and Guern, J. Immunological evidence for the existence of carrier protein for sucrose transport in tonoplast vesicles from red beet (*Beta vulgaris* L.) root storage tissue, *Plant Physiol.* **102**, 751, 1993.

Ghosh, M., Shen, J. and Rosen, B.P., Pathways of As(III) detoxification in *Saecharomyces cerevisiae*, *Proc. Natl. Acad. Sci. USA*, **96**, 5001, 1999.

Giannini, J.L., Holt, J.S. and Briskin, D.P., The effect of glyceollin on proton leakage in *Phytophthora megasperma* f. sp. glycinea plasma membrane and red beet tonoplast vesicles, *Plant Sci.*, **68**, 39, 1990.

Giannini, J.L., Holt, J.S. and Briskin, D.P., The effect of glyceollin on soybean (*Glycine max* L.) tonoplast and plasma membrane vesicles, *Plant Sci.*, **74**, 203, 1991.

Gifford, E.M., and Stewart, K.D., Inclusions of proplastids and vacuoles in the shoot apices of *Bryophyllum* and *Kalanchoe*, *Am. J. Bot.*, **53**, 269, 1968.

Gilroy, S., Read, N.D. and Trewavas, Elevation of cytoplasmic calcium by caged calcium or caged inositol triphosphate initiates stomatal closure, *Nature*, **346**, 769, 1990.

Godbold, D.L., Horst, W.J., Collins, J.C., Thurman, D.A. and Marschner, H., Accumulation of zinc and organic acids in roots of zinc tolerant and non-tolerant ecotypes of *Deschampsia caespitosa*, *J. Plant Physiol*, **116**, 59, 1984.

Goerlach, J. and Willims-Hoff, D., Glycine uptake into barley mesophyll vacuoles is regulated but not energized by ATP, *Plant Physiol.*, **99**, 134, 1992.

Goodman, R.N. and Plurad, S.B., Ultrastructural changes in tobacco undergoing the hypersensitive reaction caused by plant pathogenic bacteria, *Physiol. Plant Pathol.*, **1**, 11, 1971.

Gordon-Weeks, R., Koren'kov, V.D., Steele, S.H. and Leigh, R.A., Tris is a competitive inhibitor of K^+ activation of the vacuolar H^+-pumping pyrophosphatase, *Plant Physiol.*, **114**, 901, 1997.

Granell, A., Belles, J.M. and Conejero,V., Induction of pathogenesis-related proteins in tomato citrus exocortis viroid, silver ions and ethophon, *Physiol. Mol. Plant Pathol*, **31**, 83, 1987.

Green, P.B., Erickson, R.O., and Buggy, J., Metabolic and physical control of cell elongation rate, *Plant Physiol.*, **47**, 423, 1971.

Greutert, H. and Keller, F., Further evidence for stachyose and sucrose/H^+ antiporters on the tonoplast of Japanese artichoke (*Stachys siebolodi*) tubers, *Plant Physiol.*, **101**, 1317, 1993.

Gries, G.E. and Wagner, G.J., Association of nickel versus transport of cadmium and calcium in tonoplast vesicles of oat roots, *Planta*, **204**, 390, 1998.

Griesbach, R. and Sink, K., Evacuolation of mesophyll protoplasts, *Plant Sci. Lett.*, **30**, 297, 1983.

Groover, A., and Jones, A.M., Tracheary element differentiation uses a novel mechanism coordinating programmed cell death and secondary cell wall synthesis, *Plant Physiol.*, **119**, 375, 1999.

Groover, A., DeWitt, N., Heidel, A., and Jones, A., Programmed cell death of plant tracheary elements differentiating *in vitro*, *Protoplasma*, **196**, 197, 1997.

Grotewald, E., Chamberlin, M., Snook, M., Siame, B., Butter, L., Swenson, J., Moddock, S., St. Clair, G. and Bowen, B., Engineering secondary metabolism in maize cells by ectopic expression of transcription factors, *Plant Cell*, **10**, 721, 1998.

Gruhnert, C., Biehl, B. and Selmer, D., Compartmentation of cyanogenic glucosides and their degrading enzymes, *Planta*, **195**, 36, 1994.

Guern, J., Mathieu, Y., Kurkdjian, A., Manigault, P., Manigault, J., Gillet, B., Beloeil, J.C. Lallemand, J.Y., Regulation of vacuolar pH of plant cells. II. A ^{31}PNMR study of the modifications of vacuolar pH in isolated vacuoles induced by proton pumping and cation/H^+ exchanges, *Plant Physiol*, **89**, 27, 1989.

Gupta, H.S. and De, D.N., Functional analogy of plant vacuoles with animal lysosomes, *Curr. Sci.*, **52**, 680, 1983a.

Gupta, H.S. and De, D.N., Uptake and accumulation of acridine orange by plant cells, *Proc. Ind. Nat. Sci. Acad., India* **B 49**, 553, 1983b.

Gupta, H.S., Acridine orange-induced formation of myelin-like structures in plant cell vacuoles, *Indian J. Genet*, **45**, 133, 1985a.

Gupta, H.S., Plant lysosomes: aspects and prospects, *Curr. Sci.*, **54**, 554, 1985b.

Gupta, H.S. and De, D.N., Mechanism of drug detoxification by plant cells as studied by fluorescence microscopy, *J. Assam. Sci. Soc.*, **28**, 1, 1985c.

Gupta, H.S. and De, D.N., Acridine-orange induced vacuolar uptake of cytoplasmic organelles in plant cells: an ultrastructural study, *J. Plant Physiol.*, **132**, 254, 1988.

Guy, M. and Kende, H., Conversion of 1-aminocyclopropane-1-carboxylic acid to ethylene by isolated vacuoles of *Pisum sativum* L., *Planta*, **160**, 281, 1984.

Hager, A., and Helme, M., Properties of an ATP-fueled, Cl-dependent proton pump localised in membranes of microsomal vesicles from maize coleoptiles, *Z. Naturforsch.*, **36c**, 997, 1981.

Hajibagheri, M.A. and Flowers, T.J., X-ray micro analysis of ion distribution within root cortical cells of the halophyte *Suaeda maritima* (L.) Dum., *Planta*, **177**, 131, 1989.

Hall, M.D., and Cocking, E.C., The response of isolated *Avena coleoptile* protoplasts to indole-3-acetic acid, *Protoplasma*, **79**, 225, 1974.

Hanower, P., Brzozowska, I., and Niamien N'Goran, M., Absorption des acides amines par les lutoides du latex d'*Hevea brasiliensis*, *Physiol. Plant*, **39**, 299, 1977.

Hara-Nishimura, I., Nishimura, M. and Akazawa, T., Biosynthesis and intracellular transport of 11S globulin in developing pumpkin cotyledons, *Plant Physiol.*, **77**, 747, 1985.

Hara-Nishimura, I., Takeuchi, Y. and Nishmura, M., Molecular characterization of a vacuolar processing enzyme related to a putative cysteine proteinase of *Schistosoma mansoni*, *Plant Cell*, **5**, 1651, 1993.

Harder, D.E., and Chong, J., Structure and physiology of haustoria, in 'The Cereal Rusts', Bushnell, W.R., and Roelfs, A.P., Eds, Academic Press, New York, 1984, 431.

Harley, S.M. and Lord, J.M., *In vitro* endoproteolytic cleavage of castorbean lectin precursors, *Plant Sci.*, **41**, 111, 1985.

Haupt, W., Physiology of movement, *Prog. Botany*, **44**, 222, 1982.

Hawker, J.S., Smith, G.M., Phillips, M. and Wiskich, J.T., Sucrose phosphatase associated with vacuole preparations from red beet, sugar beet and immature sugarcane stem, *Plant Physiol.*, **84**, 1281, 1987.

Hedrich, R. and Becker, D., Green circuits: the potential of plant specific ion channels, *Plant Mol. Biol.*, **26**, 1637, 1994.

Hedrich, R. and Neher, E., Cytoplasmic calcium regulates voltage dependent ion channels in plant vacuoles, *Nature*, **329**, 833, 1987.

Hedrich, R., Flügge, V.I., and Fernandez, J.M., Patch-clamp studies of ion transport in isolated plant vacuoles, *FEBS. Letters*, **204**, 228, 1986.

Hedrich, R., Barbier-Brygoo, H., Felle, H., Flügge, U.I., Lüttge, U. Maathuis, F.J.M., Mark, S., Prins, H.B.A., Raschke, K., Schnabl, H., Schroeder, J.I., Struve, I., Taiz, L., and Zeigler, P., General mchanisms for solute transport across the tonoplast of plant vacuoles: a patch-clamp survey of ion channels and proton pumps, *Bot. Acta*, **101**, 7, 1988.

Heineke, D., Wildenberger, K., Sonnewald, U., Willmitzer, L., and Heldt, H.W., Accumulations of hexoses in leaf vacuoles: Studies with transgenic tobacco plants expressing yeast-derived invertase in the cytosol, vacuole or apoplasm, *Planta*, **194**, 29, 1994.

Hensel, W., Movement of pulvinated leaves, *Prog. Botany*, **49**, 171, 1987.

Hepler, P.K., and Wayne, R.O., Calcium and plant development, *Ann. Rev. Plant Physiol.*, **36**, 397, 1985, 439.

Herman, E.M. and Lamb, C.J., Arabinogalactan-rich glucoproteins are localised on the cell surface and in intravacuolar multivesicular bodies, *Plant Physiol.*, **98**, 264, 1992.

Heuillet, E., Moreau, A., Halpern, S., Jeanne, N. and Puiseux-Dao, S., Cadmium binding to a thiol-molecule in vacuoles of *Dunaliella bioculata* contaminated with $CdCl_2$: electron probe microanalysis. *Biol.Cell*, **58**, 79, 1986.

Higinbotham, N., Electropotential of plant cells, *Ann. Rev. Plant Physiol.*, **24**, 25, 1973.

Hillmer, S., Depta, H. and Robinson, D.G., Confirmation of endocytosis in higher plant protoplasts using lectin-gold conjugates. *Eur. J. Cell Biol.* **41**, 142, 1986.

Hinder, B., Schellenberg, M., Rodoni, S., Ginsburg, S., Vogt, E., Martinoia, E., Matile, P. and Hörtensteiner, S., How plants dispose of chlorophyll catabolites: Directly energized uptake of tetrapyrrolic breakdown praducts into isolated vacuoles, *J. Biol. Chem.*, **271**, 27233, 1996.

Hiraiwa, N., Takeuchi, Y., Nishimura, M. and Hara-Nishimura, I., A vacuolar processing enzyme in maturing and germinating seeds: its distribution and associated changes during development, *Plant Cell Phyiol.* **34**, 1197, 1993.

Hiraiwa, N., Kondo, M., Nishimura, M. and Hara-Nishmura, I., An aspartic endopeptidase is involved in the breakdown of propeptides of storage proteins in protein storage vacuoles of plants, *Eur. J. Biochem.*, **246**, 133, 1997.

Hodges, T.K., ATPase, associated with membranes of plant cells, *Encl. Plant Physiol.*, **2A**, 260, 1976.

Hoffman, L.M., Donaldson, D.D., and Herman, E.M., A modified storage protein is synthesised, processed and degraded in the seeds of transgenic plants, *Plant Mol. Biol.*, **11**, 717, 1988.

Höfte, H., Hubbard, L., Reizer, J., Ludevid, D., Herman, E.M. and Chrispeels, M.J., Vegetative and seed-specific isoforms of a putative solute transporter in the tonoplast of *Arabidopsis thaliana*, *Plant Physiol.*, **99**, 561, 1992.

Homeyer, U. and Schultz, G., Transport of phenylalanine into vacuoles isolated from barley mesophyll protoplasts, *Planta*, **176**, 378, 1988.

Hopp, W. and Seitz, H.U., The uptake of acylated anthocyanin into isolated vacuoles from a cell suspension culture of *Daucus carota*, *Planta*, **170**, 74, 1987.

Hopp, W., Hinderer, W., Petersen, M. and Seitz, H.U., Anthocyanin containing vacuoles isolated from the protoplasts of *Daucus carota* cell cultures, in 'The Physiological Properties of Plant Protoplast', Pilet, P.E. Ed., Springer Verlag, Berlin, 1985, 122.

Horn, M.A., Meadows, R.P., Apastol, I., Jones, C.R., Gorenstein, D.G., Heinstein, P.F. and Low, P.S., Effect of elicitation and changes in extracellular pH on the cytoplasmic and vacuolar pH of suspension-cultured soybean cells, *Plant Physiol.*, **98**, 680, 1992.

Hrazdina, G. and Wagner, G.J., Compartmentation of plant phenolic compounds: sites of synthesis and accumulation, in 'Ann. Proc. Phytochem. Soc. Europe', Vol. 25, Sumere, C.P., Lea, P.J., Eds, Oxford Uni Press, Oxford, 1985, 119.

Hüber-Wälchle, V. and Wiemken, A., Differential extraction of soluble pools from the cytosol and the vacuole of yeast (*Candida utilis*) using DEAE-dextran, *Arch. Microbiol.*, **120**, 141, 1979.

Hüve, K., Dittrich, A., Kinderman, G., Slovik, S. and Heber, U., Detoxification of SO_2 in conifers differing in SO_2-tolerance: A comparison of *Picea abies*, *Picea pungens* and *Pinus sylvestris*, *Planta*, **195**, 578, 1995.

Inz, D. and Van Montagu, M., Oxidative stress in plants, *Curr. Opin. Biotechnol.*, **6**, 153, 1995.

Ishikawa, T., The ATP-dependent glutathione S-conjugate export pump, *Trends Biochem. Sci.*, **17**, 463, 1992.

Iwasaki, I., Arata, H. and Nishimura, M., Ionic balance during malic acid accumulation in vacuoles of a CAM plant *Graptopetalum paraguayense*, *Plant Cell Physiol.*, **29**, 643, 1988.

Iwasaki, I., Arata, H., Kijima, H. and Nishimura, M., Two types of channels involved in the malate ion transport across the tonoplast of a Crassulacean acid metabolism plant, *Plant Physiol.*, **98**, 1494, 1992.

Jeschke, W.D., Roots : Cation selectivity and compartmentation, involvement of protons and regulation, in 'Plant Membrane Transport: Current Conceptual Issues', Spanswick, R.M., Lucas, W.J. and Dainty, J., Eds, Elsevier/North Holland, Amsterdam, 1980, 17.

Joachem, P., Rona, J.P., Smith, J.A.C. and Lüttge, U., Anion-sensitive ATPase activity and proton transport in isolated vacuoles of species of the CAM genus Kalanchoe, *Physiol. Plant*, **62**, 410, 1984.

Johannes, E., Brosnan, J.M. and Sanders, D., Parallel pathways for intracellular Ca^{2+} release from the vacuole of higher plants, *Plant J.*, **2**, 97, 1992.

Joyce, D.C., Cramer, G.R., Reid, M.S. and Bennett, A.B., Transport properties of the tomato fruit tonoplast, III. Temperature dependence of calcium transport, *Plant Physiol.*, **88**, 1097, 1988.

Kaestner, K.H., and Sze, H., Potential-dependent anion transport in tonoplast vesicles from oat roots, *Plant Physiol.*, **83**, 483, 1987.

Kaiser, G. and Heber, U., Sucrose transport into vacuoles isolated from barley mesophyll protoplasts, *Planta*, **161**, 562, 1984.

Kaiser, G., Martinoia, E. and Wiemken, A., Rapid appearance of photosynthetic products in the vacuoles isolated from barley mesophyll protoplasts by a new fast method, *Z. Pflanzenphysiol.*, **107**, 103, 1982.

Kaiser, G., Martinoia, E., Schröppel-Meier, G. and Heber, U., Active transport of sulfate into the vacuole of plant cells provides halotolerance and can detoxify SO_2, *J. Plant Physiol.*, **133**, 756, 1989.

Kamiya, N. and Kuroda, K., Artificial modification of the osmotic pressure of the plant cell, *Protoplasma*, **46**, 423, 1956.

Kappus, H., Lipid peroxidation: mechanisms, analysis, enzymology and biological relevance, in 'Oxidative Stree', Ed. Sies, H., Acad. Press. London, 1985, 273.

Kasamo, K. and Nouchi, I., The role of phospholipid in plasma membrane ATPase activity in *Vigna radiata* L. (Mung bean) roots and hypocotyls, *Plant Physiol.*, **83**, 823, 1987.

Kauss, H., Some aspects of calcium-dependent regulation in plant metabolism, *Ann. Rev. Plant Physiol.*, **38**, 47, 1987.

Keller, F., Transport of stachyose and sucrose by vacuoles of Japanese artichoke (*Stachys sieboldi*) tubers, *Plant Physiol.*, **98**, 442, 1992.

Kikuyama, M., and Tazawa, M., Tonoplast action potential in *Nitella* in relation to vacuolar chloride concentration, *J. Membr. Biol.*, **92**, 95, 1987.

King, G.J., Turner, V.A., Hussey, C.E., Wurtele, E.S. and Lee, M., Isolation and characterization of a tomato cDNA clone which codes for a salt-induced protein, *Plant Mol. Biol.*, **10**, 401, 1988.

Kinoshita,T., Nishmura, M. and Hara-Nishimura, I., Homologues of a vacuolar processing enzyme that are expressed in different organs in *Arabidopsis thaliana, Plant Mol. Biol.* **29**, 81,1995.

Kiyosawa, K., and Tazawa, M., Hydraulic conductivity of tonoplast-free *Chara* cells. *J. Membr. Biol.*, **37**, 157, 1977.

Klein, N., Wissenboeck, G., Dufaud, A., Gaillard, C., Kreuz, K., and Martinoia, E., Different energization mechanisms drive the vacuolar uptake of a flavonoid glycoside and a herbicide glucoside, *J. Biol. Chem.*, **271**, 29666, 1996.

Kliewer, W.M. Changes in the concentration of glucose, fructose and total soluble solids in flowers and berries of *Vitis vinifera*. *Am. J. End. Vitic.* **16**, 101, 1965.

Klionsky, D.J., Nonclassical protein sorting to the yeast vacuole, *J. Biol. Chem.*, **273**, 10807, 1998.

Knight, M.R., Campbell, A.K., Smith, S.M. and Trewavas, A.J., Transgenic plant aequorin reports the effects of touch and cold shock and elicitors on cytoplasmic calcium, *Nature*, **353**, 524, 1991.

Kochian, L.V., Cellular mechanisms of aluminium toxicity and resistance in plant, *Annu. Rev. Plant Physiol. Plant Mol. Biol.*, **46**, 237, 1995.

Kononowicz, A. K., Nelson, D. E., Singh, N. K., Hasegawa, P. M. and Bressan, R.A., Regulation of the osmotin promoter, *Plant Cell*, **4**, 513, 1992.

Krasowski, M.J. and Owens, J.N., Seasonal changes with apical zonations and ultrastructure of coastal Douglas-fir seedlings (*Pseudotsuga manziesii*), *Amer J. Bot*, **77**, 245, 1990.

Kreuz, K., Herbicide safeners: recent advances and biochemical aspects of their mode of action, in 'Proc. Brighton Crop Protect. Conf. Weeds', Brighton, UK, 1993, 1249.

Kreuz, K., Tommasini, R. and Martinoia, E., Old enzymes for a new job: Herbicide detoxification in plants, *Plant Physiol.*, **111**, 349, 1996.

Krotz, R.M., Evangelou, B.P. and Wagner, G.J., Relationship between cadmium, zinc, Cd-peptide and organic acid in tobacco suspension cells, *Plant Physiol.* **91**, 780, 1989.

Kurkdjian, A., Quiquampoix, H., Barbier-Brygoo, H., Pean, M., Manigault, P., and Guern, J. Critical evaluation of methods for estimating the vacuolar pH of plant cells, in 'Biochemistry and Function of Vacuolar Adenosine Triphosphatase in Fungi and Plants, Marin, B.P., Ed., Springer Verlag, Berlin, Heidelberg, New York, Tokyo, 1985, 98.

Kylin, A., and Hansson, G., Transport of sodium and potassium, and properties of (sodium + potassium) activated adenosine triphosphatases: possible connection with salt tolerance in plants, in 'Proc. 8th Colloq. Intern. Potash. Inst., Bern, Int. Potash Inst., 1971, 64.

Laisk, A., Pfanz, H., and Heber, U., Sulfur-dioxide fluxes into different cellular compartments of leaves photosynthesizing in a polluted atmosphere, *Planta*, **173**, 241, 1988.

Lazof, D.B., Goldsmith, J.G. and Linton, R.W., The *in situ* analysis of intracellular aluminium in plants, in 'Progress in Botany', **58**, 112, 1997.

Leigh, R.A. and Walker, R.R., ATPase and acid-phosphatase activities associated with vacuoles isolated from storage roots of red beet (*Beta vulgaris* L.), *Planta*, **150**, 222, 1980.

Leigh, R.A., and Wyn Jones, R.G., Cellular compartmentation in plant nutrition: the selective cytoplasm and the promiscuous vacuole, in 'Advances in Plant Nutrition', Tinker, P.B. and Läuchli, A., Eds, Praeger, New York, 1986, 249.

Leigh, R.A., Rees, T., Fuller, W.A. and Banfield, J., The location of acid invertase activity and sucrose in the vacuoles of storage roots of beet root (*Beta vulgaris*), *Biochem. J.*, **178**, 539, 1979.

Levanony, H., Rubin, R., Altschuler, Y. and Galili, G., Evidence for a novel route of a wheat storage proteins to vacuoles, *J. Cell Biol*, **119**, 1117, 1992.

Lew, R.R., and Spanswick, R.M., Characterization of anion effects on nitrate sensitive ATP-dependent proton pumping activity of soyabean (*Glycine max.* L.) seedling root microsomes, *Plant Physiol.*, **77**, 352, 1985.

Li, Z-S, Alfenito, M., Rea, P., Walbot, V. and Dixon, R.A., Vacuolar uptake of the phytoalexin medicarpin by glutathione-conjugate pump, *Phytochemistry*, **45**, 689, 1997.

Li, Z-S, Zhen, R.G. and Rea, P.A., 1-Chloro-2,4-Dinitrobenzene-elicited increase in vacuolar glutathione S-conjugate transport activity, *Plant Physiol.*, **109**, 177, 1995.

Li, Z-S., Lu, Y-P., Zhen, R-G., Szczypka, M., Thiele. D.J. and Rea, P.A., A new pathway for vacuolar cadmium sequestration in *Saccharomyces cerevisiae*: YCF 1- catalyzed transport of bis (aglutathionoto) cadmium, *Proc. Natl. Acad. Sci. USA*, **94**, 42, 1997.

Li, Z-S, Zhao, Y. and Rea, P.A., Magnesium adenosine 5'-triphosphate-energized transport of glutathione S-conjugates by plant vacuolar membrane vesicles, *Plant Physiol.*, **107**, 1257, 1995.

Lin, W., Wagner, G.J., Siegelman, W., and Hind, G., Membrane-bound ATPase of intact vacuoles and tonoplasts isolated from mature plant tissue, *Biochem. Biophys. Acta*, **465**, 110, 1977.

Linthorst, H.J.M., Pathogenesis related proteins of plants, *Crit. Rev. Plant Sci.* **10**, 123, 1991.

Löffelhardt, W. and Kopp, B., Subcellular localization of glucosyltransferases involved in cardiac glycoside glucosylation in leaves of *Convallaria majales*, *Phytochemistry*, **20**, 1219, 1981.

Lommel, C. and Felle, H., Transport of Ca^{2+} across the tonoplast of intact vacuoles from *Chenopodium album* L. suspension cells: ATP-dependent import and inositol-1, 4,5-triphosphate induced release, *Planta*, **201**, 477, 1997.

Lu, Y.P., Lia, Z-S., Drozdowicza, Y.M., Hörtensteiner, S.H., Martinoia, E. and Rea, P.A., ATMRP2, an *Arabidopsis* ATP binding cassette transporter able to transport glutathione S-conjugates and chlorophyll catabolites: Functional comparisons with ATMRP1, *Plant Cell*, **10**, 1, 1998.

Lunevsky, V.Z., Zherelova, O.M., Vostrikov, I.Y. and Berestovsky, G.N., Excitation of characeal cell membranes as a result of activation of calcium and chloride channels, *J. Membr. Biol.*, **72**, 43, 1983.

Lüttge, U., Carbon dioxide and water demand: Crassulacean acid metabolism (CAM), a versatile ecological adaptation exemplifying the need for integration in ecophysiological work, *New Phytol*, **106**, 593, 1987.

Lüttge, U. and Ball, E., Electrochemical investigation of active malic acid transport at the tonoplast into the vacuoles of the CAM plant *Kalanchoe daigremontiana*, *J. Membr. Biol.*, **47**, 401, 1979.

Lüttge, U., and Higinbotham, N., Transport in Plants, Springer-Verlag, New York, 1979, Chap. 12.

Lüttge, U., Smith, J.A.C. and Marigo, G., Membrane transport, osmoregulation, and the control of CAM, in 'Crassulacean Acid Metabolism', Ting, I.P. and Gibbs, M., Eds, Waverly Press, Baltimore, 1982, 69.

Lüttge, U., and Schnepf, E., Organic Substances, *Enc. Plant. Pysiol.*, **2B**, 244, 1976.

Lüttge, U., and Smith, J.A.C., Mechanisms of passive malic acid efflux from vacuoles of the CAM plant *Kalanchoe daigremontiana*, *J. Membr. Biol.*, **81**, 149, 1984.

Lüttge, U., Ball, E., and Tromballa, H.W., Potassium independence of osmoregulated oscillations of malate-levels in the cells of CAM leaves, *Biochem. Physiol. Pflanz.*, **167**, 267, 1975.

Lüttge, U., Fischer-Schliebs, E., Ratajczak, R., Kramer, D., Berndt, E. and Kluge, M., Functioning of the tonoplast in vacuolar C-storage and remobilization in crassulacean acid metabolism, *J.Exp. Bot.* **46**, 1377, 1995.

Mackenbrock, V., Gunia, W. and Barz, W., Accumulation and metabolism, of medicarpin and maackinian malonylglucosides in elicited chickpea (*Cicer arietinum* L.) cell suspension cultures, *J. Plant Physiol.*, **142**, 385, 1993.

Macklon, A.E.S., Calcium fluxes at plasmalemma and tonoplast, *Plant Cell Environ*, **7**, 407, 1984.

Maclean, D.J., Sargent, J.A., Tommerup, I.C. and Ingram, D.S., Hypersensitivity as the primary event in resistance to fungal parasites, *Nature*, **249**, 186, 1974.

Mandala, S. and Taiz, L., Proton transport in isolated vacuoles from corn coleoptile, *Plant Physiol.*, **78**, 104, 1985.

Mansour, M.M., Van Hasselt, P.R. and Kuiper, P.J.C., Plasma membrane lipid alterations induced by NaCl in winter wheat roots, *Physiol. Plant*, **92**, 473, 1994.

Maretzki, A. and Thom, M., UDP-glucose-dependent sucrose translocation in tonoplast vesicles from stalk tissue of sugar cane, *Plant Physiol.*, **83**, 235, 1987.

Maretzki, A. and Thom, M., High performance liquid chromatography-based re-evaluation of disaccharides produced upon incubation of sugarcane vacuoles with UDP-glucose, *Plant Physiol.*, **88**, 266, 1988.

Marger, M.D. and Saier, J.M.H., A major superfamily of transmembrane facilitators that catalyse uniport, symport and antiport. *Trends Biochem Sci.*, **18**, 13, 1993.

Marienfeld, S., Lehmann, H. and Stelzer, R., Ultrastructural investigations and EDX-analyses of Al-treated oat (*Avena sativa*) roots, *Plant Soil*, **171**, 167, 1995.

Marin, B.P., Evidence for an electrogenic adenosine triphosphatase in *Hevea* tonoplast vesicles, *Planta*, **157**, 324, 1983.

Marin, B.P., and Komor, E., Isolation, purification and subunit structure of H^+-translocating ATPase from *Hevea* latex, *Plant Physiol.*, **75**, 163, 1984.

Marin, B., Preisser, J., and Komor, E., Solubilization and purification of the ATPase from the tonoplast of *Hevea*, *Eur. J. Biochem.*, **151**, 131, 1985.

Marrs, K.A., The functions and regulation of glutathione S-transferase in plants, *Annu. Rev. Plant. Physiol. Plant Mol. Biol.*, **47**, 127, 1996.

Marrs, K.A., Alfenito, M.R., Lloyd, A.M., and Walbot, V., A glutathione S-transferase involved in vacuolar transfer encoded by the maize gene Bronze-2, *Nature*, **375**, 397, 1995.

Martinoia, E., Kaiser, G., Schramm, M.J. and Heber, U., Sugar transport across the plasmalemma and the tonoplast of barley mesophyll protoplasts: evidence for different transport systems., *J. Plant Physiol.*, **131**, 467, 1987.

Martinoia, E., Flügge, U.I., Kaiser, G., Heber, G., and Heldt, H.W., Energy-dependent uptake of malate into vacuoles isolated from barley mesophyll protoplast, *Biochem. Biophys. Acta*, **806**, 311, 1985.

Martinoia, E., Grill, E., Tommasini, R., Kreuz, K. and Amrhein, N., An ATP-dependent glutathione S-conjugate 'export' pump in the vacuolar membrane of plants, *Nature*, **364**, 247, 1993.

Martinoia, E., Schramm, M.J., Kaiser, G., Kaiser, W.M., and Heber, U., Transport of anions in isolated barley vacuoles I. Permeability to anions and evidence for a Cl^- uptake system, *Plant Physiol.*, **80**, 895, 1986.

Martinoia, E., Thume, M., Vogt, E., Rentsch, D. and Deitz, K-J., Transport of arginine and aspartic acids into isolated barley mesophyll vacuoles, *Plant Physiol.*, **97**, 664, 1991.

Marty, F., Role du systeme membranaire vacuolaire dans la differenciation des laticiferes d'*Euphorbia characias* L., *Comp. Rend. Acad. Sci.* (Paris), **271**, 2301, 1970.

Marty, F., Differenciation des plastes dans les laticiferes d'*Euphorbia characias*, L., *Comp. Rend. Acad. Sci.*, (Paris) **272**, 223, 1971.

Marty, F., Cytochemical studies on GERL, provacuoles, and vacuoles in root meristematic cells of *Euphorbia*, *Proc. Nat. Acad. Sci.*, USA, **75**, 852, 1978.

Marty, F., Branton, D. and Leigh, R.A., Plant vacuoles, in 'The Biochemistry of Plants: A Comprehensive Treatise', Tolbert, N.E., Ed., Academic Press, New York, 1980, 625.

Matern, V., Reichenbach, C. and Heller, W., Efficient uptake of flavonoids into parsley (*Petroselinum hortense*) vacuoles requires acylated glucosides, *Planta*, **167**, 183, 1986.

Mathieu, Y., Guern, I., Kurkdjian, A., Manigault, P., Manigault J., Zielinska, T., Gillet, B., Beloeil, J.C. and Lallemand, J.Y., Regulation of vacuolar pH of plant cells. I. Isolation and properties of vacuoles suitable for ^{31}P-NMR studies, *Plant Physiol*, **89**, 19, 1989.

Mathys, W., The role of malate oxalate and mustard oil glucosides in the evolution of zinc resistance in herbage plants, *Plant Phyiol.*, **40**, 130, 1977.

Matile, P., 'Crop Photosynthesis: Spatial and Temporal Determinants', Eds Baker, N.R., and Thomas, H., Elsevier, Amsterdam, 1992, 413.

Matile, P., and Moor, H., Vacuolation: origin and development of the lysosomal apparatus in root tip cells, *Planta*, **80**, 159, 1968.

Matile, P. and Winkenbach, F., Function of lysosomes and lysosomal enzymes in the senescing corolla of the morning glory *Ipomoea purpurea. J. Exp. Bot.*, **22**, 759, 1971.

Matile, P., Ginsburg, S., Schellenberg, M., and Thomas, H., Catabolites in senescing barley leaves are localised in the vacuole of mesophyll cells, *Proc. Natl. Acad. Sci, USA*, **85**, 9529, 1988.

Matsuoka, K., Higuchi, T., Maeshima, M. and Nakamura, K., A vacuolar-type H^+ATPase in a non-vacuolar organelle is required for the sorting of soluble vacuolar protein precursors in tobacco cells, *Plant Cell*, **9**, 533, 1997.

Mauch, F. and Staehelin, L.A., Functional implications of the subcellular localization of ethylene-induced chitinase and beta-1,3-glucanase in bean leaves, *Plant Cell*, **1**, 447, 1989.

Mauch, F., Meehl, J.B. and Staehlin, A.B., Ethylene-induced chitinase and beta-1,3-glucanase accumulate specifically in the lower epidermis and along vascular strands of bean leaves, *Planta*, **186**, 367, 1992.

Maurel, C., Kado, R.T., Guern, J. and Chrispeels, M.J., Phosphorylation regulates the water channel activity of the seed-specific aquaporin γ-TIP, *EMBO. J.*, **14**, 3028, 1995.

Maurel, C., Reizer, J., Schroeder, J.I., Chrispeels, M.J. and Saier, M.H. Jr., Functional characterization of *Escherichia coli* glycerol facilitator GlpF in *Xenopus occytes*, *J. Biol. Chem.*, **269**, 11869, 1994.

Maurel, C., Tacnet, F., Josette, G., Guern, J. and Ripoche, P, Purified vesicles of tobacco cell vacuolar and plasma membranes exhibit dramatically different water permeability and water channel activity, *Proc. Natl. Acad. Sci. USA*, **94**, 7103, 1997.

McAinsh, M.R., Brownlee, C. and Hetherington, A.M., Abscisic acid-induced elevation of guard cell cytosolic Ca^{2+} precedes stomatal closure, *Nature*, **343**, 186, 1990.

McClintock, M., Higinbotham, N., Uribe, E.G. and Cleland, R.E., Active, irreversible accumulation of extreme levels of H_2SO_4 in the brown alga, *Desmarestia*, *Plant Physiol*, **70**, 771, 1982.

Mehdy, M., Active oxgen species in plant defense against pathogens, *Plant Physiol*, **105**, 467, 1994.

Meijer, A.H., de Waal, A. and Verpoorte, R., Purification of the cytochrome P-450 enzyme geraniol-10 hydroxylase from cell cultures of *Catharanthus roseus.*, *J. Chromatogr.*, **635**, 237, 1993a.

Meijer, A.H., Verpoorte, R. and Hoge, J.H.C., Regulation of enzymes and genes involved in terpenoid indole alkaloid biosynthesis in *Cathananthus roseus*, *J. Plant Res.*, **3**, 145, 1993b.

Mende, P. and Wink, M., Uptake of quinolizidine alkaloid lupanine by protoplasts and isolated vacuoles of suspension cultured *Lupinus polyphyllus* cells. Diffusion of carrier-mediated transport?, *J. Plant Physiol.* **129**, 229, 1987.

Merchant, R. and Robards, A.W., Membrane systems associated with the plasmalemma of plant cells, *Ann. Bot.*, **32**, 457, 1968.

Mikus, M., Boba'k, M. and Lux, A., Structure of protein bodies and elemental composition of phytin from dry germ of maize (*Zea mays* L.) *Bot. Acta.*, **105**, 26, 1992.

Milicic, D., Viruskörper in Zellsafte, *Protoplasma*, **57**, 602, 1963.

Miller, A.J. and Smith, S.J., The mechanism of nitrate transport across the tonoplast of barley root cells, *Planta*, **187**, 554, 1992.

Miller, A.J., Brimelow, J.J. and John, P., Membrane potential changes in vacuoles isolated from storage roots of red beet (*Beta vulgaris* L.), *Planta*, **160**, 59, 1984.

Mimura, T., and Tazawa, M., Effect of intracellular Ca^{2+} on membrane potential and membrane resistance in tonoplast free cells of *Nitellopsis obtusa*, *Protoplasma*, **118**, 49, 1983.

Mimura, T., Dietz, K.J., Kaiser, W., Schramm, M.J., Kaiser, G. and Heber, U., Phosphate transport across biomembranes and cytosolic phosphate homeostasis in barley leaves, *Planta*, **180**, 139, 1990.

Minami, A., and Fukuda, H., Transient and specific expression of cysteine endopeptidase during autolysis in differentiation tracheary elements from *Zinnia* mesophyll cells, *Plant Cell Physiol.*, **36**, 1599, 1995.

Minorsky, P.V., A heuristic model of chilling injury in plants: A role for calcium as the primary physiological transducer of injury, *Plant Cell Environ.*, **8**, 75, 1985.

Mitchell, P., Vectorial chemistry and the molecular mechanics of chemiosmotic coupling: power transmission by protocity, *Biochem. Soc. Trans.*, **4**, 399, 1976.

Moreno, P.R.H., Van der Heijden and Verpoorte, R., Cell and Tissue cultures of *Catharanthus roseus*: A literature survey II Updating from 1988 to 1993, *Plant Cell Tiss. Org. Cult.*, **42**, 1, 1995.

Moriyama, Y., Maeda, M. and Futai, M., Involvement of a non-proton pump factor (possibly Donnan-type equilibrium) in maintenance of an acidic pH in lysosomes, *FEBS Lett.* **302**, 18, 1992.

Moriyasu, Y. and Ohsumi, Y., Autophagy in tobacco suspension-cultured cells in response to sucrose starvation, *Plant Physiol.*, **111**, 1233, 1996.

Moriyasu, Y. and Tazawa, M., Distribution of several proteases inside and outside the central vacuole of *Chara australis*, Cell Struct. Funct., **11**, 81, 1986a.

Moriyasu, Y. and Tazawa, M., Plant vacuole degrades exogenous proteins, *Protoplasma*, **130**, 214, 1986b.

Moriyasu, Y., and Tazawa, M., Degradation of proteins artificially introduced into vacuoles of *Chara australis*, *Plant Physiol.*, **88**, 1092, 1988.

Moriyasu, Y., Shimmen, T., and Tazawa, M., Vacuolar pH regulation in *Chara australis*, Cell Struct. Funct., **9**, 225, 1984a.

Moriyasu, Y., Shimmen, T., and Tazawa, M., Electrical characteristics of the vacuolar membrane of *Chara* in relation to pHv regulation, Cell Struct. Funct., **9**, 235, 1984b.

Morse, M.J. and Satter, R.L., Relationships between motor cell ultrastructure and leaf movements in *Samanea saman*, *Physiol. Planta*, **46**, 338, 1979.

Moysset, L. and Simon, E., Secondary pulvinus of *Robinia pseudoacacia* (Leguminosae): structural and ultrastructural features, *Am. J. Bot.*, **78**, 1467, 1991.

Müller, M.L., Irkens-Kiesecker, V., Kromer, D. and Taiz, L., Purification and reconstitution of the vacuolar H^+-ATPases from lemon fruits and epicotyls, *J. Biol. Chem.*, **272**, 12762, 1997.

Müller, M.L., Irkens-Kiesecker, V., Rubinstein, B. and Taiz, L., On the mechanism of hyperacidification in lemon - comparison of the vacuolar H^+-ATPase activities of fruits and epicotyls, *J. Biol. Chem.*, **271**, 1916, 1996.

Münch, E., Die Staffbewegungen in der Pflanze, Gustav Fisher, Jena, 1930.

Nakagawa, S., Kataoka, H. and Tazawa, M., Osmotic and ion regulation in *Nitella*, *Plant Cell Physiol.*, **15**, 457, 1974.

Nakanishi, Y. and Maeshima, M., Molecular cloning of vacuolar H^+-pyrophosphatase and its developmental expression in growing hypocotyl of mung bean. *Plant Physiol.* **116**, 589, 1998.

Nelson, N., Structure, function and evolution of proton-ATPases, *Plant Physiol.*, **86**, 1, 1988.

Neuman, D., Kraus, G., Heike, M. and Gröger, D., Indole alkaloid formation and storage in cell suspension cultures of *C. roseus*, *Planta Med.* **48**, 20, 1983.

Nicholson, R.L., van Scoyoc, S., Williams, E.B. and Kuc, J., Host-pathogen interaction preceeding the hypersensitive reaction of *Malus* sp. to *Venturia*, *Phytopathology*, **67**, 108, 1977.

Niemietz, C.M. and Tyerman, S.D., Characterization of water channels in wheat root membrane vesicles, *Plant Physiol.*, **115**, 561, 1997.

Niemietz, C., and Willenbrink, J., The function of tonoplast ATPase in intact vacuoles of red beet is governed by direct and indirect ion effects, *Planta*, **166**, 545, 1985.

Nishida, K., and Tominaga, O., Energy-dependent uptake of malate into vacuoles isolated from CAM plant, *Kalanchoe daigremontiana*, *J. Plant Physiol.*, **127**, 385, 1987.

Noda, T., Matsuura, A., Wada, Y. and Ohsumi, Y., Novel system for monitoring autophagy in the yeast *Saccharomyces cerevisiae*, *Biochem. Biophys. Res. Comm.*, **210**, 126, 1995.

Norberg, P. and Liljenberg, P., Lipids of plasma membranes prepared from oat root cells: Effect of induced water deficit tolerance. *Plant Physiol.*, **96**, 1136, 1991.

Nozue, M. and Yasuda, H., Occurrence of anthocyanoplasts in suspension culture of sweet potato, *Plant Cell Rep.*, **4**, 252, 1985.

Nozue, M., Kubo, H., Nishimura, M. and Yasuda, H., Detection and characterization of a vacuolar protein (VP24) in anthocyanin producing cells of sweet potato in suspension culture, *Plant Cell Physiol.*, **36**, 883, 1995.

Nozue, M., Yamada, K., Nakamura, T., Kubo, H., Kondo, M. and Nishimura, M., Expression of vacuolar protein (VP24) in anthocyanin-producing cells of sweet potato in suspension culture, *Plant Physiol.*, **445**, 1065, 1997.

Nozzolillo, C. and Ishikura, N., An investigation of the intracellular site of anthocyanoplasts using isolated protoplasts and vacuoles, *Plant Cell Rep.*, **7**, 389, 1988.

Obermeyer, G., Sommer, A. and Bentrup, F.W., Potassium and voltage dependence of the inorganic pyrophosphatase of intact vacuoles from *Chenopodium rubrum*, *Biochim. Biophys. Acta*, **1284**, 203, 1996.

Okazaki, Y., and Tazawa, M., Involvement of calcium ion in turgor regulation upon hypotonic treatment in *Lamprothamnium succintum*, *Plant Cell Environ*, **9**, 185, 1986.

Oparka, K. J., Murant, E.A., Wright, K. M., Prior, D.A.M. and Harris, N., The drug probenecid inhibits the vacuolar accumulation of fluorescent anions in onion epidermal cells, *J. Cell Sci.*, **99**, 557, 1991.

Oritz, D.F., McCue, K.F. and Ow, D.W., *In vitro* transport of phytochelatins by the fission yeast ABC-type vacuolar membrane protein (abstract no.25), *Plant Physiol.*, **105**, S-17, 1994.

Pantoja, O., Gelli, A. and Blumwald, E., Characterization of vacuolar malate and K^+ channels under physiological conditions, *Plant Physiol.*, **100**, 1137, 1992a.

Pantoja, O., Gelli, A. and Blumwald, E., Voltage-dependent calcium channels in plant vacuoles, *Science*, **225**, 1567, 1992b.

Park, H.M., Hakamatsuka, T., Sankawa, U. and Ebizuka, Y., Rapid metabolism of isoflavonoids in elicitor-treated cell suspension cutlures of *Pueraria lobata*, *Phytochemistry*, **38**, 373, 1995.

Pecket, C.R. and Small, C.J., Occurrence, location and development of anthocyanoplasts, *Phytochemistry*, **19**, 2571, 1980.

Perry, C.A., Leigh, R.A., Tomos, A.D., Wyse, R.E., and Hall, J.L., The regulation of turgor pressure during sucrose mobilisation and salt accumulation by excised storage-root tissue of red beet, *Planta*, **170**, 353, 1987.

Pfanz, H., and Heber, U., Buffer capacities of leaves, leaf cells and leaf cell organelles in relation to fluxes of potentially acidic gases, *Plant Physiol.*, **81**, 597, 1986.

Pfanz, H., Martinoia, E., Lange, O.L. and Heber, U., Flux of SO_2 into leaf cells and cellular acidification by SO_2., *Plant Physiol.*, **85**, 928, 1987.

Pfeiffer, W. and Hager, A. Ca^{2+}-ATPase and a Mg^{2+}/H^+-antiporter present on tonoplast membranes from roots of *Zea mays* L., *Planta*, **191**, 377, 1993.

Pick, U. and Weiss, M., Polyphosphate hydrolysis within acidic vacuoles in response to amine-induced alkaline stress in the halotolerant alga *Dunaliella salina*, *Plant Physiol.*, **97**, 1234, 1991.

Pick, U., Bental, M., Chitlaru, E. and Weiss, M., Polyphosphate hydrolysis - a protective mechanism against alkaline stress? *FEBS Lett.* **274**, 15, 1990.

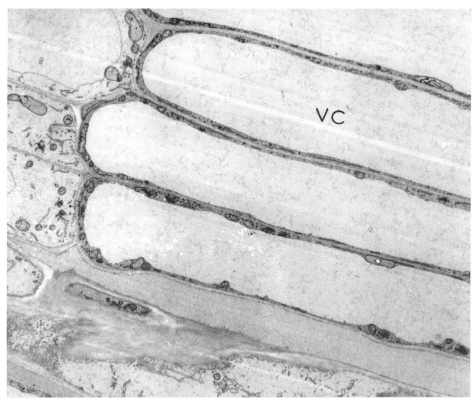

Figure 3.2 Micrograph shows the extremely vacuolated cells of the cambium in the root of *Medicago sativa*. (Drs. S. C. Maitra and D. N. De, published in *Bull. Bot. Soc.* Bengal, **24**, 115, 1970.) With permission of Botanical Society of Bengal.

by a rich cytoplasm, large nucleus and a few small vacuoles. Due to the presence of storage materials such as proteins and lipids in membrane-bound bodies and starch in plastids, the resting cambium cells show reduced vacuolation. However, the differentiating cells rapidly become vacuolated as the cellular activity necessary for cell wall synthesis and related metabolism sets in (Robards and Kidwai 1969). Sometimes, the large vacuole is penetrated by a system of anastomosing protoplasmic strands in high summer heat. When the temperature is low, the vacuole tends to divide into a number of small independent vacuoles. Electron microscope studies have shown that active growing cambium contains one or two large vacuoles, rough endoplasmic reticulum and polyribosomes. In winter the vacuoles are small and numerous, the plasmalemma is thrown into folds and the endoplasmic reticulum occurs mostly in the form of smooth vesicles. Recent observations on the seasonal changes in the ultrastructure of the cambium in the trunk of *Robinia pseudoacacia* by Farrant and Evert (1977) indicate that during dormancy, the fusiform cells are densely cytoplasmic with many small vacuoles containing proteinaceous materials. During reactivation, the proteinaceous material disappears and the vacuoles begin to fuse. During the period of cambial activity, fusiform cells become highly vacuolate and subsequently many large invaginations of the plasma membrane intrude into the vacuole, pushing the tonoplast inward and pinching off into the vacuole. The phellogen or cork cambium, is in general less vacuolar. The protoplast of the phellogen cells contains variously sized vacuoles, some of which may contain tannin.

Xylem is a complex tissue of tracheids, vessels and fibres which are all dead cells at maturity. The associated cells of xylem parenchyma are living and moderately vacuolated. Young xylem cells have several small vacuoles which expand as the cell differentiates and eventually may fuse to form a large vacuole. This fusion usually occurs at a late stage in the development of the xylem elements. The vacuolar membrane may sometimes be resolved as a triple-layered structure, and the two dense layers are of equal thickness. Following the deposition of the secondary wall, degeneration of the cytoplasm starts with a breakdown of the vacuolar membrane (Cronshaw and Bouck 1965).

3.7.6 Reproductive tissues

With respect to the life cycle, the most important tissue of any plant is its reproductive tissue, which shows the most precise cellular architecture and activities in a programmed fashion. Thus, the presence and distribution of vacuoles in the tissues of anthers, pollens and ovules are of great interest.

In the young developing anther lobes, the microspore mother cell at the very early stage may or may not contain detectable vacuoles. In *Phaseolus mungo* these cells contain a large nucleus, a dense cytoplasm with various organelles and no vacuole (Fig. 3.3). Subsequently, as the cells enlarge and prepare for meiosis, a number of small vacuoles are detectable. Although the pollen contains an adequate amount of water for subsequent activity of germination and pollen tube growth, the construction of the pollen is such that its protoplasmic content is protected against possible desiccation during pollination by a thick wall. In general, all angiosperm pollens contain a vacuole but its development and degree varies. A few examples are provided here.

No detectable vacuole exists in microspores of wheat or barley right after meiosis. When the microspores are released from the tetrad, and exine formation has begun, some vacuoles are evident in the cytoplasm. Subsequently, the small vacuoles fuse and a large central vacuole is formed before DNA synthesis (Bin 1986). The large vacuole of wheat microspore undergoes a cycle of changes during the whole uninucleate phase. At the early uninucleate stage the nucleus is spherical and is situated at the centre of the microspore. It migrates to the border of the cell at mid-uninucleate stage and further migrates to the opposite side of the germ pore and becomes oblate at late uninucleate stage. It becomes spherical again and increases in volume, forcing part of the vacuole membrane to cave in, thus entering the premitotic stage. The large vacuole begins to appear at mid-uninucleate stage, expanding its volume to the maximum at the late uninucleate stage, then beginning to decrease its volume when the microspore begins to enter the premitosis stage (He and Ouyang 1984). On the other hand, in spiderwort (*Tradescantia paludosa*) when the four microspores separate to form independent pollen grains, a large eccentrically placed vacuole is formed which displaces the nucleus to one end of the cell. As the nucleus prepares to undergo pollen mitosis it shifts to the convex inner part of the cell and the single vacuole becomes two small vacuoles on two sides of the developing spindle lying across the short axis of the cell. The daughter nuclei gradually differentiate into vegetative and generative nuclei and the vacuoles gradually diminish in size. The mature pollen with a potential for germination contains a virtually 'vacuoleless' cytoplasm with a high density of various organelles (Bal and De 1961). Similarly, the absence of vacuoles in mature pollen grains has been reported in a number of different plants by Larson (1965). His electron microscope studies have also shown that vacuolation is initiated only during pollen tube emergence and begins with a number of small vacuoles arising distal to the emerging tube. These vacuoles increase in size, coalesce and, with continued tube growth, become a single vacuole almost completely filling the pollen grain. Vacuolation of the pollen tube proximal to the pollen grain begins before total vacuolation of the grain is completed. In the germinating pollens of *Parkinsonia aculeata*, the vacuoles arise due to the swelling of the cisternae of the endoplasmic reticulum and vacuolation increases by the growth of the original body and the incorporation of additional cisternae. In the germinating pollens of *Hippeastrum belladonna*, *Hymenocallis occidentalis* and *Ranunculus macranthus*, vacuoles develop from the 'prevacuolar' bodies. These membrane-bound bodies lose their electron opacity (probably through the utilisation of the storage product), undergo enlargement and begin to fuse.

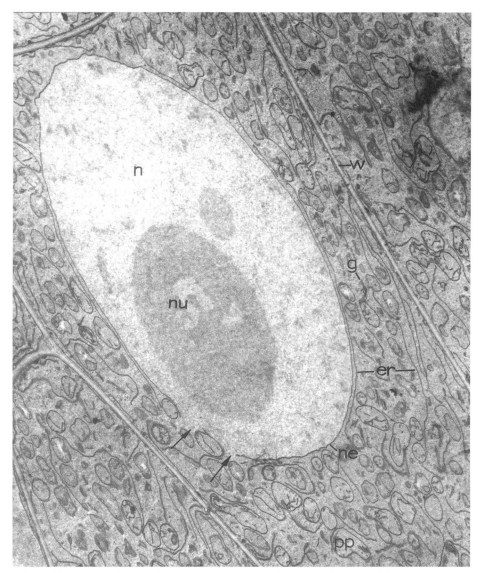

Figure 3.3 Micrograph showing the absence of any vacuole in the cytoplasm of the sporogenous cells in the anther of *Phaseolus mungo* (Drs. S.C. Maitra and D.N. De, unpublished).

The embryo sac of the angiosperms is rather watery and vacuolate. Depending on the number of megaspores involved in its formation it could be monosporic, bisporic or tetrasporic. Irrespective of the number of spores involved the cell destined to be the embryo sac undergoes rapid enlargement with large vacuoles, and the nuclei undergo free nuclear division to produce generally 8-nucleate or 4- or 16- nucleate embryo sacs. The mature embryo sac is vacuolate, or even watery, with or without a discernible tonoplast. Even when the nuclei are maintained as cells delimited by a fine membrane they remain vacuolate.

The ultrastructure of the cotton embryo sac has been described by Jensen (1965a, 1965b, 1965c) (Fig. 3.4). The major portion of the egg cell is occupied by a large vacuole, placing the egg nucleus at one side of the embryo sac. Present in the cytoplasm, particularly at the micropylar end

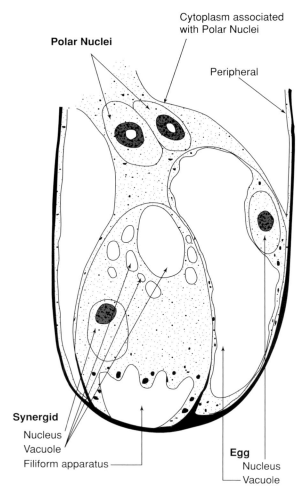

Figure 3.4 Structure of the micropylar end of the mature megagametophyte of cotton (from Jensen 1965). With permission of the Botanical Society of America.

and near the nucellus, are other structures which may be small vacuoles. These are surrounded by an extremely dense membrane and are irregular in shape. The central cell containing the two polar nuclei is a large highly vacuolate cell. Similarly, the synergid cell contains one vacuole of considerable size, and one-third of the cell at the chalazal end is occupied by numerous vacuoles. The vacuoles are bound by a single smooth membrane. Their shape is generally elongated and slightly irregular. The nucellar tissue surrounding the embryo sac is also interesting. The cells near the micropylar end, the so-called collar cells, contain dense cytoplasm and are almost free of typical vacuoles. In contrast, the side and chalazal cells of the nucellus are characterised by single large central vacuoles.

In the angiosperms, the development of the endosperm, especially in the early stages, is accompanied to a great extent by vacuolation. In the *nuclear type* of endosperm development, after double fusion, the triploid nucleus undergoes free nuclear division in the embryo sac which enlarges gradually and remains highly vacuolate (Fig. 3.5). In certain cases, the nuclei of the endosperm may occupy a parietal position and a large vacuole forms in the centre of the embryo sac (Fig. 3.6), or they may be held in the central vacuole by transvacuolar strands. In the *cellular*

Occurrence and distribution of vacuoles 53

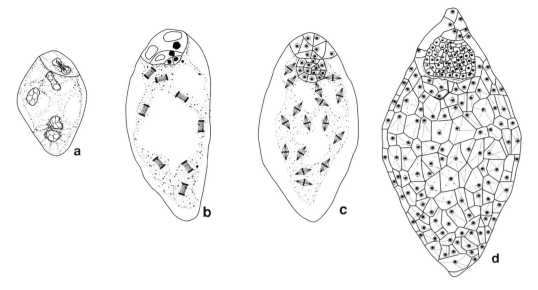

Figure 3.5 Stages in the development of the embryo and cellular endosperm in *Acalypha indica*. **a** both the dividing zygote and the 4-nucleate endosperm are vacuolate; **b** embryo 4-celled, many free nuclei dividing in highly vacuolate endosperm; **c** multicellular embryo developing within free-nucleate endosperm; **d** two-lobed embryo within vacuolate and walled endosperm cells (from Johri and Kapil 1953).

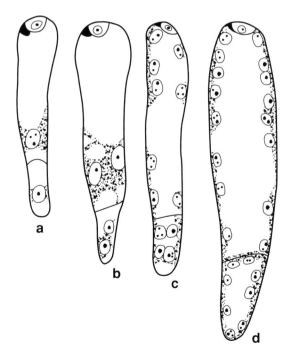

Figure 3.6 Stages in the development of the endosperm in the embryo sac of *Eremurus himalaicus*. The free nuclear division takes place in the enlarging highly vacuolate endosperm cells (from Maheshwari 1950).

type endosperm development, the endosperm remains cellular and the cells continue to maintain vacuoles of various dimensions.

The role of the vacuole is very important in the development of the embryo. Jensen (1963) described the changes that take place in vacuoles in cotton after fertilisation. The large vacuole of the egg cell diminishes in volume and becomes much smaller at the base of the zygote. This decrease in vacuolar volume accompanies the decrease in volume of the egg cell. The cytoplasm, which formerly surrounded the large vacuole, now surrounds the nucleus as a dense mass. The vacuole continues to decrease in volume as the zygote matures, but the volume of the cell ceases to decrease after it reaches a point about one-half its original volume. The result of the continued decrease in size of the vacuole under these conditions is that the cytoplasm surrounding the nucleus appears less dense. In general, after fertilisation a large vacuole in the embryo sac pushes the zygote away from the micropylar end and a transverse division cuts the highly vacuolate cell near the micropylar end, which becomes the basal cell. In cotton most of the vacuoles of the basal cell appear to be surrounded by a thick membrane. Associated with these vacuoles are halos of small vesicles which seem to be fusing with the vacuolar membrane (Jensen 1963). Generally, the basal cell is invariably vacuolate and large and forms the suspensor by a series of transverse divisions. The end cell closest to the micropyle enlarges and becomes a highly vacuolate sac which functions as a haustorium. The other cells of the row of the suspensor are much less vacuolated and the cells of the embryonal mass are characterised by dense cytoplasm with little or no detectable vacuolation. Again, the suspensor may develop into a large haustorium which penetrates between the cells of the endosperm and nucellar tissue. In certain genera of Papilionaceae, the suspensor may consist of two pairs of highly vacuolate multinucleate cells or two rows of highly vacuolate uninucleate cells. In the Rubiaceae, the suspensor at first develops as a multicellular thread, and later the cells closest to the micropyle develop lateral projections which become highly vacuolate and act as haustoria (Maheshwari 1950).

In the club-shaped proembryo of pearl millet, Taylor and Vasil (1995) noticed the existence of a continuous gradient of vacuoles from the highly vacuolated basal suspensor cells to the densely cytoplasmic apical cells (Fig. 3.7). In the late coleoptile stage embryo, the cells of the root are more vacuolated than those of the shoot, with most vacuoles located around the cell periphery. On the other hand, the shoot meristem cells have dense cytoplasm and a few small vacuoles. The scutellar cells are highly vacuolated and contain numerous lipid droplets that vary considerably in size. The epidermal scutellar cells are much smaller and less vacuolated than other cells. Some vacuoles contain electron-dense material along the tonoplast. The vacuoles of the coleorhiza cells form a gradient, from the root cap cells which have the fewest vacuoles, to the cells at the tip of the coleorhiza which have the most. The suspensor cells are highly vacuolated with electron-dense material lining the tonoplast. Taylor and Vasil (1995) stated that ultra-structurally embryogenesis in pearl millet is typical of that reported in other grasses.

3.7.7 Specialised cells

The distribution of vacuoles in particular cells of angiosperms has been described above. It remains to mention some interesting cases of the distribution of vacuoles in certain special cells and tissues. The nectariferous tissues of the angiosperms are generally composed of small thin-walled cells with large nuclei, dense granular cytoplasm and small vacuoles. Some of these vacuoles are barely identifiable under an electron microscope. During secretion the secretory cells contain a dense cytoplasm rich in organelles and the vacuolar system is reduced. Such structural features are generally believed to be associated with high metabolic activity. It has been suggested that vesicles derived from ER cisternae or Golgi bodies which release their contents by fusion with the plasmalemma secrete sugar solution. In *Colchicum ritchii*, the nectary consists of transparent epidermal cells which contain large central vacuoles surrounded by a thin layer of cytoplasm and rounded parenchyma cells with a granular content. The vacuole of the parenchyma cells is

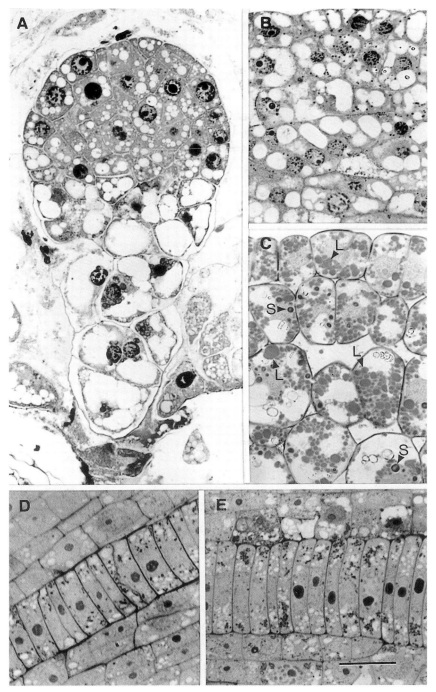

Figure 3.7　Bar = 20 μm. **a** Club-shaped proembryo stage. Note dense apical cells and vacuolated basal suspensor cells. **b** Part of a scutellum from an early coleoptile stage embryo. Note vacuoles and epidermal layer. **c** Mature embryo scutellar cells, note vacuoles and intercellular spaces. **d** Root cells of the late coleoptile stage embryo. **e** Central metaxylem cells of a mature embryo root (from Taylor and Vasil 1995). With permission of the Botanical Society of America.

relatively small and the cytoplasm is rich in organelles. It is therefore assumed that only the parenchyma cells of the nectary function in nectar secretion (Fahn 1989).

In the esential oil-secreting cells of the Labiatae, the vacuoles contain osmiophilic substances. At the stage of secretion, the cytoplasm retracts irregularly from the cell wall which becomes split between the pectic and cuticular layers. During the process of secretion, the vacuoles lose their contents. In the old cells they are empty, and the extraplasmatic space becomes greatly enlarged.

Prominent vacuoles are a constant feature in salt glands. In many xeromorphic species the vesiculate trichomes may cover the entire epidermis of the leaves, as in *Atriplex portulacoids*. In these hairs the salt is secreted by the cytoplasm into the vacuole which enlarges and becomes a bladder (Osmond et al. 1969). Ultrastructural studies by Balsamo and Thomson (1993) in *Frankenia grandifolia* show that the salt glands consist of two collecting cells and six secretory cells. The collecting cells are highly vacuolated, containing a densely staining osmiophilic substance along the periphery of the tonoplast. The secretory cells also contain several medium to large vacuoles. In the adaxial salt glands of a mangrove plant *Avicennia germinans*, two collecting cells, eight secretory cells and one single stalk cell contain vacuoles of a decreasing order of total volume, with the collecting cells showing the highest degree (55%) of vacuolation.

All motor cells responsible for curvature movement in plant parts contain vacuoles which are directly involved in a rise and fall in turgor pressure. Usually the central vacuole in a typical turgid motor cell fragments when it contracts and reforms from the coalescence of the small vacuoles when the cell expands as reported in carnivorous plants *Dionaea* and *Drosera* (Toriyama 1973). The process is similar to the changes in the guard cells during opening and closing of stomata as reported in *Vicia faba*, *Lillum candidum* and *Anemia rotundifolia* (Humbert and Guyot 1982). The pulvinar motor cells of a number of leguminous species, e.g. *Samanea saman* (Morse and Satter 1979), *Albizzia* sp. (Campbell and Garber 1980), *Mimosa pudica* (Fleurat-Lessard 1988) and *Robinia pseudoacacia* (Moysset and Simon 1991) show changes both in number and size of non-tannin vacuoles during leaflet movement. Although Toriyama and Jaffe (1972) have observed changes in both size and shape of tannin vacuoles during seismonastic movements of the main pulvinus of *Mimosa pudica*, the tannin vacuoles containing phenolic and polysaccharide compounds in the *Robinia* motor cells do not exhibit any change in size and number.

When cells and tissues are cultured they become more vacuolate than the cells of the explant or comparable cells and tissues *in planta*. As the cell division proceeds and the cells stay together in a mass the degree of vacuolation decreases. A compact callus mass or a differentiated embryo has less vacuolation in its cells. In single cell cultures in liquid medium the cells are highly vacuolated and abnormally swollen. In suspension cultures the cells have no constraints on shape or surface. The high degree of vacuolation is possibly the result of partially controlled absorption and various osmotica, growth regulators and cultural conditions, since compared to the callus cells the cells in suspension are in direct contact with the medium. When suspension cells are plated on solid or semi-solid medium, the degree of vacuolation is reduced to some extent.

3.8 SHAPE, NUMBER AND VOLUME

As early as 1876, Charles Darwin recorded that in a given tissue the shape, number and volume of vacuoles in a cell may vary, based on the anthocyanin vacuoles of *Drosera* (vide Chapter I). A consideration of the distribution of vacuoles in the plant kingdom, presented earlier, also indicates the diversity of shape, number and volume in different cell types.

A physico-chemical consideration of the cytoplasm points to its nature as a heterogenous system in which the various organelles, particles, membranes, etc., are the *dispersed phase* and water is the *continuous phase*. Since all the particles and organelles, including vacuoles, in the dispersed phase are not alike, cytoplasm is said to be a *polyphasic colloid*. According to the second law of thermodynamics, the free energy of a system tends to decrease; therefore, the surface free energy also tends to decrease and the dispersed particles in a heterogerous system assume a shape with minimal surface area. Consequently, a vacuole limited by the tonoplast, like a drop of liquid suspended in another liquid or watery ground protoplasm, always tends to assume a spherical shape, since a sphere has a smaller surface than any differently shaped object of the same volume (vide Giese 1962). Thus, very often in a typical turgid cell a vacuole is more or less spherical, spheroidal or ellipsoidal. Otherwise, a vacuole may take any conceivable shape due to its origin, development and function within the physical constraints imposed by the presence of other organelles and structures present in the cell. Again, this consideration that a single large globule has less surface area than many small ones suggests that the small ones may coalesce. This consideration can explain the fusion of vacuoles, unless other factors or surface properties interfere.

The accuracy of determination of the number of vacuoles (copy number) and their volume per cell depends on the method used as well as on fixation, staining and subsequent steps both for light and electron microscopy. Thus, some of the earlier estimates of vacuolar volume may be erroneous. However, a general consensus has emerged on the estimates of vacuolar volume fraction near the root apex. Some recent investigations present reliable data. Munnich and Zoglauer (1979) obtained a vacuolar fraction of 33% of the tissue volume in the first 2-mm segment of maize roots and 38% in the fourth 1-mm segment. Gerson and Poole (1972) recorded that in the apical 1-mm segment of mung been root the vacuolar volume is 22% of the tissue volume. Similarly Patel et al. (1990) recorded the vacuolar volume fraction in pea roots to be 22 and 33% in the first and second 1-mm segments, but 57 and about 74% in the fourth and seventh 1-mm segments respectively.

Using a sophisticated method of stereological evaluation of cellular and subcellular volumes from light and electron micrographs, Winter et al. (1993) stated that in barley leaf epidermal cells 99% of the volume is occupied by the vacuole and in the mesophyll cells it is about 73%. They present similar data for spinach leaves, where vacuoles occupy 89% of the epidermal cells and 79% of the mesophyll cells (Winter et al. 1994).

The total volume of vacuoles in a cell may not remain constant, but may change reversibly. The best example of reversible change in vacuolar volume is provided in stomatal guard cells. The volume of the guard cell vacuole increases during stomatal opening and decreases during closing, paralleling the volume changes measured for the entire guard cell (Fricker and White 1990). Similar reversible change in vacuolar volume has been noted in the non-tannin vacuoles of the extensor and flexor cells of the secondary pulvinus of *Robininia psueudoacacia* by Moysset and Simon (1991). Most extensor motor cells show a large central vacuole that occupies most of the cell when the pulvinus is open, but various smaller vacuoles appear when it is closed. In contrast, the great majority of flexor cells are univacuolate or have a few vacuoles in the bent pulvinus, whereas they are more multivacuolate in the straight pulvinus. The analysis of the vacuolar area/cellular area ratio shows that during dark-induced closure it decreases from 0.82 to 0.67 in the extensor cells, but increases from 0.59 to 0.65 in the flexor cells.

An excellent example of enormous increase in the tonoplast surface is provided by the vacuole system of pea cotyledon. Early in its development the cotyledon contains typical plant cells with a large central vacuole, mostly containing water. As protein synthesis and transport continue, the vacuole begins to develop as a protein storage vacuole. Craig et al. (1979) estimated that the surface area of the vacuole system increased 100 times from 5500 to 552 000 μ^2 during the transition from central vacuole to protein storage vacuole.

In specialised cells, vacuolar volume can vary widely. For example, in the secretory cells of the salt glands of *Frankenia grandiflora* it is 29% of the cell volume, and in *Avicennia germinans* it is 33%. In the collecting cells of *A. germinans* it is 52% and in the stalk cells it is 23% (Balsamo and Thomson 1993). Various physiological alterations and stress are known to cause changes in vacuolar volume. A standard feature of suspension culture cells is that compared to the cells of the tissue explants their vacuolar volume increases rapidly. Reuveni et al. (1991) reported that in suspension culture, cells of beetroot show a vacuolar volume of about 70% and those of carrot cells comprise 80% of the total cell volume. In the process of cold-hardiness in the stem tissues of peach, Wisniewski and Ashworth (1986) noticed a reduction in the size of vacuoles. However, in

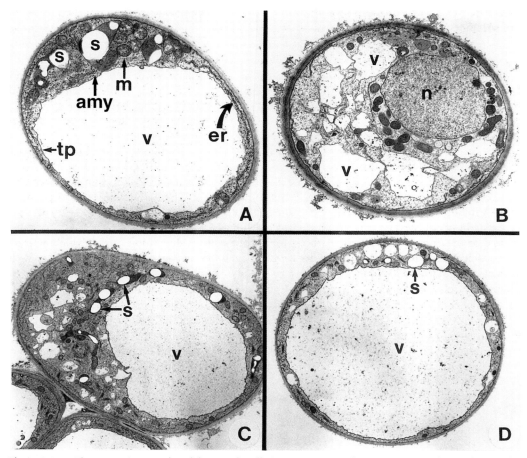

Figure 3.8 Electron micrographs of the peach cells in suspension culture. **a** A non-acclimated control cell characterised by a large central vacuole and peripheral cytoplasm. Numerous amyloplasts containing starch grains are visible. **b** Cold-acclimation by 3°C in dark for 10 days causes replacement of the large central vacuole by microvacuoles, cytoplasmic augmentation and disappearance of starch grains. **c** When cold-acclimated cells are transferred to 24°C in dark for 3 days, the central vacuole and the starch grains reappear in the deacclimated cells. **d** In cells treated with 75 μM abscisic acid, no ultrastructural alteration was induced by cold-acclimation.
amy: amyloplast; er: endoplasmic reticulum; m: mitrochondria; n: nucleus; s: starch; tp: tonoplast; v: vacuole. (Courtesy Arora, R. and Wisniewski, M.E., *Plant Cell Tissue Org.Cult*, **40**, 17, 1995.) With permission of Kluwer Academic Publishers.

suspension culture cells of peach associated with low temperature-induced cold acclimation, the large central vacuole is replaced by smaller vacuoles (Fig. 3.8). Metal toxicity is another example of physiological stress. An exposure to subtoxic concentrations of zinc can induce a 2.93-fold and 6.78-fold increase in the total vacuolar volume fraction in rice and wheat root meristematic cells, whereas no significant increase is noted in rye roots (Davies, *et al.* 1992).

Another example of increase in vacuolar number and volume as a response to physiological stress is noted in marine green alga, *Chlorococcum littorale* which is resistant to extremely high concentrations of CO_2. Kurano *et al.* (1998) noted that high CO_2 conditions induce an increase in the number and size of vacuoles in the alga. Lately, Sasaki *et al.* (1999) have confirmed these observations with the aid of differential interference fluorescence microscopy and immunoassay of

V-ATPase. Their time-course studies show that vacuolar number and volume, as well as the level, and specific activity of V-ATPase, increase after transferring *C. littorale* cells from air to 40% CO_2.

3.9 SUMMARY

Procaryotes

No true vacuoles are found in procaryotes. Certain members of cyanobacteria and purple sulfur bacteria show the presence of **gas vacuoles** which are gas-filled chambers and have no similarity with typical vacuoles found in other groups.

Algae

- As a rule, one or more vacuoles are present in all algal cells. In many cases, the vacuole may be huge, occupying most of the cell volume and leaving a thin film of protoplasm lining the cytoplasm or suspending the nucleus and a little cytoplasm at the centre of the cell by protoplasmic strands.
- In addition to the normal vacuoles in all cells which maintain water at relatively constant turgor pressure, in certain algal groups there exist contractile vacuoles (CV) which actively participate in osmorgulation. The CVs are generally found in flagellates, zoospores, etc., which lack a rigid cell wall. The CVs, outlined by a single membrane, absorb water and enlarge to reach the diastole stage when the membrane fuses with the plasma membrane and releases its contents to the exterior reaching the systole stage. Then the small vacuoles at the base of the main CV again accumulate water and fuse to form a large vacuole, and the process is repeated.
- Pusule is a specialised contractile vacuole found in the dinoflagellates. It consists of an interconnected system of vesicles that are outlined by a double membrane system composed of plasma membrane closely appressed with the vacuolar membrane. Pusule is connected with the exterior of the cell by a slender canal and does not exhibit any contractile activity.

Fungi

- Vacuoles are widespread in all groups of fungi. In the aquatic fungi, Phycomycetes, vacuoles may appear as a fine network of canaliculi along the length of the hyphae. In certain phycomycetes contractile vacuoles present in the biflagellate zoospores are called water expulsion vacuole (WEV). WEVs are surrounded by several types of vacuoles and golgi bodies to form what is termed an WEV apparatus.
- In Ascomycetes and Basidiomycetes vacuoles are chiefly spherical or elongated in shape. In certain basidiomycetes tubular or nodular vacuole systems have been found along the length of hyphae. A very special vacuolar system exists in yeasts.

Bryophyta

In the bryophytes, including liverworts and mosses, vacuoles are ubiquitous and abundant, apart from the antherozoids where vacuoles are not present.

Pteriophyta

In the pteridophytes, including club mosses, horse tails and ferns, vacuoles occur in all types of cells and diverse vacuolar inclusions are noted. The most interesting feature is the formation of a huge central vacuole during the development of the megagametophyte from the megaspore of the heterosporus pteridophytes. The antherozoids of certain groups appear as non-vacuolate.

Gymnosperm

The gymnosperms exhibit typical distribution of vacuoles in all types of cells. However, a high degree of vacuolation is a major phemomenon in the reproductive structure, especially in the development of the megagametophyte, the egg cell, and the proembryo as typified by *Cycas*.

Angiosperm

A general survey of occurrence and distribution of vacuolation in angiosperms show that the meristematic cells have dense cytoplasm with no or very few small vacuoles. Most of the differentiated cells like mesophyll cells, cortical cells of stems and roots, have a large single vacuole. As in the gymnosperms, vacuoles are invariably involved in the organisation of all tissues at all stages of the reproductive system.

Specialised cells

In various specialised cells, vacuoles play different roles. Salt glands in many plants are characterised by prominent vacuoles.
- All cells responsible for curvature movements contain vacuoles which directly participate in the rise and fall of turgor pressure. Reversible coalescence of small vacuoles after absorption of water to form a large vacuole causes turgidity, and fragmentation of the large vacuole with loss of water causes flaccidity. This process is observed in stomatal guard cells, pulvinar cells, as well as in the tentacles of carnivorus plants. Tissue culture cells, especially those in suspension culture, are more vacuolated than the original tissue explants.
- Regarding the shape of the vacuole, no generalisation can be made, apart from noting that in a typical turgid cell a vacuole tends to be spheroidal or ellipsoidal. No generalisation can be made on vacuolar volume and size, which depend on cell type and the physiological state of the cell.

3.10 REFERENCES

Arora, R. and Wisniewski, M.E., Ultrastructural and protein changes in cell suspension cultures of peach associated with low temperature-induced cold acclimation and abscisic acid treatment, *Plant Cell Tissue Org, Cult.*, **40**, 17, 1995.

Bal, A.K. and De, D.N., Developmental changes in the submicroscopic morphology of the cytoplasmic components during microsporogenesis in Tradescantia, *Develop. Biol.*, **3**, 241, 1961.

Balsamo, R.A. and Thomson, W.W., Ultrastructural features associated with secretion in the salt glands of *Frankenia grandifolia* (Frankeniaceae) and *Avicennia germinans* (Avicenniaceae), *Am. J. Bot.*, **80**, 1276, 1993.

Banta, L.M., Robinson, J.S., Klionsky, D.J. and Emr, S.D., Organelle assembly in yeast: characterization of yeast mutants defective in vacuolar biogenesis and protein sorting, *J. Cell Biol.*, **107**, 1369, 1988.

Bartinicki-Garcia, S. and Hemmes, D.E., Some aspects of the form and function of Oomycete species, in 'The Fungal Spore', Weber, D.J. and Hess, W.W. Eds., John Wiley, New York, 1976, 593.

Bessey, E.A., Morphology and Taxonomy of Fungi, 1st Ed., Blakiston Co. Philadelphia, 1950, 366.

Bin, H., Ultrastructural aspects of pollen embryogenesis in *Hordeum*, *Triticum* and *Paeonia*, in 'Haploids of Higher Plants', Han, H. and Hongyuan, Y., Eds. China Academic Pub, Beijing, 1986, 91.

Bold, H.C., and Wynne, M.J., Introduction to the Algae, Prentice-Hall Inc. Englewood Cliffs, 1978, 72.

Campbell, N.A. and Garber, R.C., Vacuolar reorganization in the motor cells of *Albizzia* during leaf movement, *Planta*, **148**, 251, 1980.

Cecich, R.A., An electron microscopic evaluation of cyto-histological zonation in the shoot apical meristem of *Pinus banksiana.*, *Am. J. Bot.*, **64**, 1263, 1977.

Cho, C.W. and Fuller, M.S., Observations of the water expulsion vacuole of *Phytophthora palmivora*, *Protoplasma*, **149**, 47, 1989.

Clegg, J. and Filosa, M.F., Trehalose in the cellular slime mold, *Dictyostelium mucoroides*, *Nature*, **192**, 1077, 1961.

Cohen-Bazire, G., The photosynthetic apparatus of procaryotic organisms, in 'Biological Ultrastructure: The origin of cell organelles', Harris, P. Ed. Oregon State University Press, Corvallis, 1971, Chapter 5.

Colt, W.M. and Endo, R.M., Ultrastructural changes in *Pythium aphanidermatum* zoospores and cysts during encystment, germinations and penetrations of primary lettuce roots. *Phytopathology*, **62**, 751, 1972.

Craig, S., Goodchild, D.J. and Hordham, A.R., Structural aspects of protein accumulation in developing pea cotyledons. I. Qualitative and quantitative changes in parenchyma cell vacuoles, *Aust. J. Plant Physiol.*, **6**, 81, 1979.

Cronshaw, J., Phloem differentiation and development, in 'Dynamic Aspects of Plant Ultrastructure', Robards, A.W., Ed., McGraw Hill Book Co. U.K., 1974, Chapter 11.

Cronshaw, J., and Bouck, G.B., The fine structure of differentiating xylem elements, *J. Cell Biol.*, **24**, 415, 1965.

Davies, K.L., Davies, M.S. and Francis, D., Zinc-induced vacuolation in root meristematic cells of cereals, *Ann. Bot.*, **69**, 21, 1992.

Dodge, J.D., The ultrastructure of *Chroomonas mesostigmatica* Butcher. (Cryptophyceae). *Arch. Mikrobiol.*, **69**, 226, 1969.

Dodge, J.D., The ultrastructure of the dinoflagellate pusule: a unique osmo-regulatory organelle. *Protoplasma*, **75**, 285, 1972.

Dykstra, M., Quoted by L.S. Olive, The Mycetozoa, Acad. Press, New York, 1975, 189.

Esau, K., Minor veins in Beta leaves: Structure related to function, *Amer. Phil. Soc.* **111**, 119, 1967.

Fahn, A. Plant Anatomy, Pergamon Press, Oxford, 1989.

Farrant, J.J. and Evert, R.F., Seasonal changes in the ultrastructure of the vacuolor structure and distribution in cambium of *Robinia pseudoacacia*.,*Trees*, **11**, 191, 1997.

Fleurat-Lessard, P., Structural and ultrastructural features of cortical cells in motor organs of sensitive plants, *Biol. Rev.*, **63**, 1, 1988.

Fogg, G.E., Stewart, W.D.P., Fay, P., and Walsby, A.E., The blue green Algae, Acad. Press, London, 1973, 459.

Fricker, M.D. and White, N., Volume measurement of guard cell vacuoles during stomatal movements using confocal microscopy. *Micro*, **90**, 345, 1990.

Gangulee, H. C. and Kar, A.K., College Botany, Vol 2, 1st Edition, New Central Book Agency, Calcutta, 1968. 971.

Giese, A.C., Cell Physiology, W.B. Saunders, Philadelphia. London, 1962, 66.

Gregg, J.H. and Badman, W.S. Morphogenesis and ultrastructure in *Dictyostelium*, *Develop. Biol*, **22**, 96, 1970.

Greson, D.F. and Poole, R.J., Chloride accumulation by mung been root tips, *Plant Physiol.*, **50**, 603, 1972.

Grove, S.N., Fine structure of zoospore encystment and germination in *Pythium aphanidermatum*. *Am. J. Bot.*, **57**, 745, 1970.

Hahl, H.R. and Hamamoto, S. Ultrastructure of spore differentiation in *Dictyostelium*: The prespore vacuole. *J. Ultrastruct.Res.* **26**, 442, 1969.

Hausmann, K. and Patterson, D.J., Involvement of smooth and coated vesicles in the function of the contractile vacuole complex of some Cryptophycean flagellates., *Exp. Cell Res.*, **135**, 449, 1981.

He, D. G. and Ouyang, J.W., Callus and plantlet formation from cultured wheat anthers at different developmental stages, *Plant Sci, Lett*, **33**, 71, 1984.

Ho, H.H., Hickman, C.J. and Telford, R.W., The morphology of zoospores of *Phytophthora megasperma* var. Sojae and other phycomycetes. *Can. J. Bot.*, **46**, 88, 1968.

Hoch, H.C. and Mitchell, J.E., The ultrastructure of zoospores of *Aphanomyces eutiches* and their encystment and subsequent germination. *Protoplasma*, **75**, 113, 1972.

Hoffman, L.R. Fertilization in *Oedogonium*. I Plasmogamy. *J. Phycol.*, **9**, 62, 1973.
Humbert, C. and Guyot, M., Vacuome des cellules de garde et mouvements des stomates, *Physiol. Veg.*, **20**, 239, 1982.
Ikeda, T. and Takeuchi, I., Isolation and characterization of a prespore specific structure of the cellular slime mold, *Dictyostelium discoideum*, *Develop. Growth and Differentiation* **13**, 221, 1971.
Jensen, W.A., Cell development during plant embryogenesis, *Brookhaven Symp. Biol.*, **16**, 179, 1963.
Jensen, W.A., The ultrastructure and composition of the egg and central cell of cotton, *Am. J. Bot*, **52**, 781, 1965(a).
Jensen, W.A., The ultrastructure and histochemistry of the synergids of cotton, *Am. J. Bot.*, **52**, 238, 1965(b).
Jensen, W.A., The composition and ultrastructure of the nucellus in cotton, *J. Ultrastructure Res.* **13**, 112, 1965(c).
Kitching, J.A., Osmoregulation and ionic regulation in animals without kidneys, *Symp. Soc. Exp. Biol.*, **8**, 63, 1954.
Knauf, G.M., Welter, K., Mueller, M. and Mendgen, K., The haustorial host parasite interface in rust-infected bean leaves after high-pressure freezing, *Physiol. Mol. Plant Pathol.*, **34**, 519, 1989.
Kofoid, C.A., and Swezy, The free-living unarmoured Dinoflagellate. *Mem. Univ. Calif.*, **5**, 1, 1921.
Kurano, N., Sasaki, T. and Miyachi, S., Carbon dioxide and microalgae, in 'Advances in Chemical Conversion for Mitigating Carbon dioxide Studies in Surface Science and Catalysis', Inui, T., Anpo, M., Izui, K., Yanagida, S., and Yamaguchi, T. Eds., Elsevier Science, Amsterdam, 1998, 55.
Larson, D.A., Fine structural changes in the cytoplasm of germinating pollen, *Am. J. Bot.*, **52**, 139, 1965.
Lovtrup, S. and Honsson Mild, K., Permeation, diffusion and structure of water in living cells, in 'International Cell Biology', H.G. Schweiger. Ed., Springer-Verlag, Berlin, 1981, 889.
Ma, Y. and Steeves, T.A., Characterizations of stelar initiation in shoot apices of ferns, *Ann. Bot.*, **75**, 105, 1995.
Maeda, Y. and Takeuchi, I., Cell differentiation and fine structure in the development of cellular slime molds, *Develop. Growth and Differentiation*, **11**, 231, 1969.
Maheshwari, P., An Introduction to the Embryology of Angiosperms, McGraw Hill, New York, 1950.
McCully, M.E. and Canny, M.J., The stabilization of labile configurations of plant cytoplasm by freeze-substitution., *J. Microsc.* **139**, 27, 1985.
Morse, M.J. and Satter, R.L., Relationships between motor cell ultrastructure and leaf movements in *Samanea saman*, *Physiol. Planta.*, **46**, 338, 1979.
Moysset, L. and Simon, E., Secondary pulvinus of *Robinia pseudoacacia* (Leguminosae): structural and ultrastructural feature, *Am, J,. Bot.*, **78**, 1467, 1991.
Munnich, H. and Zoglauer, M., Bestimmung des Volumenanteils der Vacuolen in Zellen der Wurzelsitze vn Zea mays L., *Biologische Rundschau*, **17**, 119, 1979.
Norris, R.E., Unarmoured marine dinoflagellates. *Endeavour*, **25**, 124, 1966.
Osmond, C.B., Luttge, V., West, K.R., Pallaghy, C.K., and Sacher-Hill, B., Ion absorption in *Atriplex* leaf tissue: II. Secretion of ions to epidermal bladders, *Aust. J. Biol. Sci.*, **22**, 797, 1969.
Patel, D.D., Barlow, P.W. and Lee, R.B., Development of vacuolar volume in the root tips of pea, *Ann. Bot.*, **65**, 159, 1990.
Pickett-Heaps, J.D., Green Algae. Sinauer Assoc. Sunderland, 1975, 372.
Porter, J., and Jost, M., Physiological effects of the presence and absence of gas vacuoles in blue-green alga *Microcystis aeruginosa* Kuetz. emmend. Elenkin., *Arch. Microbiol.*, **110**, 225, 1976.
Reichle, R.E., Fine structure of *Phytophthora parasitica* zoospores, *Mycologia*, **61**, 30, 1969.

Reuveni, M., Lerner, H.R. and Poljakoff-Mayber, A., Osmotic adjustment and dynamic changes in distribution of low molecular weight solutes between cellular compartments of carrot and beetroot cells exposed to salinity, *Am. J. Bot.* **78**, 601, 1991.

Robards, A.W. and Kidwai, P., A comparative study of the ultratructure of resting and active cambium of *Salix fragilis* L. *Planta*, **84**, 239, 1969.

Robertson, R.W. and Fuller, M.S., Ultrastructural aspects of the hyphal tip of *Sclerotium rolfsi* preserved by freeze-substitution, *Protoplasma*, **146**, 143, 1988.

Sasaki, T., Pronina, N.A., Maeshima, M., Iwasaki, I., Kurano, N. and Miyachi, S., Development of vacuoles and vacuolar H^+-ATPase activity under extremely high CO_2 conditions in *Chlorococcum littorale* cells. *Plant Biol.*, **1**, 68, 1999.

Schnepf, E. and Koch, W., Über die Entstehung der pulsierenden Vacuolen von *Vacuolaria viresens* aus dem Golgi-Apparat., *Arch. Mikrobiol.*, **54**, 229, 1966.

Shear, H., and Walsby, A.E., An investigation into the possible light shedding role of gas vacuole in a planktonic blue-green alga. *Br. Phycology J.*, **10**, 241, 1975.

Shepherd, V.A., Orlovich, D.A. and Ashford, A.E., A dynamic continuum of pleiomorphic tubules and vacuoles in growing hyphae of a fungus. *J. Cell Biol.*, **104**, 495, 1993.

Smith, B., Ultrastructure of fresh water Phycomycetes, in 'Recent Advance in Aquatic Mycology', Gareth Jones, E.B. Ed., Elek Science, London, 1976, 603.

Smith, G.M., Cryptogamic Botany, Vol. 1, 1st Edition, McGraw Hill Book Co, New York, 1938, 157.

Smith, G.M., Cryptogomic Botany, Vol. 2, 2nd Edition, McGraw Hill Book Co., New York, 1955, 96.

Taylor, M.G. and Vasil, I.K., The ultrastructure of zygotic embryo development in pearl millet (*Pennisetum glaucum*; Poaceae), *Am. J. Bot.*, **52**, 205, 1995.

Toriyama, H., Tannins and tannin vacuole in the motor organ of higher plants, *Sci. Report*. Tokyo Women's Christian College, 29–31, 386, 1973.

Toriyama, H. and Jaffe, M.J., Migration of calcium and its role in the regulation of seismonasty in the motor cell of *Mimosa pudica* L., *Plant Physiol.*, **49**, 72, 1972.

Walsby, A.E., Structure and functions of gas vacuoles, *Bact. Revs.*, **36**, 1, 1972.

Winter, H., Robinson, D.G. and Heldt, H.W., Subcellular volumes and metabolite concentrations in barley leaves, *Planta*, **191**, 180, 1993.

Winter, H., Robinson, D.G. and Heldt, H.W., Subcellular volumes and metabolite concentrations in spinach leaves, *Planta*, **193**, 530, 1994.

Wisniewski, M. and Ashworth, E.N., A comparison of seasonal ultrastructural changes in stem tissues of peach (*Prunus persica*) that exhibit contrasting mechanisms of cold hardiness, *Bot. Gaz.* **147**, 407, 1986.

IV
Ultrastructure and chemical composition of tonoplast

4.1 ULTRASTRUCTURE

The ultrastructure of the tonoplast differs from other membranes. It has a great capacity to undergo expansion. Salyaev (1985) records that it is capable of extraordinary stretching—up to 90%—which cannot be explained by the current concept of membrane structure. The high fluidity and resistance to great shearing forces may be explained by the presence of large quantities of the unsaturated fatty acids, linoleic and linolenic acid. The tonoplast has not been shown under light microscope, except as a phase boundary between two optical phases of cytoplasm and vacuole sap. It is possible to view it indirectly by depositing precipitates on the tonoplast by enzyme histochemical methods (vide Methodology).

Kaufmann and De (1956) were the first to detect a distinct tonoplast with the aid of electron microscopy. The tonoplast is single-layered, i.e. made of a unit membrane, with a thickness of 8–12 nm, according to various authors. The thickness may vary depending on the material, as well as on the type of fixation and staining procedure for electron microscopy. The tonoplast is continuous without any interruption or pore. It is not normally connected with any other membrane or membranous structure. Typically the ribosomes are not known to be present on any surface of the tonoplast.

Densitometric studies by Salyaev (1985) show that the outer face (PF) of the tonoplast has higher electron density than the inner (vacuolar) or exosplasmic face (EF). However, Marty (1982) reports that the EF is thicker and darker, probably because of osmiophilic compounds like polyphenolics and carbohydrates bound to the peripheral elements of the membrane. The asymmetry of the two surfaces of the membrane is also shown by experiments in which colloidal gold particles coated with ConA were incubated with unsealed fragments derived from the vacuole membrane. Marty (1985) demonstrated that the gold particles were attached to the EF of the tonoplast and concluded that the receptors for ConA, or other lectins, are localised on the EF. The carbohydrate chains of the glycoconjugates are thought to be exoplasmic macromolecular extensions of the tonoplast.

Volkmann (1981) showed from first freeze-fracture electron microscopic (FFEM) studies that in the root cells of garden cress, the PF of the tonoplast can easily be distinguished from the EF by its very dense packing of intramembranous particles (IMPs). Marty (1985) and Salyaev (1985) studied the particulate fine structure of the tonoplast of beetroot vacuoles, both from intact cells and from isolated membrane fragments with the aid of FFEM. Marty's micrographs of intact cells revealed that the EF appears to be uneven due to a mixture of pits and particles often associated in aggregates or in rows (1600 particles/μm^2; average diam, 11.2 nm). The PF contains bigger particles (average diam, 12.0 nm) which are more densely packed (2400 particles/μm^2). The electron micrographs by Salyaev show slightly different measurement for the IMPs—10.6 nm in PF and 10.1 nm in EF layers. The distribution density of the IMPs differs widely—2006/μm^2 in PF and 725/μm^2 in EF. The background matrix of the two faces is relatively smooth. In the case of isolated tonoplast fragments Marty could not detect much difference in diameter of the IMPs nor density on the two faces. In contrast, the diameter of the particles and their density on the plasma membrane are very different.

A method has been devised to differentiate sterol-rich membranes from sterol-poor ones. A treatment of sterol-containing membranes with filipin induces the appearance of 20 nm filipin-induced lesions (FIL) which appear as protuberances or invaginations of a very characteristic structure (Verkleij et al. 1973, Severs 1981). After FFEM combined with filipin-treatment, the tonoplast and the plasmalemma can be easily characterised by the frequency of intramembranous particles and the arrangement of the FILs. Compared with the plasmalemma, the tonoplast is characterised by a lesser number of FILs which are typically arranged in clusters of different size (on plasmalemma they are evenly distributed) (Vom Dorp et al. 1986). The filipin treatment has little effect on ER or nuclear membrane which never exhibits FIL. FILs are rarely found on amyloplasts and only a few can be detected on dictyosomes.

An additional feature, i.e. of the presence of IMPs in a group of four or 'tetrads' on the tonoplast of the mucilage vacuoles of *Gigartina teedii* and *Ceramium rubrum*, has been demonstrated by Tsekos et al. (1985). In later detailed FFEM studies of the tonoplast of mucilage vacuoles of 27 species of red algae, Tsekos and Reiss (1993) recorded IMPs with average diameters between 8.5 and 14.4 nm. The tetrads are most clearly and more frequently located in the EF than the PF. They lie individually within the vacuole membrane or form clusters in several species and are randomly distributed (Fig. 4.1). In two species, *Ceramium diaphanum* var. *strictum* and *Laurencia obtusa*, the tetrads span both the outer and inner leaflets as transmembrane particles of the tonoplast. The tetrad frequency in EF face varies from 2 to 87/μm^2, and that of single particles from 102 to 695/μm^2. The authors suggest that since mucilage vacuoles in red algae are involved in the synthesis and export of cell surface polysaccharides, the membrane tetrad is a multienzyme complex participating in the synthesis of polysaccharides of the amorphous extracellular matrix.

The presence of IMPs has also been reported in the tonoplast of yeast by Baba et al. (1995). Whereas the PF and EF of the membranes of the autophagic bodies of yeast cells are free of IMPs, both faces of the tonoplast, especially the PF, show many randomly-distributed IMPs. The researchers record starvation leading to reduction of IMPs, as counted on EF, to 70% of the control frequency. There is obviously a wide divergence in the occurrence, dimension and density of IMPs in different cells as reported by various workers.

A similar line of investigation by Balsamo and Thomson (1995) on the salt-effects on tonoplast of the hypodermis and mesophyll cells of *Avicennia germinans* shows that IMPs are present in both PF and EF of both plasma membrane and tonoplast. These researchers concluded a possible relationship between the increase in the tonoplast ATPase concentration and activity and IMP density and size with increased exposure to salt.

Bowman et al. (1989) were the first to visualise ATPase particles on the tonoplast. The negative-staining electron microscopy of the tonoplast vesicles isolated from *Neurospora crassa* revealed the presence of 9 nm stalked F_1-like ATPase particles. Similar studies by Morrè et al. (1991) on highly purified tonoplast fractions, isolated by preparative free-flow electrophoresis from hypocotyls of etiolated soybean, exhibited head and stalk structures resembling those reported by Bowman et al. (1989). The stalked particles are unevenly distributed and locally concentrated in some membrane regions, with a more sparse distribution elsewhere.

More recently, Ward and Sze (1992) have visualised knob-like structures of 10 to 12 nm in diameter on the vacuolar membrane vesicles by negative staining. These structures are present in patches and clearly represent a major component of the membrane. From a top view some appear to have a trigonal shape; while others appear tetrahedral or tetragonal. Since the treatments that inactivated ATPase and solubilised the peripheral sector were also effective in removing these particles from native membrane vesicles, the authors concluded that these knob-like structures are the peripheral sectors of the vacuolar membrane H^+-ATPase. Lüttge et al. (1995) recorded the

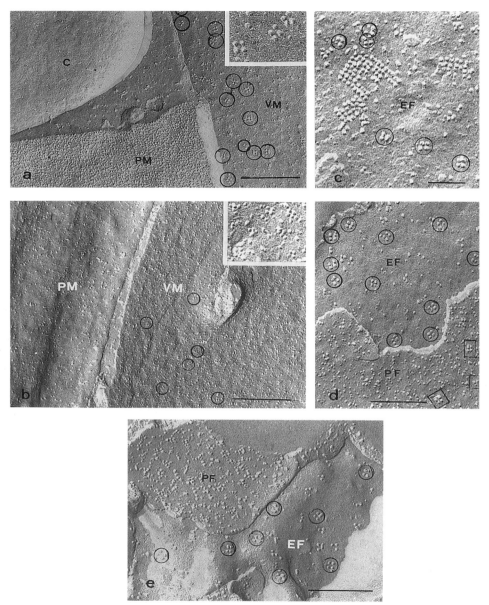

Figure 4.1 Fracture through freeze-etched cells of some red algae. **a, b, c** *Ceramium diaphanum* var. *strictum*, cortical cell; **a** a part of a chloroplast (C), the protoplasmic fracture face of the plasma membrane (PM) with high density of particles and the exoplasmic fracture face of the vacuole membrane (VM) with tetrads (circles). Bar = 0.3 µm. Inset, tetrads at high magnification × 135 000. **b** EF of the plasma membrane (PF) with low particle density and PF of the vacuole membrane (VM) with tetrads (circles). Bar = 0.3 µm. Inset, tetrads at high magnification × 148 000. **c** EF of the vacuole membrane with tetrads and tetrads in clusters. Bar = 0.1µm. **d, e** *Spermothamnion johannis*, vegetative cell; **d** EF of vacuole membrane of tetrads and PF of plasma membrane which, apart from single particles, also contains 'membrane tetrads' (rectangles). Bar = 0.2 µm. **e** Two different vacuoles; EF face with many tetrads; PF face with higher particle density than the EF face of the other vacuole is also visible. Bar = 0.2 µm. (Courtesy: I. Tsekos and H.D. Reiss, *Ann. Bot.* 72, 213, 1993.) With permission of Academic Press.

ultrastructure of V-ATPase from the negatively stained, as well as from the freeze fracture replicas, of the tonoplast vesicles of *Mesembryanthemum crystallinum*. Detailed measurements of the head and stalk (V_1 domain) and the membrane integral proteolipid (V_0 domain) have also been obtained (vide Lüttge and Ratajczak 1997). The average diameter of the head is about 10–12 mm and the length of the head and stalk is about 12–14 mm.

4.2 Chemical analysis of tonoplast

4.2.1 Lipids

Like most biomembranes, the tonoplast is composed of proteins, including glycoproteins and enzymes, and lipids including glycolipids. However, the relative amount of the constituents appears to vary considerably. In beetroot, Marty and Branton (1980) reported a tonoplast phospholipid: protein ratio of 0.68 and a sterol: phospholipid ratio of 0.21. In a subsequent publication by Marty (1985), these ratios appear as 0.58 and 0.30 respectively. In thoroughly washed tonoplasts, Salyaev (1985) obtained a protein: total lipid ratio of 0.26 and a carbohydrate: protein ratio of 0.38. These ratios appear to be too low compared to the calulated values 1.7 and 0.79 respectively in the same tissue (Marty 1985).

In the analysis of lipid and protein components of yeast tonoplast conducted by Kramer *et al.* (1978) and Van der Wilden and Matile (1978), the lipid: protein ratio appeared to be about 1.5. The membrane contains about 60–70% neutral lipid and 30–40% phospholipid, chiefly composed of phosphatidylcholine (33% of the total lipid phosphate), phosphatidylethanolamine (15–28%), phosphatidylserine and phosphatidylinositol (jointly 33–43%). The two-dimensional thin layer chromatography led to the identification of 17 lipid classes (Marty 1980). The complex polar lipids included phosphatide and glycolipids. The phosphatides contain phosphatidylcholine (54%), phosphatidylethanolamine (24%), phosphatidylserine (1%), phosphatidylglycerol (4%), phosphatidylinositol (5%) and phosphatidic acid (12%). The glycolipids consist of five classes including two ceramide glycoside-like and one sulfoglycoside-like fraction. The neutral lipids, which are low in quantity, consist of sterols, monoglycerides, diglycerides, triglycerides, steryl esters and fatty acids. Although the protein: phospholipid ratio and phosphatidylcholine: phosphatidylethanolamine ratio of tonoplast of beetroot is similar to that of yeast, the actual composition is different, especially in the amount of phosphatidylserine. Moreover, yeast tonoplast contains more lysolecithin than beetroot (DuPont *et al.* 1976). Phosphatidic acid is absent in yeast and oat tonoplast (Verhoek *et al.* 1983), but present in beetroot tonoplast. In this respect, the membrane of lutoids of rubber tree laticifers is very distinctive, its phosphatidic acid accounting for 82% of total lipid phosphate (DuPont *et al.* 1976). In beetroot tonoplast Salyaev (1985), detected a good amount (5%) of free fatty acids.

4.2.2 Proteins

The protein composition of the purified tonoplast varies in different plants. In *Hippeastrum*, Wagner (1981) reported only 5 major peptides (69–53 kD) and 9 minor ones. In beetroot tonoplast, Salyaev (1985) reported 10 major and 10–12 minor bands, all below 70 kD in molecular weight. However, in the same tissue 15 major bands and at least 15 minor bands with molecular weights ranging from 91 kD to 12 kD were reported by Marty (1985). In order to comprehend the supramolecular organisation of proteins Marty (1985) conducted some selective elution studies and recorded that in addition to some peripheral proteins, a few polypeptides are either electrostatically or hydrophobically interacting with membrane core components. In a similar approach, Salyaev (1985) concluded that, in addition to 1.5% of the peripheral proteins, 50% of the proteins are loosely bound integral proteins (IP) and 42% are strongly bound IPs. Actual identification of transmembrane proteins (carriers, channels, pumps, etc.) involved in various tonoplast activities is described later.

Although the various peptides from the tonoplast of *Hippeastrum* (Wagner 1981) failed to show the presence of any carbohydrate ligand, it is now established that many of the tonoplast proteins are carbohydrate-containing polypeptides. The tonoplast contains receptor proteins for at least three lectins, concanavalin A (ConA), wheat germ agglutinin (WGA) and *Ricinus communis* aggutinin (RCA), and these proteins possess residues of galactose, glucose/mannose, as well as galactosamine and glycosamine (Marty 1985; Salyaev 1985).

With the development of techniques for isolating tonoplast and greater sophistication in high resolution electrophoresis, precise detection of various enzymes has become possible. The most important are the proton pumping enzymes, ATPase and pyrophosphatase. In addition, Ca^{2+}-ATPase, ABC transporters, NADH cytochrome C oxidoreductase and phosphoenolpyruvate carboxylase are present in many tonoplasts. A number of enzymes of biosynthesis including sucrose group-translocator enzymes, α-mannosidase, permeases and proteinases also occur widely (Table 5.3).

Besides the various enzymes which confer specific properties of the vacuole membrane, a number of the constituent proteins of the tonoplast have been purified and characterised. After the one-dimensional gel-electrophoresis studies on the polypeptides from the tonoplasts of barley by Kaiser *et al.* (1986) and of two CAM plants by Kenyon and Black (1986), the two-dimensional gel electrophoresis by Deitz *et al.* (1988) provided a detailed description of the polypeptide composition of the tonoplast of barley mesophyll cells. They identified 14 tonoplast specific polypeptides with the aid of Coomassie brilliant blue R staining of the gel. Ni and Beevers (1991), using silver staining, detected as many as 68 polypeptides in the tonoplast of corn seedling root cells. Immunoblot analysis with antibodies against a tonoplast holoenzymes ATPase revealed the presence of 72, 60 and 41 kD polypeptides. Washing with salt and NaOH indicated that a total of 38 tonoplast specific polypeptides were peripherally associated and 30 peptides were identified to be integral proteins. Affinity blotting with concanavalin A and secondary antibodies indicated that 8 out of 30 integral polypeptides and 13 out 38 peripheral polypeptides contained detectable carbohydrate moieties.

In this connection, it is interesting to note that with *Mesembryanthemum crystallinum* the change from C_3 photosynthesis to CAM is accompanied by the appearance of 55, 41 and 36 kD proteins which are non-H^+-ATPase fractions. Again, one polypeptide of about 24 kD appeared in the tonoplast of salt-treated plants (Bremberger and Lüttge 1992). With the aid of monoclonal antibody technique Liedtke *et al.* (1993) identified a 71 kD protein (TOP 71 antigen) in the tonoplast of a wide range of plants including a number of monocots and dicots, green alga *Chara* and the fern *Matteucia*. They suggested that the protein has a general function and not necessarily an enzymatic one.

The vacuole plays an important role in calcium homoeostasis and considerable evidence indicates that the vacuole membrane has mechanisms for accumulation and release of calcium. In their search for calcium-binding proteins, Seals *et al.* (1994) identified a specific protein in the isolated vacuole membrane fraction of celery. They identified a 42 kD, calcium-dependent protein (VCaB42) which binds on the cytosolic side of the tonoplast. This annexin-like protein has Ca-affinity comparable to calmodulin. The researchers hypothesised that VCaB42 is involved in maintaining the structural integrity of the vacuole, regulating and forming membrane channels or regulating calcium levels or vacuolar processes, e.g. catalysis of membrane fusion.

Many observations on plasma membrane and vacuole membrane of plant and animal cells have indicated the existence of a water-selective channel in these membranes. The animal membrane integral proteins (MIP) belong to an ancient family of proteins characterised by six membrane-spanning domains and a short chain of 22 amino acids that are absolutely conserved, including the signature sequence NPAXT, of which NPA is found in outside loops on the opposite sites of the membrane (Reizer *et al.* 1993, Fig. 4.2). Site-directed mutagenesis of the loops containing the NPA motifs further indicated that these segments probably line the path of water

Ultrastructure and chemical composition of tonoplast   69

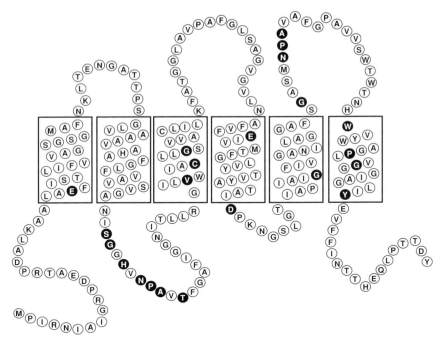

Figure 4.2   Schematic representation of aquaporin γ-TIP in the tonoplast showing the six putative transmembrane domains. The orientation of the protein is based primarily on extensive studies with the bovine lens protein MIP, the orientation of the sixth transmembrane domain has been confirmed for the seed protein α-TIP. The residues that are conserved in all members of the MIP family are shown as dark circles (from Chrispeels and Maurel 1994). With permission of the American Society of Plant Physiologists.

permeation. These results led to the hourglass model in which the NPA loops project from the outer and inner leaflets towards the centre of the membrane (Jung et al. 1994, Fig. 4.2).

The MIP family contains the major integral membrane proteins of protein storage vacuoles (PSV), called tonoplast integral protein (TIP), which was first identified in the cotyledons of *Phaseolus vulgaris* (Johnson et al., 1989, 1990). This protein, called α-TIP, is present in seed tissue of every plant thus far examined and is highly conserved. The amino acid sequence of TIP is similar to that of pore proteins found in other organisms. This indicates that TIP may be involved with the transport of small molecules into or out of PSV.

Maurel et al. (1993) found γ-TIP, another homologue of MIP, in the vacuolar membrane of cells of shoots, roots and flowers of *Arabidopsis*. These water channel proteins, called aquaporins, (Agre et al. 1993, Chrispeels and Maurel 1994), have a molecular mass of approximately 27 kD, which is close to the size of the functional unit of the water channel as determined by radiation inactivation method. Macey (1984) postulated that the radial diameter of the water channel may lie between 1.5 A° (the radius of water molecule) and 2.0 A° (the radius of urea molecule), and water is constrained to move as a single file of molecules. The archetypal aquaporin AQP1 is a partly glycosylated water selective channel and is a homotetramer containing four independent aqueous channels. Each AQP1 monomer has six tilted, bilayer-spanning alpha-helices which form a right-handed bundle surrounding a central mass. The central mass is formed by the two NPA (Asn-Pro-Ala)-containing loops which carry the highly conserved residues involved in water permeation (Walz et al. 1997).

Although γ-TIP is constitutively expressed (Daniels et al. 1994), some of the aquaporins may be expressed in a tissue-specific manner, e.g. a-TIP is exclusively specific for seed protein storage vacuoles, Rtob7 is found in roots of tobacco, and NOD26 in peribacteroid membranes of root nodules. Certain other aquaporins are induced by physiological conditions, as, for example, desiccation induces formation of aquaporins RD-28 and TRG-31 in *Arabidopsis thaliana* and *Pisum sativum* respectively. Another aquaporin, Dip, is expressed in mature seeds and dark-grown seedlings of *Antirrhinum majus*.

Besides the lipids and integral proteins, what is most important is the information on biochemical characterisation and molecular architecture of the various enzymes, pumps, transporters and ion channels in the tonoplast.

4.2.3 Tonoplast enzymes

4.2.3.1 *Proton ATPase*

One of the most ubiquitous enzyme complexes of the tonoplast is the ion motive ATPase. Three major classes of ATPases have been identified, designated as 'P', 'F' and 'V' (Pedersen and Carafoli, 1987). The P-type ATPases are found in the plasma membrane of plants, fungi, eubacteria and animals, and are inhibited by vanadate. P-type ATPase is usually made of a 100 kD single polypeptide (Nelson and Taiz 1989). F-type ATPases occur in the plasma membrane of aerobic bacteria and the inner membrane of mitochondria and chloroplasts. Their molecular mass of 50 kD consists of two multi-subunit components F_1 and F_0 and they are inhibited by DCCD (N, N'-dicyclohexylcarbodiimide) and azide. The third category, V-type ATPases (EC.3.6.1.3), although inhibited by DCCD, nitrate and bafilomycin, are insensitive to azide and vanadate. They are located in the vacuolar systems of eukaryotic cells, viz. vacuoles of plants, lysosomes, endosomes, Golgi vesicles, as well as clathrin-coated vesicles. They are large multimeric masses of about 400 to 750 kD, comprising a hydrophilic catalytic complex and an integral membrane moiety (Forgac 1992; Sarafian *et al.* 1992).

Target-size analysis by radiation inactivation is a well established method to study structure–function relationship in biologically active macromolecules without prior purification or solubilisation (Kepner and Macey 1986; Jung 1984). By this method, the functional molecular mass of the tonoplast ATPase from corn coleoptiles (Mandala and Taiz 1985) and mung bean seedlings (Wang *et al.* 1989) was found to be about 400 kD, whereas those of maize roots (Chanson and Pilet 1989) and *Neurospora crassa* (Bowman *et al.* 1986) were 542 and 520 kD, respectively. Such analysis of ATPase from tonoplast of *Acer pseudoplatanus* indicated its molecular mass to be about 400 kD (Magnin *et al.* 1995). However, the authors agreed that this value was an underestimate because if one adds up the identified subunits and takes into account that A and B subunits are present in triplicate, the molecular mass should be above 600 kD. Ward and Sze (1992) estimated its mass in oat root as about 650 kD.

In general the V-ATPases, which are composed of 8 to 10 different subunits, are assembled into a peripheral sector V_1, located on the cytosolic face of the membrane, and an integral membrane sector V_0. Whereas V_1 represents the hydrophilic catalytic complex for ATP hydrolysis, V_0 is the hydrophobic non-catalytic membrane-spanning regulatory complex. V_1 is composed of two major subunits, A and B, and three minor subunits, C, D and E, with a stochiometry of A_3B_3CDE (Gogarten *et al.* 1989). V_0 contain six copies of 16 kD proteolipid and one copy each of minor subunits of 100, 38 and 19 kD (Nelson 1991). Later research showed large variations in the composition of V_0 and V_1 subunits.

The complete subunit composition of vacuolar H^+-ATPase from oat roots has been determined by Ward and Sze (1992). This enzyme, with a relative molecular weight of 650 kD, is composed of 10 different subunits: 70, 60, 44, 42, 36, 32, 29, 16, 13, and 12 kD. Of these, 6 different polypeptides; viz 70, 60, 44, 42, 36 and 29 kD units, are peripheral subunits and are responsible for ATP hydrolysis and ATP-dependent H^+- pumping activities. The remaining four polypepitdes of 32, 16, 13, and 12 kD make up the integral sector.

Table 4.1 Comparison of subunit composition in kD of tonoplast H^+-ATPase from angiosperms.

Dicot				Monocot	
Kalanchoe Bremberger et al. 1998	Red beet Parry et al. 1989	Mung bean Matsuura-Endo et al. 1990	Maple Magnin et al. 1995	Barley DuPont & Morrissey, 1991	Oat Ward & Sze, 1992
	100		95	115	
72	67	68	66	68	70
56	55	57	56	53	60
48	52		54		
	44	44		45	44
42	42	38	40	42	42
		37	38	34	36
	32	32	31	32	32
28	29				29
16	16	16	16	17	16
		13		13	13
		12		12	12

The V_1 complex includes three copies each of nucleotide-binding catalytic 70 kD and the regulatory 60 kD subunits, known as subunits A and B respectively. The membrane integral sector or V_o which forms the proton channel is made up of six copies of the 16 kD, c proteolipid together with 1 to 3 other subunits (Sze et al. 1992). Mung bean H^+-ATPase consists of nine different polypeptides, which include two major subunits of 68 and 57 kD, a DCCD-binding subunit of 16 kD and six other minor subunits (Matsuura-Endo et al. 1990). The subunits of 68, 57, 44, 38, 37, 32, 13 and 12 kD were released from the membranes by treatment with chaotropic anion. These eight subunits may be components of the peripheral part of ATPase. The 16-kD subunit is an integral membrane proteolipid.

The primary amino acid sequence of 16 kD proteolipid subunit of eukaryotes contains four membrane-spanning domains (Mandel et al. 1988). It is interesting to note that 16 kD transcripts differentially accumulate in different tissues and increase dramatically in tissues undergoing rapid expansion, particularly in anthers, ovules and petals (Hasenfratz et al. 1995).

Unlike the oat enzymes, vacuolar H^+-ATPases purified from red beet (Parry et al. 1989) and barley (DuPont and Morrissey 1991) contain a large subunit of 100 to 115 kD.

Recently Magnin et al. (1995) studied the H^+-ATPase purified from isolated vacuoles of *Acer pseudoplatanus*. The H^+-ATPase consists of at least eight subunits of 95, 66, 56, 54, 40, 38, 31 and 16 kD. The 66 kD polypeptide cross-reacted with monoclonal antibodies raised to the 70 kD subunit of the vacuolar H^+-ATPase of oat roots. The available data on the subunit composition of the tonoplast H^+-ATPase of 4 dicots and 2 monocots have been compared (Table. 4.1). The monocot plants barley and oats show more homology in subunit compositions possibly due to the fact that they belong to the family Poaceae (Graminaceae). In contrast, the four dicot plants, which happen to belong to different families, show more diversity. A recent review on vacuolar H^+-ATPases is provided by Lüttge and Ratajczak (1997). They concluded that, in addition to species difference, the variation of the subunits in the composition of V-ATPase is due to different physiological functions and ecophysiological responses. They have suggested that such an enzyme, which is directly involved in ecophysiological adaptations at the molecular level, may be called an eco-enzyme. In response to stress the enzyme may be post-translationally modified and may show a moderate change in gene expression.

4.2.3.2 Proton-PPase

In addition to H$^+$-ATPase, plant cells contain a second proton pump on the tonoplast, H$^+$-pyrophosphatase (V-PPase) (EC 3.6.1.1.) which could be an additive or potentially serve as a backup system if V-ATPase is impaired (Rea and Sanders 1987). V-PPase is an abundant membrane protein constituting about 1–10% of total tonoplast protein and capable of contributing substantially to the generation of an inside-acid inside-positive transtonoplast H$^+$-electro-chemical potential difference (Rea et al. 1992; Rea and Poole 1993). The estimate of the target size or functional mass of V-PPase by radiation inactivation is between 91 kD and 96 kD for uncoupled substrate hydrolysis. Depending on the methodology and the plant species, the relative molecular weight varies between 64 kD and 73 kD. Rea et al. (1992) established a common identity of the enzyme from four different plants, which are between 64.5 and 67 kD. Whereas 700-amino acid V-PPase polypeptide has a predicted mass of 81 kD (Sarafian et al. 1992), structural studies indicate that the enzyme is made of a single 70 kD polypeptide species. Recent cDNA cloning of the enzyme from four different species shows that it is made of 760–775 amino acids and has a molecular mass of 79–81 kD, whereas radiation inactivation studies indicate its mass as 88–96 kD. The minimum unit competent for enzyme activity is the homodimer of the polypeptids (Zhen et al. 1997).

On the basis of computer-assisted hydropathy plots of V-PPases from *Arabidopsis* and *Beta*, Rea et al. (1992) established that the emzyme is an extremely hydrophobic integral membrane protein consisting of 13–16 amphipathic transmembrane spans which may serve to generate and stabilise a relatively compact structure through the formation of inter-helical hydrogen bonds and salt bridges (Fig. 4.3). In addition, several of the putative hydrophilic domains are characterised by the presence of clusters of charged residues which may participate in cation or anion binding. Thus, the hydrophilic segment linking membrane spans I and II contains four contiguous glutamate residues, the segment linking spans IV and V contains four acidic residues in a stretch of eight amino acids, the segment between spans VIII and IX contains the sequence RXRXR and the penultimale hydrophilic domain, linking spans XII and XIII, contains a preponderance of lysine. The overall orientation of the 81 kD polypeptide is depicted in accordance with the 'positive-inside rule' wherein, for the majority of the spanning domains, positively charged amino acids are located predominantly towards the cytosol. Recent research has shown that a total of 16 transmembrane spans exist in V-PPase, i.e. three spans in addition to the 13 predicted previously (Zhen et al. 1997).

4.2.3.3 Ca^{2+}-ATPase

Two classes of active Ca^{2+} transporters are present in vacuolar membranes (i) Ca^{2+}-ATPase and (ii) nH$^+$/Ca^{2+} antiporter. There are two types of Ca^{2+}-ATPases: one which is stimulated by calcium-binding protein calmodulin (CaM) and the other which is not. Recently, Askerlund (1999) showed that the 111 kD CaM-stimulated, CaM-binding Ca^{2+} ATPase is definitely associated with the tonoplast. Recent isolation of genes encoding two high- and low-affinity Ca^{2+}/H$^+$ antiporters, CAX1 and CAX2, from *Arabidopsis*, indicate that each has a central acidic motif and 11 membrane-spanning domains, similar to those detected in yeast and E. coli (Hirschi et al. 1996).

ATP-binding cassette transporters

The group of organic solute transporters known as ATP-binding cassette (ABC) transporters (also referred to as traffic ATPases) is directly energised by MgATP. All ABC transporters comprise one or two copies each of two basic structural elements: hydrophobic, integral transmembrane domains (TMD) containing multiple (usually four or six) transmembrane helices, and a cytoplasmically oriented ATP-binding domain (nucleotide binding fold, NBF). The TMDs span the

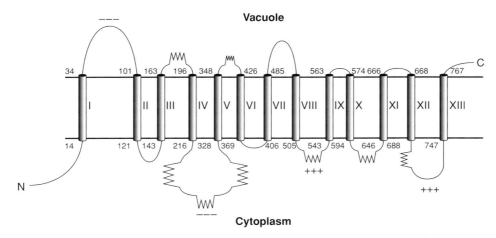

Figure 4.3 Tentative topographic model of vacuolar H^+-PPase from computer-assisted hydropathy plots of the deduced amino acid sequence of cDNAs encoding the substrate-binding polypeptide of the H^+-PPase isolated from *Arabidopsis*.
WWW α-helix; ---- clusters of negative charge; +++ clusters of positive charge; N amino terminus; C carboxyl terminus (from Rea *et al.* 1992). With permission of Elsevier Science.

membrane to form the pathway for solute movement across the bilayer and determine the substrate specificity of the transporter. The cytosolically oriented NBFs couple ATP hydrolysis to solute movement by an unknown mechanism (Higgins 1992). The topology of the *Arabidopsis* tonoplast glutathione S-conjugate pump, a member of the ABC-transporter family, has been worked out in detail (Rea *et al.* 1998).

Apart from the above-mentioned transporter enzymes, very little is known about the molecular identification of numerous porters or channels of the tonoplast. Some information is available on the molecular biology of the K^+ channel, the channel for the most abundant cation in higher plants. Low affinity K^+ transport is carried out by voltage-gated channels that allow the passive flow of K^+ down their electrochemical gradient. The channel is made of six transmembrane domains, S1 to S6, an amphipathic S4 domain involved in voltage sensing and a hydrophilic hairpin region between S4 and S5, possibly form the channel pore and contain ion binding sites. Both the C– and N– termini are located in the cytosol (vide Fox and Guerinot 1998).

The Na^+/H^+ tonoplast antiport, on the other hand, is possibly made of a 170 kD polypeptide. (Barkla and Blumwald 1991). Except the malate transporter, so far no carboxylate transporter of the tonoplast has been identified. Recently, Lahjouji *et al.* (1996) reported the labelling of a 37 kD tonoplast polypeptide of *Catharanthus roseus* by a photolysable malate analogue. Polyclonal antibodies raised against this polypeptide suggested that 37 kD polypeptide is involved in vacuolar malate transport.

4.3 SUMMARY

4.3.1 Ultrastructure

Electron microscopy has revealed that the tonoplast is a **single unit membrane** whose two faces may be assymetrical. Freeze fracture electron microscopy has demonstrated the presence of intramembranous particles of different dimensions, and different distibution pattern and density on both the cytosolic face and the inner face of the tonoplast. The most distinctive of the intramembranous particles is V-ATPase, which is seen as a stalked structure of total length 12–14 nm with a head 10–12 nm in diameter.

4.3.2 Chemical composition

- The types of **lipids** and **proteins** in the tonoplast may vary between species and different tissues of the same species. The lipid/protein ratio and the ratio of neutral lipid/phospholipid also show variation. In general, neutral lipids constitute 60–70% and phospholipids 30–40%. The tonoplast may contain as many as 17 lipid classes: 6 classes of phospholipids, 5 classes of glycolipids, and 6 classes of neutral lipids.
- Depending on the plant, tissue and the method of isolation and purification, the protein composition of various tonoplasts appears very diverse. Proteins may be **peripheral**, loosely or strongly bound **integral proteins**. For example, tonoplast of corn seedling root cells may contain as many as 68 types of polypeptides, of which 30 are integral proteins and 38 are peripheral proteins. Again, 8 out of 30 integral proteins and 13 of 38 peripheral proteins contain detectable carbohydrate moieties.
- A number of receptor proteins, as well as numerous proteins/enzymes which act as transporters of ions and molecules and/or catalyse specific reactions, are also present in the tonoplast. The most important integral protein of the tonoplast is **aquaporin**, the water-selective channel. The amino acid sequence of the 27 kD polypeptide suggests an hourglass model in which the 6 hydrophobic segments are the trans-membrane domains (TMDs) and the peptide domain containing Asn-Pro-Ala (NPA) motifs paticipate in forming the aqueous channel.
- The tonoplast **proton pump V-ATPase** comprises 8 to 10 subunits assembled into a peripheral cytosolic sector V_1 and an integral membrane sector V_0. Again, in different plants, V_1 may be composed of up to 8 subunits and V_0 composed of up to 5 units. Similarly the molecular mass of the enzyme may be 400 kD, 520 kD or 650 kD.
- Another proton pump, **V-PPase (H^+-pyrophosphatase)** is a homodimer of 88–96 kD protein made of 13–16 TMDs with a total of 760–775 amino acids.
- Little information is available for the structural characteristics of numerous other transporters. Ca^{2+}-ATPase is a 111 kD protein, Na^+/H^+ antiporter is a large protein of 170 kD and the malate transporter is a small 37 D polypeptide. Ca^{2+}/H^+ antiporter has 11 TMDs, the voltage-gated K^+ channel has 6 TMDs and the ABC transporter is made of 4–6 TMDs.

4.4 REFERENCES

Agre, P., Sasaki, S. and Chrispeels, M. J., Aquaporins — a family of water channel proteins, *Am. J. Physiol.*, **265**, F461, 1993.

Askerlund, P., Calmodulin-stimulated Ca^{2+}-ATPases in the vacuolar and plasma membranes in cauliflower, *Plant Physiol.*, **114**, 999, 1997.

Baba, M., Osumi, M. and Ohsumi, Y., Analysis of membrane structures involved in autophagy in yeast by freeze-replica method, *Cell Struct. Function*, **20**, 465, 1995.

Balsamo, R. A. and Thomson, W. W., Salt-effects on membranes of the hypodermis and mesophyll cells of *Avicennia germinans* (Avicenniaceae): A freeze-fracture study, *Am. J. Bot.* **82**, 435, 1995.

Barkla, B.J. and Blumwald, E. Identification of a 170 kDa protein associated with the vacuolar Na^+/H^+ antiport of *Beta vulgaris*, *Proc. Natl. Acad. Sci. USA* **88**, 11177, 1991.

Bowman, B.J., Dschida, W.J., Harris, T. and Bowman, E.J., The vacuolar ATPase of *Neurospora crassa* contains an F1-like structure, *J. Biol. Chem.*, **264**, 15606, 1989.

Bowman, E.J., Mandala, S., Taiz, L. and Bowman, B.J., Structural studies of the vacuolar membrane ATPase from *Neurospora crassa* and comparison with the tonoplast membrane ATPase from *Zea mays: Proc. Natl. Acad. Sci. USA*, **83**, 165, 1986.

Bremberger, C., Haschke, H.P. and Lüttge, U., Separation and purification of the tonoplast ATPase and pyrophosphatase from plants with constitutive and inducible Crassulacean acid metabolism, *Planta*, **175**, 465, 1988.

Bremburger, C. and Lüttge, U., Dynamics of tonoplast protein pumps and other tonoplast proteins in *Mesembryanthemum crystallinum* L. during the induction of Crassulacean acid metabolism, *Planta*, **188**, 575, 1992.

Chanson, A. and Pilet, P.E., Target molecular size and sodium dodecyl sulfate polycrylamide gel electrophoresis analysis of the ATP- and pyrophosphate-dependent proton pumps from maize root tonoplast, *Plant Physiol.* **90**, 934, 1989.

Chrispeels, M.J. and Maurel, C., Aquaporins: the molecular basis of facilitated water movement through living plant cells?, *Plant Physiol.*, **105**, 9, 1994.

Daniels, M., Mirkov, T.E. and Chrispeels, M.J., The plasma membrane of *Arabidopsis thaliana* contains a mercury-insensitive aquaporin that is a homolog of the tonoplast water channel protein TIP, *Plant Physiol.*, **106**, 1325, 1994.

Dietz, K.J., Kaiser, G. and Martinoia, E., Characterization of vacuolar polypeptides of barley mesophyll cells by two-dimensional gel electrophoresis and by their affinity to lectins, *Planta*, **176**, 362, 1988.

DuPont, F.M. and Morrissey, P.J., Purification of a vacuolar ATPase from barley roots (abstract No.67). *Plant Physiol.*, **96**, S-13, 1991.

DuPont, J., Moreau, F., Lance, C. and Jacob, J.L., Phospholipid composition of the membrane of lutoids from *Hevea brasiliensis* latex. *Phytochemistry*, **15**, 1215, 1976.

Forgac, M., Structure and properties of the coated vesicle H^+-ATPase, *J. Bioengrg, Biomembr*, **24**, 339, 1992.

Fox, T.C., and Guerinot, M.L., Molecular biology of cation transport in plants, *Ann. Rev. Plant Physiol. Plant Mol. Biol.*, **49**, 669, 1998.

Gogarten, J.P., Kibak, H., Dittrich, P., Taiz, L., Bowman, E.J., Bowman, B.J., Manolson, M., Poole, R.J., Date, T., Oshima, T., Konishi, J., Denda, K. and Yoshida, M., The evolution of vacuolar H^+-ATPase: implications for the origin of eukaryotes, *Proc. Natl. Acad. Sci. USA*, **86**, 6661, 1989.

Hasenfratz, M.P., Tsou, C. and Wilkins, T.A., Expression of two related vacuolar H^+-ATPase 16-kilodalton proteolipid genes is differentially regulated in a tissue-specific manner, *Plant Physiol.*, **108**, 1395, 1995.

Higgins, C.F., ABC transporters: from microorganisms to man, *Ann. Rev. Cell Biol.* **8**, 67, 1992.

Hirschi, K.D., Zhen, R-G., Cunningham, K.W., Rea, P.A., and Fink, G., CAX1, an H^+/Ca^{2+} antiporter from Arabidopsis, *Proc. Natl. Acad. Sci. USA*, **93**, 8782, 1996.

Johnson, K.D., Herman, E.M. and Chrispeels, M.J., An abundant, highly conserved tonoplast protein in seeds, *Plant Physiol.*, **91**, 1006, 1989.

Johnson, K.D., Hofte, H., Chrispeels, M.J., An intrinsic tonoplast protein of protein storage vacuoles in seeds is structurally related to a bacterial solute transporter (GlpF), *Plant Cell*, **2**, 525, 1990.

Jung, C.H., Molecular weight determination by radiation inactivation, in 'Molecular and Chemical Characterization of Membrane Receptors', Vol 3. Venter, J.C. and Harrison, L. C., Eds, Alan R. Liss, New York, 1984, 193.

Jung, J.S., Preston, G.M., Smith, B.L., Guggino, W.B., and Agre, P., Molecular structure of the water channel through aquaporin CHIP: the tetrameric-hourglass model; *J. Biol. Chem.* **269**, 14648, 1994.

Kaiser, G., Martinoia, E., Schmitt, J.M., Himcha, D.K. and Heber, U., Polypeptide pattern and enzymatic character of vacuoles isolated from barley mesophyll protoplast, *Planta*, **169**, 345, 1986.

Kaufmann, B.P. and De, D.N., Fine structure of chromosome, *J. Biophys. Biochem. Cytol*, **2** Suppl. 419, 1956.

Kenyon, W.H. and Black, C. C., Electrophoretic analysis of protoplast, vacuole, and tonoplast vesicle protein in Crassulacian acid metabolism plants. *Plant Physiol.*, **82**, 916, 1986.

Kepner, G.R. and Macey, R.I., Membrane enzyme systems: molecular size determinations by radiation inactivation. *Biochem. Biophys. Acta*, **163**, 188, 1986.

Kramer, R., Kopp. F., Niedermeyer, W. and Fuhrmann, G.F., Comparative studies of the structure and composition of the plasmalemma and the tonoplast in *Saccharomyces cerevisiae.*, *Biochim. Biophys. Acta*, **507**, 369, 1978.

Lahjouji, K., Carrasco, A., Bouyssou, H., Cazaoux, L., Marigo, G., and Canut, H., Identification with a photoaffinity reagent of tonoplast protein involved in vacuolar malate transport of *Catharanthus roseus*, *Plant Journal*, **9**, 799, 1996.

Liedtke, C., Martiny-Baron, G., Volkmann, D. and Scherer, G.F.E., Monoclonal antibody TOP71 recognises a tonoplast protein of wide distribution in the plant kingdom, *Planta*, **190**, 491, 1993.

Lüttge, U., and Ratajczak, R., The physiology, biochemistry and molecular biology of the plant vacuolar ATPase, in 'The Plant Vacuole', Eds. Leigh, R.A., Sanders, D., Academic Press, London, 1997, 253.

Lüttge, U., Fischer-Schliebs, E., Ratajczak, R., Kramer, D., Berndt, E., and Kluge, M., Functioning of the tonoplast in vacuolar C-storage and remobilization in crassulacean acid metabolism, *J. Exp. Bot.*, **46**, 1377, 1995.

Macey, R. I., Transport of water and urea in red blood cells. *Am. J. Physiol.*, **246**, C195, 1984.

Magnin, T., Fraichard, A., Trassat, C. and Pugin, A., The tonoplast H^+-ATPase of *Acer pseudoplatanus* is a vacuolar-type ATPase that operates with a phosphoenzyme intermediate, *Plant Physiol.*, **109**, 285, 1995.

Mandala, S. and Taiz, L., Partial purification of a tonoplast ATPase from corn coleoptiles, *Plant Physiol*, **78**,104, 1985.

Mandel, M., Moriyama, Y., Hulmes, J.D., Pan, Y.C., Nelson, H. and Nelson, N., cDNA sequence encoding the 16kD proteolipid of chromaffin granules implies gene duplication in the evolution of H^+-ATPase. *Proc. Natl. Acad. Sci. USA*, **85**, 5521, 1988.

Marty, F., Isolation and freeze-fracture characterization of vacuole membrane fragments, in 'Plasmalemma and Tonoplast: Their Functions in the Plant Cell', Marmé, D., Marré, E. and Hertel, R., Eds., Elsevier Biomedical Press, Amsterdam, 1982, 179.

Marty, F., Analytical characterization of vacuolar membranes from higher plants, in 'Biochemistry and Function of Vacuolar Adenosinetriphosphatase in Fungi and Plants', Marin, B.P. Ed., Springer Verlag, Berlin, 1985, 14.

Marty, F. and Branton, D., Analytical characterization of beet root vacuole membrane, *J. Cell. Biol.* **87**, 72, 1980.

Matsuura-Endo, C., Maeshima, M. and Yoshida, S., Subunit composition of vacuolar membrane H^+-ATPase from mung bean, *Eur. J. Biochem.*, **187**, 745, 1990.

Maurel, C., Reizer, J., Schroeder, J.I., Chrispeels, M.J., The vacuolar membrane protein γ-TIP creates water specific channels in *Xenopus oocytes*, *EMBO J.*, **12**, 2241, 1993.

Morre, D.J., Liedtke, C., Brightman, A.O. and Scherer, G.F.E., Head and stalk structures of soybean vacuolar membranes, *Planta*, **184**, 343, 1991.

Nelson, N. and Taiz, L.,The evolution of H^+-ATPases. *Trends Biochem. Sci* **14**, 113, 1989.

Nelson, N., Structure and pharmacology of the proton-ATPase. *Trends Pharmacol. Sci.*, **12**, 71, 1991.

Ni, M. and Beevers, L., Characterization of polypeptides isolated from corn seedling roots, *Plant Physiol.*, **97**, 264, 1991.

Parry, R.V., Turner, J.C. and Rea, P.A., High purity preparations of higher plant vacuolar H^+-ATPase reveal additional subunits, *J. Biol. Chem*, **264**, 20025, 1989.

Pedersen, P.L. and Carafoli, E., Ion-motive ATPases II Energy coupling and work output. *Trends Biochem. Sci.* **12**, 146, 1987.

Rea, P.A. and Poole, R.J., Vacuolar H^+-translocating pyrophosphatase, *Ann. Rev. Plant Physiol. Plant Mol. Biol.*, **44**, 157, 1993.

Rea, P.A. and Sanders, D., Tonoplast energization: Two H^+ pumps, one membrane. *Physiol. Plant.*, **71**, 131, 1987.

Rea, P.A., Britten, C.J. and Sarafian, V., Common identity of substrate-bending subunit of vacuolar H^+-translocating inorganic pyrophosphase of higher plant cells, *Plant Physiol.*, **100**, 723, 1992.

Rea, P.A., Kim, Y., Sarafian,V., Poole, R.J., Davies, J.M. and Sanders, D., Vacuolar H^+-translocating pyrophosphatases: a new category of ion translocase, *Trends Biochem. Sci.* **17**, 348, 1992.

Rea, P.A., Li, Z-S., Lu, Y-P., and Drozdowicz, Y.M., From vacuolar GS-X pumps to multispecific ABC transporters, *Ann. Rev. Plant Physiol. Plant Mol. Biol.*, **49**, 727, 1998.

Reizer, J., Reizer, A. and Saier, M.H. Jr., The family of integral membrane channel proteins: sequence comparisons, evolutionary relationships, reconstructed pathway of evolution, and proposed fuctional differentiation of the two repeated halves of the protein. *Crit. Rev. Biochem. Mol. Biol.*, **28**, 235, 1993.

Salyaev, R.K., Plant vacuole membrane: structure and properties, in 'Biochemistry and Function of Vacuolar Adenosinetriphosphatase in Fungi and Plants', Ed. Marin, B. P., Springer Verlag, Berlin, 1985, 3.

Sarafian, V., Kim, Y., Poole, R.J. and Rea, P.A., Molecular cloning and sequence of cDNA encoding the pyrophosphate energized vacuolar membrane proton pump (H^+-PPase) of *Arabidopsis thaliana*, *Proc. Natl. Acad. Sci. USA*, **89**, 1775, 1992.

Sarafian, V., Potier, M. and Poole, R.J., Radiation inactivation analysis of vacuolar H^+-ATPase and H^+-pyrophosphatase from *Beta vulgaris* L. *Biochem J.*, 283, 493, 1992.

Seals, D.F., Porrish, M.L. and Randall, S.K., A 42- kilodalton annexin-like protein is associated with plant vaculoes, *Plant Physiol.*, **106**, 1403, 1994.

Severs, N.J., Warren, R.C. and Barnes, S.H., Analysis of membrane structure in the transitional epithelium of rat urinary bladder.3 Localization of cholesterol using filipin and digitonin, *J. Ultrastuc. Res.* **77**, 160, 1981.

Sze, H., Ward, J.M. and Lai, S., Vacuolar H^+-ATPase from plant: structure, function and isoforms. *J. Bioenerg. Biomembr.* **24**, 371, 1992.

Tsekos, I. and Reiss, H.D., The supramolecular organization of red algal vacuole membrane visualized by freeze-fracture, *Ann. Bot.*, **72**, 213, 1993.

Tsekos, I., Reiss, H.D. and Schnepf, E., Occurrence of particle tetrads in the vacuole membrane of the marine red algae *Gigartina teedi* and *Ceramium rubrum*, *Naturwiss.* **72**, 489, 1985.

Van der Wilden, W. and Matile, P., Isolation and characterization of yeast tonoplast fragments, *Biochem. Physiol. Pflanz*, **19**, 285, 1978.

Verhoek, B. Haas, R., Wrage, K., Linscheid, M. and Heinz, E., Lipids and enzymatic activities in vacuolar membranes isolated via protoplasts from oat primary leaves. *Z. Naturforsch*, **38C**, 770, 1983.

Verkleij, A.J., De Kruiff, B., Gerritsen, W.F., Demel, R.A., Van Deenen, L.L.M. and Ververgaert, P.H.J., Freeze-fracture electron microscopy of erythrocytes, *Acholeplasma laidlawii* cells and liposomal membranes ofter the action of filipin and amphotericin B., *Biochim. Biophys. Acta.* **291**, 577, 1973.

Volkmann, D., Structural differentiation of membranes involved in the secretion of polysaccharide slime by root cap cells of garden cress (*Lepidium sativum* L.), *Planta*,**151**, 180, 1981.

Vom Dorp, B., Volkmann, D. and Scherer, G.E.E., Identification of tonoplast and plasma membrane in membrane fractions from graden cress (*Lepidium sativum* L.) with and without filipin treatment, *Planta*, **168**, 151, 1986.

Wagner, G.J., Enzymic and protein character of tonoplast from *Hippeastrum* vacuoles, *Plant Physiol.*, **68**, 499, 1981.

Walz, T., Hirai, T., Murata, K., Heymann, J.B., Mitsuoka, K, Fujioshi, Y., Smith, B.L., Agre, P., and Engel, A., The three-dimensional structure of aquaporin-1, *Nature*, **387**, 624, 1997.

Wang, Y., Cin, H.Y., Chou, W.M., Chung, T.P. and Pan, R.L., Purification and characterization of tonoplast ATPase from etiolated mung bean seedlings, *Plant Physiol.* **90**, 475, 1989.

Ward, J. M. and Sze, H., Subunit composition and organization of the vacuolar H^+-ATPase from oat roots, *Plant Physiol.* **99**, 170, 1992.

Zhen, R-G., Kim, E.J., and Rea, P.A., The molecular and biochemical basis of pyrophosphate-energized proton translocation at the vacuolar membrane, in 'The Plant Vacuole, Eds Leigh, R.A, and Sanders, D., Acadamic Press, London, 1997, 298.

V
Vacuolar contents

Vacuoles are generally filled with an aqueous medium known as 'vacuole sap', 'cell sap' or 'sap', which is distinctly separated from the rest of the aqueous phase of the cytoplasm by a single membrane, the tonoplast (Fig. 5.1). Vacuole sap, in addition to a lot of dissolved substances and colloidal matter, occasionally contains solid or microscopically visible particulate matter, crystalloid substances and concretions of various sorts which are sometimes referred to as ergastic substances. The solid or insoluble vacuolar contents are considered generally as metabolic end-product, reserve, secretory or excretory materials.

5.1   SOLIDS AND PARTICULATE INCLUSIONS
5.1.1 Crystals
A number of inorganic crystals are found in diverse types of cells (Guilliermond 1941). Crystals of various salts of calcium, especially calcium oxalate, are very wide-spread. Silicate or gypsum occurs less frequently, while crystals of many organic substances such as betanine, various

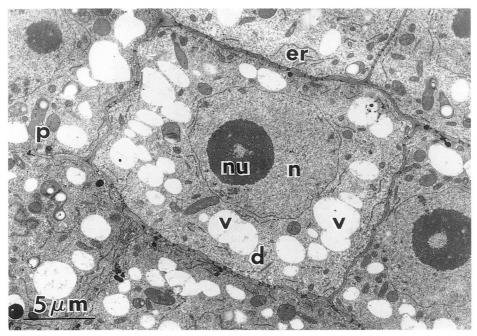

Figure 5.1   Micrograph of a young cortical cell of the root tip of *Rubus chamaemorus* L. showing many small vacuoles (V) which are empty. (Courtesy Dr. A.K. Bal.)

80 Plant cell vacuoles: an introduction

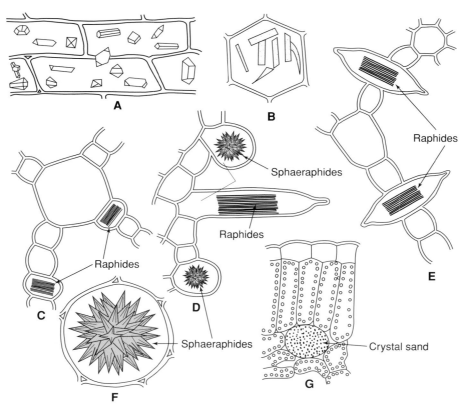

Figure 5.2 Calcium oxalate crystals. **a** solitary crystals in the tunic of *Allium* (onion); **b** solitary crystals in a cell of *Leonurus*; **c** raphides in the petiole of *Colocasia* (arum); **d** raphides and sphaeraphides in the leaf of *Pistia*; **e** raphides in the petiole of *Eichornia* (water hyacinth); **f** sphaeraphides in the petiole of *Carica* (papaw); **g** Crystal sand in the leaf of *Atropa belladona* (from Gangulee *et al.* 1961).

anthocyanin pigments and proteins are also known to be present. The solitary rhombohedral crystals, sheaf-like bundles of long acicular crystals, are called raphides. The clustered crystals in globose masses are called sphaeraphides or druses. Solitary needle-like crystals, small prismatic crystals and minute crystals known as *crystal sands*, are also found in many types of cells (Fig. 5.2) (Gangulee *et al.* 1961). In certain members of Ficaceae, Cucurbitaceae and Acanthaceae, crystal aggregates of calcium carbonate are attached with a stalk and appear as a bunch of grapes in the vacuoles. These are known as cystoliths (Fig. 5.3). The statolith of *Chara* rhizoid is a dense structure made chiefly of barium sulphate (Schröter *et al.* 1975; Van Stevenick and Van Stevenick 1978). Again there are certain bodies in the vacuoles of the cells of *Chondria* (Rhodophyceae) which are responsible for an optical interference phenomenon known as iridescence of the thallus (Feldman 1970).

5.1.2 Aleurone grains

Many seeds, embryos, endosperms and cotyledons contain crystalline or amorphous proteinaceous bodies known as **aleurone grains** or **proteid grains** (Ashton 1976). These exist in dehydrated vacuoles. At germination, the aleurone grains are rehydrated and become vacuoles. Thus they exist as reserve food. In cereals, a layer of cells just beneath the grain coat, known as the aleurone layer, remains filled with grains. In leguminous seeds they remain as small grains and occur

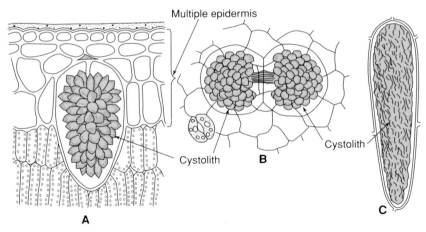

Figure 5.3 Calcium carbonate crystals. **a** in the leaf of *Ficus elastica* (India rubber plant); **b** double cystolith in *Momordica* sp.; **c** elongated cystolith in *Ruellia* sp. of Acanthaceae (from Gangulee et al. 1961).

alongside starch grains. In the endosperm of castor bean a distinct proteinaceous crystalloid structure, along with a globoid structure made of phytin, is seen in the composite aleurone grain. The transformation of aleurone vacuole to aleurone grain takes place by gradual dehydration which causes salts of myoinositol hexaphosphate (phytic acid) to precipitate in a globoid mass. Simultaneously, the proteins, which are corpuscularly dispersed, begin to arrange themselves to the lattice order of crystalloid. Finally the remaining proteins solidify in a homogeneous matrix embedding both globoid and crystalloid structure (Frey-Wyssling 1953). By the use of specific fluorescent antibodies, Graham and Gunning (1970) have demonstrated the exclusive localisation of legumin and vicilin in the aleurone grains of bean cotyledon cells. Cytochemical localisation studies by Pon (1965) have shown that the globoid bodies definitely contain insoluble phosphates. The isolation of globoid bodies from cotton seed aleurone grains has yielded conclusive evidence that phytic acid is one of its constituents (Lue and Altschul 1967). The seed storage proteins can be conveniently classified on the basis of their extraction and solubility of water (albumins), dilute saline (globulins), alcohol/water mixtures (prolamins) and dilute acid or alkali (glutelins) (Osborne 1924). The protein storage vacuoles of maturing seeds of many species are composed of crystalloids, matrix and membrane. 2S albumin is located in the vacuolar matrix (Hara-Nishimura et al. 1993) and 11S globulin is the primary constituent of the vacuolar crystalloid (Hara-Nishimura et al. 1985; 1987). The vacuolar processing enzymes responsible for processing proproteins to mature proteins are also localised in the matrix. In certain leguminous plants the 7S vicilin-type and 11S legumin-type globulins appear to be in the same protein bodies with no spatial separation (see Harris et al. 1993). Whereas prolamin and globulin storage proteins are present in separate protein bodies in rice, they are located within the same protein bodies in other cereals. Prolamin inclusions are present in a globulin matrix in oats (Lending et al. 1989) and globulin (triticin) inclusions are present within a prolamin matrix in wheat (Bechtel et al. 1991). Immunogold labelling methods have shown that in maize α-zeins form the protein body core, with β- and γ-zeins at the periphery (Lending et al. 1992). The δ-zeins may also be present in the protein body core (Esen and Stetler, 1992). Rechinger et al. (1993) proposed that in barley the quantitatively minor S-rich γ_1 and γ_2 hordeins form a peripheral layer surrounding a core of β hordeins (S-rich) and C hordeins (S-poor).

As well as the seed storage proteins, a number of vegetative storage proteins have been characterised. The vacuoles of soybean leaves and immature organs contain three vegetative storage

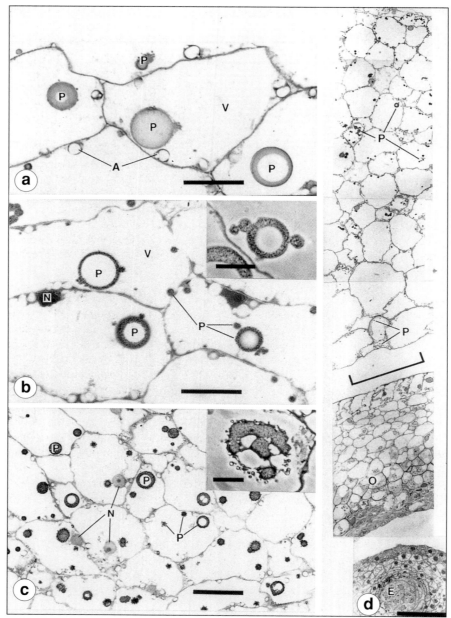

Figure 5.4 Immunocytochemical localisation of proteinase inhibitors I and II by light microscopy in fruits of *Lycopersicon peruvianum*. A, amyloplast; N, nucleus; P, protein aggregate; V, vacuole. **a** Preimmune control, Bar = 20μm. **b** Localisation of proteinase inhibitor I. The protein aggregates are covered with silver-enhanced gold particles, Bar = 20 μm. Inset: enlargement of a labelled protein aggregate. Bar = 10μm. **c** Localisation of proteinase inhibitor II. Protein aggregates are uniformly and heavily labelled. Bar = 40 μm. Inset: enlargement of a labelled protein aggregate. Bar = 10 μm. **d** Montage of radial section through the fruit labelled to identify inhibitor I protein demonstrating the range of tissues examined. Ovule (O) and embryo (E) tissues are not labelled. Most cells of the pericarp (indicated by bracket) exhibit labelled protein aggregates. Bar = 100 μm (from Wingate *et al. Plant Physiol.*, **97**, 490 1991). With permission of the American Society of Plant Physiologists.

proteins (vsp) of approximately 27, 29 and 94 kD. The 94 kD vsp is an enzyme, lipoxygenase which catalyses the hydroperoxidation of polyunsaturated lipids (Tranbarger et al. 1991). In soybean, the single-cell layer of aleurone cells contain a unique xyloglyco-protein in the storage vacuoles which also contain conglycinin and agglutinin (Yaklich and Herman 1995). As well as aleurone grains, many species contain proteins as droplets or minute structures. In the leaves of tomato the occurrence of chymotrypsin inhibitor protein can be observed as spherical conspicuous droplets in the vacuoles by electron microscopy of mesophyll cells (Ryan and Shumway 1971). Proteinase inhibitor proteins, which are considered as defensive chemicals in plant tissues, are also localised in the vacuoles. Shumway et al. (1976) and Walker-Simmons and Ryan (1977) demonstrated by immunochemical labelling in situ and biochemical analysis of isolated vacuoles of wounded and non-wounded tomato leaves that the proteinase inhibitor proteins are accumulated as non-membrane-bound protein aggregates within the central vacuoles of non-wounded mesophyll cells. Using immuno-cytochemical labelling, Wingate et al. (1991) recorded by light and electron microscopy that parenchyma cells of the tomato pericarp contain inhibitor I and II proteins in non-membrane bound dense vacuolar aggregates, while the funiculus, ovule and early embryonic cells do not (Fig. 5.4). Vacuoles are also used for seasonal storage of proteins in various tissues, e.g. during late summer wood-ray cells of *Populus* store proteins in small vacuoles (Sauter et al. 1989).

Many plants contain ribosome-inactivating proteins (RIPs) which are components of immunotoxins or antiviral agents. Some of these RIPs are known to be stored in vacuoles—ricin, for example, is accumulated in the soluble matrix of vacuolar protein bodies in the endosperm of *Ricinus communis* (Tully and Beevers 1976; Youle and Huang 1996). Ultrastructural immunolocation studies by Carzaniga et al. (1994) indicated that another RIP, saporin, is present as individual deposits and large aggregates within the parenchyma-cell central vacuoles, the coenocytic lining of the perisperm 'vacuole' and the extracellular spaces of the seed of *Saponaria officinalis*.

Besides proteins and phytates, certin vacuoles contain a number of carbohydrate inclusions. Recent studies indicate that inulin, a reserved carbohydrate, is synthesised in the vacuole from sucrose and stored there. Not all reserve foods are produced or stored in vacuoles. Starch is produced essentially in chloroplasts and in various types of plastids, especially the amyloplasts. Unlike vacuoles, these are all surrounded by a double membrane and contain other membranous structures. However, a search in various species may reveal some interesting cases of vacuole. For example, leucosin, presently known as chrysolaminarin (a β-linked glucan), is present in some members of *Ochromonas sp.* (Chrysophyceae) as a large central leucosin vacuole outlined by a single membrane (Bold and Wynne 1978).

5.1.3 Lipid bodies

The important storage materials in many plant cells are lipids, fats and oils. These may occur freely, either suspended in the vacuole or in the ground cytoplasm as small bodies or droplets. They are stainable by fat-soluble dyes like Sudan IV, are highly osmiophilic and thus appear as dense bodies. Suter and Majno (1965) proved by specific fat staining that the osmiophilic particles were lipid particles. Maitra and De (1972) reported that the cells of the central zone of the root cap of *Medicago sativa* contain lipid bodies, but the outermost of the root cap cells contain both lipid bodies and lipid vacuoles which are produced by partial utilisation of lipids. Spherosomes, on the other hand, are considered a separate type of vacuole and are characterised by high refractivity, a globular shape and high triglyceride content (Sorokin 1967). They stain well with lipophilic dyes such as Sudan red and Nile blue. Sorokin and Sorokin (1968) and Walek-Czernecka (1965) demonstrated acid phosphatase and a number of hydrolases in the spherosomes of the epidermal cells of onion bulb scales. The presence of various hydrolytic enzymes in such particles qualifies them as special small vacuoles (Armentrout et al. 1968; Wilson et al. 1970). Oil droplets may be present without any outlining membrane in the cytoplasm. In liverworts distinct vacuoles, often known as oil ideoblasts, occur in body cells. The oil is included in a special vacuole together with some hydrophilic material, mainly polysaccharides. Oil is thought to be synthesised in the vacuole on its membrane (Schnepf 1981). Even in the presence of prominent vacuoles, oil drops may occur in the cytoplasm of young root cells of *Valeriana* sp.

5.1.4 Isoprenoids, polyterpenes, rubber, resin, and essential oil

Various derivatives of the 5-carbon compound isoprene, C_5H_8 occur in plant cells as essential oils, resins, carotenoids and rubber. Whereas terpenes of 10- to 30-carbon compounds make essential oils and resin, carotenoids are solid structures with 40 carbon atoms and more double bonds. Rubber and gutta-percha are high polymer derivatives. The various essential oils are not synthesised or produced in vacuoles but are stored in special cells transformed as oil sacs. The carotenoids occur as plastid pigments and are not generally known to occur in vacuoles. Vacuoles are not involved in synthesis and storage of terpenoids or resins either. In the investigated species of *Pinus*, the synthesis of terpene resin in the epithelial cells of ducts is carried out by endoplasmic reticulum. The nuclear envelope, plastids and mitochondria take part in resin synthesis (Vassilyev 1970; Werker and Fahn 1968; Loomis and Croteau 1973), and afterwards the droplets of resin being surrounded by plasmalemma invaginations are eliminated from the protoplast. The isolated portion of plasmalemma and cytoplasm surrounding the droplet undergo lysis, and the resin reaches the apoplast (Fahn and Benayoun 1976). The pattern of synthesis and release of gum-resins of *Rhus glabra* (Fehn and Evert 1974) and *Mangifera indica* (Joel and Fahn 1980) are similar and do not involve the vacuole at any stage. In contrast, rubber exists as discrete particles throughout the vacuole and cytoplasm of latex vessels or laticifers of various plants. In most cases there is very little differentiation between vacuole and cytoplasm.

Rubber particles may originate in three possible ways:
a) Particles occur *de novo* in the cytoplasmic matrix and stay there (Dickenson 1969; Pujarniscle 1971; Hebaul 1981).
b) Particles originate *de novo* in the cytoplasm and are subsequently transferred to the vacuole (Schulze *et al.* 1967; Marty 1974).
c) Particles are synthesised and held in the vacuole (Groneveldon 1976; Fineran 1983). The studies by Roy and De (1992) indicate that in *Calotropis gigantea*, rubber particles are synthesised in the small vacuoles which are first formed in the cytoplasm. Finally the disorganised vacuolar lumen of the laticifer houses the rubber particles.

5.1.5 Volutin and polyphosphate

The cytoplasm or vacuoles of many organisms show many particulate structures which, when stained with metachromatic dyes like toluidine blue, develop red colour instead of blue. They may contain polyphosphates and are sometimes called volutin granules.

Fungal vacuoles contain polyphosphates which give metachromatic staining with toluidine blue and are responsible for vital staining of vacuoles with neutral red. Guillermond (1941), like Dangeard (1916), observed that vital stains induce a precipitation of coloured particles in vacuoles, indicating that in the natural state, polyphosphates (which they called 'metachromatine') are colloidally dispersed. The so-called volutin granules (metachromatic granules) are highly concentrated deposits of polyphosphates (Waime 1947) which can exist next to the vacuole (Jordanov *et al.* 1962). If yeasts are starved by aeration in distilled water, followed by the addition of phosphate, the number of metachromatic basophilia rises markedly, provided energy sources are present. These structures are located exclusively in vacuoles. In *Scenedesmus*, the accumulation of polyphosphate is associated with the appearance of electron dense granules in small vacuoles (Sundberg and Nishammar-Holmvall 1975). The halo-tolerant unicellular alga *Dunaliella salina* accumulates large amounts of polyphosphates in association with K^+ and Mg^{2+} in the acid vacuoles (Pick *et al.* 1991; Pick and Weiss 1991).

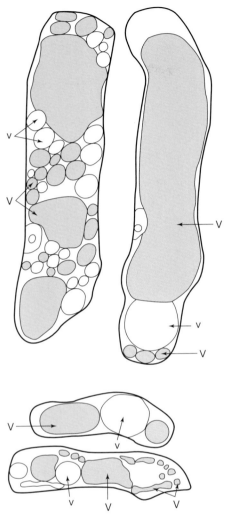

Figure 5.5 Living cells from the inner part of the fleshy receptacle. Two types of vacuoles, one (*V*) varying in size and shape with colloidal contents and anthocyanin pigment; the other (*v*), varying in size but always spherical, without colloidal substance (from Guilliermond 1941).

5.2 Colloidal or dissolved substances

5.2.1 Phenolics

The phenolics, e.g. tannin, anthocyanin and anthoxanthin, are a diverse group of compounds, which are widespread as dissolved substances in vacuoles of plant cells. Tannins (M.W. 500–3000) generally have a large number of phenolic hydroxy groups to form effective crosslinks between proteins and other macromolecules. They are, in general, amorphous astringent substances which combine with ferric salts to produce blue, black or green colour in sap, precipitate gelatin from solution, and dissolve in hot water. The presence of tannin and other phenolics has been used to designate vacuoles as 'A' type or full cell sap, in contrast to cells with 'B' type or empty cell sap which do not contain phenolics (see Chapter II) (Fig. 5.5 and 5.6). The presence of tannins has been demonstrated with the aid of electron microscopy, specifically in the shoot apex of *Oenothera* (Diers *et al.* 1973), in cultured cells of white spruce (Chafe and Durzan 1973) and slash pine (Baur

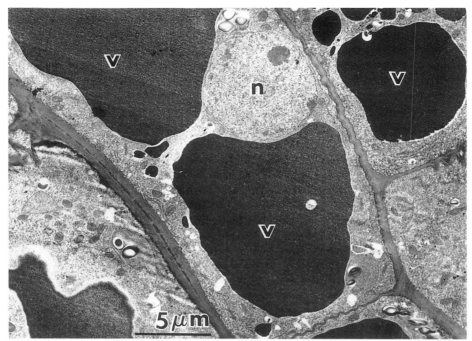

Figure 5.6 Micrograph of middle root cap layer of cells of *Rubus chamaemorus* L. showing vacuoles (V) loaded with polyphenolic compounds stained densely with osmium and lead. (Courtesy Dr. A.K. Bal.)

and Walkinshow 1974). In horseradish root cells phenolics are exclusively located in vacuoles (Grob and Matile 1980). Large deposits of phenolics have also been noted in the infected and uninfected cells of the root nodules of some plants, produced due to infection by the symbiotic actinomycete *Frankia* (Newcomb and Wood 1987).

5.2.2 Anthocyanin and anthoxanthin pigments

A large part of the colouration of plant parts is due to the presence of water-soluble anthocyanin and anthoxanthin pigments in the vacuole sap. Carotenoids, by contrast, are insoluble and are present in particulate form in the cytoplasm. Anthocyanins are derivatives of polyhydroxyflavylium with a sugar compound — e.g. glucose, galactose, rhamnose or gentiobiose. The presence and number of hydroxyl groups, methylation and sugar moiety produce red, violet and blue coloration. Other substances, including tannins and anthoxanthin, cause intensification of colour and the presence of metal ions like iron or aluminium may change the shade. Acidity of the cell sap is also involved in changing the colour. The anthoxanthins are yellow or orange coloured pigments, which are derivatives of flavone. The flavones may also be variously substituted with hydroxyl or methoxyl groups and may exist in free form or as glycosides. The presence of pigments is obvious in most vacuoles by their colouration. One of the best documented cases is the occurrence of pigments in the young rose leaflet, which show anthocyanin-containing vacuoles at various stages of development. At the tips of the leaflets, thread-like vacuoles containing pigments enlarge and coalesce to produce a single large pigmented vacuole. In many cases the concentration of the pigment may be too low for visual detection in living cells and in other cases, distinct crystals of the pigment can be found (Fig. 5.7). Using advanced biochemical methods, some workers have precisely detected a few pigments and their metabolites in the vacuole. For example, Hopp *et al.* (1985) isolated anthocyanin containing vacuoles from protoplasts of *Daucus carota*

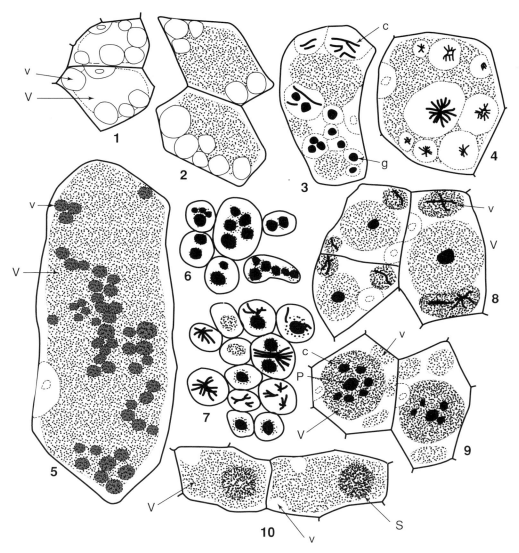

Figure 5.7 1–7, Fruit of *Rubus fruticosus*. Two types of vacuoles. 1–4, exocarp. 1, young green fruit; large and small colourless vacuoles (V, v). 2, ripening fruit; large vacuole with raspberry-red anthocyanin pigment and small colourless vacuoles. 3, ripe fruit; three vacuoles with red pigment, numerous small vacuoles either with colloidal bodies (g) or needle-shaped crystals (c), both coloured violet by anthocyanin. 4, ripe fruit; one large vacuole with red pigment, smaller vacuoles with crystals of dark violet pigment, isolated or in bundles. 5, mesocarp, ripening fruit; large vacuole (V), with dilute solution of raspberry-red pigment, numerous small vacuoles (v), with brick-red pigment and raspberry-red colloidal granules. 6, 7, mesocarp, ripe fruit; small vacuoles with one or more violet colloidal bodies, sometimes also with needle-shaped crystals of pigment, isolated or in groups, sometimes with the crystals only. 8, *Wisteria sinensis*; epidermis of petal; large central vacuole (V), with red-violet pigment and small peripheral vacuoles (v), with a concentrated solution of blue-violet pigment and crystalline needles of dark blue pigment. 9, *Hibiscus syriacus*; epidermis of petal; large central vacuole (V), with red pigment and dark red tannin bodies (P); small peripheral vacuoles (v), with mauve pigment. 10, *Canna indica*; epidermis of leaf; large vacuole (V), with red pigment and a large spherical crystal (S); one or more small colourless peripheral vacuoles. (*in vivo*) (from Guilliermond 1941).

cell cultures. Anhalt and Weissenböck (1992) recorded that two flavones, luteolin 7-0-diglucuronyl-4'-0-glucuronide and luteolin 7-0-diglucuronide, were entirely located in the vacuoles of the mesophyll protoplasts of young primary leaves of rye. Dhurrin, a cyanogenic glucoside, is accumulated in the vacuoles of leaves of *Sorghum* seedlings (Saunders and Conn 1978). Another cyanogenic glucoside, linamarin, was also found to be exclusively localised in the central vacuole of mesophyll and epidermal layers of *Hevea brasiliensis* by Gruhnert et al. (1994). The presence of a red anthocyanin pigment has been vividly demonstrated in the vacuoles of *Euphorbia millii* (Morikawa et al. 1988).

5.2.3 Alkaloids and steroids

These complex cyclic compounds, slightly soluble in water and basic in reaction, are typical secondary products of certain families of higher plants, especially Solanaceae and Papaveraceae. Most alkaloids occur in vacuoles or vesicles. Nicotine accumulates in the vacuoles of leaf cells of tobacco (Saunders 1979) and capsaicin is present in the vacuoles of placental cells of capsicum fruits (Fujiwake et al. 1986). Experiments on uptake in cultured cells indicate that all of the alkaloids may be located in vacuoles (Deus-Newmann and Zenk 1984, 1985). Using cytochemical and autoradiographic techniques at the ultrastructural level, Neumann and Muller (1967) demonstrated the vacuolar localisation of the alkaloids protopine, sanguinarine and allocryptopine in the tissue culture cells of *Macleaya cordata*. The addition of hexachloroplatinic acid to the fixative causes large electron-dense clumps of precipitated alkaloids to be formed in the vacuoles. In *Cheliodonium majus*, vacuoles isolated from the laticifers show the presence of a number of alkaloids like sanguinarine, chelerythrine, berberine and dihydrocopticine. Yellow berberine pigments had been clearly demonstrated in the vacuoles of *Coptis japonica* (Morikawa et al. 1988). Similarly, small vacuoles obtained from the poppy fruit latex contain all the morphine (Matile 1978). More often than not it is extremely difficult to detect the site of synthesis and accumulation of alkaloids (Böhm 1980). Various physiological conditions affect whether or not an alkaloid is stored in the vacuole. In *Catharanthus roseus*, alkaloids are stored in vacuoles with a pH of 3.0, in contrast to the 'normal' vacuolar pH of 5.0 (Anderson et al. 1985). Löffelhardt et al. (1979) concluded that the cardenolides of the leaf of *Convallaria majalis* are primarily stored in vacuoles.

5.2.4 Sugars

Various types of sugars occur in plant cells in free or conjugated form. Cell sap may contain mono-, di-, tri- or even tetra-saccharides in free form in the dissolved state. Sucrose, glucose and fructose appear to be the chief sugars found in cell sap in most plants. Mannitol is found in brown algae and floridoside is found in many red algae. Large amounts of sugars are accumulated in trees, shrubs and xerophytic plants compared to herbs. Various physiological factors, the type of cell, time of day and season greatly affect the amount of dissolved sugar in a cell, since the sugars are used as ready energy for many metabolic activities.

Certain tissues, like parenchymatous cells of sugar cane, store a huge quantity of sucrose, away from general metabolic activity. Similarly, beetroot tissue contains 90% of its sucrose in its vacuoles. Besides general storage, the vacuoles are also involved in biosynthesis of higher saccharides from mono- or disaccharides. For example, *Pisum* mesophyll cell vacuoles are capable of energy-dependent uptake of D-glucose. In addition to biosynthesis and mobilisation of saccharides, vacuoles are the likely site for glycosilation and production of various metabolites. Large amounts, up to 80% of dry weight, of stachyose are found in the vacuoles of tuber parenchyma of *Stachys sieboldii* (Keller and Matile 1985) and gentianose in the storage roots of *Gentiana lutea* (Keller and Wiemken 1982). A number of glycosides and glucosides also exist in vacuolar sap (Alibert et al. 1985; Werner and Matile 1985). In the heterotrophic suspension cells of sugar cane, Preisser et al. (1992) found an accumulation of hexoses, but not of sucrose, in the vacuoles. Heineke et al. (1994) made similar observations in the leaf cells of tobacco, where sucrose was located mainly in the cytosol, and up to

98% of hexoses were accumulated in the vacuole. Winter et al. (1993), using non-aqueous fractionation techniques combined with stereological evaluation of cellular and subcellular volumes from light and electron micrographs, found that in barley leaves sucrose concentration was as high as 232 mM in the cytosol, compared to ≥ 21 mM in the vacuole. In spinach leaves the concentration of sucrose in the cytosol is five times higher than in the vacuole (Winter et al. 1994).

5.2.5 Organic acids

The most prevalent of the many aliphatic organic acids present in plants are the dicarboxylic acids — oxalic, malic and tartaric; and the tri-carboxylic acid, citric. Many different organic acids occur in individual plant cells, either as short-lived metabolic intermediates or as special accumulates. Thus oxalic acid may comprise as much as 50% dry weight of certain leaves. Similarly, citric acid may be as much as 60% dry weight of lemon, making its pH as low as 2.5. On the basis of pH of cell sap, general appearance of the cell and consideration of compartmentation, these acids are known to occur in vacuoles, away from the centres of active metabolism.

Through the analysis of the kinetics of isotope-labelling, MacLennan et al. (1963) were the first to demonstrate that carboxylic acids are stored in vacuoles. In the latex of *Hevea* the small vacuoles, called lutoids, appear to contain citric acid at a concentration up to 24 times higher than the cytoplasm, but malic acid remains at equal concentration in both vacuoles and cytoplsm (Reballier et al. 1971). The development of reliable techniques to isolate intact vacuoles has enabled better definition of organic acids. Researchers have shown that oxalic (Wagner 1981), ascorbic (Grob and Matile 1980), malic, citric and isocitric acids (Buser and Matile 1977; Kenyon et al.1979; Wagner 1979) are predominantly or exclusively present in vacuoles. The Crassulcean acid metabolism (CAM) plants are characterised by fixation of CO_2 and accumulation of malic acid at night and utilisation of this acid as an endogenous source of CO_2 during the day. Vacuoles contain all the malic acid and show a diurnal fluctuation of malic acid, whereas isocitric acid, which is also abundant in CAM plants, remains at a constant level (Buser and Matile 1977; Kenyon et al. 1979).

In contrast to the aliphatic acids, the aromatic acids are not widely distributed as free acids. They occur as esters in oils and resins or as glycosides and tannins. Their occurrence in vacuoles is not well documented.

5.2.6 Amino acids

Early physiologists recognised a number of free amino acids in plant cells. These may occur both in the cytoplasm as well as in the vacuoles as separate pools and their composition may be significantly different, reflecting a degree of compartmentation (Wagner 1979; Martinoia et al. 1981). The amino acid pools may be differentially responsive to various metabolic activity including protein synthesis. In sweet clover more than 50% of free amino acids are found in vacuoles (Boudet et al. 1981). Similarly, 70% of the neurotoxin, 2,4-diaminobutyric acid, a non-protein amino acid is localised in the vacuoles of leaf tissues of *Lathyrus sylvestris* (Foster et al. 1987). In apple cotyledons, more than 80% of the free amino acids histidine, arginine, tryptophan and valine are contained in the vacuole, with most of the other free amino acids being present in vacuoles at levels greater than 50% of their cellular level (Yamaki 1982). Wagner (1977) suggested that the vacuolar pool is slow or inactive compared to the cytoplasmic pool. Vacuoles of different tissues contain different amino acids — both in type and concentration. He also detected 20% of cellular serine, 25% of glutamate, 21% of glutamine and 80% of tryptophan in isolated vacuoles of *Hippeastrum* petals. The constitution of the amino acid pool not only varies from tissue to tissue, as expected, but also varies with supply of nitrogen. For example, if carrot explants are grown on different concentrations of NH_4NO_3, the level of amino acids in the vacuoles varies (Mott and Steward 1972).

In vacuoles aqueously isolated from mesophyll protoplasts of barley, Dietz et al. (1990) noted a relatively high total amino acid concentration of 77 mM with alanine, leucine and glutamine dominant. A non-aqueous fractionation procedure showed a much lower amino acid level in the vacuoles of spinach leaves (Riens *et al.* 1991). Winter *et al.* (1993), using similar techniques combined with stereological evaluation of cellular and subcellular volumes from micrographs, found a very low concentration of total amino acids (≤ 2 mM) in barley leaf vacuoles compared to the cytosolic compartment (275 mM). Fricke *et al.* (1994) obtained a similar result in the epidermal vacuoles of barley. Studies on the amino acid pool of fungal cells show that the vacuoles contain a large amount of amino acids. Fungal cells may contain a pool of amino acids that can be eliminated only with difficulty by starvation. A study of the kinetics of amino acid uptake in *Neurospora* by Zalokar (1961) showed at least two pools of amino acids; one, an intermediate pool of relatively constant size from which amino acids are taken for protein synthesis, and the second, an expansible pool into which excess amino acids are shunted as a reserve. Since the old vacuolated cells are less active in amino acid uptake than the young ones, it has been suggested that the vacuoles may be a site for such a pool. Wiemken and Nurse (1973) reported that the vacuoles of *Candida* accounted for at least 90% of the total amino acid pool and contained mainly the nitrogen-rich amino acids. The basic amino acids, especially arginine and glutamine, are accumulated preferentially. The concentration in the vacuoles may be as much as 20 times that of the cytoplasm. However, glutamic acid, which accounts for one-third of the amino acid pool, may be almost absent in the vacuoles (Wiemken and Dürr 1974). It is thus possible that the amino acids represent a reserve for soluble nitrogen.

It has been demonstrated that amino acids are actively transported into the vacuole. They may also passively enter the vacuole by an exchange system. An amino acid pool may also be created by lytic breakdown of proteins inside a vacuole.

5.2.7 Hormones

The presence of plant growth hormones in vacuoles is not very obvious. Abscisic acid (ABA) is present in cytoplasm and chloroplasts as well as in the vacuoles of mesophyll cells. The vacuolar amount may almost double at night yet remain low compared to the other components (Hartung *et al.* 1982). From a study of biotransformation of ABA in cell suspension cultures of *Lycopersicon esculentum*, Lehman and Glund (1986) showed that conjugated forms of ABA were located in the vacuoles whereas ABA and its acidic metabolites were found mainly in the extra-vacuolar fractions. They proposed that ABA-ß-D-glucopyranosyl ester (ABA-Glc) is irreversibly compartmented in the vacuoles of plant cells. It is interesting to note that ABA-Glc is a non-convertible metabolite of ABA metabolism and is exclusively located in the vacuole. Similar compartmentation of glucoside conjugates of ABA has also been reported by Bray and Zeevaart (1985).

Guy and Kende (1984) have demonstrated that in *Pisum sativum*, the conversion of 1-amino cyclopropane-1-carboxylic acid (ACC) to ethylene (the last step in ethylene biosynthesis), takes place in the vacuole. The isolated vacuoles contained most of the ACC present in the protoplast and converted the stored ACC to ethylene. Gibberellin-A_1 and its glucosides occur in the vacuoles of cowpea and barley (Garcia-Martinez *et al.* 1981). The accumulation of synthetic auxin, 2,4-dichlorophenoxy acetic acid in a conjugated form has been demonstrated in soybean vacuoles (Schmitt and Sanderman 1982). Fusseder and Zeigler (1985) have demonstrated that when dihydrozeatin is supplied to photoautotrophically growing cells of *Chenopodium rubrum*, the cytokinin is compartmentalised in the vacuoles after O-glycosylation. Thus it may be inferred that vacuoles are the subcellular storage compartment for glucosidic conjugates of plant hormones in general.

5.2.8 Inorganic ions

It has long been known that inorganic ions are present in almost all vacuoles. Numerous studies have revealed the presence of a number of ions including potassium, sodium, calcium, magnesium, chlorides, sulphates, nitrates and phosphates in vacuole sap. Osterhout (1936) showed that

Table 5.1 Comparison of ionic concentration (in milliequivalents) in cell sap of fresh water *Nitella* and surrounding pond water (after Hoagland and Davis 1923).

Ion	Cell sap	Pond water	Ratio cell sap:pond water
Cl	9.7	0.9	11
Na	10.0	0.2	50
K	54.3	–	–
Ca	20.5	1.6	13
Mg	35.4	3.4	10
SO_4	16.7	0.7	24
PO_4	3.7	0.004	925
NO_3	0	0.5	–

Table 5.2 K^+, Na^+ and Cl^- concentration in various compartments in algal cells and their bathing solution.

Algae	Compartment	Ion concentration (mM)			Reference
		K^+	Na^+	Cl^-	
1. *Nitella translucens*	Solution	0.1	1.0	1.3	MacRobbie 1971
	Cytoplasm	119	14	65	
	Vacuole	75	65	150–170	
2. *Acetabularia mediterrania*	Solution	10	470	550	Saddler 1970
	Cytoplasm	400	57	480	
	Vacuole	355	65	480	
3. *Valonia ventricosa*	Solution	11	485	590	Gutknecht 1968
	Cytoplasm	434	40	138	
	Vacuole	625	44	643	

Valonia (a marine alga with coenocytic vesicular thallus of about 3–5 cm in diameter containing about 20 ml of cell sap), accumulates potassium at a much higher concentration in the vacuole than in sea water, whereas sodium, calcium and sulphate were lower. Like *Valonia*, the cylindrical internodal cells of the fresh-water alga *Nitella clavata*, which are about 15 cm long and 1 cm wide, readily provide sap for analysis (Table 5.1).

A comparison of salt water and fresh water algae obtained by various workers indicates that the sap composition and concentration in these two groups are similar, whereas external solution concentration was much lower for the fresh water forms. Thus the ratio of cell sap concentration to that of the bathing solution was much higher in fresh-water plants than in salt-water plants (Table 5.2). The vacuolar ionic content may differ from cytoplasmic as well as bathing fluid.

This type of differential ion accumulation is also noted in a totally different type of vacuole. The small vacuoles in the *Hevea* latex contain different ion concentrations than the cytoplasm. Mg^{2+} was accumulated 90-fold, Ca^{2+} 6-fold and Cu^{2+} 2-fold in the vacuole compared to the cytoplasm. In contrast, equal concentrations of K^+ were found in the vacuole and cytoplasm (Ribaillier et al. 1971). Efflux analysis of *Atriplex spongiosa* leaf slices suggests that Na^+ is restricted to vacuoles, whereas K^+ has a more uniform distribution throughout the cell. In *Suaeda maritima*, Rb^+, which can be detected under the electron microscope, when absorbed in large quantities appears to be restricted to the vacuoles.

Such differential ionic make ups may not be universal. Quantitative microprobe analysis of thin freeze-dried sections of baker's yeast cells yielded evidence that a number of univalent ions are about equally distributed between vacuoles and cytoplasm (Roomans and Seveus 1976).

Table 5.3 Vacuole/extravacuole partitioning of plant cell metabolites determined from studies of isolated vacuoles, as compiled by Wagner (1982).

Substance	Percentage in vacuole	Tissue	References
Amino acids	50; 52; 85	*Hippeastrum*, tulip; barley;	Martinoia *et al.* 1981; Wagner 1981; Heck *et al.* 1981
Sucrose	44–100; most; 62	*Hippeastrum*, tulip; beetroot; castorbean	Wagner 1979; Leigh *et al.* 1979; Nishimura & Beevers 1978
Glucose, fructose	50–100	*Hippeastrum*, tulip	Wagner 1979
Malic acid	most	*Bryophyllum*	Buser & Matile 1977
Ascorbate-derived oxalate	70	barley	Wagner 1981
Na^+, Mg^+, Ca^{2+}, Cl^-, K^+	most	*Hippeastrum*, tulip	Lin *et al.* 1977
Anthocyanin	100	*Hippeastrum*, tulip	Wagner 1979
Betanin	all	beetroot	Leigh *et al.* 1979
Ascorbate	100	horseradish root	Grob & Matile 1980
Phenolics	100	horseradish root	Grob & Matile 1980
Glucosinolates	100	horseradish root	Grob & Matile 1980
Dhurrin	all	*Sorghum*	Saunders & Conn 1978
Nicotine	93	tobacco	Saunders 1979
Proteinase inhibitor	all	tomato	Walker-Simmons & Ryan 1977
Gibberellin	30–100	barley, cowpea	Ohlrogge *et al.* 1980
Nitrate	99	barley	Martinoia *et al.* 1981

Similarly, significant differences between the ion contents in vacuoles and cytoplasm could not be detected when whole protoplasts and intact vacuoles of *Tulipa* petal cells were analysed (Lin and Wagner 1977).

The relative concentration of various inorganic ions is frequently controlled by the composition of the bathing solution. On the other hand, when plants are grown in identical synthetic nutrient environments there are often clear species differences in the accumulation of various elements. For example, members of Chenopodiaceae accumulate very high levels of Mg^{2+}. *Astragalus missouriensis* with 2.1 ppm of selenium completely fails to accumulate selenium from the soil, whereas *A. biseulatus* contains as much as 8400 ppm of the ions. The secretory (vesicular) cells of Rhodophyceae are colourless and have a large central vacuole. These cells may contain enough iodine to produce a blue colour in herbarium paper which contains starch as filler. The accumulation of iodine in kelp, which are members of Phaeophyceae, is well known. It is interesting to note here that bromine is involved in the formation of refractile inclusions in the vesicle cells of the members of Bonnemaisonaceae (Rhodophyceae). Wolk (1968) has shown that if these algae are grown in the absence of bromine, then the secretory cells become vestigial, lacking the normal large vacuole with its refractile contents. The most spectacular case of anion accumulation has been demonstrated in *Desmarestia*, a large marine brown alga (Phaeophyceae) which inhabits cold waters. The cells may contain, together with some malic acid, free sulfuric acid at a concentration of 0.44 N in the vacuoles, resulting in a pH value of 0.8 to 1.8 (Meeuse 1956, 1962). The cells of the desmid *Closterium lunular* contain vacuoles at each end of the cell. These vacuoles are fluid-filled with one to several hundred tiny crystals of barium sulphate, as has been determined by scanning proton microprobe analysis.

Finally, it has been realised that ions may occur in vacuoles in various states. They may be ionised, dissolved but not ionised, adsorbed, readily soluble or hardly soluble. On the basis of an extensive investigation, Kinzel (1987) noticed a certain pattern of accumulation of calcium in vacuoles of various plant families. Chenopodiaceae generally contain calcium in an undissolved form of Ca-oxalate whereas Crassulaceae and Brassicaceae have dissolved calcium. Some members of Brassicaceae contain super-saturated non-ionised solutions of Ca-sulphate, possibly as colloids. Pectin bound calcium is present in the Boraginaceae. X-ray analytical methods indicate that tanniniferous cells in the mesophyll of mangrove (*Aegiceras corniculatum*) contain two types of vacuole; one containing large amounts of osmiophilic organic solutes with little or no Cl^-, and the other free of osmiophilic organic solute but containing significant quantities of Cl^-. Wagner (1982) compiled a quantitative estimate of various metabolites in vacuoles derived from a number of authors (Table 5.3). This table underlines the role of vacuole as a storage compartment.

5.2.9 Miscellaneous substances

A number of secondary metabolites and toxic or potentially toxic substances have been located in vacuoles. The storage roots of horseradish for example, contain, along with phenolics, two glucosinolates, viz. sinigrin and gluconasturtin, which are exclusively localised in the vacuoles (Grob and Matile 1980). Vacuoles are also a repository for glucosides like dhurrin, a cyanogenic glucoside in *Sorghum* (Saunders *et al.* 1977), cardiac glucoside in *Convallaria* (Löffelhardt *et al.* 1979), O-coumaryl-glucoside in *Melilotus* (Oba *et al.* 1981).

Measurements of ultraviolet absorption show that S-adenosyl phosphate, an important intermediate in glucose metabolism, accumulates in the vacuoles of yeasts. As well as their natural contents, vacuoles can be forced to accumulate different substances. For example, other sulphonium compounds, like S-adenosyl methionine, can be detected in yeast cells (Svihla and Schlenk 1960) and saponin has been found in the vacuoles of *Avena sativa* (Urban *et al.* 1983). Anthraquinones are located in the vacuoles of cultured cells of *Gallium mollugo* (Anderson *et al.* 1985). Utilising indirect post-embedding immunogold electron microscopy, Herman *et al.* (1988) demonstrated the presence of specific lectins exclusively sequestered in the protein storage vacuoles of leaf and bark tissue cells of *Sophora japonica*. Many of the protein bodies of the parenchyma cells of mature cotyledons of *Dioclea lehmani* contain lectins. In immature cotyledons and in mature embryo axis, these lectins are exclusively localised in the vacuoles. Düggelin *et al.* (1988) reported that two fluorescent lipofuscin-like compounds, formed due to the breakdown of chlorophyll in the senescent primary leaves, are located exclusively in the vacuole. Proline derivatives also occur there and the presence of nucleosides has also been reported. Leinhos *et al.* (1986) have recorded that a part of cellular uridine is located in the vacuoles of tomato cells in culture. As the search for vacuolar localisation continues, more and more chemical substances are expected to be found in vacuoles.

5.3 ENZYMES

Vacuoles contain soluble proteins and enzymes. The vacuoles of *Hippeastrum*, *Tulipa* and wheat may contain 4–15% of the protoplast's soluble proteins (Wagner *et al.* 1981). The protein content of isolated vacuoles of tobacco cultured cells are estimated to be 5–11% that of protoplasts, whereas vacuoles from mature leaves of tobacco contain as much as 29% (Boller and Kende 1979). The relative amount of a given protein or enzyme in the cytosol or vacuole may vary. For example, Lin and Wittenbach (1981) studied the subcelllluar location of protease and concluded that all the proteolytic activity of corn and wheat leaf mesophyll cell protoplasts can be assigned to vacuoles. Similarly, an enzyme may be strictly localised, e.g. V-ATPase and V-PPase, in most tonoplasts and α-mannosidase in yeast tonoplast (Van der Wilden and Matile 1973). Table 5.4 shows the quantitative estimation of various enzymes in vacuoles by different workers.

A general survey of the various enzymes present in the vacuole (Table 5.5) indicates that enzymes may be found in vacuoles of all groups of plants. The method of investigation, whether it is *in situ* localisation by light or electron microscopic techniques or biochemical isolation from whole tissue or isolated vacuoles, dictates or limits the sensitivity and specificity. Obviously, different types of cells or tissues used for detection may show the presence of different enzymes.

Table 5.4 Vacuole/extravacuole partitioning of active enzymes as determined from studies of isolated vacuoles (compiled by Wagner 1982).

Enzyme	Percentage in vacuole	Number of tissues	References
Acid phosphatase	30–100	8	Heck et al. 1981; Saunders 1979; Grob & Matile 1980; Boller and Kende 1979; Nishimura & Beevers 1978; Butcher et al. 1977
Phosphodiesterase	55–100	4	Grob & Matile 1980; Boller & Kende 1979; Nishimura & Beevers 1978
Acid nuclease	50–100	3	Nishimura & Beevers 1978; Butcher et al. 1977
Acid protease	80–100	4	Heck et al. 1981; Lin & Wittenbach 1981; Boller & Kende 1979; Wagner et al. 1981
Carboxypeptidase	most	1	Nishimura & Beevers 1978
α-Mannosidase	100	4	Heck et al. 1981; Boller & Kende 1979
α-Galactosidase	80–100	2	Boller & Kende 1979
β-Glucosidase	most	1	Nishimura & Beevers 1978
β-N-acetylglucosaminidase	90–100	4	Heck et al. 1981; Boller & Kende 1979
Acid invertase	most–100	2	Leigh et al. 1979
Phytase	most	1	Nishimura & Beevers 1978
Myrosinase	29	1	Grob & Matile 1980
Peroxidase	50–70	3	Grob & Matile 1980; Boller & Kende 1979

For example, whereas the small vacuoles of root meristem of *Zea mays* show a range of hydrolytic enzymes, in the scutellar vacuoles lipase predominates and in the leaf mesophyll it is the proteases. Vacuoles, even those residing in the same cell, may not be alike. Isolated lysosomes for example, are very different from the typical large central vacuole of a mature cell. Thus, aleurone grains, lutoids, etc. are not typical vacuoles. In many cases it has not been possible for researchers to state the precise location of enzymes. However, many reports conclusively demonstrate the presnce of different enzymes on tonoplast and vacuole sap. The identification or nomenclature of enzymes has been inadequate, especially in the early studies, but most recent reports have been very precise.

Of six broad categories of enzymes, the hydrolases are the only ones present in vacuolar sap. Apart from a few well-documented reports of oxido-reductases in the tonoplast, enzymes of most other categories, namely transferases, isomerases and ligases, are possibly absent from vacuoles. Hydrolases are the enzymes which catalyse the splitting of C–O, C–N, C–C, and some other bonds by the addition of water. The three main groups of hydrolases are *peptidases*, which hydrolyse peptide bonds between amino acid residues in proteins, *esterases*, which hydrolyse ester bonds of phosphoric acid, sulfuric acid and carbonic acid and *glycosidases*, which hydrolyse glycosidic bonds between sugars in glycogen, starch, etc.

Both types of peptidases, viz. *endopeptidases*, which cleave internal peptide bonds, and *exopeptidases*, which cleave terminal peptide bonds, are present in vacuoles. It is highly significant that, at least in certain plants, endopeptidases are exclusively located in the vacuole sap, whereas other peptidases are widely distributed throughout the cell (Canut et al. 1985). The endopeptidases, known as endoproteinases or simply proteinases, are classified not on the basis of substrate, but on the basis of active site residue. The proteinases are divided into four major groups: cysteine proteinases, serine proteinases, aspartic acid proteinases and metalloproteinases. They all occur in vacuoles or protein bodies in various plant cells (Callis 1995; see Runeberg-Roos et al. 1994). Of the two types of exopeptidases, the *aminopeptidases* acting on

Table 5.5 Presence of enzymes in various types of vacuoles in plant species. (Enzyme nomenclature or vacuole type has been presented as given by the respective authors. No attempt is made to standardise them.

Sl. No.	Plant species	Tissue	Methods: Biochemical/ Cytochemical	Vacuole: Type/ Structure	Enzyme	References
	ALGAE:					
1.	*Asteromonas gracilis*	Cell	Biochem Cytochem	Sap	Acid phosphatase	Swanson & Floyd 1979
2.	*Gleomonas kupfferi*	Cell	Cytochem	Sap	Cytidine 5'-monophosphatase, β-nicotinamide adenine diphosphatase, β-glycerophosphatase	Domozych 1989
3.	*Acetabularia mediterranea*	Stalk	Biochem	Subcell cytoplast	Acid phosphatase, RNase	Luscher & Matile 1974
4.	*Nitella axilliformis*	Internodal	Biochem	Sap	Phosphatase, Carboxypeptidase	Doi et al. 1975
	PHYTOFLAGELLATE:					
5.	*Euglena gracilis*	Cell	Cytochem	Lysosome	Hydrolases	Brandes & Bertini 1964
6.	*Euglena gracilis*	Cell	Biochem	Particulate	Acid phosphatase, Esterase	Bennum & Blum 1966
7.	*Ochromonas malhamensis*	Cell	Biochem	Particulate	Hydrolase	Lui et al. 1968
	FUNGI:					
8.	*Saccharomyces cerevisiae*	Cell	Biochem	Sap	Protease A, Protease B, Carboxypeptidase, RNase, α-mannosidase, acetylesterase, leucyl-aminopeptidase, invertase, β-glucosidase, acid phosphatase (Ca^{2+}), alkaline phosphatase (Mg^{2+})	Wiemken et al. 1979
9.	*Saccharomyces cerevisiae*	Cell	Biochem	Tonoplast	NADH: Cytochrome C, α-mannosidase, oxidoreductase NADH: DIP Oxidoreductase, ATPase (Mg^{2+})	Van der Wilden et al. 1973
10.	*Saccharomyces cerevisiae*	Cell	Biochem	Tonoplast	Permease (arginine-specific)	Boller 1977
11.	*Saccharomyces cerevisiae*	Cell	Biochem	Sap	Trehalase	Keller et al. 1982
12.	*Saccharomyces cerevisiae*	Cell	Biochem	Tonoplast	H^+-ATPase	Kakinuma et al. 1981
13.	*Neurospora crassa*	Conidia	Biochem	Vacuole	Endopeptidase, aminopeptidase, RNase, Phosphatases, phosphodiesterase, invertase	Matile 1971

Table 5.5 Presence of enzymes in various types of vacuoles in plant species. (Enzyme nomenclature or vacuole type has been presented as given by the respective authors. No attempt is made to standardise them. (*Continued*)

Sl. No.	Plant species	Tissue	Methods: Biochemical/ Cytochemical	Vacuole: Type/ Structure	Enzyme	References
14.	*Neurospora crassa*	Hypha	Biochem	Tonoplast	H^+-ATPase	Bowman & Bowman 1982
15.	*Botrytis cinerea*	Hypha	Cyto	Lysosome	Acid phosphatase, (thioacetic acid) esterase, acid deoxyribonuclease II, β-galactosidase, arylsulfatases	Pitt 1968
16.	*Sclerotinia fructigena*	Hypha	Cyto	Lysosome	Acid phosphatase, α-L-arabinofuranosidase	Hislop *et al.* 1974
17.	*Ceratocystis fagacearum, Ceratocystis fagacearum, C. fimbriata, Fomes annosus, Agaricus campestris, Alternaria tenuis, Piptocephalis virginiana, Microtypha microspora*	Hypha	Cyto	Spherosome	Acid β-glycerophosphatase, arylsulfatase, acid DNase	Wilson *et al.* 1970
18.	*Coprinus lagopus*	Hypha	Biochem	Sap	Endopeptidase, RNase, phophatase, β-glucosidase, chitinase	Iten & Matile 1970
	MYXAMOEBA:					
19.	*Dictyostelium discoideum*	Amoeboid	Biochem	Lysosome	Acid phosphatase, Ribonuclease, Deoxyribonuclease, Proteinase, amylase, β-N-acetylglucosaminidase, α-mannosidase, β-glucosidase	Ashwoth & Weiner 1973
20.	*Polysphondylium pallidum*	Amoeboid	Biochem	Lysosome	Phosphatase, RNase, DNase, proteinase, amylaase, β-N-acetylglucosaminidase	O'Day 1973
	FERN:					
21.	*Asplenium fontanum*	Meristem	Biochem	Small Vacuole	Protease, RNase, DNase, phosphatase, phosphodiesterase, β-galactosidase	Coulomb 1971
	GYMNOSPERM:					
22.	*Psuedotsuga sp.*	Seed	Biochem	Spherosome	Lipase	Ching 1968

Table 5.5 Presence of enzymes in various types of vacuoles in plant species. (Enzyme nomenclature or vacuole type has been presented as given by the respective authors. No attempt is made to standardise them. *(Continued)*

Sl. No.	Plant species	Tissue	Methods: Biochemical/ Cytochemical	Vacuole: Type/ Structure	Enzyme	References
23.	*Larix decidua*	Roots	Cyto	Lysosome	Acid phophatase, acid DNase, esterase, β-glucuronidase, aryl sulfatase	Walek-Czerneeka 1965
	MONOCOTS:					
24.	*Zea mays*	Root	Cyto	Tonoplast	ATPase	Sexton & Hall 1978
25.	*Zea mays*	Root	Biochem	Small vacuole	Endo and exopeptidases, RNase, DNase, phosphatase, phosphodiesterase, acetyl esterase, β-amylase, α- and β-glucosidase, β-galactosidase	Matile 1966; Matile 1968a
26.	*Zea mays*	Root	Cyto	Tonoplast	Peroxidase	Parish 1975
27.	*Zea mays*	Scuttellum	Biochem	Sap	Lipase	Matile 1976
28.	*Zea mays*	Mesophyll	Biochem	Sap	Protease	Lin & Wittenbach 1981
29.	*Triticum aestivum*	Mesophyll	Biochem	Sap	Protease	Lin & Wittenbach 1981
30.	*Triticum aestivum*	Leaf	Biochem	Sap	Peptide hydrolase	Waters *et al.* 1982
31.	*Triticum aestivum*	Leaf (senescing)	Biochem	Sap	Protease	Wittenbach *et al.* 1982
32.	*Triticum vulgare*	Mesophyll	Biochem	Tonoplast	Ca^{2+}-ATPase	Diaz de Leon & Wyn Jones 1985
33.	*Hordeum vulgare*	Mesophyll	Biochem	Sap	Acid proteinase, α-Mannosidase, acid phosphatase	Heck *et al.* 1981
34.	*Hordeum vulgare*	Leaf	Biochem	Tonoplast	α-mannosidase, proteinase	Martinoia *et al.* 1981
35.	*Hordeum vulgare*	Mesophyll	Biochem	Sap	Endoproteinases EP_1, EP_2	Thayer & Huffaker 1984
36.	*Hordeum vulgare*	Seed	Biochem	Aleurone grain	Endopeptidase, phytase, acid proteinase	Ory & Heningsen 1969; Tronier *et al.* 1971
37.	*Hordeum vulgare*	Seed	Cyto	Aleurone grain	Acid phosphatase	Ashford & Jacobson 1974

Table 5.5 Presence of enzymes in various types of vacuoles in plant species. (Enzyme nomenclature or vacuole type has been presented as given by the respective authors. No attempt is made to standardise them. (*Continued*)

Sl. No.	Plant species	Tissue	Methods: Biochemical/ Cytochemical	Vacuole: Type/ Structure	Enzyme	References
38.	*Sorghum bicolor*	Leaf	Biochem	Sap	Dhurrin β-glucosidase	Kojima *et al.* 1979
39.	*Sorghum bicolor*	Mesophyll	Biochem	Sap	Dhurrin β-glucosidase, hydronitrile lyase	Saunders *et al.* 1977
40.	*Sorghum bicolor*	Seed	Biochem	Aleurone grain	Protease, RNase, phosphatase, pyrophosphatase, phytase, α- and β-glucosidase, β-galactsidase	Adams & Novellie 1975
41.	*Oryza sativa*	Seedling	Biochem	Tonoplast	H^+-ATPase, H^+-PPase	Carystinos *et al.* 1995
42.	*Hippeastrum sp.*	Petal	Biochem	Tonoplast	ATPase (Mg^{2+})	Wagner & Mulready 1985
43	*Hippeastrum sp.*	Petal	Biochem	Sap	RNase, DNase, acid phosphatase	Butcher *et al.* 1977
44.	*Tulipa sp.*	Petal	Biochem	Sap	α-galactosidase	Boller & Kende 1979
45.	*Tulipa sp.*	Petal	Biochem	Tonoplast	ATPase (Mg^{2+}), Pyrophosphatase (Mg^{2+})	Wagner & Mulready 1983
46.	*Tulipa sp.* *Ananas comosus*	Leaf	Biochem	Sap	Proteinase, α-galactosidase, β-N-acetylglucosamine, phosphodiesterase acid phosphatase, α-manno-sidase	Boller & Kende 1979
47.	*Allium cepa*	Scale	Cyto	Lysosome	Acid phosphatase, esterase	Walek-Czernecka 1962
48.	*Allium cepa*	Root	Cyto	Lysosome	Acid phosphatase	Novikoff 1961
49.	*Allium cepa*	Root	Cyto	Particulate	Acid phosphatase	Jensen 1956
50.	*Allium cepa*	Cambium	Cyto	Lysosome	Acid naphthol AS-BI-phosphatase, acid-β-glycerophosphatase, naphthol ASD esterase	Gahan & McLean 1969
51.	*Allium cepa*	Scale	Biochem	Sap	Allinase	Lancaster & Collin 1981
52.	*Saccharum officinarum*	Stem	Biochem	Tonoplast	GTPase and GDPase	Thom & Komer 1984
53.	*Tradescantia bracteata*	Pollen	Cyto		Lysosome Acid phosphatase, acid DNase, esterase, β-glucuronidase, aryl sulfatase	Gorska-Brylass 1965
54.	*Phleum pratense*	Root	Cyto	Particulate	Acid phosphatase, sulfatase	Avers 1961

Table 5.5 Presence of enzymes in various types of vacuoles in plant species. (Enzyme nomenclature or vacuole type has been presented as given by the respective authors. No attempt is made to standardise them. (*Continued*)

Sl. No.	Plant species	Tissue	Methods: Biochemical/ Cytochemical	Vacuole: Type/ Structure	Enzyme	References
55.	*Panicum virginatum*	Root	Cyto	Particulate	Acid phosphatase, sulfatase	Avers 1961
	DICOTS:					
56.	*Pisum sativum*	Root tip	Cyto	Lysosome	Acid phosphatase	Dyar 1950
57.	*Pisum sativum*	Root	Cyto	Sap	Peroxidase	Sexton & Hall 1978
58.	*Pisum sativum*	Root	Cyto	Tonoplast	Peroxidase	Hall & Sexton 1972
59.	*Pisum sativum*	Root	Cyto	Sap	β-glycero phosphatase	Sexton *et al.* 1971
60.	*Pisum sativum*	Cotyledon	Biochem	Aleurone grain	Endopeptidase, RNase, Phosphatase, acetylesterase, α-glucosidase, β-amylase	Matile 1968
61.	*Phaseolus vulgaris*	Leaf	Biochem	Sap	Chitinase	Boller & Vögeli 1984
62.	*Phaseolus vulgaris*	Leaf	Biochem	Sap	Chitinase, β-1,3-glucanase	Mauch & Staehelin 1989
63.	*Vicia faba*	Root meristem	Cyto	Lysosome	Acid phosphatase	Gahan 1965; Novikoff 1961
64.	*Vicia faba*	Meristem Cambium	Cyto	Lysosome	Acid naphthol ASBI phosphatase, Cambium acid β-glycerophosphatase, naphthol ASD esterase	Gahan & McLean 1969
65.	*Vicia faba*	Seed	Biochem	Aleurone grain	Endopeptidase, Phosphatase, grain Phytase	Morris *et al.* 1970
66.	*Vigna sp.*	Seed	Biochem	Aleurone	Caseolytic activity, carboxypeptidase, α-mannosidase, N-acetylglucosaminidase	Harris & Chrispeels 1975
67.	*Vigna radiata*	Cotyledon	Biochem	Protein body	α-mannosidase, N-acetyl-β-glucosaminidase, ribonuclease, acid phosphatase, phosphodiesterase, carboxypeptidase	Van der Wilden *et al.* 1980
68.	*Vigna sp.*	Leaf	Biochem	Sap	Chitinase	Boller & Vögeli 1984
69.	*Meliotus alba*	Leaf mesophyll	Biochem	Sap	Protease	Canut *et al.* 1985
70.	*Meliotus alba*	Mesophyll	Biochem	Sap	Serine proteinase, cysteine proteinase, aspartic proteinase, metalloproteinase	Canut *et al.* 1987

Table 5.5 Presence of enzymes in various types of vacuoles in plant species. (Enzyme nomenclature or vacuole type has been presented as given by the respective authors. No attempt is made to standardise them. (*Continued*)

Sl. No.	Plant species	Tissue	Methods: Biochemical/ Cytochemical	Vacuole: Type/ Structure	Enzyme	References
71.	*Glycine max*	Root nodule	Biochem	Membrane envelope	ATPase	Verma et al. 1978
72.	*Glycine max*	Hypocotyl	Biochem	Tonoplast	NADH-oxidoreductase	Barr et al. 1986
73.	*Glycine max*	Cultured cell	Cytochem	Sap & tonoplast	Peroxidase	Griffing & Fowke 1989
74.	*Glycine max*	Hypocotyl	Cytochem	Tonoplast	NADH-oxidoreductase	Morré et al. 1987
75.	*Glycine max*	Mesophyll	Immunocyto	Sap	Lipoxygenase	Tranbarger et al. 1991
76.	*Beta vulgaris*	Storage root	Biochem	Sap	Acid phosphatase, Acid invertase	Leigh et al. 1979
77.	*Beta vulgaris*	Storage root	Biochem	Tonoplast	ATPase	Leigh & Walker 1980
78.	*Beta vulgaris*	Storage root	Biochem	Tonoplast	Pyrophosphatase (Mg^{2+})	Walker & Leigh 1981
79.	*Beta vulgaris*	Storage root	Biochem	Tonoplast	ATPase, Pyrophosphatase	Leigh & Walker 1985
80.	*Nicotiana rustica*	Leaf	Biochem	Sap	Acid phosphatase, β-N-acetylglucosaminidase	Saunders 1979
81.	*Nicotiana rustica*	Leaf	Biochem	Tonoplast	α-Mannosidase	Saunders & Gillespie 1984
82.	*Nicotiana tabacum*	Cultured cell	Biochem	Sap	Nuclease, β-fructosidase (invertase), α-mannosidase, acid phosphatase, phosphodiesterase	Boller & Kende 1979
83.	*Nicotiana tabacum*	Seed endosperm	Biochem	Spherosome	Endopeptidase, RNase, DNase, phosphatase, acetylesterase, Lipase	Matile & Spichiger 1968
84.	*Nicotiana tabacum*	Leaf	Biochem	Sap	Basic peroxidase	Schoss et al. 1987
85.	*Lycopersicum esculentum*	Meristem cambium	Cyto	Lysosome	Acid naphthol AS-BI phosphatase, acid β-glycerophosphatase, naphthol AS-D esterase	Grahan & McLean 1969
86.	*Lycopersicum esculentua*	Leaf	Biochem	Sap	Acid phosphatase, carboxypeptidase	Walker-Simmons & Ryan 1977
87.	*Lycopersicum esculentum*	Cultured cell	Biochem	Sap	RNase I	Abel & Glund 1986, 1987
88.	*Lycopersicum esculentum*	Fruit immature	Cyto	Tonoplast	Peroxidase	Thomas & Jen 1980

Table 5.5 Presence of enzymes in various types of vacuoles in plant species. (Enzyme nomenclature or vacuole type has been presented as given by the respective authors. No attempt is made to standardise them. (*Continued*)

Sl. No.	Plant species	Tissue	Methods: Biochemical/ Cytochemical	Vacuole: Type/ Structure	Enzyme	References
89.	*Lycopersicum esculentum*	Fruit	Immunocyto	Sap	Carboxypeptidase	Mehta *et al.* 1996
90.	*Solanum tuberosum*	Young shoot	Biochem	Small vacuole	RNase, phosphatase, phosphodiesterase, acetylesterase	Pitt & Galpin 1973
91.	*Campanula persicifolia*	Leaf stomatal (guard cell)	Cyto	Small vacuole, spherosome	Acid phosphatase, acid β-glycerophosphatase, acid naphthol phosphatase	Sorokin & Sorokin 1968
92.	*Campanula sp. Bryonia dioica Vinca minor*	Pollen grain, pollen tube	Cyto	Lysosome	Acid phophatase, acid DNase, esterase, β-glucuronidase, aryl sulfatase	Gorska-Brylass 1965
93.	*Cucumis sativus*	Root	Cyto	Tonoplast	Pyrophosphatase	Poux 1967
94.	*Cucumis sativus*	Root	Cyto	Sap	Peroxidase	Poux 1969
95.	*Cucumis sativus Linum usitatissimum*	Root meristem	Cyto	Small vacuole	Acid phosphatase	Poux 1965
96.	*Hevea brasiliensis*	Latex	Biochem	Lutoid	Endopeptidase, RNase, DNase, phosphatase, phosphodiesterase, β-glucosidase, β-gactosidase, β-N-acetylglucosaminidase	Pujarniscle 1968
97.	*Hevea brasiliensis*	Latex	Biochem	Lutoid tonoplast	NADH-cytochrome C reductase & NADH-ferricyanide reductase	Moreau & Jacob 1975
98.	*Hevea brasiliensis*	Latex	Biochem	Tonoplast	ATPase	D'Auzac 1977
99.	*Ricinus communis*	Seed endosperm	Biochem	Sap	Acid phosphatase, acid protease, phosphodiesterase, RNase, phytase, β-glucosidase	Nishimura 1978
100.	*Ricinus communis*	Seed endosperm	Biochem	Spherosome	Lipase	Ory *et al.* 1968
101.	*Chelidonium majus*	Latex	Biochem	Lutoid	Endopeptidase, RNase, phosphatase	Matile *et al.* 1970
102.	*Cannabis sativa*	Seed ungerminated	Biochem	Aleurone grain	Endopeptidase	St. Angelo *et al.* 1969

Table 5.5 Presence of enzymes in various types of vacuoles in plant species. (Enzyme nomenclature or vacuole type has been presented as given by the respective authors. No attempt is made to standardise them.) (*Continued*)

Sl. No.	Plant species	Tissue	Methods: Biochemical/ Cytochemical	Vacuole: Type/ Structure	Enzyme	References
103.	*Crambe abyssinica*	Cotyledon immature	Cyto	Spherosome	β-glycerophosphatase	Smith 1974
104.	*Helianthus annus*	Cotyledon germinating	Biochem	Aleurone grain	Acid proteinase	Schrarrenberger *et al.* 1972
105.	*Gossypium hirsutum*	Cotyledon germinating	Biochem	Aleurone grain	Endopeptidase, phosphatase	Yatsu *et al.* 1971
106.	*Gossypium hirsutum*	Seed fibres	Electron Histo	Tonoplast	H$^+$-ATPase	Joshi *et al.* 1988
107.	*Bryophyllum daigremontianum*	Leaf	Biochem	Sap	Acid phosphatase, RNase	Buser & Matile 1977
108.	*Armoracia lapathifolia*	Root	Biochem	Sap	Acid phosphatase	Grob & Matile 1979
109.	*Pinguicula grandiflora, P. vulgaris, P. lusitanica, P. caudata*	Sessile gland cell	Cyto	Sap	Phosphatase, esterase	Heslop-Harrison & Knox 1971
110.	*Convallaria majalis*	Leaf	Biochem	Tonoplast	Glucosyltransferase	Löffelhardt & Kopp 1981
111.	*Prunus Serotina*	Stem & Leaf	Immunocyto	Sap	Prunasine hydrolase, mandelonitrile lyase	Swain & Polton 1994
112.	*Cyara cordunculus*	Stigma	Immunocyto	Sap	Aspartic proteinase	Ramalho-Santos *et al.* 1997

the peptide bond of the amino-terminal amino acid appear to be less prevalent than the *carboxypeptidases* acting on the peptide bond of the carboxy-terminal amino acid. The carboxypeptidases have been reported from the vacuoles of *Nitella*, yeast, tomato leaf, etc.

The esterases, comprising *phosphomonoesterases*, *phosphodi-esterases* and *acetylesterases*, are the most prevalent group of enzymes in vacuoles. The phosphomonoesterases, generally assayed by their action on β-glycerophosphate, are non-specific phosphatases. They are typified by acid phosphatase with pH optimum about 5, which is widely present in vacuoles and is the easiest enzyme for cytochemical demonstration *in situ*, both by light and electron microscopy. *Phytase* is also a phosphomonoesterase which hydrolyses myoinositol hexa-phosphate and is reported in a number of aleurone grain type vacuoles. Phosphodiesterases, which include the *nucleases*, cleave the ester bonds of phosphate linking two sugar residues, and are widely distributed in vacuolar sap. The two types of nucleases, *exonucleases* and *endonucleases*, are found in all groups of plants. Acetylesterases, which cleave esters of carboxylic acids, are less widespread than the other two classes of esterases. *Lipases*, which belong to this class, are recorded in storage vacuoles of oily seeds. *Arylsulfatases* are generally found in vacuoles of fungi.

The *glycosidases* are involved in breakdown of di- and oligo-saccharides, various structural and reserve polysaccharides, as well as a number of glycosides. They are widespread in vacuoles of all groups of plants. Various storage vacuoles invariably contain glycosidases. The vacuoles in fungi and cells with certain storage material, like pollen grains, lutoids, etc., show an abundance of endo- and exo-glycosidases.

As well as the hydrolytic enzymes, vacuolar sap also contains non-hydrolytic cleaving enzymes known as *lyases*.

5.4 Summary

5.4.1 Solids and particulate inclusions

The lumen of the vacuole may cantain diverse inorganic, organic or conjugated solid substances. Crystals of these substances appear as raphide, sphaeraphide, cystolith, statolith, etc., inside the vacuole.

- Crystalline or amorphous solid proteinaceous bodies known as aleurone grains are present in storage cells of many seeds, embryos, cotyledons and endosperms. Aleurone grains are diverse in chemical composition and may contain phytin, a polyphosphate, as well as large amounts of protein, for example, albumin, globulin, prolamin and/or glutelin. Moreover, specific proteins like zein in maize or hordein in barley may form the bulk of the aleurone grain. Vacuoles may also store lipoxygenase, chymotrypsin inhibitor or proteinase inhibitor proteins and ribosome-inactivating proteins like ricin and saporin.
- Although starch is not stored, certain reserve carbohydrates like inulin and fructans are stored in vacuoles. Many vacuoles contain **lipids**, fats and oils. Apart from rubber, it is doubtful that isoprene derivatives like resin and essential oil are stored in vacuoles. Phytate and other polyphosphates like volutin are found in the vacuoles of some cells.

5.4.2 Colloidal and dissolved substances

- Diverse phenolic compounds like tannin and anthocyanin are found in the vacuoles of many plants. Water soluble anthocyanins and anthoxanthins are extensively present in vacuole sap to produce colouration of leaves, floral parts, etc.
- Most of the alkaloids are known to be accumulated in vacuoles. Commercially important alkaloids like terpenoid indole alkaloids, cardenolides and morphine are stored in vacuoles.
- **Sugars** in free or conjugated forms are widely found in vacuoles, especially in storage parenchyma of sugar cane, beetroot and ripening fruit cells. But in many cases, sucrose concentration in the cytosol is much higher than in the vacuole.

- **Organic acids**, mainly the dicarboxylic and tricarboxilic acids, are present in vacuolar sap. Oxalic, ascorbic, malic, citric and isocitric acids are exclusively located in the lumen. In CAM plants malic acid is stored in the mesophyll vacuole at night at a very high concentration, whereas isocitric acid is maintained at a constant level throughout the day.
- Although **amino acids** are known to occur in most vacuoles, their concentration varies widely. In many cases it may be extremely low compared to the cytosolic concentration. However, fungal cells generally show a higher concentration of amino acids in the vacuole, a possible reserve pool of soluble nitrogen.
- Free **hormones** are not known to be present in vacuoles. However, conjugates of certain hormones may occur in certain vacuoles.
- Vacuolar sap houses numerous **inorganic ions**, including potassium, sodium, calcium, magnesium, chloride, sulfate, nitrate and phosphate, in concentrations that are sometimes much higher than the cytoplasm, apoplast or other surrounding medium. This difference in ion concentration may vary between ions in the same vacuole, depending on the metabolic and osmotic requirements of the cell.
- A number of **miscellaneous** substances such as lectins, saponin and anthraquinones also occur in vacuoles. The autophagic process and the presence of hydrolytic enzymes in the vacuole leads to a number of break-down products of chlorophyll, nucleic acids, membrane lipids, etc., especially in senescent cells.

5.4.3 Enzymes

As has been discussed in Chapter IV, the tonoplast invariably contains V-ATPase and frequently PPase, along with other enzymes and carriers for transport of diverse ions and molecules across the tonoplast. Of six broad categories of enzymes, vacuolar sap contains only the **hydrolysing enzymes**. A **master table** (Table 5.5) presents the distribution of various enzymes from more than 100 publications on all groups of the plant kingdom. All the three main groups of hydrolases, viz. peptidases including endopeptidases and exopeptidases, esterases including phosphomono-, phosphodi- and acetyl-estarases and the glucosidases, are represented in diverse vacuoles.

5.5 References

Abel, S. and Glund, K., Localization of RNA-degrading enzyme activity within vacuoles of cultured plant cells, *Physiol. Plant.*, **66**, 79, 1986.

Abel, S. and Glund, K., Ribonuclease in plant vacuoles: Purification and molecular properties of the enzyme from cultured tomato cells, *Planta*, **172**, 71, 1987.

Adams, C.A. and Novellie, L., Acid hydrolases and autolases and autolytic properties of protein bodies and spherosomes isolated from ungerminated seeds of *Sorghum bicolor* (Linn.) Moench., *Plant Physiol.*, **55**, 7, 1975.

Alibert, G., Boudet, A.M., Canut, H. and Rataboul, P., Protoplast in studies of vacuolar storage compounds, in 'The Physiological Properties of Plant Protoplasts', Ed. Pilet, P.P., Springer Verlag, Berlin, 1985, 105.

Anderson, L.A., Phillipson, J.D. and Roberts, M.F., Biosynthesis of secondary products by cell cultures of higher plants, in 'Plant Cell Culture: Advances in Biochemical Engineering/Biotechnology Vol. 31', Ed. Fletcher, A., Springer Verlag, Berlin, 1985, Chap. 1.

Anhalt, S. and Wissenböck, G., Subcellular localization of luteolin glucuronides and related enzymes in rye mesophyll, *Plant*, **187**, 83, 1992.

Armentrout, V.N., Smith, G.G. and Wilson, C.L., Spherosomes and mitochondria in the living fungal cell. *Amer. J. Bot.*, **55**, 1062, 1968.

Ashford, A.E. and Jacobson, J.V., Cytochemical localization of phosphatase in barley aleurone tissue: the pathway of gibberellic acid induced enzyme release, *Planta*, **120**, 81, 1974.

Ashton, F., Mobilization of Storage Proteins of Seeds., *Ann. Rev. Plant Physiol.*, **27**, 95, 1976.

Ashwoth, J.M. and Weiner, E., The lysosomes of the cellular slime mould *Dictyostellium discoideum*, in 'Frontiers of Biology, 29, Lysosomes in Biology and Pathology, Vol. 3', Ed. Dingle, J.T., North-Holland, Amsterdam, 1973, Chap. 3.

Avers, C.J., Histochemical localization of enzyme activities in root meristem cells, *Amer. J. Bot*, **48**, 137, 1961.

Barr, R., Sandelius, A.S., Crane, F.L. and Morré, D.J., Oxidation of reduced pyridine nucleotides by plasma membranes of soybean hypocotyl, *Biochem. Biophys. Res. Commun.*, **131**, 943, 1986.

Baur, P. S. and Walkinshaw, C.H., Fine structure of tannin accumulations in callus cultures of *Pinus elliotti.*, *Can. J. Bot.* **52**, 615, 1974.

Bechtel, D . B., Wilson, J. D. and Shewry, P. R., Immunocytochemical localisation of the wheat storage protein triticin in developing endosperm tissue, *Cereal Chem.* **68**, 573, 1991.

Bennum, A., and Blum, J.J., Properties of the induced acid phosphatase and of the constitutive acid phosphatase of *Euglena*, *Biochem. Biophys. Acta*, **128**, 106, 1966.

Böhm, H, The formation of secondary metabolites in plant tissue and cell cultures, *Int. Rev. Cytol. Suppl.* **11B**, 183, 1980.

Bold, H.C. and Wynne, M.J., Introduction to the Algae, Prentice Hall, Englewood Cliffs, 1978.

Boller, T. Die Arginin-Permease der Hefevakuole, Ph.D. thesis No. 5928, Swiss. Fed. Inst. Technol., Zurich, 224, 1977.

Boller, T, and Kende, H., Hydrolytic enzymes in the central vacuole of plant cells, *Plant Physiol.*, **63**, 1123, 1979.

Boller, T. and Vögeli, U., Vacuolar localization of ethylene induced chitinase in bean leaves, *Plant Physiol.*, **74**, 442, 1984.

Boudet, A.M., Canut, H. and Alibert, G., Isolation and characterization of vacuoles from *Melilotus alba* mesophyll, *Plant Physiol*, **68**, 1354, 1981.

Bowman, E.J. and Bowman, B.J., Identification and properties of an ATPase in the vacuolar membranes of *Neurospora crassa*, *J. Bacteriol*, **151**, 1326, 1982.

Brandes, D. and Bertini, F., Role of golgi apparatus in the formation of cytolysosomes, *Exptl. Cell Res.*, **35**, 194, 1964.

Bray, E.A. and Zeevaart, A.D., The compartmentation of abscisic acid and ß-D-glucopyranosyl abscisate in mesophyll cells, *Plant Physiol*, **79**, 719, 1985.

Buser, Ch. and Matile, P., Malic acid in vacuoles isolated from *Bryophyllum* leaf cells., *Z. Pflanzenphysiol.*, **82**, 462, 1977.

Butcher, H.C., Wagner, G.J. and Siegelman, H.W., Localization of acid hydrolases in protoplasts: Examination of the proposed lysosomal function of mature vacuole. *Plant Physiol.*, **59**, 1098, 1977.

Callis, J. Regulation of protein degradation, *Plant cell*, **7**, 845, 1995.

Canut, H., Dupre, M., Carrasco, A. and Boudet, A.M., Proteases of *Melilotus alba* mesophyll protoplasts, *Planta*, **170**, 541, 1987.

Carystinos, G.D., MacDonald, H.R., Monroy, A.F., Dhindsa, R.S. and Poole, R.J., Vacuolar H^+-translocation pyrophosphatase is induced by anoxia or chilling in seedlings of rice, *Plant Physiol.*, **108**, 614, 1995.

Carzaniga, R., Sinclair, L., Fordham-Skelton, A.P., Harris, N. and Croy, R.R.D., Cellular and subcellular distribution of saporins, type-1 ribosome-inactivating proteins in soapwort (*Saponaria officinallis* L.), *Planta*, **194**, 461, 1994.

Chafe, S. C. and Durzan, D.J., Tannin inclusions in cell suspension cultures of white spruce, *Planta*, **113**, 251, 1973.

Ching, T.M., Intracellular distribution of lipolytic acvtivity with female gametophyte of germinating Douglas fir seeds. *Lipids*, **3**, 482, 1968.

Chrispeels, M.J., Biosynthesis, processing and transport of storage proteins and lectins in legume cotyledons. *Phil. Trans. R.Soc. Lond.*, **B304**, 309, 1984.

Coulomb, P., Phytolysosomes dans les frondes d'*Asplenium fontanum* (Filicinées, Polypodiacées). Isolement sur gradient dosages de quelques hydrolases et contròl des culots obtenus, par la microscopie electronique, *J. Microsc.* (Paris), **11**, 299, 1971.

D'Auzac, J., ATPase membranaire de vacuoles lysosomales : les lutoids du latex d'*Hevea brasiliensis*, *Phytochemistry*, **16**, 1881, 1977.

Dangeard, P.A., Sur les corpuscles métachromatiques des Levures, *Bull. Soc. Myc.*, France, **32**, 27, 1916.

Deus-Newmann, B. and Zenk, M.H., A highly selective alkaloid uptake system in vacuoles of higher plants, *Planta*, **162**, 250, 1984.

Deus-Newmann, B. and Zenk, M.H., Accumulation of alkaloids in plant vacuoles do not involve an ion-trap mechanism, *Planta*, **166**, 121, 1985.

Diaz de Leon, J.L. and Wyn Jones, R.G., Ca^{2+}-ATPase and their occurrence in vacuoles in higher plants in 'Biochemistry and Function of Vacuolar Adenosine triphosphatase in Fungi and Plants', Ed. Marin, B.P., Springer Verlag, Berlin, 1985, 57.

Dickenson, P.B., Electron microscopical studies of the latex vessel system of *Hevea brasiliensis*, *J. Rubber Res. Inst.* (Malaya) **21**, 543, 1969.

Diers, L.F., Schoetz, F. and Meyer, B., Über die Ausbildung von Gerbstaff Vacuolen bei Oenotheren., *Cytobiol*, **7**, 10, 1973.

Dietz, K.J., Jäger, R., Kaiser, G. and Martinoia, E., Amino acid transport across the tonoplast of vacuoles isolated from barley mesophyll cells. *Plant Physiol.*, **92**, 123, 1990.

Doi, E., Ohtsuri, C. and Matoba, T., Lysosomal enzyme activities in the central vacuole of the internodal cells of *Nitella*, *Plant Sci. Lett.*, **4**, 243, 1975.

Domozych, D.S., The endomembrane system and mechanism of membrane flow in the green alga, *Gleomonas kupfferi* (Volvocales, Chlorophyta) II. A cytochemical analysis, *Protoplasma*, **149**, 108, 1989.

Duggelin, T., Schellenberg, M., Borttik, K., and Matile, P., Vacuolar location of lipofuscin- and proline-like compounds in senescent barley leaves, *J. Pl. Physiol.*, **133**, 492, 1988.

Dyar, M.T., Some observations on starch synthesis in pea root tips, *Amer. J. Bot*, **37**, 786, 1950.

Esen, A. and Stetler, D.A., Immunocytochemical location of gamma-zein in the protein bodies of maize endosperm,. *Amer. J. Bot.* 79,243,1992

Fahn, A. and Benayoun, J., Ultrastructure of resin ducts in *Pinus halepensis* : Development, possible sites of resin synthesis and mode of its elimination from the protoplast, *Ann. Bot.* **40**, 857, 1976.

Fahn, A. and Evert, R.F., Ultrastructure of the secretory ducts of *Rhus glabra* L., *Amer. J. Bot.*, **61**, 1, 1974.

Faye, L. and Chrispeels, M.J., Transport and processing of the glycosylated precursor of concanavalin A in jack bean, *Planta*, **170**, 217, 1987.

Faye, L., Greenwood, J.S., Herman, E.M., Sturm, A. and Chrispeels, M.J., Transport and post-translational processing of the vacuolar enzyme α-mannosidase in jack bean cotyledons, *Planta*, **174**, 271, 1988.

Feldman, G., Sur le ultrastructure descorps irisants des Chondria (Rhodophyceae). *C.R. Acad. Sci.*, Paris, **270**, 1244, 1970.

Fineran, B.A., Differentiation of non-articulated laticifers in Poinsettia (*Euphorbia pulcherrima* Wild.), *Ann. Bot.* **52**, 279, 1983.

Foster, J.G., Cress, W.D., Wright, S.F. and Hess, J.L., Intracellular localization of neurotoxin 2,4-diaminobutyric acid in *Lathyrus sylvestris* L. leaf tissue, *Plant Physiol.*, **83**, 900, 1987.

Frey-Wyssling, A., Submicroscopic Morphology of Protoplasm, Elsevier Pub. Co., Amsterdam, 1953, 194.

Fricke, W., Leigh, R.A. and Thoma, A.D., Concentrations of inorganic and organic solutes in extracts from individual epidermal, mesophyll and bundle sheath cells of barley leaves, *Planta*, **192**, 310, 1994.

Fuesseder, A. and Ziegler, P., Metabolism and compartmentation of dihydrozeatin exogenously supplied to photoautotrophic suspension cultures of *Chenopodium rubrum, Planta*, **173**, 104, 1988.

Fujiwake, H., Suzuki, T. and Iwai, K., Intracellular localization of capsaicin and its analogues in *Capsicum* fruit, II The vacuole as the intracellular accumulation site of capsaicinoid in the protoplast of *Capsicum* fruit, *Plant Cell Physiol*, **21**, 1023, 1980.

Gahan, P.B., Histochemical evidence for the presence of lysosome-like particles in root meristem cells of *Vicia faba, J. Exptl. Botany*, **16**, 350, 1965.

Gahan, P.B. and McLean, J., Subcellular localization and possible functions of acid glycerophosphatases and naphthol esterases in plant cells. *Planta*, **89**, 126, 1969.

Gangulee, H.C., Das, K.S. and Datta, C., College Botany, Vol.1., Central Book Agency, Calcutta, India, 1996, 301.

Garcia-Martinez, J.L., Ohlrogge, J.B. and Rappaport, L., Differential compartmentation of gibberellin A1 and its metabolites in vacuoles of cowpea and barley leaves. *Plant Physiol.*, **68**, 865, 1981.

Gorska-Brylass,A., Hydrolases in pollen grains and pollen tubes, *Acta, Soc. Botan. Polon.*, **34**, 589, 1965.

Graham, T.A. and Gunning, B.E.S., The localization of legumin and vicillin in bean cotyledon cells using fluorescent antibodies, *Nature*, **228**, 81, 1970.

Griffing, L.R. and Fowke, L.C., Cytochemical localization of peroxidase in soybean suspension culture cells and protoplasts: intracellular vacuole differentiation and presence of peroxidase in coated vesicles and multivesicular bodies, *Protoplasma*, **128**, 22, 1989.

Grob, K. and Matile, P., Vacuolar location of glucosinolates in horseradish root cells, *Plant Sci. Lett.*, **14**, 317, 1979.

Grob, K. and Matile, P., Compartmentation of ascorbic acid in vacuoles of horseradish root cells: Note on vacuolar peroxidase, *Z. Pflanzenphysiol.*, **98**, 235, 1980.

Groeneveld, H.W., Biosynthesis of *Latex triterpenes* in Euphorbia: evidence for a dual synthesis. *Acta. Bot. Neerlandica*, **25**, 459, 1976.

Gruhnert, C., Biehl, B. and Selmar, P. Compartmentation of cyanogenic glucosides and their degrading enzymes, *Planta*, **195**, 36, 1994.

Guilliermond, A., The Cytoplasm of Plant Cell, *Chronica Botanica*, USA, 1941, 72.

Gutknecht, J., Permeability in *Valonia* to water solutes: apparent absence of aqueous membrane pores, *Biochem. Biophys. Acta.*, **163**, 20, 1968.

Guy, M. and Kende, H., Conversion of l-aminocyclopropane-1- carboxylic acid to ethylene by isolated vacuoles of *Pisum sativum* L., *Planta*, **160**, 281, 1984.

Hall, J.L. and Sexton, R., Cytochemical localization of peroxidase activity in root cells, *Planta*, **108**, 103, 1972.

Hara-Nishimura, I., Nishimura, M. and Akazawa, T., Biosynthesis and intracellular transport of 11 S globulin in developing pumpkin cotyledons, *Plant Physiol.*, **77**, 747, 1985.

Hara-Nishimura, I., Hayashi, M., Nishimura, M. and Akazawa, T., Biogenesis of protein bodies by budding from vacuoles in developing pumpkin cotyledons, *Protoplasma*, **136**, 49, 1987.

Hara-Nishimura, I., Takeuchi, Y., Inoue, K. and Nishimura, M., Vesicle transport and processing of the precursor to 2S albumin in pumpkin, *Plant J.* **4**, 793, 1993.

Harris, N. and Chrispeels, M.J., Histochemical and biochemical observations on storage, protein metabolism and protein body autolysis in cotyledons of germinating mung bean, *Plant Physiol.*, **56**, 292, 1975.

Harris, N., Henderson, J., Abbot, S.J., Mulcrone, J. and Davies, J.T., Seed development and structure, in 'Seed Storage Compounds', Eds Shewry, P.R. and Stobart, A.K. Oxford Sci. Pub, Oxford, 1993, 3.

Hartung, W., Gimmler, H. and Heilmann, B., The compartmentation of abscisic acid (ABA), of ABA-biosynthesis, ABA-metabolism and ABA-conjugation, in 'Plant Growth Substances', 1982, Ed. Wareing, P.F., Academic Press, London, 1982, 325.

Hébant, C., Ontogenie de laticiféres du systéme primaire de l' *Hevea brasiliensis*: une étude ultrastructural et cytochimique, *Canad. J. Botany*, **59**, 974, 1981.

Heck, U., Martinoia, E. and Matile, P., Subcellular localization of acid proteinase in barley mesophyll protoplasts, *Planta*, **151**, 198, 1981.

Heineke, D., Wildlenberger, K., Sonnewald, U., Willmitzer, L. and Heldt, H. W., Accumulation of hexoses in leaf vacuoles: studies with transgenic tobacco plants expressing yeast-derived invertase with cytosol, vacuole or apoplasm, *Planta*, **194**, 29, 1994.

Herman, E.M., Hankins, C.N. and Shannon, L.M., Bark and leaf lectins of *Sophora japonica* are squestered in protein-storage vacuoles, *Plant Physiol.*, **86**, 1027, 1988.

Heslop-Harrison, Y. and Knox, R.B., A cytochemical study of the leaf gland enzymes of insectivorous plants of the genus Pinguicula, *Planta*, **96**, 183, 1971.

Hislop, E.C., Barnaby, V.M., Shellis, C. and Laborda, F., Localization of a-L-arabinofuranosidase and acid phosphatase in mycelium of *Sclerotinia fructigena*, *J. Gen. Microbiol.*, **81**, 79, 1974.

Hoagland, D.R. and Davis, A.R., The composition of cell sap in relation to absorption of ions, *J. Gen. Physiol*, **5**, 629, 1923.

Hopp, W., Hinderer, W., Peterson, M. and Seitz, H.U., Anthocynin-containing vacuoles isolated from protoplasts of *Daucus carota* cell cultures, in 'The Physiological Properties of Plant Protoplasts', Ed. Pilet, P.E., Springer, Berlin, Heidelberg, New York, Tokyo, 1985, 122.

Iten, W. and Matile, P., Role of chitinase and other lysosomal enzymes of *Coprinus lagopus* in the autolysis of fruiting bodies, *J. Gen. Microbiol.*, **61**, 301, 1970.

Jensen, W.A., The cytochemical localisation of acid phosphatase in root tip cells, *Amer. J. Bot.*, **43**, 50, 1956.

Joel, D.M. and Fahn, A., Ultrastructure of the resin ducts of *Mangifera indica* L. (Anacardiaceae). I. Differentiation and senescence of the shoot ducts., *Ann. Bot.*, **46**, 225, 1980.

Jordanov, J., Grigorov, I. and Bojadzieva-Michaijlova, A., Über die Zytochemie der Polyphosphate bei *Aspergillus oryzae.*, *Acta. Histochem.* **13**, 165, 1962.

Joshi, P.A., Stewart, J. Mc.D. and Graham, E.T., Ultrastructural localization of ATPase activity in cotton fiber development during elongation, *Protoplasma*, **143**, 1, 1988.

Kakinuma, Y., Oshumi, Y. and Anraku, Y., Propeties of H^+ translocating adenosine triphosphatase in vacuolar membranes of *Saccharomyces cerevisiae*, *J. Biol. Chem.*, **256**, 10859, 1981.

Keller, F. and Matil, P., The role of vacuole in storage and mobilization of stachyose in tubers of *Stachys siebolii*, *J. Plant Physiol*, **119**, 369, 1985.

Keller, F. and Wiemken, A., Differential compartmentation of sucrose and gentianose in the cytosol and vacuoles in storage root protoplasts from *Gentiana lutea* L., *Plant Cell Res.*, **1**, 274, 1982.

Keller, F., Schellenberg, M. and Wiemken, A., Localization of trehalase in vacuoles and cytosol of yeast, *Saccharomyces cerevisiae*, Arch, *Microbiol*, **131**, 298, 1982.

Kenyon, W. H., Kringstad, R. and Black, C. C., Diurnal changes in the malic acid content of vacuoles isolated from leaves of the crassulacean acid metabolism plant, *Sedum telephium*, *FEBS. Lett.* **94**, 281, 1979.

Kinzel, H., Forms and fractions of vacuolar calcium, *Proc. XIV. Int. Bot. Congress.* (Abstr.) **19**, 1987.

Kojima, M., Poulton, J.E., Thayer, S.S. and Conn, E.E., Tissue distributions of dhurrin and of enzymes involved in its metabolism in leaves of Sorghum bicolor, *Plant Physiol.*, **63**, 1022, 1979.

Lancaster, J.E. and Collin, H.A., Presence of allinasae in isolated vacuoles and alkyl cysteine sulphoxides in the cytoplasm of bulbs of onion (*Allium cepa*), *Plant Sci. Lett.* **22**, 169, 1981.

Lehmann, H. and Glund, K., Abscisic acid metabolism- Vacuolar/extravacuolar distribution of metabolites, *Planta*, **168**, 559, 1986.

Leigh, R.A. and Walker, R.R., ATPase and acid phosphatase activities associated with vacuoles isolated from storage roots of red beet *Beta vulgaris*, *Planta*, **150**, 222, 1980.

Leigh, R.A. and Walker, R.R., Salt stimulated ATPase and PPase activities associated with vacuoles from higher plants, in 'Biochemistry and Function of Vacuolar Adenosine Triphosphatase in Fungi and Plants', Ed. Marin, B.P., Springer Verlag, Berlin, 1985, 45.

Leigh, R.A., Rees, T., Fuller, W.A. and Banfield, J., The location of acid invertase activity and sucrose in the vacuoles of storage roots of beetroot (*Beta vulgaris*), *Biochem J.*, **178**, 539, 1979.

Leinhos, V., Krauss, G.J. and Glund, K., Evidence that a part of cellular uridine of a tomato (*Lycopersicum esculentum*) cell suspension culture is located in the vacuole, *Plant Sci.*, **47**, 15, 1986.

Lending, C.R., Chesnut, R.S., Shaw, K.L. and Larkin, B.A. Immunolocalisation of avenin and globulin storage proteins in developing endosperm of *Avena sativa* L., *Planta*, **178**, 315, 1989.

Lending, C.R., Chesnut, R.S., Shaw, K.L. and Larkins, B.A., Synthesis of zeins and their potential for amino acid modification, in 'Plant Protein Engineering', Eds Shewry, P.R. and Gutteridge S., Cambridge Uni. Press, Cambridge, 1992, 209.

Lin, W., and Wittenbach, V.A., Subcellular localization of protease in wheat and corn mesophyll protoplasts, *Plant Physiol.*, **67**, 969, 1981.

Lin, W., Wagner, G.J., Siegelman, H.W. and Hind, G., Membrane-bound ATPase of intact vacuoles and tonoplasts isolated from mature plant tissue, *Biochem. Biophys. Acta*, **464**, 110, 1977.

Löffelhardt, W., Kopp, B. and Kubellea, W., Intracellular distribution of cardiac glycosides in leaves of *Convallaria majalis*. *Phytochemistry*, **18**, 1289, 1979.

Löffelhardt, W. and Kopp, B., Subcellular localization of glucosyl transferases involved in cardiac glycoside glucosylation in leaves of *Convallaria majalis*, *Phytochemistry*, **20**, 1219, 1981.

Loomis, W.D. and Croteau, R., Biochemistry and physiology of lower terpenoids, in 'Terpenoids: Structure, Biogenesis and Distribution, Recent Advances in Biochemistry', Eds Runeckles, V.C. and Mabry, T.J., Academic Press, New York and London, Vol 6, 1973.

Lui, N.S.T. and Altschul, A.M., Isolation of globoides from cotton seed aleurone grain, *Arch. Biochem. Biophys.*, **121**, 678, 1967.

Lui, N.S.T., Roels, O.A., Trout, M.E. and Anderson, O.R., Subcellular distribution of enzymes in *Ochromonas malhamensis*, *J. Protozool*, **15**, 536, 1968.

Luscher, A. and Matile, P., Studies on the localization of RNase and other enzymes in *Acetabularia*, *Planta*, **118**, 323, 1974.

MacLennan, D. H., Beevers, H. and Harley, J. L., Compartmentation of acids in plant tissues, *Biochem. J.*, **89**, 316, 1963.

MacRobbie, E.A.C., Fluxes and compartmentation in plant cells, *Ann. Rev. Plant Physiol.* **22**, 75, 1971.

Maitra, S.C. and De, D.N., Ultrastructure of root cap cells: formation and utilization of lipid, *Cytobios*, **5**, 111, 1972.

Martinoia, E., Heck, U. and Wiemken, A., Vacuoles as storage compartments for nitrate in barley leaves, *Nature*, **289**, 292, 1981.

Marty, F., Vesicules autophagiques des laticiferes differenciés d'*Euphorbia characias* L., *Comp. Rend. Acad. Sci (Paris)* **272**, 399, 1971.

Matile, P. Enzyme den Vakuolen aus Wurzelzellen von Meiskeimlingen, Ein beitrag zur funktionellen Bedeutung der Vakuole beider intrazellularen, Verdauung. *Z. Naturfrosch.*, **B21**, 871, 1966.

Matile, P., Lysosomes of root tip cells in corn seedlings, *Planta*, **79**, 181, 1968a.

Matile, P., Aleurone vacuoles as lysosomes, *Z. Pflanzenphysiol.*, **58**, 365, 1968b.

Matile, P., Vacuoles, lysosomes of *Neurospora*, *Cytobiologie*, **3**, 324, 1971.

Matile, P., Vacuoles, in Plant Biochemistry, 3rd Ed. Eds Bonner, J. and Varner, J. E. Academic Press, New York, 1976, Chap 8.

Matile, P., Biochemistry and function of vacuoles, *Ann. Rev. Plant Physiol.*, **29**, 193, 1978.

Matile, P. and Spichiger, J., Lysosomal enzymes in spherosomes (oil droplets) of tobacco endosperm, *Z. Pflanzenphysiol.*, **58**, 277, 1968.

Matile, P., Jans, B. and Rikhenbacher, R., Vacuoles of *Chelidonium latex*: lysosomal property and accumulation of alkaloids, *Biochem. Physiol. Pflanzen.*, **161**, 447, 1970.

Mauch, F. and Staehelin, L.A., Functional implications of the subcellular localisation of ethylene-induced chitinase and ß-1,3- glucanase in bean leaves, *Plant Cell*, **1**, 447, 1989.

Meeuse, B.J.D., Free sulfuric acid in the brown alga, *Desmarestia*, *Biochem. Biophys. Acta*, **19**, 372, 1956.

Meeuse, B.J.D., Storage products, in 'Physiology and Biochemistry of Algae', Ed. Lewin, R.A. Academic Press, New York, 1962, 289.

Mehta, R.A., Warmbardt, R.D. and Mattoo, A.K., Tomato fruit carboxypeptidase: properties, induction upon wounding and immunocytochemical localisation, *Plant Physiol.*, **110**, 883, 1996.

Moreau, F., Jacob, J.L., Dupont, J. and Lance, C., Electron transport in the membrane of lutoids from the latex of *Hevea brasiliensis*, *Biochem. Biophys. Acta*, **396**, 116, 1975.

Morikawa, H., Hayashi, Y., Hirabyashi, Y., Asada, M. and Yamada, Y., Cellular and vacuolar fusion of protoplasts electrofused using platinum microelectrodes., *Plant Cell Physiol*, **29**, 189, 1988

Morré, D.J., Audeset, G., Penel, C. and Canut, H., Cytochemical localization of NADH-Ferricyanide oxido- reductase in hypocotyl segments and isolated membrane vesicles of soybean, *Protoplasma*, **140**, 133, 1987.

Morris, G.F.I., Thurman, D.A. and Boulter, D., The extraction and chemical composition of aleurone grains (protein bodies) isolated from seeds of *Vicia faba*, *Phytochemistry*, **9**, 1707, 1970.

Mott, R.L. and Steward, F.C., Solute accumulation in plant cells: I. Reciprocal relations between electrolytes and non electrolytes, *Ann. Bot.* **36**, 915, 1972.

Neumann, D. and Müller, E., Intrazellulärer Nachweis von Alkaloiden in Pflanzenzellen in licht und electronmikroskopischen Masstab, *Flora (Jena) Abt. A.*, **158**, 479, 1967.

Newcomb, W. and Wood, S. M., Morphogenesis and fine structure of *Frankia* (Actinomycetes): the microsymbiont of nitrogen fixing actinorhizal root nodules, *Int. Rev. Cytol.*, **109**, 1, 1987.

Nishimura, M. and Beevers, H., Hydrolases in vacuoles from castor bean endosperm. *Plant Physiol.*, **62**, 44, 1978.

Novikoff, A.B. and Goldfischer, S., Nucleoside diphosphatase activity in the golgi apparatus and its usefulness for cytological studies, *Proc. Natl. Acad. Sci. US.*, **47**, 802, 1961.

O'Day, D.H., Intracellular localization and extracellular release of certain lysosomal enzyme activities from amoebae of the cellular slime mould *Polysphondylium pallidum*, *Cytobios*, **7**, 223, 1973.

Oba, K., Conn, E.E., Canut, H. and Boudet, A.M., Subcellular localisation of 2-(ß-D-glucosyloxy)-cinammic acids and the related ß-glucosidase in leaves of *Melilotus alba* Dest., *Plant Physiol.* **68**, 1359, 1981.

Ohlrogge, J.B., Garcia-Martinez, D., Adams, D. and Rappaport, L., Uptake and subcellular compartmentation of gibberellin A_1 applied to leaves of barley and cowpea, *Plant Physiol.*, **66**, 422, 1980.

Ory, R.L. and Heningsen, K.W., Enzymes associated with protein bodies isolated from ungerminated barley seeds, *Plant Physiol.*, **44**, 1488, 1969.

Ory, R.L., Yatsu, L.Y. and Kircher, H.W., Association of lipase activity with the spherosomes of *Ricinus communis*, *Arch. Biochem. Biophys.*, **123**, 255, 1968.

Osborne, T.B., The Vegetable Proteins, Longmans Green, London, UK, 1924.

Osterhout, W.J.V., The absorption of electrolytes in large plant cells, *Bot. Rev.*, **2**, 283, 1936.

Parish, R.W., The lysosome concept in plants I. Peroxidases associated with subcellular and wall fractions of maize root tips: implications for vacuole development, *Planta*, **123**, 1, 1975.

Pick, U. and Weiss, M., Polyphosphate hydrolysis within acidic vacuoles in response to amine-induced alkaline stress in the halotolerant alga *Dunaliella salina*. *Plant Physiol.*, **97**, 1234, 1991.

Pick, U., Zeelon, O. and Weiss, M., Amine accumulation in acidic vacuoles protects the halotolerant alga *Dunaliella satina* against alkaline stress, *Plant Physiol.*, **97**, 1226, 1991.

Pitt, D., Histochemical demonstration of certain hydrolytic enzymes within cytoplasmic particles of *Botrytis cinerea* Fr., *J. Gen. Microbiol.*, **52**, 67, 1968.

Pitt, D. and Galpin, M., Isolation and properties of lysosomes from dark-grown potato shoots, *Planta*, **109**, 233, 1973.

Poux, N., Localization de l'activite' phosphatasique acide et des posphates dans les grains d' Aleurone. I. Grains d'Al. renferment a las fois globoides et cryslalloides, *J. Microscopie* (Paris), **4**, 771, 1965.

Poux, N., Localization d'activités enzymatiques dans les cellules du mèristéme radiculaire de *Cucumis sativus*, L., *J. Microsc* (Paris) **2**, 485, 1967.

Poux, N., Localization d'activités enzymatiques dans les cellules du mèristéme radiculaie de *Cucumis sativus* L. II. Activité proxidasique, *J. Microscopie*, **8**, 855, 1969.

Preisser, J., Sprügel, H. and Komor, E., Soluble distribution between vacuole and cytosol of sugarcane suspension cells: Sucrose is not accumulated in the vacuole, *Planta*, **186**, 203, 1992.

Pujarniscle, S., Caractère lysosomal des lutoides du latex d'*Hevea brasiliensis*, Mull. Arg., *Physiol. Veget.*, **6**, 27, 1968.

Pujarniscle, S., Étude biochimique des lutoides du latex *d'Hevea brasiliensis*. Mull. Arg. differences et analogies avec les lysosomes. Memoire Office de la Recherche Scientifique et Technique Outre-Mer., **48**, 100, 1971.

Ramalho-Santos, M., Pisarra, J., Verissimo, P., Pereira, S., Salema, R., Pires. E. and Faro, C.J., Cardosin A, An abundant aspartic proteinase, acuumulates in protein storage vacuoles in the stigmatic papillae of *Cyara cardunculus* L., *Planta*, 203, 204, 1997.

Rechinger, K.B., Simpson, D.J., Svendson, I. and Cameron-Mills, V., A role for γ-3 hordein in the transport and trageting of prolamin peptides to the vacuole of developing barley endosperm, *Plant J.*, **4**, 841, 1993.

Ribaillier, D., Jacob, J.L. and D'Auzac, J., Sur certaines caractéres vacuolaires des lutoids du latex d'*Hevea brasiliensis*, Mull. Agr., *Physiol. Vég.* **9**, 423, 1971.

Riens, B., Lohaus, G., Heineke, D. and Heldt, H.W., Amino acid and sucrose content determined in the cytosolic, chloroplastic and vacuolar compartment and in the phloem sap of spinach leaves, *Plant Physiol.* **97**, 227, 1991.

Roomans, G.M. and Seveus, L.A., Subcellular localization of diffusible ions in the yeast *Saccharomyces cerevisiae*: quantitative microprobe analysis of thin freeze-dried sections. *J. Cell Sci.* **21**, 119, 1976.

Roy, A. T. and De, D. N., Studies on differentiation of laticifers through light and electron microscopy in *Calotropis gigantea* (Linn.) R.Br., *Ann. Bot.* **70**, 443, 1992.

Runeberg-Roos, P., Kervinen, J., Kovaleva, V., Raikhel, N.V. and Gal, S., The aspartic proteinase of barley is a vacuolar enzyme that processes probarley lectin in vitro, *Plant Physiol.*, **105**, 321, 1994.

Ryan, C.A. and Shumway, L.K., Studies on the structure and function of chymotrypsin inhibitor I in the Solanaceae family, in 'Proc. Int., Res. Conf. on Proteinase Inhibitors', Munich, Nov. 1970, Eds Fritz, H. and Tshescher, H., Walter de Gruyter, Berlin- New York, 1971, 175.

Saddler, H.D.W., The ionic relations of *Acetabularia mediterranea*, *J. Exp. Bot.* **21**, 345, 1970.

Saunders, J.A., Investigations of vacuoles isolated from tobacco: Quantitation of nicotine, *Plant Physiol.* **64**, 74, 1979.

Saunders, J.A., Conn, E.E., Lin, C.R. and Stocking, C.R., Subcellular localization of the cyanogenic glucoside of *Sorghum* by autoradiography, *Plant Physiol*, **59**, 647, 1977.

Saunders, J.A. and Conn, E.E., Presence of the cyanogenic glucoside dhurrin in isolated vacuoles from *Sorghum*, *Plant Physiol.*, **61**, 154, 1978.

Saunders, J.A. and Gillespie, T.M., Localization and substrate specificity of glycosidases in vacuoles of *Nicotiana rustica*, *Plant Physiol.*, **76**, 885, 1984.

Sauter, J.J., van Cleve, B. and Apel, K., Protein bodies of ray cells of *Populus canadensis* Moench 'robusta', *Planta*, **173**, 31, 1988.

Schmitt, R. and Sandermann, H., Specific localization of ß-D- glucoside conjugates of 2,4-dichlorophenoxyacetic acid in soyabean vacuoles, *Z. Naturforsch*, **37C**, 772, 1982.

Schnarrenberger, C., Oeser, A. and Tolbert, N.E., Isolation of protein bodies on sucrose gradients, *Planta*, **104**, 185, 1972.

Schnepf, E., Special cytology: Differentiated cells and cell development in higher plants, *Progr. in Botany*, **43**, 13, 1981.

Schoss, P., Walter, C. and Mäder, M., Basic peroxidases in isolated vacuoles of *Nicotiana tabacum*, *Planta*, **170**, 225, 1987.

Schröter, K., Läuchli, A. and Sievers, A., Mikroanalytische Identifikation von Bariumsulfat-Kristallen in den Statolithen der Rhizoide von *Chara fragilis*, Desv., *Planta*, **122**, 213, 1975.

Schulze, Ch., Schnepf, E. and Mothers, K., Über die Lokalization der Kautschukpartikel in verschiedenen Typen von Milchröhren, *Flora*, **158**, 458, 1967.

Sexton, R. and Hall, J.L., Enzyme Cytochemistry, in 'Electron Microscopy and Cytochemistry of plant Cells'. Ed. Hall, J.L, Elsevier North Holland, Amsterdam, 1978, Chap 2.

Sexton, R., Cornshaw, J. and Hall, J.L., A study of biochemistry and cytochemical localisation of β-glycerophosphatase activity in root tips of maize and pea, *Protoplasma*, **73**, 417, 1971.

Shumway, L.K., Yang, V.V. and Ryan, C.A., Evidence for the presence of proteinase inhibitor I in vacuolar protein bodies of plant cells, *Planta*, **129**, 161,1976.

Smith, C.G., The ultrastructural development of spherosomes and oil bodies in the developing embryos of *Crambe abyssinica*, *Planta*, **119**, 125, 1974.

Sorokin, H.P., The spherosomes and the reserve fat in plant cells, *Amer. J. Bot.* **54**, 1008, 1967.

Sorokin, H.P. and Sorokin, S., Fluctuations in the acid phosphatase activity of spherosomes in the guard cells of *Càmpanula persicifolia*. *J. Histochem. Cytochem.*, **16**, 791, 1968.

St. Angelo, A.J., Ory, R.L. and Hansen, H.J., Localization of an acid proteinase in hemp seed, *Phytochemistry*, **8**, 1135, 1969.

Sundberg, I. and Nishammar-Holmvall, M., The diurnal variation in phosphate uptake and ATP level in relation to deposition of starch, lipid and polyphosphate in synchronized cells of *Scenedesmus*, *Z. Pflanzenphysiol.*, **76**, 270, 1975.

Suter, E.R. and Majno, G., Passage of lipid across vacuolar endothelium in new born rats, J. Cell Biol., **27**, 163, 1965.

Svihla, G. and Schlenk, F., S-Adenosylmethionine in the vacuole of *Candida utilis.*, *J. Bacteriol.* **79**, 841, 1960.

Swain, E. and Polton, J.E., Immunocytochemical localisation of prunasin hydrolase and mandelonitrile lyase in stems and leaves of *Prunus serotina*, *Plant Physiol.*, **106**, 1285, 1994.

Swanson, J. and Floyd, G.L., Acid phosphatase in *Asteromonas gracilis* (Chlorophyceae, Volvocales): a biochemical and cytochemical characterization, *Phycologia*, **18**, 362, 1979.

Thayer, S.S. and Huffaker, R.C., Vacuolar localization of endoproteinases EP1 and EP2 in barley mesophyll cells, *Plant Physiol.*, **75**, 70, 1984.

Thom, M. and Komer, E., Role of the ATPase of sugarcane vacuoles in energization of the tonoplast, *Eur. J. Biochem.*, **138**, 93, 1984.

Thom, M., Willenbrink, J. and Maretzki, A., Characteristics of ATPase from sugarcane protoplast and vacuole membranes, *Physiol. Plant.*, **58**, 497, 1983.

Thomas, R.L. and Jen, J.J., The cytochemical localization of peroxidase in tomato fruit cells, *J. Food Biochem.*, **4**, 247, 1980.

Tranbarger, T.J., Franceschi, R., Hildebrand, D.F. and Grimes, H.D., The soybean 94-kilodalton vegetative storage protein is a lipoxygenase that is localised in paraveinal mesophyll cell vacuoles, *Plant Cell*, **3**, 973, 1991.

Tronier, B., Ory, R.L. and Heningsen, K.W., Characterization of the fine structure and proteins from barley protein bodies, *Phytochemistry*, **10**, 1207, 1971.

Tully, R.E. and Beevers, H., Protein bodies of the castor bean endosperm, *Plant Physiol.*, **58**, 710, 1976.

Urban, B., Laudenbach, U., and Kesselmeier, J., Saponin distribution in the etiolated leaf tissue and subcellular localization of steroidal saponins in etiolated protoplasts of oat (*Avena sativa* L.), *Protoplasma*, **118**, 121, 1983.

Van der Wilden, W., Herman, E.M., Chrispeels, M.J., Protein bodies of mung bean cotyledons as autophagic vacuoles, *Proc. Nat. Acad. Sci. U.S.*, **77**, 428, 1980.

Van der Wilden, W., Matile, P., Schellenberg, M., Meyer J., and Wiemken, A., Vacuolar membrane isolation from yeast cells, *Z. Naturoforsch.*, Teil C, **28**, 416, 1973.

Van Steveninck, R.F.M. and Van Steveninck, M.E., Ion localization, in 'Electron Microscopy and Cytochemistry of Plant Cells', Ed. Hall, J.L., Elsevier, Amsterdam, 1978, Chap. 4.

Vassilyev, A.E., On the localization of synthesis of terpenoids in the plant cell (in Russian), *Rastitelni resursi.* **6**, 29, 1970.

Verma, D.P.S., Kazazian, V., Zogbi, V. and Bal, A.K., Isolation and characterization of the membrane envelope enclosing the bacteroids in soybean root nodules, *J. Cell. Biology*, **78**, 919, 1978.

Wagner, G.J., Intracellular localization of vacuolar and cytosol components of protoplasts after vacuole isolation, *Plant Physiol*, **9**, Suppl, 104, 1977.

Wagner, G.J., Content of vacuole-extravacuole distribution of neutral sugars, free amino acids and anthocyanin in protoplasts., *Plant Physiol.* **64**, 88, 1979.

Wagner, G. J., Vacuolar deposition of ascorbate-derived oxalic acid in barley, *Plant Physiol.*, **67**, 591, 1981.

Wagner, G.J., Mulready, P., and Cutt, J., Vacuole /extravacuole distribution of soluble protease in *Hippeastrum* petal and *Triticum* leaf protoplasts, *Plant Physiol.*, **68**, 1081. 1981.

Wagner, G.J. and Mulready, P., Characterization and solubilization of nucleotide specific Mg^{2+}-ATPase and Mg^{2+}-pyrophosphatase of tonoplast, *Biochem. Biophys. Acta* **728**, 267, 1983.

Waime, J.M., Étude d'une substance polyphosphorée basophile et métachromatique chez les levures. *Biochim. Biophys. Acta* **l**, 234, 1947.

Walek-Czernecka A., Mise en évidence de la phosphatase acide (monophosphostérase II) dans les sphérosomes des cellules epidermiques des écailles bulbaires d' *Allium cepa*, *Acta. Soc. Bot. Polon.*, **31**, 539, 1962.

Walek-Czernecka, A., Histochemical demonstration of some hydrolytic enzymes in the spherosomes of plant cells. *Acta. Soc. Bot. Polon.* **34**, 573, 1965.

Walker, R.R. and Leigh, R.A., Mg^{2+}-dependent, cation- stimulated inorganic pyrophosphatase associated with vacuoles isolated from storage roots of red beet *Beta vulgaris*, *Planta*, **153**, 150, 1981.

Walker-Simmons, M. and Ryan, C.A., Immunological identification of proteinase inhibitors I and II in isolated tomato leaf vacuoles, *Plant Physiol.*, **60**, 61, 1977.

Waters, S.P., Nobel, E.R. and Dalling, M.J., Intracellular localization of peptide hydrolases in wheat *Triticum aestivum* L. leaves, *Plant Physiol.* **69**, 575, 1982.

Werker, E.and Fahn, A., Site of resin synthesis in cells of *Pinus halepensis* Mill., *Nature*, **218**, 388,1968.

Werner, C. and Matile, Ph., Accumulation of coumaryl glucosides in vacuoles of barley mesophyll protoplasts., *J. Plant Physiol*, **118**, 237, 1985.

Whaley, W.G., Mollenhauer, H.H. and Leech, H.J., The ultrastructure of the meristematic cell. *Amer. J. Bot.*, **47**, 401, 1960.

Wiemken, A. and Dürr M., Characterization of amino acid pool in vacuolar compartment of *Saccharomyces cerevisiae*, *Arch. Microbiol.* **101**, 45, 1974.

Wiemken, A. and Nurse, P., The vacuole as a compartment of amino acid pools in yeast, in 'Proc. Third Int. Specialized Symp. on Yeast', Otarieni/Helsinki, part II, 1973, 33.

Wiemken, A., Schellenberg, M. and Urech, K., Vacuoles, the sole compartments of digestive enzymes in yeast *Saccharomyces cerevisiae, Arch. Microbiol.*, **123**, 23, 1979.

Wilson, C.L., Stiers, D.L. and Smith, G.G., Fungal lysosomes or spherosomes, *Phytopathology*, **60**, 216, 1970.

Wingate, V.P.M., Franceschi, V.R. and Ryan, C.A., Tissue and cellular localization of proteinase inhibitors I and II in the fruit of wild tomato, *Lycopersicon peruvianum* (L.) Mill., *Plant Physiol.*, **97**, 490, 1991.

Winter, H., Robinson, D.G. and Heldt, H.W. Subcellular volumes and metabolic concentrations in barley leaves, *Planta*, **191**, 180, 1993.

Winter, H., Robinson, D. G. and Heldt, H.W. Subcellular volumes and metabolic concentrations in spinach leaves, *Planta* **193**, 530, 1994.

Wittenbach, V.A., Lin, W. and Hebert, R.R., Vacuolar localization of proteases and degradation of chloroplasts in mesophyll protoplasts from senescing primary wheat leaves, *Plant Physiol.* **69**, 98, 1982.

Wolk, C.P., Role of bromine in the formation of the refractile inclusions of the vesicle cells of the Bonnemaisonaceae (Rhodophyta), *Planta*, **78**, 371, 1968.

Yaklich, R.W. and Herman, E.M., Protein storage vacuoles of soybean aleurone cells accumulate a unique glycoprotein, *Plant Sci.*, **107**, 57, 1995.

Yamaki, S., Distribution of sorbitol, neutral sugars, free amino acids, malic acid and some hydrolytic enzymes in vacuoles of apple cotyledons, *Plant Cell Physiol*, **23**, 881, 1982.

Yatsu, L.Y., Jacks, T.J. and Hensarling, T.P., Isolation of spherosomes (oleosomes) from onion, cabbage and cotton seed tissues, *Plant Physiol.*, **48**, 675, 1971.

Youle, R.J. and Huang, A.H., Protein bodies from the endosperm of castor bean, *Plant Physiol.*, **58**, 703, 1976.

Zalokar, M., Kinetics of amino acid uptake and protein synthesis in *Neurospora.*, *Biochem. Biophys. Acta*, **46**, 423, 1961.

VI
Biogenesis and development of vacuoles

6.1 EARLY VIEWS

The earliest view on the origin of vacuoles in plants was that they arise *de novo* in the protoplasm to form visible droplets when water is adequate (Nägeli 1855). Early workers supported this view till de Vries (1885) proposed that vacuoles originate from individualised bodies which he called 'tonoplasts'. As the 'tonoplast' secretes cell sap within itself, it gradually enlarges and becomes the vacuolar membrane. Pfeffer (1890) was the first worker to induce vacuole formation. He introduced granules of asparagin into the plasmodium of a myxomycete, observed formation of membrane around the granule and supported the contention of the *de novo* origin of vacuoles. Went (1890) considered vacuoles as a permanent feature of the protoplasm and suggested that a vacuole can arise only from a pre-existing one. On the basis of his belief in the alveolar nature of protoplasm, Strasburger (1898) suggested that the enlargement and fusion of alveoles give rise to vacuoles. With the aid of various fixatives, Bensley (1910) studied the process of vacuolation in root tips of a number of plants. He recorded that vacuoles, although under certain fixation appearing as fine separate droplets in the cytoplasm, actually exist as units. In very young cells a system of fine canals 'canaliculi' enlarge and gradually coalesce with the growth of the cells leading to the formation of the large central vacuole (Fig. 6.1).

6.2 *DE NOVO* ORIGIN

Guilliermond (1941) and Frey-Wyssling (1953) considered the origin of vacuoles from the viewpoint of the physico–chemical properties of colloids. Guilliermond suggested that small particles of hydrophilic colloidal materials with greater imbibitional forces than the other cytoplasmic colloids are deposited in the cytoplasm and as these particles become hydrated they swell, become fluid and form vacuoles. Formation of the tonoplast membrane was explained as resulting from surface forces operating at the interfaces between an aqueous and a non-aqueous phase. Dainty (1968) speculated that a vacuole might start in a small region centered on a particularly hydrophilic (and water structure-ordering) macromolecule. This developing vacuole would need a fairly rapid increase in the local solute concentration and a phospholipid membrane to surround it. Phospholipid would need to be synthesised in the cytoplasm and 'adsorbed' as a bimolecular layer, or equivalent, at the surface of the developing vacuole.

The question of *de novo* origin arises from the fact that most meristematic cells are apparently devoid of vacuoles. Even the development of electron microscopic techniques did not solve the problem readily. Competent electron microscopists like Porter and Machado (1960), and Frey-Wyssling and Muhlethaler (1965) could not detect vacuoles in meristematic cells.

6.2.1 *De novo* liposome

Amphipathic lipids, depending on their concentration and ionic composition, may spontaneously exist as micelles, sheets of bilayers and also liposomes. A vacuole can be regarded as a large

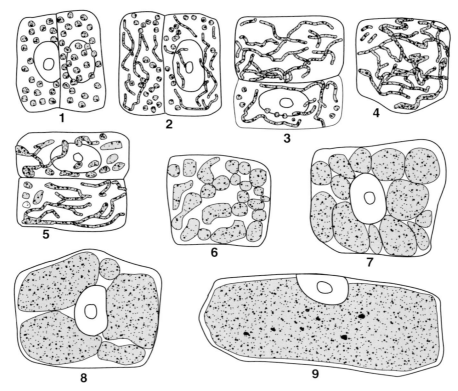

Figure 6.1 Barley root. Stages in development of the vacuolar system. 1–5, meristem; 6–8, adjacent region of differentiation; 9, mature cells of cortical parenchyma. Vital staining with neutral red (from Guilliermond 1941).

liposome where the membrane contains a number of proteins and enzymes and the lumen contains a variety of substances dissolved or undissolved in water. Such a scenario makes it easy to visualise the *de novo* origin of vacuole from polar lipids which readily and spontaneously form very thin bilayers separating two aqueous compartments.

6.2.2 *De novo* Golgi bodies

Although certain experiments with animal cells indicate that Golgi apparatus may arise *de novo* (Zorn *et al.* 1979; Maniotis and Schilwa 1991), it is widely accepted that membrane-bound organelles cannot originate *de novo*. Nunnari and Walter (1996) stated that *de novo* organelle synthesis is never required; organelles always grow by proliferation of other pre-existing ones.

6.2.3 Growth of tonoplast

Under certain conditions, plant cells elongate very rapidly at rates ranging from 20 to 75 µm per hour, with concomitant enlargement of their vacuoles. Such cases, even if the tonoplast is temporarily disrupted, would require an enormous amount of tonoplast-specific phospholipids and proteins. Compared to the biogenesis of the double-layered envelope of chloroplast or mitochondrion, the biogenesis or growth of the single lipid bilayer of tonoplast is rather simple. It only involves insertion and anchorage of specific lipids and proteins, like the outer envelope of the other organelles.

A direct delivery of tonoplast proteins may take place in the case of formation of the tonoplast of protein storage vacuoles (PSV). The tonoplast of the central vacuole of the developing pea cotyledon undergoes a 100-fold expansion of its surface area. Serial sections of developing protein bodies show that prior to the formation of truly individual PSVs, the central vacuole develops a highly convoluted surface with finger-like projections, indicating great expansion of the tonoplast (Craig *et al.* 1980).

6.2.3.1 Regeneration of tonoplast

The complete removal of vacuoles and their subsequent regeneration provides a unique system for biogenesis of vacuoles *in vivo*. Griesbach and Sink (1983) could remove all the vacuoles from isolated mesophyll protoplasts by subjecting them to density-gradient centrifugation. Freshly isolated mini-protoplasts were devoid of vacuoles, as judged by electron microscopy and the absence of acid hydrolase activity. Electron microscopy revealed the appearance of small vocuoles inside the endoplasmic reticulum within 20 hours of culture in mannitol, and the vacuolar soluble enzymes α-mannosidase, β-N-acetylglucosaminidase and acid phosphatase could be recovered after 48 hours (Hörtensteiner *et al.* 1992). On the other hand, during the regeneration of vacuoles the tonoplast enzymes ATPase and PPase increased. Again, the ability to glycosylate proteins does not appear to be essential for the formation of new vacuoles since the addition of tunicamycin, an inhibitor of asparagine-linked glycosylation, had no effect on the regeneration (Hörtensteiner *et al.* 1994). These results were essentially confirmed by Newell *et al.* (1998) who studied vacuole development in cultured evacuolated oat mesophyll protoplasts with the aid of confocal laser scanning microscopy (CLSM) and synthesis of V-ATPase, PPase and 23 kD vacuolar membrane protein. Their CLSM studies indicated that developing vacuoles took the form of a tubular network which expanded and fused to form a series of interconnected vacuoles. This is exactly what was stated by Guilliermond about 60 years ago! Thus the biogenesis of tonoplast or whole vacuole may occur in certain cases without the intermediary of Golgi bodies or vesicles which appear to be the common method in most cells.

6.2.3.2 Synthesis and direct transfer of lipids to tonoplast

Membranes can grow by expansion when lipids and proteins are inserted and anchored into the existing membrane elements. Such a process of membrane growth may essentially consist of simultaneous and/or sequential (i) synthesis of lipids and (ii) synthesis of proteins and their insertion and anchorage.

Cellular fatty acids are synthesised in smooth endoplasmic reticulum (SER), with the exception of chlorophyllous cells where all the fatty acids are made in chloroplast stroma. SER provides the necessary machinery on its outer cytosolic layer: the fatty acids along with glycerophosphate, and the enzymes for the synthesis of specific lipid molecules. The newly synthesised lipids in the outer layer are known to be equilibrated with the inner lumen layer by phospholipid translocators (flippase), specific for each type of phospholipid. The addition of new lipids on both sides of the SER leads to its growth. The tonoplast, in one way, may be compared with the outer envelope of mitochondria and chloroplasts, since all three are exposed to the environs of the cytosol. The fact that the outer envelope of chloroplast plays a major role in the biosynthesis of various lipids is very supportive (Douce and Joyard 1990).

If it is assumed that the tonoplast has no machinery for its own lipid synthesis and expansion, then the transfer of lipids by phospholipid exchange proteins or lipid transfer proteins (LTPs), which remove lipids from one source and deliver to a target organelle or membrane, is the only possible mechanism. In mammals, several classes of LTPs, including phosphatidylcholine transfer proteins, phosphatidylinositol transfer proteins and non-specific lipid transfer proteins (nsLTPs), have been detected (Wirtz 1991). Only nsLTPs have been detected in plants, but they are not yet shown to be involved in the actual transfer of lipids between organelle membranes. The occurrence and gene expression of nsLTPs is localised in the surface layers of various plant organs, such as seedlings, leaves and stem, flowers and peduncles and developing fruits. Osafune *et al.* (1996)

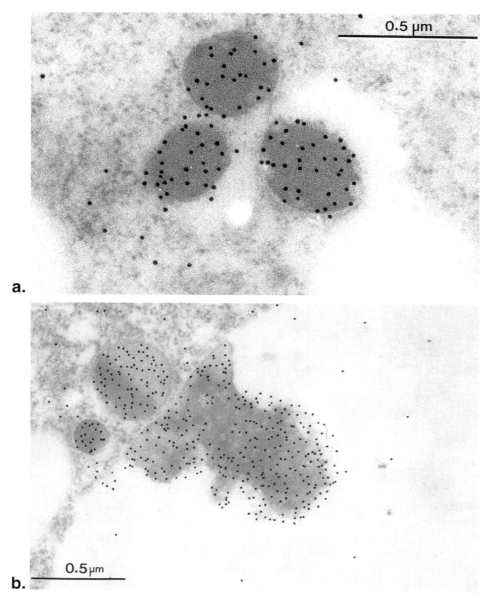

Figure 6.2   **a** Electron micrograph of a section of the endosperm cell from a developing castor bean fruit at 14 days after flowering. The section was labelled with the antibody against castor bean nsLTP. An arrow indicates nsLTPs (gold particles) localised in the dense vesicles which occur in the cytoplasm. ER, endoplasmic reticulum. **b** Electron micrograph of a section of the endosperm cell from a developing castor bean fruit at 28 days after flowering. nsLTPs (gold particles) localised in the dense vesicles approach the central vacuole (V) in the cell and assemble in the vacuole (from Osafune *et al.* 1996). With permission of Elsevier Science.

recorded nsLTPs localised in the outer layer of the endosperm and epicarp in developing castor bean fruits. They are stored in vacuoles in early stages of development but disappear later (Fig. 6.2). Thus it is yet to be demonstrated that nsLTPs are actually involved in the growth of tonoplast, or for that matter, of any plant cell membrane. Possibly the plant nsLTPs act as a carrier of long chain fatty acids which are precursors of suberin, lignin and flavonoids (Thoma *et al.* 1994).

6.2.3.3 *Synthesis and direct transfer of proteins to tonoplast*

All proteins are synthesised by the ribosomes either free in cytosol or attached to the endoplasmic reticulum (ER). Those which are synthesised in the cytosol either reside there or are delivered to their destination. Generally, specific proteins are destined to enter specific membranes or organelles and the targeting is done by signals on the protein itself. Such signals can be specific amino acid sequence or carbohydrate/phosphate residues attached to the protein. Hence immediately after synthesis the proteins may or may not need processing for their delivery. Certain membrane proteins are directly inserted in the outer membrane of the mitochondria from the cytosol without any requirement for ATP. Such proteins may have a signal–anchor sequence and stop–transfer sequence ordering them not to pass through the membrane. The process may be a general feature of protein targeting to the outer membrane of some organelles which are exposed to the cytosol. The tonoplast may grow by such insertion of proteins, most prabably in cases where it grows fast.

6.3 Origin through lipid utilisation

One possible mode of origin and development of vacuoles has been demonstrated in the root cap cells of alfalfa (*Medicago sativa*) by Maitra and De (1972). As the root elongates, the cells of the central zone of the root cap move from the innermost region to the outermost cell layer and are finally sloughed off. The cells of the central zone of the root cap contain numerous small electron dense lipid granules. The cells further away contain lipid particles and minute vacuoles. As these cells mature and move towards the periphery of the root cap the minute vacuoles enlarge and the amount of electron dense lipid decreases. Thus, as lipid digestion progresses the vacuoles enlarge. At a still later stage, as the root cap cells move to the outermost layer, large vacuoles occupy most of the cell and most of the lipid disappears. All the stages in the origin and development of vacuoles concomitant with the utilisation of lipid could be traced (Fig. 6.3). It is interesting to note that although these cells contain Golgi cisternae no Golgi vesicles are discernible. Thus, in these cells Golgi vesicles are not involved in the vacuolation process. Dawes and Barilotti (1969) have also recorded a number of small vacuoles in the growing tip of a blade of *Caulerpa* leaf. These vacuoles are numerous and contain lipid particles. Lipids are the most concentrated source of energy available in the cell, yielding over twice as many calories per gram as either carbohydrate or protein, and provide twice as much metabolic water from complete oxidation. Thus, as the lipid is oxidised, the metabolic water is not dispersed in the cytoplasm but is contained within the membrane formed by the lipid itself.

In this connection, certain observations made in fungi are of interest. The light micrographs of a cleaving zoosporangium of *Saprolegnia ferax* show the presence of a central vacuole which enlarges and delimits uninucleate masses of cytoplasm. Finally the developing zoospores are clothed with a membrane derived in part from both the vacuoles and cell membranes. Since the sporangium does not normally expand significantly during this process, the vacuoles must expand at the expense of the cytoplasm. Such expansion could easily result if the osmotic potential of the vacuoles became higher than the cytoplasm causing flow of water from cytoplasm to vacuoles. Additional membrane synthesis is required simultaneously to envelop the enlarging vacuole. It has been suggested that both processes are brought about by the fusion of the 'dense body' vesicles with the central vacuole (Gay and Greenwood 1970; Gay *et al.* 1974). These 'dense body' vesicles produced only on induction of sporulation, are rich in phospholipids and some appear to fuse with the cleaving central vacuole. Some of the phospholipids may be utilised for expansion of the vacuolar membrane whilst others are degraded into smaller vacuoles, thus increasing the osmotic pressure of the vacuole. Phospholipidase A and B and a diesterase could be the cleaving enzymes. The lipid containing dense bodies are also known to be involved in the generation of hyphal vacuoles. In certain aquatic phycomycetes these dense bodies remain in zoospores throughout their prior development. In *Pythium* (Colt and Endo 1972; Grove 1970), *Saprolegnia* (Gay *et al.* 1971)

120 Plant cell vacuoles: an introduction

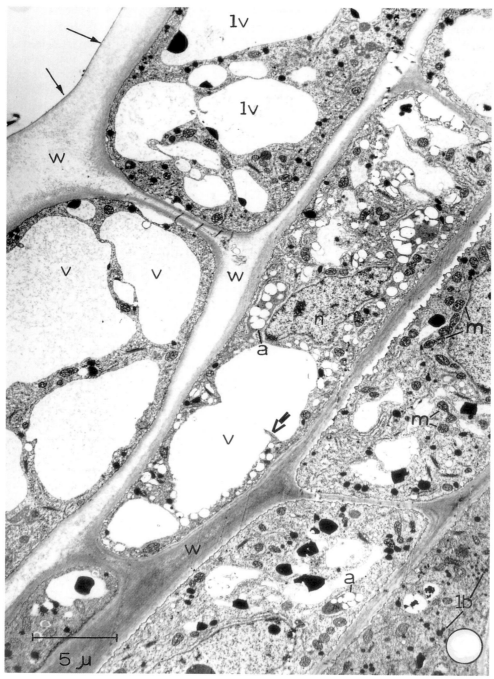

Figure 6.3 A part of the outer zone of the root cap of *Medicago sativa*. The cells are elongated and parallel to the arc of the root. The gradual enlargement of lipid vacuole (lv) from the inner cells to the outermost cells can be followed from the right lower corner to the left upper corner. The outermost cells show some very big vacuoles containing small lipid bodies. The thick arrow indicates the coalescence of vacuoles. Thin arrows indicate the outermost cell wall. Primary cell wall (W), amyloplast (a), lipid body (lb) and mitochondrion (m) (after Maitra and De 1972). With permission of Faculty Press.

and *Aphanomyces* (Hoch and Mitchell 1972), the dense body vesicles enlarge and coalesce with concomitant shrinkage of the osmiophilic granules so that they form the vacuole of the germling. Again, the action of lipase on oil bodies in germinating maize kernel supports the above observations. Huang (1996) observed that newly synthesised lipase binds specifically to oil bodies. The recognition signal on the oil bodies for lipase is likely to be oleosins. During or after lipolysis, the phospholipid layers of the oil bodies fuse with the vacuolar membrane, eventually forming the large central vacuole.

6.4 ORIGIN OF ALEURONE VACUOLES

A number of workers have studied the origin and development of special vacuoles known as aleurone vacuoles which store aleurone grains. The storage of various proteins, polyphosphates, etc., has already been discussed in Chapter V. The protein bodies (PB) in seed storage parenchayma cells may be formed by different routes. In cereals, the PBs develop directly from ER by dilation and vesiculation. In the endosperm of maize, Khoo and Wolf (1970) showed that protein is deposited at the enlarged end of the ER, which develops into numerous provacuoles. They have shown association of polyribosomes with the tonoplast of developing aleurone vacuoles. More critical studies have shown that the prolamins of rice (Krishnan *et al.* 1986) and maize (Larkins and Hurkman 1978) are retained in the lumen of ER which becomes distended to form PBs. In addition to the prolamin-containing PBs, the rice endosperm also contains another type of PB containing glutenin which is transported from the ER to the vacuole via the Golgi appraratus. The origin of protein storage vacuoles (PSV), as they are called in dicots, also appears to be diverse. In certain cases, the large vacuoles present in the young parenchymatous cotyledonary cells disappear during development to be replaced by a second group of vacuoles which are converted into PSVs. In cowpea, along with subdivision of pre-existing vegetative vacuoles, new PSVs may arise from Golgi vesicles (Harris and Boulter 1976). A similar double origin of PSVs from subdivision of a vacuole, along with swelling of ER, has been recorded by Adler and Müntz (1983) in broad beans. Craig *et al.* (1980), on the other hand, have shown that large vacuoles of developing pea cotyledon parenchyma cells fragment to form the small protein bodies found in mature seeds. Prior to this, seed storage proteins accumulate in the vacuoles at the periphery of the vacuole where they are associated with tonoplast membranes. Critical studies using ultrastructural-immunocytochemical methods (Hoh *et al.* 1995) have detected two distinct populations of vacuoles in the cotyledons of maturing pea seeds. The original vegetative vacuoles appear to become surrounded by a cisternal, tubular membrane system that already contains deposits of storage proteins. Deposits of the storage proteins vicilin and legumin in the lumen, and the presence of α-TIP in the membranes of the expanding membrane system provided evidence of its identity as a precursor to the PSV.

The aleurone vacuoles of barley contain droplets of triglycerides embedded between the half-unit membranes of the tonoplast. Huang (1992) incorrectly called these droplets oleosomes. (According to Bergfields *et al.* (1978) oleosomes are plant spherosomes that are rich in lipids, but devoid of acid phosphatases and other lytic enzymes. They are specifically found in fat-storing cells of developing seeds and fruits.) According to Napier, *et al.* (1996) the newly formed lipid droplets are detached from the ER and fuse with the tonoplast. However, Bethke *et al.* (1998) propose that the ER membrane that synthesises the neutral lipid also serves as the site of storage protein accumulation.

The structure and utilisation of the stored food in the aleurone grain demonstrate an interesting feature of the lysosomal nature of this specialised vacuole. As well as proteins and phytic acid, the presence of acid phosphatase has been demonstrated in the aleurone grains of wheat embryo and cotyledonary cells of *Cucumis* and *Linum* (Poux 1963, 1965). The aleurone grains of pea cotyledons have been found to contain various hydrolases including protease, RNase and acid phosphatase (Matile 1968a). In *Cannabis sativa* seeds the grains contain reserve protein edestin and its specific enzyme edestinase (St. Angelo *et al.* 1968, 1969). Ory and Hennigsen, (1969) showed the presence of protease and specific phytase in the grains of ungerminated barley seeds. Hydrolytic activity has also been demonstrated in the protein bodies of bean cotyledons (Van der Wilden *et al.* 1980). Since aleurone grains are outlined by a single membrane and contain hydrolytic enzymes, they are equivalent to lysosomes. What makes them different is a presence of substrate and its specific enzyme. In germinating seeds, the utilisation of the storage proteins is reflected in the gradual disappearance of the protein matrix and the swelling of the aleurone vacuoles. The utilisation and mobilisation of the contents of the aleurone grain without the disintegration of the outlining membrane or tonoplast indicates the grain's true vacuolar nature. In plants whose cotyledons develop into green leaves, the aleurone vacuole becomes the central vacuole (vaan der Eb and Nieuwdorp 1967; Treffry *et al.* 1967; Jones and Price 1970). In pea the cotyledons eventually undergo senescence and the digestion of the reserve proteins is followed by the digestion of the surrounding cytoplasm (Hinkelmann 1966). The ontogeny of vacuoles from protein bodies during germination of seeds has also been reported by a number of workers (Pernolett 1978; Brown *et al.* 1982; Bollini *et al.* 1983). In the germinating seeds of *Prunus serotina*, Swain and Poulton (1994) recorded that numerous small protein bodies within a given cotyledonary cell coalesced to form larger protein bodies. Later these protein bodies enlarged and became less dense in the process of transforming into vacuoles. The cotyledons of mature black cherry seeds contain the cyanogenic diglucoside (R)-amygdalin and monoglucoside (R)-prunasin. During 3 weeks of germination, the reduction in amygdalin level was paralleled by declines in the levels of amygdalin hydrolase (AH), prunasin hydrolase (PH), mandelonitrile lyase (MDL), and β-cyanoalanine synthase. At all stages of seedling development, AH and PH could be detected by immunocytochemistry within the vascular tissues. In contrast, MDL occurred mostly in the cotyledonary parenchyma cells and also in the vascular tissues. Soon after imbibition of water AH, PH and MDL were found within protein bodies and were later detected in the vacuoles derived from these organelles.

Recently Bethke *et al.* (1998) reviewed how the cereal aleurone grain is metamorphosed from a storage compartment to a lytic organelle. When the aleurone cells imbibe water or are subjected to a growth regulator like gibberellic acid a cascade of events is initiated. These include coalescence of protein storage vacuoles, acidification of vacuole lumen, activation of enzyme, hydrolysis of stored reserves and release of hydrolysates and minerals from the aleurone cells (Bethke *et al.* 1998). Thus the storage structures become lytic organelles, rapidly hydrolysing the stored polymers in their lumen. The proteins are degraded by proteases and completely hydrolysed to amino acids, phytates are hydrolysed by phytases and other enzymes to inositol phosphates, free inositol and phosphate. The solubilised minerals are released from the aleurone cell as cations and anions. Triglycerides are degraded by lipases and the resulting free acids are released. Finally, the vacuole becomes filled only with soluble substances.

6.5 ORIGIN THROUGH LYTIC PROCESSES

As stated above, the earliest detectable vacuole under an electron microscope is a tiny vesicle known as a provacuole, which enlarges and coalesces to produce a typical vacuole. Provacuoles possibly contain hydrolases and are lysosomal in nature. Berjak (1972) first demonstrated that acid phosphatase activity is localised in provacuoles, which are pinched off from the endoplasmic reticulum. Thus the differentiation of ER into provacuoles involves the synthesis of lysosomal enzymes. Presumably, as provacuoles undergo fusion and increase in size, enzyme accumulation takes place and the membrane of the vacuole may assume specific properties (Matile 1968b).

In animal cells, a lysosome may engulf cell structures like mitochondria or endoplasmic reticulum and digest them. With leftover cell debris or partial cytoplasmic lysis, lysosomes form a cytolysosome. Villiers (1967) was the first to report on the occurrence of cytolysosomes in plants,

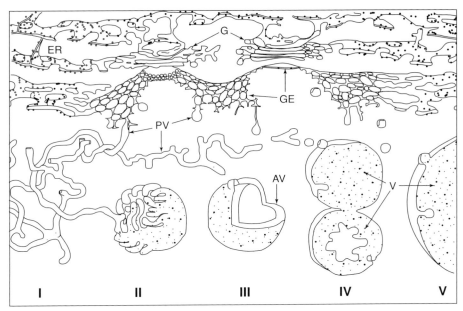

Figure 6.4 Stages in the formation of vacuoles. AV, autophagic vacuole; ER, endoplasmic reticulum; G, Golgi stack; GE, GERL; M, mitochondrion; PV, provacuole; V, vacuole. ER, Golgi apparatus, and GERL are extensively drawn in the upper part of the diagram because they are present at any step of vacuole maturation. The formation of vacuoles during cell differentiation is shown from left to right in the lower portion of the drawing. Steps I-V as explained in the text (from Marty 1978).

in the embryos of *Fraxinus excelsior*, which were kept under stress conditions. Large portions of cytoplasm containing various organelles were delimited by a double membrane and in some cases the content showed signs of breakdown. Eventually, with the breakdown of the contents and the inner membrane, a vacuole was formed. The demonstration of acid-β-glycerophosphatase activity in these structures suggested the lytic activity of the compartment. It was also shown that the membranes of the ER are involved in formation of the compartment and subsequent breakdown of the enclosed cytoplasm (Villiers 1971). Subsequently a number of workers have demonstrated the presence of a cytolysosome-like digestive compartment in various other plant cells, both under stress and in normal conditions. In the meristematic cells of *Hordeum* roots, Buvat (1968) traced the origin of vacuoles that isolate cytoplasmic regions before destroying them. At first, closed double membranes are observed forming one or two concentric systems with profiles similar to the saccules of smooth ER. These saccules swell and form small vesicles, partially confluating in cupuliform vacuoles enclosing a part of the cytoplasm. Similar sequestration of a part of the cytoplasm by the ER envelope leading to vacuolation has been demonstrated in many other plant cells. Sometimes concentric ER membranes may undergo fusion to form multilayered cytolysosomes or vacuoles containing multiple vesicles (Marty 1970; Coulomb and Coulomb 1972; Cresti *et al.* 1972; Mesquita 1972).

Novikoff and coworkers (1967, 1973) noted similar membrane-bound hydrolytic activity in neurons of rat ganglion nodosum. In order to designate an interrelationship between the Golgi apparatus, endoplasmic reticulum and lysosomes, they used the acronym GERL to describe the region of the smooth ER located at the trans aspect of the Golgi apparatus. In neurons of the dorsal root ganglia GERL possesses cytochemically demonstrable acid phosphatase activity and appears to produce primary lysosomes. On the basis of studies using high voltage electron microscopy of thick and thin sections and enzyme cytochemistry, Marty and coworkers (1978, 1980) extended the GERL concept to plant cells. They showed that in *Euphorbia* meristems large plant cell vacuoles result from a number of steps. The GERL produces numerous provacuoles which drive a programmed cellular autophagy leading to formation of young vacuoles which undergo enlargement and/or fusion with provacuoles to produce the large mature vacuole. They indicated various stages of vacuole formation in a diagram (Fig. 6.4). The provacuoles may fuse or elongate to produce fine tubes throughout the cytoplasm (step I). The lysosomal tubes undergo a program of cellular autophagy by wrapping themselves around portions of cytoplasm (step II) and merge laterally to surround a portion of the cytoplasm (step III) with a continuous exoplasmic space loaded with digestive enzymes. The provacuoles contain high levels of acid phosphatase and thiolacetic acid esterase activity at all steps of development into autophagic vacuoles. These small vacuoles may undergo fusion (step IV). In the vacuolated cells, the provacuoles are transferred directly from GERL to the pre-existing large vacuoles (step V).

6.6 Origin from membranous components

6.6.1 Origin from endoplasmic reticulum

Buvat (1958) recorded that parallel profiles of smooth endoplasmic reticulum in the cells of *Elodea canadensis* formed saccules from a series of dilations. These dilations give rise to vacuoles. Similar dilations have also been noticed by Poux (1962), although Manton (1962) was not convinced of the connection between vacuoles and endoplasmic reticulum. The possible origin of vacuoles from minute spherical bodies, not related to any existing organelle of the cell, has been proposed by Whaley *et al.* (1964). They regarded the dense spherical bodies as the precursors of vacuoles in meristematic cells. These bodies gradually become less electron dense, develop an irregular shape, increase in number by fragmentation, and subsequently coalesce to form vacuoles.

According to Bowes (1965) accumulation of metabolites in certain localised regions of the ER in meristematic cells may lead to vacuole formation. Again, in degenerating or ageing cells, a breakdown or hydrolysis of cellular organelles may result in the production of vacuoles. From his studies on germinating pollen grain, Larson (1965) contended that the cisternae of ER undergo swelling to produce vacuoles. In certain cases, highly osmiophilic inclusions or prevacuolar bodies may become vacuoles, possibly with some incorporation of ER as they enlarge. Matile and Moor (1968) used freeze-etching technique for the root cap cells and radicular meristem of maize and showed that small vesicles are produced from the ER. These vesicles, called *provacuoles*, amalgamate repeatedly, to form large vacuoles. Similarly, on the basis of studies on root-tip cells of lupine, Mesquita (1969) concluded that vacuoles originate from local dilations of ER in the meristematic cells. The small vesicles pinched off from the ER generally show higher electron density. This may indicate that after separation from ER the vesicle has undergone a degree of differentiation. Mesquita suggested that such an increase in electron density can be noted in ER at the sites where vacuolation begins. Moreover, dilated regions of the ER are smooth or carry only a few ribosomes, while the rest of the ER is rough. In *Chara*, the young internodal cell which is produced from a sub-apical cell undergoes rapid vacuolation to attain gigantic dimensions. Pickett-Heaps (1967b, 1975) noted that in these cells the dilating ER produces numerous vacuoles which in turn fuse to produce the large central vacuole.

Fineran (1973) recorded from EM sections that distentions of portions of ER form numerous 'provacuolar' vesicles which then coalesce and further expand to produce vacuoles in root tip cells. A combined thin section and freeze-etch study of later stages in vacuole formation showed that each expanding provacuole had a sheet of fenestrated ER lying close to its bounding membrane over the whole external surface.

Chafe and Durzan (1973) studied the production of tannin vacuoles in cultured white spruce cells. The electron dense tannin material can be first detected in the lumen of the smooth ER which undergoes dilation and finally produces tannin vacuoles. A similar phenomenon has been reported in the cells of slash pine by Baur and Walkinshaw (1974). Parham and Kaustinen (1977) showed that tannin synthesis takes place in small provacuoles which may fuse later with the central vacuole. A similar phenomenon occurs with alkaloid synthesis. Electron microscope studies of root meristem of *Berberis parviflora* show that small vesicles, as well as vacuoles, contain dense precipitates of protoberberine alkaloids, jatrorrhizine, palmatine and berberine (Amman *et al.* 1986). These vesicles fuse with the small vacuoles and release their osmiophilic contents in them (Deliu *et al.* 1994).

Hilling and Amelunxen (1985) have also shown that SER tubules dilate to form small vacuoles. Continued fusion of small vacuoles with nearby tubules and subsequent dilation result in the formation of large vacuoles. Thus the entire SER becomes the vacuole.

A reversible development of vacuole ER has been documented by Gamelei (1990). During leaf maturation the ER enlarges with the onset of photosynthesis and forms the central vacuole in the mesophyll cells. The reverse transformation of the vacuole into ER in mesophyll, bundle-sheath and intermediary cells was observed in autumn leaves of evergreen species after photosynthesis had stopped. A similar reversible transformation from ER to vacuole was reported to be a normal process in the development of trichomes (Wilson *et al.* 1990). This agrees with observations by Buvat (1971) in the early days of electron microscopy.

6.6.2 Origin from plasma membrane

The involvement of plasma membranes in the formation or enlargement of vacuoles has come about from studies on pinocytosis. Poux (1962) showed that small globular vacuoles are formed by pinocytosis in barley root cells. These globular vacuoles or enclaves probably unload their contents into the pre-existing vacuoles. Again, the process of endocytosis can be induced in certain cells. Reizmann (1985) records that when yeast cells are treated with lucifer yellow CH, small vacuoles may originate from the plasma membrane by the process of endocytosis. The plasma membrane thus forms the membrane of the vacuole which contains the dye. Such prevacuolar compartments or endosomes, as they are commonly called, are also produced by endocytosis of many substances, including mating pheromone alpha-factor (see p. 138). Many workers believe that the process of endocytosis cannot be a regular process in the life of a plant cell and its occurrence is possibly limited.

At the end of telophase of the meristematic cells, a cytoplasmic mass is distinctly formed between two daughter nuclei to form what is known as a phragmoplast. From the earliest stage of phragmoplast formation, together with profiles of ER and other undefined structures, tiny vacuoles of irregular shape are noticeable (Whaley *et al.* 1960). In permanganate-fixed preparations Buvat (1971) observed that these vacuoles appear to contain electron-dense material. He suggested that, in the phragmoplast, these vacuoles come from pre-existing closed plasma membranes secreting a manganophilous material that may be identified with the electron-negative or hydrophilous vacuolar colloid typical of the vacuole system.

6.6.3 Origin from Golgi bodies

The possible ontogeny of vacuoles from Golgi bodies has been suggested on the basis of early electron microscopic studies. According to Marinos (1963) individual Golgi cisternum can undergo extensive swelling and turn into vacuoles in cells of *Hordeum*. In the stomatal guard cells and hair cells of wheat, Pickett-Heaps (1967a) showed that vesicles derived from Golgi discharge into the vacuoles. Another approach to this aspect has been taken by Koncalova (1965) and Ahmadian-Tehrani (1969) in their studies on ultrastructural aspects of vital staining by neutral red. They recorded that the Golgi bodies participate in the transport of dye which enters the cell by pinocytosis and is led to the vacuoles. Besides the reports on Golgi cisternae directly becoming vacuoles, the first report of actual incorporation of Golgi vesicles into developing vacuoles was made by Berjak and Villiers (1970).

Comprehensive work on the ultrastructural interrelationship of various organelles carried out by Matile and Moor (1968) showed that *provacuoles*, produced by rough ER, undergo repeated fusion and enlargement, leading to the formation of typical vacuoles. Moreover, developing vacuoles may incorporate Golgi vesicles through tonoplast invagination. Similar invagination may also introduce portions of cytoplasm into the vacuole, leading to formation of autophagic vacuoles.

Biochemical studies of vacuole structure and function are even more convincing. The occurrence of a number of substances in vacuoles, especially enzymes, needs to be understood in terms of their synthesis or transport to the vacuole. Since the vacuole has no mechanism of protein synthesis, all enzymes present in the vacuole sap must be imported in newly synthesised or processed form. Most lysosomal proteins in eucaryotic cells are synthesised on polysomes found on the endoplasmic reticulum and are immediately translocated into the lumen of ER. These proteins are then sorted and targeted to various destinations for use or storage. On the basis of extensive organelle fractionation, electron microscopic, histochemical and autoradiographic studies, Novikoff (1976) and Marty (1978) were the first to show that many proteins are routed through the Golgi bodies for processing, including glycosylation.

6.6.3.1   Golgi-vesicle-vacuole

Electron microscopic studies on the endomembrane system of a chlamydomonad flagellate, *Gloeomonas kupfferi* (Chlorophyta) by Domozych (1989a) show an intimate relationship between peripheral and contractile vacuoles with ER network and Golgi apparatus. In this organism, the endomembrane system consists of the ER network, a perinuclear complex of 14 to 18 Golgi bodies and 8 to 12 vacuoles and an anterior contractile vacuole complex. The individual Golgi is polar with distinct cis- and trans-faces. The cis-face is closely associated with transition vesicles emerging from the adjacent ER. The large vesicles emerge from peripheral swellings of terminal cisternae. The Golgi-associated ER is connected to the peripheral vacuolar system. The electron cytochemistry of the endomembrane system reveals remarkable compartmentalisation of the various enzymes (Domozych, 1989b). Inosine 5'-diphosphatase is located throughout all Golgi cisternae and vesicles associated with the contractile vacuole. Other alkaline phosphatases like thiamine pyrophosphatase, ATPase and inosine 5'-triphosphatase are localised within the trans-face cisternae and vesicles of the contractile vacuole. Inosine 5'-monophosphatase is localised at the plasma membrane and not within the endomembrane system. The acid phosphatases, including cytidine 5'-monophosphatase, NADPase and β-glycerophosphatase are localised in vesicles emerging from the central terminus of the trans-face of the Golgi and in the peripheral vacuolar network (Fig. 6.5).

6.6.3.2   ER-Golgi-vacuole

Hara-Nishimura *et al.* (1985) have clearly demonstrated that transport of many cellular substances can take place with the aid of vesicles, in the case of developing cotyledon cells of *Cucurbita* sp. They showed that trimeric *proglobulin* molecules are synthesised in the endoplasmic reticulum and transported to the vacuoles via dense vesicles, the membranes of which fuse with the tonoplast. Globulin molecules in the vacuoles form crystalloids which become larger in size during the development of cotyledons and finally bud out of the vacuoles giving rise to protein bodies.

A critical work by Herman *et al.* (1994) shows that the central vacuolar membrane originates ultimately from the ER via the Golgi vesicles. They studied the subcellular localisation of vacuolar H^+-ATPase (V-ATPase) by electron microscopic and immunological methods in root cells of oat seedlings. In immature root tip cells, which lack large vacuoles, most of the V-ATPase was localised with ER. The peripheral sub-unit B of the proton pump, was present in tonoplast, Golgi and ER, as well as numerous small Golgi-derived vesicles (0.1 to 0.3 μm diameter), presumably the provacuoles. Electron micrographs suggest that the vacuole precursors, i.e. the provacuoles, could originate from the trans-Golgi network (TGN) and mature vacuoles are formed as a result of

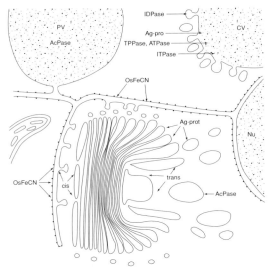

Figure 6.5 Diagrammatic representation of compartmentalisation within the endo-membrane system of *Gloeomonas*. Information is based upon enzyme and silver proteinate cytochemistry presented in this paper and OsFeCN (osmium ferricyanide) labelling. IDPase activity is found throughout all cisternae of the dictyosome and vesicles of the contractile vacuole (CV). TTPase, ATPase, and ITPase activity is found in association with the contractile and trans-face cisternae of the dictyosome. AcPase (incl. CMPase, β-glycerophosphatase, NADPASE) activity is found in vesicles emerging from the transface of the dictyosome and peripheral vacuolar network (*PV*). OsFeCN labelling is restricted to ER and the nuclear envelope. Silver proteinate labelling is found in cisternal peripheries, especially those at the transface. The nucleus (*Nu*) displays no labelling (from Domozych 1989b). With permission of Springer Verlag.

fusion of many small ones. An excellent review on vesicle sorting, targeting and budding through Golgi compartments has been presented by Staehlin and Moore (1995). It is now realised that the cellular and subcellular mechanisms of the secretory system are possibly the same in all eucaryotes, and the Golgi apparatus is the pivotal organelle for the basic organisation of the system. Biochemical, histochemical and immunocytochemical studies have shown that the Golgi apparatus in mammalian cells is composed of four types of cisternae: *cis*, medial, *trans* and trans-Golgi network (TGN), which have definite positions in the stack and characteristic enzyme constituents. The experiment, using a number of antibodies against glycan epitopes on glycoproteins and non-cellulosic polysaccharides, has shown that plant Golgi is similarly compartmentalised (Staehlin and Moore 1995). After due processing of the secretory substance is completed, the TGN serves as the compartment that sorts and packages proteins, lipids, glycoproteins and non-cellulosic polysaccharides in small vesicles. The details of the mechanism will be presented later.

6.7 INHERITANCE OF VACUOLES IN YEAST

Although many workers do not consider yeast cells as typical plant cells, it is still useful to review recent work on the development, inheritance and segregation (division) of vacuoles in *Saccharomyces cerevisiae*. There is no question of *de novo* origin of vacuoles in these cells. The division or inheritance of vacuoles takes place by a distinctly different mode from that of typical plant cells. Whereas the copy number of vacuoles in yeast is one or, at most, a few, the highly watery typical plant cell may show a high copy number, and cytokinesis leads to a *stochastic* inheritance of vacuoles. In yeast, vacuoles undergo what is called *ordered* inheritance during budding and are restricted spatially to an axis between the maternal vacuole and the bud. Detailed cytological

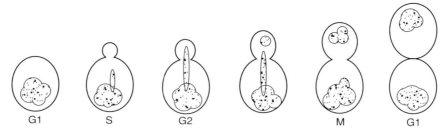

Figure 6.6 Division of vacuole during the cell cycle in *Saccharomyces cerevisiae*. (After Conradt *et al.* 1992.) With permission of Rockefeller University Press.

studies by Weisman and Wickner (1988) revealed that in the newly developing bud, vacuoles are founded by a stream of vacuole-derived membranous vesicles and tubules which are transported from the mother cell to the bud. These vesicular structures fuse in the new bud to form one or a few vacuoles. The vacuole inheritance follows a strict sequence of events during cell cycle, as shown by Conradt *et al.* (1992) in a wild type strain. In the G_1 phase vacuoles are clustered together. During the early S phase, when cells possess a small bud, the maternal vacuoles form a tubulo-vesicular structure which is directed toward the emerging bud (Fig. 6.6). Throughout the S and G_2 phase, materials are transported from the mother cell vacuole into the bud along the tubulo-vesicular pathway. During mitosis this transport is disrupted. The vacuolar material that enters the bud from the mother cell forms the new bud vacuole. From their *in vitro* experiments on vacuolar inheritance, Conradt *et al.* (1994) could detect four biochemically distinct sequential stages of vacuolar fusion in vitro. Stage I requires exposure of vacuoles to solutions of moderate ionic strength. Stage II requires stage I vacuoles and cytosol. In stage III, stage II vacuoles react with ATP to form striking tubular extensions, reminiscent of those observed *in vivo*. At stage IV the vesicles, when present in a certain minimal concentration, fuse homotypically without any further requirement for any soluble component.

The above considerations point to three major pathways by which different types of vacuoles can be formed in plant cells. Besides *de novo* biogenesis in certain cells, vacuoles may develop from (i) ER; (ii) Golgi vesicles and provacuoles; and (iii) lytic processes.

The present information on the biochemical aspects of biogenesis unifies all the apparently distinct pathways.

6.8 BIOCHEMICAL ASPECTS OF BIOGENESIS

6.8.1 Vesicular transport

In conventional electron microscopy, various terms like vesicles, small vacuoles, provacuoles, etc., have been used for indicating spherical single membrane-bound structures containing materials of different electron capacity, which may give rise to vacuoles. Although it may not be possible to correlate the diverse spherical structures observed in different types of cells, it is known that these vesicular structures are involved in the biogenesis of vacuoles. For example, the vegetative storage proteins vsp27 and vsp29 are associated with the Golgi apparatus in the paraveinal mesophyll cells of soybean and can be observed in Golgi-derived vesicles adjacent to the tonoplast (Franceschi and Giaquinta 1983). This indicates that the proteins, synthesised in membrane-bound ribosomes, move through the Golgi cisternae and are packaged into vesicles that fuse with the tonoplast to deliver the proteins (Staswick 1990). Yaklich and Herman (1995) used EM immunocytochemistry to demonstrate that the vacuoles of soybean aleurone cells accumulate a xyloglycoprotein which passes through the Golgi apparatus. They interpreted this to indicate that xylose transfer to the recepient polypeptides occurs in the medial-trans Golgi and the proteins are packaged and eventually deposited in the protein storage vacuole. The vesicles which are involved in the transport of proteins are produced from specialised coated regions of membranes and so bud off as **coated vesicles**. The coating is a cage of proteins on the cytosolic surface of the membranes.

Most of the information on the biochemistry and genetics of different types of coated vesicles has been obtained first from animal and yeast cells, and later from plant cells. In general, all coated vesicles have similar sizes (50–100 mm) but contain different coatings and cargo proteins. As a first step to vesicle formation, the coat proteins physically deform a patch of the donor membrane into a vesicle, sequester the membrane protein and package the cargo inside it. The vesicle then transports the membrane and the packaged protein to the target organelle. Thus the coats govern the traffic of the vesicles between the membrane and the organelles. Once coated vesicles are formed, disassembly of the coat allows vesicle fusion with the target membrane and the coat complexes return to the cytoplasm for additional cycles of vesicle formation (vide review Rothman 1994). Reviewing the state of the art, Springer et al. (1999) present a model for general mechanism of transport vesicle budding. At first a small GTPase is recruited to the donor membrane by its cognate guanine nucleotide exchange factor (GEF). It binds to the membrane in its GTP state. Then an GTPase-activating protein (GAP) recognises both the GTPase and a membrane protein that acts as a primer to form a priming complex. These priming complexes associate laterally and more coat proteins are bound. As the cargo proteins diffuse into the bindings site, they become trapped by their interactions with the coat. On the other hand, the priming complex may decay after the GTPase is stimulated to hydrolyse GTP, yielding free GTPase and a polymeric coat on the membrane. Finally the membrane deforms into a coated bud, giving rise to a vesicle. Three well-characterised classes of coated vesicles are known: clathrin-coated vesicles (CCV), coatomer-coated I (COPI) and coatomer-coated II (COPII) vesicles. Whereas COPI and COPII are generally smaller (50–70 nm dia), the post-TGN CCVs are a little larger (~100 nm). The CCVs mediate transfer among the plasma membrane, endocytic and TGN compartments. The clathrin coat consists of three components: two different complexes of coat proteins (clathrin and adaptor complexes (AP)) and small GTP-binding protein, ADP-ribosylation factor (ARF). Clathrin is a three-legged molecule forming a distinct shape, a triskelion, which is well-adapted for assembly into polygonal arrays characteristic of clathrin coats. AP complexes not only provide a membrane-binding site for clathrin but also interact with trafficking membrane proteins to serve a cargo-selective function. Two related APs have been characterised for clathrin assembly: those budding from the trans-Golgi network (TGN) interact with AP-1, whereas those budding from the plasma membrane interact with AP-2. A third complex, AP-3, first identified in mammalian cell, has been detected in yeast. AP-3 has sequence and structural homology to AP-1 and AP-2 and is found in both Golgi and endosomal membranes. It is needed for an alternative Golgi-to-vacuolar pathway in yeast (Schmid 1997; Cowles et al. 1997; Pishvaee and Payne 1998). The first characterisation of CCVs in plants was made by Lin et al. (1992) in pea cotyledons. These CCVs, which carry cargo proteins from the TGN to the vacuole or a prevacuolar compartment homologous to the mammalian endosome, contain lectin and legumin precursors.

In contrast to CCVs, COPI vesicles mediate intra-Golgi and ER-to-Golgi bidirectional, rather non-specific, transport (Orci et al. 1997; Schekman and Mellman 1997). COPI contains two separable constituents: the coatomer complex and ARF. The coatomer complex consists of seven (α, β, β^1, γ, δ, ϵ and ζ) proteins which assemble into a large complex before attaching as a single unit to the donor membrane. The unit is called a coatomer. The molecular details of the coatomer-coated vesicle formation have been reviewed by Rothman (1994). In brief, the vesicle begins to develop when ARF carrying a guanosine diphosphate (GDP) contacts the membrane. Then an enzyme replaces the GDP with GTP — a change that enables ARF to recruit a number of coatomer proteins. The assembly of ARF and coatomer units on the membrane causes a bud to form and pinch off from the membrane as a coated vesicle. Before the vesicles fuse with the target membrane the whole coat is removed. The conversion of GTP to GDP, by release of a phosphate group, causes removal of ARF and coatomer molecules from the vesicle.

130 Plant cell vacuoles: an introduction

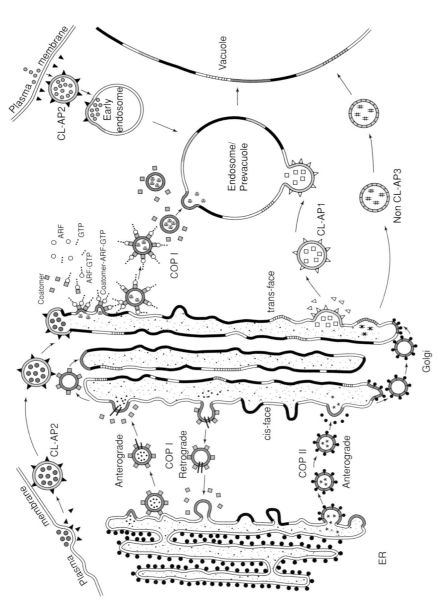

Figure 6.7 Diagrammatic representation of major vesicular trafficking of membrane and soluble proteins between endoplasmic reticulum (ER) Golgi complex, provacuole/endosome and vacuoles. Different clathrin-AP complexes are indicated by different triangles (△, ▲) and COPI and COPII coats are indicated by shaded squares (▨) and solid shperes (●) on the vesicle membrane. Lumen proteins are shown as asterisks (*), small spheres (•), hollow squares (▫) and helices (⚏). Membranes are shown as solid (▬), hollow (═) or banded (▦) to indicate different segments with various integral proteins and lipids.

In contrast to the other two classes of vesicles, the COPII vesicles mediate anterograde (cis to trans) transport from ER to Golgi apparatus. COPII consists of three parts: two coat protein complexes (Sec23/24p and Sec13/31p) and one small GTP-binding protein, Sar1p. COPII vesicles are formed by the concerted action of subunits of the coat that interact with each other and with specific proteins in the ER. Kuehn et al. (1998) investigated the role of COPII coat proteins in cargo selection and recruitment. They demonstrated that the COPII components Sar1, Sec23 and Sec24 form a specific complex with integral membrane and soluble vesicle cargo. A diagrammatic representation of major vesicular trafficking of membrane and soluble proteins between ER, Golgi complex, provacuole/endosome and vacuole is presented in Fig. 6.7.

There is no consensus on the COP-mediated traffic and it is likely that different cell systems operate different modes of trafficking. The possible models, which are not mutually exclusive are:
i) COPI vesicles mediate both anterograde and retrograde traffic.
ii) COPII vesicles direct ER-to-intermediate compartment (IC) (Hauri and Schweizer 1992) or vesicular tubular clusters (VTC) (Bannykh and Balch 1997) and subsequently COPI mediates IC- or VTC-to-Golgi transport.
iii) COPII mediates anterograde and COPI retrogade traffic.
iv) Both COPs mediate independent trafficking.

Coatomer and ARF comprise the essential and rate-limiting components of COPI vesicle formation at the Golgi complex (Orci et al. 1993). Similarly, three cytosolic and peripheral membrane proteins (Sar1p, Sec23/24p, and Sec13/31p) are necessary and sufficient to form functional COPII vesicles from isolated ER membrane fractions (Matsuoka et al. 1998). Considering these — and also earlier observations on clathrin, which forms uniformly sized empty cages at high protein concentrations in the absence of the membrane (Keen et al. 1979) — it may be concluded that the coat subunits are the principal determinant of vesicle morphogenesis. However, the sequential binding of the coat proteins to specific lipids, especially phosphorylated inositol phospholipids, is necessary for the budding of vesicles (Matsuoka et al. 1998). On the other hand, coat proteins should not be considered as indifferent to the cargo they carry. The diversity of the proteins and phospholipids of the coat suggests that the coat proteins, as well as the AP complexes, confer a degree of specificity to the cargo. On the basis of their finding of at least 5 type II cytoplasmically oriented integral membrane proteins on the ER-to-Golgi vesicles, Hay et al. (1997) suggested that if several functionally distinct vesicles exist for anterograde ER-to-Golgi traffic, each may shuttle different sets of cargo molecules, display distinct or partially distinct surface properties and be competent to dock and fuse with specific target sites.

6.8.2 SNARE hypothesis on vesicular transport

Vesicle uncoating and fusion in mammalian cells require a number of cytosolic proteins, including an N-ethylmaleimide-sensitive factor (NSF), a 76kD protein with two ATP-binding motifs, and soluble NSF attachment proteins (SNAPs) (Rothman and Orci 1992).

The current hypothesis of vesicle targeting and fusion states that the specificity required to accurately target vesicles derived from a donor organelle to a distinct acceptor organelle, is mediated by a set of membrane proteins. These proteins are generically referred to as SNARE proteins, which are the attachment proteins for SNAPs during fusion (Söllner et al. 1993). During budding, transport vesicles are tagged with v-SNAREs which serve as specific address markers for the cognate t-SNAREs in the target membrane to dock the vesicle. The general fusion machinery then assembles at the membrane scaffold. Much of the machinery, including the ATPase, NSF and the SNAP proteins, is readily available to all cellular membranes. The 20S docking and fusion particle in synaptic vesicles in mammalian brain membranes consists of NSF, SNAPs, a vesicle-associated membrane protein (VAMP) and syntaxin, another integral membrane protein. Finally the SNAPs bind to the SNARE complex, enabling NSF to bind. ATP hydrolysis by NSF is essential for fusion, leading to bilayer fusion by an unknown biophysical mechanism.

This pairing between v- and t-SNAREs defines a model that is valid for many different vesicle-trafficking steps throughout the secretory pathway, in cells as evolutionarily distant as neurons, yeast or plants. Studies in yeast identified a set of v- and t-SNAREs that mediate transport between the ER and Golgi apparatus. A different set of SNAREs underlies shuttling between the Golgi apparatus and vacuoles (Becherer *et al.* 1996). Similar sets, which are rapidly being identified in *Arabidopsis* and other plants, are discussed later.

The problem of docking specificity, however, is yet to be resolved. Many different cellular functions in eucaryotic cells are regulated by proteins that undergo a cycle of GTP-binding and hydrolysis. When the proteins are charged with GTP it is active and when GTP is hydrolysed to GDP it is inactive. Proteins are also called GTPase because of the GTP hydrolysis that they catalyse. The super family of small GTP-binding proteins contains a subfamily of Rab proteins in mammalian cells whose equivalent is Ypt proteins in yeast. Their homologues have also been identified in plant cells. In view of the presence of a wide range of Rab proteins for different target membranes it has been proposed that these are the principal source of docking specificity. In the presence of a guanine nucleotide exchange factor (GEF) and GTP, a specific Rab protein is attached to the vesicle where it may regulate v-SNARE in a way that v-SNARE would be active only when it is bound to GTP-Rab proteins (Novick and Brennwald, 1993). The Rab family is thought to add an extra layer of specificity to the vesicle fusion process (Lupashin and Waters 1997; Novick and Zerial 1997).

Recent research (Gonzalez and Scheller 1999; McBride *et al.* 1999) indicates that Rab GTPases and their effector molecules interact specifically. This interaction may contribute specificity to membrane trafficking at the levels of vesicle translocation, docking and fusion.

In animal cells, an unresolved puzzle has been the relationship between vesiculo-tubular clusters (VTC) produced by ER and the cis-Golgi network (CGN) which contain similar marker proteins derived from the ER. Presley *et al.* (1997) used green fluorescent protein conjugated to viral glycoprotein to study the ER-to-Golgi transport in living cells. They have shown that, upon export from the ER, fluorescent-tagged proteins become concentrated in many differently-shaped pre-Golgi structures and move rapidly along microtubules by using a microtubule minus-end-directed motor complex of dynein/dynactin. These structures are often large enough to deform in shape during transport, suggesting that they are non-vesicular membrane transport intermediates. Martinez-Menárguez *et al.* (1999) report that VTCs play a vital role in concentration of soluble cargo protein after exiting the ER and before reaching the Golgi. In view of the fact that in most plant cells the cytoplasm is rather thinly populated by various intracellular compartments, surface information and specificities of vesicles or membrane-bound structures may not be enough for targeting to the appropriate organelle. In such cases, movement similar to VTCs along the microtubule may be envisaged. As the methodology of using green fluorescent protein is now established, the existence or otherwise of similar transport mechanism may soon be demonstrated.

6.8.3 Vesicular transport in yeast: SNARE mechanism

As well as studies on mammalian cells, isolation of many temperature-sensitive, secretion mutants of yeast has helped identify the various molecular features of secretory transport. Cloning and sequencing of affected genes and electron microscopy of the cellular ultrastructure have provided a wealth of information to establish the molecular features of the secretory system. The various components of the SNARE hypothesis have been detected in yeast. The slow intrinsic activity of NSF protein is necessary for vesicle fusion. The sequence of yeast protein Sec18p shows that it is a homologue of NSF. Sec17p is the yeast homologue alpha-SNAP, since alpha-SNAP can restore the fusion activity of cytosolic extracts from yeast Sec17 mutants.

As stated earlier, the Ypt subfamily is the yeast-equivalent of Rab proteins in a mammalian system. Sec4p is localised on vesicles as well as on plasma membranes. Whereas Ypt1p is located on vesicles from ER to Golgi, the COPII proteins of these vesicles are made of Sar1p, Sec 13/31p, Sec 23/24p and others (vide Verma et al. 1994).

Schimmoeller and Riezman (1993) reported that the Ypt7 gene encodes the yeast homologue of mammalian Rab7 protein. The Ypt7, a small GTPase, is involved in the regulation of transport steps from late endosomes to vacuoles. Cowles et al. (1994) isolated a protein Vps45p (which is required for the efficient sorting of proteins to the vacuole) from a temperature-sensitive mutant of yeast. They suggested that Vps45p may act to stabilise the interactions of a vacuolar sorting pathway t-SNARE, with a corresponding v-SNARE and thereby facilitate the vesicle fusion event. In the absence of Vps45p, the SNARE complex may rapidly dissociate, hindering vesicle fusion. Alternatively, Vps45p may play a direct role in vesicle recognition of the target membrane. Finally, Vps45p may interact with Rab-like GTPase. It may be noted here that since Vps45 mutant cells contain a vacuole, the tonoplast components are delivered independently of Vps45p function. Vps1p, a high-molecular-weight GTPase (~80kD), is a member of a growing family that is found in a number of organisms and is involved in a variety of cellular processes.

6.8.4 SNARE mechanism in plants

The essential features of various types of vesicular transport between ER, Golgi apparatus, endosomes and vacuoles/lysosomes, as well as the SNARE mechanism as understood in mammalian and yeast systems, have been discussed above. Although the gross pattern of origin and development of vacuoles in diverse plant cells has been very well-documented by both electron microscopical and biochemical methodologies, the molecular mechanism of vesicular traffic and the validity and operation of the SNARE mechanism are yet to be established.

One of the key features of vesiculation and its fusion is the activity of GTP-binding proteins (Staehlin and Moore 1995) which occur as a sub-family of Rab/Ypt genes, first discovered in yeast. A number of homologues of this subfamily have also been identified in plant tissues, e.g. coleoptile of maize, root nodules of soybean and *Arabidopsis*. For example, Rab 6 is a small GTP-binding protein involved in the regulation of vesicle transport from TGN in mammals. Bednarek et al. (1994) cloned from *Arabidopsis*, a homologous gene AtRAB6 which shows 79% homology with mammalian gene, Ryh1 and Ypt6 counterparts. The direct detection and characterisation of GTP-binding proteins on a plant tonoplast have been made by Perroud et al. (1997). GTP-binding assays performed on the extracted tonoplast of spinach (*Spinacia oleracea*) leaves revealed an evident GTP-binding activity. Three groups of proteins of molecular weights of known GTP-binding protein categories were detected in the range 18, 30 and 40kD.

As described earlier, in mammalian and yeast systems 3 types of coated vesicles, viz. clathrin-coated vesicles (CCV) and coatomer-coated, COPI and COPII vesicles, have been identified. In plant cells, although CCVs are widely detected in diverse plant tissues (Beevers 1996), COPI and COPII vesicles are yet to be confirmed. A separate category of rather large vesicles (dia 200 to 400 nm), called precursor-accumulating vesicles (PAC), have been detected in maturing pumpkin seeds by Hara-Nishimura et al. (1998). PACs are responsible for transport of storage proteins from the ER directly to the protein storage vacuoles, bypassing the Golgi complex. Whether they belong to the class of coated-vesicles or endosome/provacuoles is yet to ascertained. They may also appear as homologous to the vesicular-tubular clusters (VTC) in animal cells.

In animal systems mannose-6-phosphate receptors (MPRs) transport newly synthesised lysosomal hydrolases from the Golgi to prelysosomes and then return to the Golgi for another round of transport. Depending on divalent cations for ligand-binding, MPRs are of two types: cation-independent (CI-MPR) and cation-dependent (CD-MPR). Both types use CCVs for export from the TGN. However, MPRs are not known to exist in plants. As discussed later, the sorting signals for proteins reside in short peptide sequences in the protein or its precursor protein. These sorting signals can occur at the amino terminus, carboxy terminus or somewhere else in the protein. In many plants investigated so far, an 80kD protein performs similar functions and is recognised as a potential receptor. Kirsch et al. (1994) identified such a vacuolar targeting receptor with a high molecular weight of 80kD (BP-80) and demonstrated that the N-terminal targeting determinant of alurain interacts specifically with BP-80 protein present in fractions enriched in clathrin-coated vesicles from developing pea cotyledons. Furthermore, the same protein also binds to NTTP involving the NPIR motif of prosporamin from sweet potato and to CTTP of pro-2S albumin from Brazil nut, but not to the CTTP of barley lectin (Kirsch et al. 1996). Finally, it has been established that BP-80 traffics in clathrin-coated vesicles between Golgi and prevacuolar compartments (Paris et al. 1997). Shimada et al. (1997) isolated two potential vacuolar sorting receptors of 72 kD and 82 kD for vacuolar proteins from dense vesicles of developing pumpkin cotyledons. As stated below, three types of adaptor proteins, viz. AP-1, AP-2 and AP-3, couple the CCVs with MPRs to serve as a cargo-selection device. So far, only AP-1 has been detected in plants (Holstein et al. 1994). In animal systems, the signals required for collection of MPRs have been identified to some extent for both CI- and CD-MPRs of AP-1 and AP-2 complexes. The signal sequences in CI-MPR in both AP-1 and AP-2 are chiefly YKYSKV (Tyr-Lys-Tyr-Ser-Lys-Val). Similar presence of tyrosine-based signal sequences that could be involved in recruiting the adaptor proteins into the vesicles have been noted in 80kD receptor-like protein from *Arabidopsis*, pumpkin, wheat germ and developing pea cotyledon (Butler et al. 1997; Beevers and Raikhel 1998). In yeast systems CCVs do not deliver the cargo directly to the vacuole but via the endosome. That there exists a prevacuolar intermediate compartment comparable to endosome in plants is supported by the fact that there is usual uncoating of the CCVs before fusion and the 80 kD protein is detected in such a compartment (Ahmed et al. 1997).

Parallel to the yeast systems, v- and t-SNAREs have also been identified in plant systems. In yeast, PEP12 gene product functions as an endosomal t-SNARE required for the transport of CPY from TGN to the endosome. A homologue of PEP12, AtPEP12 has been isolated from *Arabidopsis* by Bassham et al. (1998). It was localised in a post-Golgi compartment in the vacuolar pathway (Conceicao et al. 1997). Another *Arabidopsis* protein AtELP is a potential vacuolar targeting receptor. It is homologous to yeast transmembrane cargo receptor, is detected at the TGN of root cells and can preferentially interact with TGN-specific AP-1 clathrin-adaptor complex (Ahmed et al. 1997). In keeping with the vacuolar cargo trafficking model of yeast, these workers also observed that AtELP co-localises with AtPEP12p on the prevacuolar compartment. Another t-SNARE, AtVAM3p, is a homologue of yeast Vam3p. This protein is localised to small patches on the tonoplast at the contact between small vacuoles and large vacuoles in *Arabidopsis* system, suggesting that it is involved in membrane–membrane interaction (Sato et al. 1997). As a t-SNARE, yeast VPS45 is involved in transport between TGN and prevacuole, and its *Arabidopsis* homologue AtVPS45p is a peripheral membrane protein. What is significant is that this protein co-fractionates with the potential vacuolar targeting receptor AtELP and partially co-fractionates with AtPEP12p. These observations lead to the conclusion that AtVPS45p functions in vacuolar protein trafficking (Bassham and Raikhel 1998).

Inositol lipids are known to interact directly with proteins and control many cellular activities including intracellular vesicle transport (Simonsen et al. 1998). A 3-phosphorylated inositol lipid, phosphatidylinositol-3-phosphate (PtdIns(3)P), is formed by the action of a phosphoinositol-3-OH Kinase (P13K). In yeast, Vps34p—a peripheral membrane protein—has been identified as a PtdIns3K. It is essential for protein sorting and transport of vesicles from the late Golgi to the vacuole. A P13K gene has been isolated from *Arabidopsis* showing some sequence homology to the yeast VPS34 gene (Welters et al. 1994). Thus the parallelism of plant systems with yeast and animal systems is established and it is hoped that more details of the vesicular transport mechanism of plants will soon be available. The SNARE mechanism may operate only in cases of biogenesis of plant vacuoles which contain proteins, glycoproteins, etc., but not in vacuoles where inorganic substances or simple organic acids, sugars, tannins, etc., are not routed through Golgi cisternae.

The generalised concept of vesicle formation and targeting the vesicle to its destination involves the following steps:
  i) The donor membranes coated by certain cytosolic proteins are shaped into vesicles which contain soluble cargo in its lumen.
  ii) The vesicles are then released in the cytosol, and
  iii) the protein coats dissociated from the vesicle by GTP hydrolysis.
  iv) The targeting proteins on the surface of naked vesicles are recognised by the specific receptor proteins on the acceptor membranes to initiate docking.
  v) Subsequent fusion of the membranes is followed by discharge of the cargo in the lumen of the target organelle or out in the periplasmic space.

6.8.5 Mechanics of fusion

The available information from diverse cell systems suggests that regulated fusion of transport vesicles with organelles and vacuoles requires ATPases (Sec18/NSF fusion protein), accessory proteins (Sec17/soluble NSF-attachment proteins (SNAP)), integral membrane SNAP receptors (SNAREs), and GTPases (Rab/Ypt). To ascertain the validity of the v- and t-SNAREs in vacuolar fusion in a yeast system, Nichols *et al.* (1997) used two strains of yeast, one lacking the vacuolar alkaline phosphatase (Pho8p) and the other lacking the proteinase A (Pep4p) that processes and thereby activates pro-alkaline phosphatase. Fusion of purified vacuoles from these strains would lead to mixing of vacuole contents and hence production of phosphatase activity, which is assayed spectrophotometrically. They demonstrated that both the SNAREs are normally present and vacuoles containg the v-SNARE can fuse with those containing the t-SNARE. Vacuoles containing neither SNARE cannot fuse with those containing both, demonstrating that docking is mediated by cognate SNAREs on the two organelle membranes.

The distinct ordered phases that occur before **fusion** of vacuoles are variously called **priming** and **docking**; priming, tethering and docking (Peters and Mayer, 1998); or tethering and docking (Schekman, 1998). Docking consists of two steps: tethering and SNARE pairing (Ungermann *et al.* 1998). Despite this semantic confusion, the essential steps leading to fusion are being unravelled. Ungermann *et al.* (1998) reported that during priming, pre-existing v/t SNARE complexes containing target membrane t-SNARE Vam3, the SNAP-25 homologue Vam7 and the v-SNARE Nyv1, Sec17/a-SNAP, Sec18/NSF and stabilising factor LMA1, are disassembled by Sec17/Sec18/ATP. The primed vacuoles undergo docking which consists of two sequential reactions: a reversible **tethering** mediated by the GTPase Ypt7 and SNARE **pairing**, in which SNARE proteins from opposite membranes form a complex in *trans*.

Weber *et al.* (1998) define the minimal machinery for the fusion of functionally different membranes (heterotypic) or identical membranes (homotypic). From an evaluation of available experimental evidence they argued that cognate v- and t-SNARE proteins are most likely sufficient as the minimal recognition machinery for pairing specific membranes as partners for subsequent fusion. With the aid of quick-freeze/deep-etch electron microscopy, Hanson *et al.* (1997) have shown that SNARE complexes are 14 nm long, 2 nm wide rods. Cognate v-and t-SNAREs bind each other via membrane-proximal heptad repeat regions that may form coils or closely related helical bundles. The membrane anchors of v-and t-SNAREs emerge at the same end of the rod (Lin and Scheller 1997), suggesting that the rod must lie approximately in the plane of contact between the vesicles paired by a complex of a v-SNARE in one with a t-SNARE in the other, a structure which Weber *et al.* (1998) term a SNARE pin. They have shown that recombinant v- and t-SNARE proteins reconstituted in separate liposomes assemble into SNARE pins–SNARE complexes linking two membranes. This leads to spontaneous fusion of the docked membranes at physiological temperature. Weber *et al.* suggest that SNARE-pins are the sufficient and minimal machinery for cellular membrane fusion, and that SNAPs and ATPase NSF are 'add-ons' needed for repeated use of a bilayer fusion hairpin. At the final stage of fusion, as suggested by Peters and

136 Plant cell vacuoles: an introduction

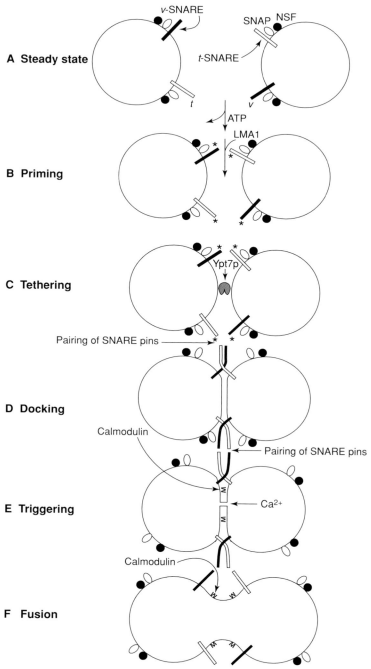

Figure 6.8 A generalised model for membrane or vesicular fusion. **a** Both the membranes are provided with cognate v- and t-SNARE pins (⌒ ⌒), α-SNAP (○) NFS (●) **b** Priming (*) of the membranes takes place with the aid of ATPase activity (NSF) and is stabilised by LMA1. **c** Primed membranes are tethered together by a small GTP-binding protein (Ypt7p). **d** Membrane docking involves pairing of the cognate SNARE pins and bridging two membranes. **e** Membrane docking triggers the local activation of Ca^{2+} channel leading to Ca^{2+} flow from vesicular pool. **f** As the Ca^{2+} sensor, calmodulin (**M**) binds tightly to vacuoles and the Ca^{2+}-calmodulin complex promotes bilayer mixing, causing fusion.

Mayer (1998), docking produces an unknown signal leading to the transient opening of Ca^{2+} channels. The resulting Ca^{2+} efflux is perceived by calmodulin which is the first protein of the post-docking phase to be identified. Calmodulin then binds to the vacuoles more tightly and activates enzymes that catalyse bilayer/contents mixing. A generalised model for fusion of vacuolar membranes is depicted in Fig. 6.8.

Other Ca^{2+} binding proteins may play a certain role in fusion. Moss (1997) noted that annexins, a family of proteins, can play a role in vesicle fusion, especially in exocytosis and endocytosis in animal systems (Creutz 1992; Burgoyne and Clague 1994). Similarly, plant annexins are also involved in fusion (for review see Clark and Roux 1995). Recently, Seals and Randal (1997) have shown that VCaB42, an annexin, is associated with vacuolar vesicles at a physiological level of Ca. From its pattern of expression relative to the expansion of cells, they suggested that VcaB42 participates in the early events of vacuole biogenesis by fusion.

6.8.6 Prevacuoles as intermediate structures

The general scheme of development of lysosomes in animal cells is as follows:

$$ER \rightarrow \text{coated vesicle} \rightarrow Golgi \rightarrow \text{Golgi vesicle} \rightarrow Endosome \rightarrow Lysosome$$

The same sequence of events also occurs in yeast cells. In addition to the coated-vesicles which are small in dimension (50–100 nm dia), repeated references to larger structures, variously called prevacuole, cytosome, small globular vacuole, etc., have been made. In mammalian cells, soluble lysosomal enzymes are sorted from secreted proteins in the trans-Golgi network and transported to the lysosome via an endosomal compartment (Griffiths and Simons 1986; Dahms *et al.* 1989). In order to appreciate the trafficking of lumen and membrane proteins, it is necessary to examine the prevacuole/endososme as an intermediate structure in yeast.

During the last decade, a large number of mutants of *Saccharomyces cerevisiae*, which affect various aspects of the vacuole, have been isolated and characterised. For example, vacuole protein targeting (*vpt*) mutants may (i) exhibit defects in proper localisation and processing of vacuolar proteins, (ii) affect biogenesis of normal vacuoles, (iii) affect structural and functional characteristics of vacuoles, (iv) affect vacuole membrane assembly, or (v) accumulate vesicles and membrane-enclosed compartments showing no resemblance to normal vacuoles (Banta *et al.* 1988). A number of yeast *vac* mutants, which are defective in vacuole inheritance, vesicle fusion or vacuolar protein targeting have also been identified (Haas *et al.* 1994; Conradt *et al.* 1994). Similarly, analyses of a number of vacuole protein sorting (*vps*) mutants have yielded a wealth of information (Rothman *et al.* 1990; Nothwehr *et al.* 1995). On the basis of cytological and genetical studies of these mutants, it may be concluded that in most cases, lipid and protein constituents of the vacuole membrane are transported together with the luminal soluble components and enzymes via a common vesicle carrier.

Direct evidence has been presented by Vida *et al.* (1993) that a vacuolar hydrolase carboxypeptidase Y is sorted away from secretory proteins in the late Golgi-cisternae (TGN) and transits to the vacuole via an endosome-like prevacuolar compartment. This structurally separate compartment contains a precursor form of carboxypeptidase Y (p2CPY), the Golgi complex endoprotease (Kex2p), and the vacuole markers alkaline phosphatase and mature CPY. Since the endocytosed mating pheromone alpha-factor co-fractionates with the vacuole markers, Vida *et al.* proposed that the endosome is the delivery compartment of proteins from both biosynthetic and endocytic routes (Fig. 6.9). Using completely different techniques, Schimmoeler and Riezman (1993) provided independent evidence that the endocytosed pheromone alpha-factor and vacuolar hydrolase came in direct contact in the prevacuolar endosome compartment. Thus the pathways of vacuole biogenesis and endocytosis in yeast intersect in the endosomal membrane system. Additional evidence is available that Kex2p, as well as the Golgi membrane protein DPAP-A, moves to the vacuole membranes via the prevacuolar or the late endosomal compartment in yeast (Nothwehr *et al.* 1995). Another type of vesicle for targeting proteins from cytoplasm to vacuoles

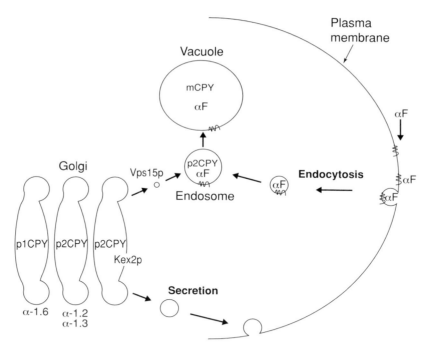

Figure 6.9 Compartmental model for protein delivery to the vacuole through biosynthetic and endocytic routes in yeast. The mating pheromone alpha-factor (αF) is shown undergoing receptor-mediated endocytosis and degradation in the vacuole after transport from an endosome. The Golgi-modified p2CPY precursor is shown undergoing the sorting step in the late Golgi before transport to an endosome followed by maturation in the vacuole (from Vida et al. 1993). With permission of Rockefeller University Press.

in yeast has been described recently by Baba et al. (1997). The precursor form of vacuolar hydrolase aminopeptidase I is localised mostly in restricted regions of the cytosol as a complex with spherical particles, termed as Cvt (cytoplasm to vacuole targeting) complex. During vegetative growth, the Cvt complex is selectively wrapped by a membrane sac forming a double membrane-bound structure of ~150 nm diam, which fuses with the vacuolar membrane. Hence, the mechanism of protein synthesis and delivery from ER to vacuole via vesicles and prevacuole in yeast is identical to that in animal cells. It is expected that a similar mechanism may operate in those plant cells, especially where the vacuole has hydrolytic properties and shows presence of vesicles and/or prevacuoles. As stated previously, many studies have pointed to the role of provacuoles in vacuole formation in plants. Even in cases where hydrolytic activity is not involved, provacuoles may be an intermediate structure. For example, Parham and Kaustinen (1977) demonstrated that tannin synthesis takes place in small provacuoles which may fuse with the central vacuole.

6.8.7 Membrane trafficking

The phenomenon of *membrane flow*, or trafficking, deals with the flow of a part of the membrane either as a small piece or as a macromolecular assemblage from one donor structure or membrane to another target membrane. It takes place as a regular feature in the life cycle of most cells. Together with this flow, lipids and proteins may undergo modification both in quantity and quality. This feature differentiates the recipient membrane from the donor membrane. The change or modification of the membrane is supplemented by deposition, intercalation or integration of lipids and proteins synthesised elsewhere or donated by other membranes. Thus, a growing organelle and its membrane component may have very specific lipid and protein composition.

Electron microscopic studies on the biogenesis of vacuoles and the development of the vacuole membrane have been presented earlier in this chapter.

6.8.7.1 Independent sorting of membrane and lumen protein

In certain cases it is possible to distinguish between membrane protein and soluble proteins, enzymes and other substances in the lumen of the vesicle. The study of mutations in the VPS45 gene in yeast by Cowles *et al.* (1994) is a case in point. The VPS45 mutants are temperature sensitive, mis-sort multiple vacuolar hydrolases and have a single vacuolar structure. The fact that these mutant cells contain a vacuole suggests that vacuolar membrane constituents are delivered independently of Vps45p function.

6.8.7.2 Membrane protein sorting

In animals integral proteins of the plasma membrane, secretory vesicles and lysosomes leave the ER and cross the Golgi stack together. The pathway of a number of membrane proteins in yeast is becoming clear. Lately, Becherer *et al.* (1996) have reported that the PEP12p integral membrane yeast protein is associated with both the endosome fraction and the Golgi marker Kex2p pool. They suggest that PEP12p may mediate the docking of Golgi-derived transport vesicles at the endosome. Again, in a totally different membrane system of root nodules formed by infection of *Rhizobium* in leguminous plants, a peribacteroid membrane (PBM) surrounds the bacteria. Since PBM is a mosaic membrane with properties common to both plasma membrane and vacuoles, there must be certain integral membrane components common to Golgi cisternae, vesicles and the other membranes (Cheon *et al.* 1993). A mutation in yeast clathrin heavy chain protein ChcLp, prevents Kxe2p endoprotease, a Golgi-membrane protein responsible for processing of the alpha-factor precursor and DPAP-A (dipeptidyl aminopeptidase A), to proceed directly to the prevacuolar compartment, but indirectly via the plasma membrane (Seeger and Payne 1992). Nothwehr *et al.* (1995) also reported that Vs1p, a GTPase, is required for all membrane traffic from the Golgi to the prevacuolar compartment. In the absence of Vps1p, a vacuolar membrane protein ALP (alkaline phosphatase) and a Golgi membrane fusion protein A-ALP are delivered to the vacuole via the plasma membrane. Thus it is possible that in yeast Chc1p and Vps1p act together in the formation and budding of vesicles from the late Golgi that are bound for the prevacuolar compartment (Fig. 6.10) (Nothwehr *et al.* 1995).

6.8.7.3 Lumen protein sorting

As a result of processing in the ER many different types of glycoproteins (identical oligosaccharide chains of 10 sugar units) arrive at the *cis* face of Golgi stock. What happens to the glycoproteins in the TGN depends on their destination in the cell. The secretory proteins and plasma membrane proteins generally lose most of their mannose units and gain sialic acid, galactose and N-acetylglucosanine units. However, the lysosomal proteins acquire a different modification in the Golgi complex; their N-linked oligosaccharide chains acquire terminal mannose-6-phosphate residues that enable the hydrolases to bind to mannose-6-phosphate-specific receptors (MPRs) with high affinity. The resulting MPR/lysosomal enzyme complex is transported from the Golgi to an acidified prelysosomal compartment where the low pH induces the complex to dissociate (Kornfield 1992). The released lysosomal enzymes are packaged into lysosomes while the receptors either return to the Golgi to repeat the process or move to the plasma membrane where they function to internalise exogenous ligands (vide review Munier-Lehmamn *et al.* 1996).

Acidification of the lysosome or vacuole is a major factor for proper sorting of the luminal proteins. If the acidification of yeast vacuoles by V-ATPase activity is inhibited by the drug bafilomycin A1, soluble vacuolar hydrolases carboxypeptidase Y (CPY) and proteinase A (PrA) are accumulated and secreted but the tonoplast protein alkaline phosphatase (ALP) remains unaffected (Banta *et al.* 1988; Klionsky and Emr 1989). This suggests that there are differences in the

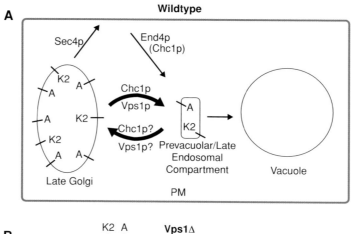

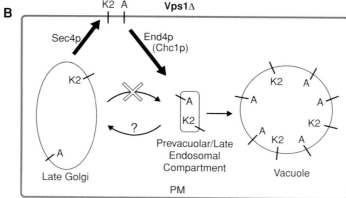

Figure 6.10 Model for the role of Vps1p in membrane trafficking events in yeast. DPAP A and Kex2p are represented by the letters *A* and *K2* respectively, whereas the plasma membrane is represented by *PM*. The *thick* and *thin* arrows represent pathways that are taken and not taken by late Golgi membrane proteins respectively. In wild type cells (*A*) Vps1p and Chc1p are required for all membrane traffic from the late Golgi to the prevacuolar compartment, and their possible involvement in the retrieval step is noted by question marks. The requirement of Sec4p for membrane traffic from the late Golgi to the plasma membrane and of End4p and Che1p for endocytosis are also indicated. In *vps1* Δ mutant cells (*B*) all membrane traffic from the late Golgi to the prevacuolar compartment is blocked and the delivery of integral membrane proteins to the vacuole occurs via the plasma membrane (from Nothwehr *et al.* 1995). With permission of Rockefeller University Press.

targeting mechanism for soluble and membrane proteins. Whereas information required for tonoplast localisation of ALP resides in the cytoplasmic tail and transmembrane region of ALP (Klionsky and Emr 1990), the tonoplast protein DPAP-B appears not to contain any definitive vacuolar targeting region (Roberts *et al.* 1992). Curiously, another vacuolar hydrolase, carboxypeptidase S, which is synthesised at the ER as a type II integral membrane protein after transit through Golgi apparatus, enters the vacuole as acid-dependent soluble protein (Morano and Klionsky 1994). The workers suggested that there are multiple pathways for vacuolar membrane protein transport, represented by classes of membrane proteins that arrive by default (DPAP-B), by rapid recognition of cytoplasmic targeting determinants (ALP) or by a pH sensitive, slower active sorting process (CPS). Lately more information has become available from the study of Pep7p proteins of yeast by Webb *et al.* (1997). They reported that Pep7p function is required for the transport of the Golgi-precursors of the soluble hydrolase CPY, PrA and PrB to the endosome

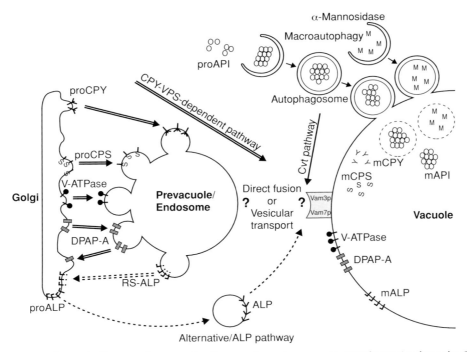

Figure 6.11 Schematic diagram of transport pathways of membrane and luminal proteins from the Golgi apparatus to the vacuole. The Cvt pathway (thin arrow) which can be distinguished from macroautophagy only by kinetic studies is represented by aminopeptidaseI (API) and alpha-mannosidase, both of which are luminal proteins. The CPY (VPS-dependent) pathway (double arrow) is responsible for a majority of vacuolar proteins, both luminal and membrane. The pathway is represented by carboxypeptidaseY (CPY) (Y), carboxypeptidase S (CPS)(S), vacuolar ATPase (V-ATPase)(♦) and dipeptidyl aminopeptidase A (DPAP-A) (▬). DPAP-A undergoes retrieval from vacuole to endosome to Golgi due to a FXFXD tail motif. The alternative ALP pathway (broken arrow) transports membrane protein alkaline phosphatase (ALP) to the vacuole without passing through the prevacuole. Transplantation of FXFXD by a retention sequence (RS) motif may cause its retrieval to Golgi. Prefix 'm' indicates mature form of the protein in the vacuole.

and concluded that Pep7p functions as a regulator of vesicle docking and/or fusion of the endosome. However, Pep7p is not needed for transport of tonoplast hydrolase ALP.

6.8.8 Transport pathways

Although there is a large amount of data on protein trafficking in yeast (vide Sato *et al.* 1998; Bryant *et al.* 1998), there appears to be no hard and fast rule on the co-transport of lumen and membrane protein trafficking. The consensus (Bryant and Stevens 1998; Klionsky 1998) is for three distinct pathways in yeast through which luminal and membrane proteins are transported to the vacuole as follows (Fig. 6.11):

 i) VPS-dependent or CPY pathway
 ii) Alternative or ALP (alkaline phosphatase) pathway
 iii) Cytoplasm-to-vacuole targeting or Cvt pathway.

6.8.8.1 VPS-dependent or CPY pathway

A number of soluble vacuolar hydrolases, e.g. carboxypeptidase Y (CPY), carboxypeptidase S (CPS) Proteinase A (PrA), Proteinase B(PrB), and tonoplast proteins, e.g. V-ATPase, dipeptidyl aminopeptidase A (DPAP), are trafficked through the seretory pathway (Klionsky *et al.* 1990,

Raymond *et al.* (1992). The CPY pathway is characterised by a receptor mediated system in which VPS10 gene products are essential. After synthesis in the ER and processing through the Golgi, CPY emerges as p2CPY in the late Golgi and binds with VPS10p. The receptor-ligand complex is then transported by Golgi-derived vesicles to the endosomal or prevacuolar compartment (PVC). As p2CPY dissociates from the VPS10p, it is released in the lumen of the PVC and the receptor is recycled back to the Golgi. The same receptor also recognises PrA and is responsible for its delivery to PVC for onward delivery to the vacuole. Inside the vacuole, p2CPY is cleaved by vacuolar proteases into its active, mature 61kD form (mCPY). Like CPY, CPS, a type II membrane protein, is transported via the secretory pathway to the vacuole as inactive precursor pCPS and upon delivery is cleaved at a site just lumenal to the transmembrane domain to yield a soluble active mature form mCPS (Burd *et al.* 1997; Cowles *et al.* 1997b). Oberbeck *et al.* (1994) found the most important transmembrane protein of the tonoplast V-ATPases, with a complete set of peripheral and membrane integral subunits in highly purified fractions of ER and Golgi, in maize root cells. It is suggested that V-ATPase holoenzyme is completely assembled in ER and is sorted via the Golgi to the tonoplast. Hillt and Stevens (1995) indicated that 100kD subunit Vph1p follows essentially the CPY pathway and is finally deposited in the vacuolar membrane. As stated earlier, DPAP-A, a resident transmembrane protein, follows the CPY pathway. It travels from TGN to the prevacuolar compartment and subsequently to the vacuole. Nothwehr *et al.* (1993) and Bryant and Stevens (1997) have shown that DPAP-A is retrieved from the PVC back to TGN due to a motif FXFXD within the cytosolic tail of the protein.

6.8.8.2   Alternative/ALP pathway

In contrast to the membrane and lumenal proteins described above, the integral membrane protein alkaline phosphatase (ALP) does not travel to PVC, but bypasses it and moves directly to the vacuole (Cowles *et al.* 1997a; Piper *et al.* 1997). The 33-residue N-terminal cytosolic tail of ALP is both necessary and sufficient for the diversion of the membrane protein from the VPS-dependent CPY pathway to the alternative route to the vacuole. The ablation of the 33-residue from ALP results in trafficking of the protein to the PVC and subsequently to the vacuole following the CPY pathway. Thus ALP defines a novel sorting pathway from the TGN to the vacuole. Adaptor protein AP-3, which is not associated with clathrin, is involved in trafficking of ALP along the alternative pathway (Cowles *et al.* 1997b). On the other hand, Bryant *et al.* (1998) have shown that a hybrid ALP with a retention sequence (RS-ALP) containing FXFXD in its cytosolic tail causes a retrograde membrane trafficking out of the vacuole to PVC to TGN. This study questions the idea that the vacuole is the terminal compartment in protein trafficking.

6.8.8.3   Cytoplasm-to-vacuole targeting (Cvt) pathway

The vacuolar proteins that are not processed through TGN may be transported directly from the cytosol to the vacuole. A great deal of work has been conducted on such a protein aminopeptidase I (API) which is a soluble vacuolar hydrolase (Klionsky 1998). Unlike most vacuolar proteins API does not travel along the CPY- or ALP-secretory pathway and it does not require trafficking out of the ER or through the Golgi apparatus. In other words, API is synthesised in the cytosol and transported directly through the Cvt pathway (Klionsky 1998). After synthesis, the precursor API oligomerises as a dodecamer of 732kD which is too large to pass through a protein channel and is not contained in any small coated vesicle. The dodecamer forms complexes that are surrounded by a sequestering double membrane (Scott *et al.* 1997), and the complex behaves like an autophagosome. The outer membrane of the Cvt vesicle fuses with the tonoplast and allows the release of the inner single membrane vesicle into the vacuole lumen. This vesicle is broken down by vacuolar hydrolases, releasing API into the lumen. A similar process occurs for transport of alpha-mannosidase I (Yoshihisa and Anraku 1990). The Cvt pathway is very similar to autophagy during which a part of the cytoplasm is enclosed by a double membrane and internalised in the vacuole. The autophagic pathway, which is generally induced by nutrient deprivation, causes degradation of large amounts of intracellular protein. Whereas the Cvt pathway is specific for designated proteins, autophagy is non-specific. The two pathways display different kinetics. Whether the process is the Cvt pathway or autophagy is determined by three criteria, viz. the conditions of import, the rate, and the yield.

The validity of the SNARE mechanism in each of the three pathways may be examined in brief. Wada et al. (1992) identified several mutants of the VAM7 gene, which are defective in vacuolar assembly and morphogenesis. The VAM7 deletion mutants lack normal vacuoles and, instead, accumulate numerous vesicular structures that contain vacuolar proteins. The VAM3 gene, on the other hand, which is essential for autophagic and biosynthetic protein transport to the vacuole, is a syntaxin homologue (Darsow et al. 1997). Lately, Sato et al. (1998) indicated that Vam7p is associated with the tonoplast and functions at a late stage in vacuole delivery of multiple proteins that transit via distinct biosynthetic pathways. Moreover, genetic and physical analysis have shown that Vam7p functionally interacts with the vacuolar t-SNARE Vam3p. Sato et al. propose that docking and/or fusion of prevacuolar transport intermediates from at least three distinct biosynthetic pathways to the vacuole (described above) requiring the SNAP-25 family member Vam7p and the syntaxin homologue Vam3p. What is still intriguing is the question of multiple **delivery** machanisms when the same **targeting** mechanism operates for all three pathways in yeast. Klionsky (1998) offers three possible explanations: (i) Multiple delivery mechanisms for resident hydrolases will ensure the passage of at least a minimal complement of vacuolar enzymes, (ii) in contrast to other eucaryotes, yeast vacuoles have to undertake diverse functions in the same cell, and (iii) some proteins like API may oligomerise for their function and/or stability and being unable to enter the vacuole through the secretory pathway have to resort to an autophagic mechanism. Although great strides have been made in recent years to comprehend the molecular mechanisms of biogenesis of vacuoles, much remains to be learned.

6.9 SORTING SIGNALS FOR PROTEINS

Proteins may be sorted to plant vacuoles by four possible signalling systems:
  i) The **positive signals** in the polypeptide domains target the proteins directly to the vacuole.
  ii) The absence of any sorting signals leads to **direct uptake** of the protein by the vacuole.
  iii) When positive signalling for some other compartment fails, the proteins enter the vacuole by a **default mechanism**.
  iv) In the absence of any signalling at the molecular or macro-level, proteins enter the vacuole by **autophagy**.

6.9.1 Positive signalling for plant proteins

The molecular mechanism of protein targeting through vesicle formation in mammalian cells and yeast is well known and its parallel in plant systems is gradually becoming understood, but there has been more emphasis on analysing the polypeptide domains that target proteins directly to the vacuoles. Such positive signals for sorting to lysosomes in animal systems are quite uniform, in that as the protein progresses through the Golgi cisternae, it undergoes precise glycosylation. As discussed earlier in detail, mannose-6-phosphate residues which are post-translationally added to the oligosaccharide chains of the lysosomal proteins in the Golgi-apparatus, are recognised by specific receptors in the trans-Golgi-network. This recognition leads to the packaging of the hydrolase-Man-6-P receptor complexes into clathrin-coated vesicles, which ultimately results in the transport of the hydrolases to the lysosome (Pfeffer and Rothman 1987; Kornfeld and Mellman 1989).

The mechanism of synthesis, transport, sorting, modification and maturation of a vacuolar hydrolytic enzyme in yeast has been studied well in the case of serine-protease carboxypeptidase Y (CPY). CPY, like many vacuolar enzymes is synthesised as a large inactive precursor and enters the

144 Plant cell vacuoles: an introduction

Figure 6.12 A cartoon by M. Toufton in Rothman et al. TIBS, 14, 347, 1989. With permission of Elsevier Science.

lumen of the endoplasmic reticulum where it receives four asparagine-linked core oligosaccharides (Hasilik and Tanner 1978a and b; Jones 1984). The glycosylated polypeptide is then moved to the Golgi complex where additional oligosaccharide modification occurs and where sorting from non-vacuolar proteins takes place. The conversion of the pro-CPY to the mature enzymatically active protease takes place upon arrival in the vacuole, by cleavage of an amino terminal 8kD propeptide (Stevens et al. 1982). Although like mammalian lysosome proteins, yeast cells do phosphorylate mannose residues on CPY, they do not require a mannose-6-phosphate determinant for targeting as has been shown by tunicamycin inhibition of glycosylation. Unlike the mechanism of protein sorting of typical mammalian lysosome, CPY targeting to yeast vacuole is not dependent on modification of its core oligosaccharides. In contrast, it has been shown by mutagenesis and hybrid protein studies that the determinant of protein-sorting resides in the polypeptide structure of pro-CPY. Valls et al. (1987) concluded that the N-terminal propeptide of CPY carries out three functions: (i) it mediates translocation across the ER, (ii) it renders the enzyme inactive during transit, and (iii) it targets the molecule to the vacuole. In agreement with the above conclusion, Johnson et al. (1987) reported that the N-terminal 20 amino acids of CPY act as a signal sequence to direct and/or assist in its translocation to ER. After cleavage of this peptide in the ER lumen, a new terminal segment is exposed (amino acids 21–50) that functions as a protein sorting signal which is both necessary and sufficient to direct vacuolar delivery of the enzyme. On the basis of region-directed mutagenesis using randomised oligonucleotides, Valls et al. (1990) concluded that four contiguous residues Gln-Arg-Pro-Leu (QRPL) constitute the core of an N-terminal topological element, immediately following the ER signal sequence, that is necessary for targeting the

Sweet potato sporamin A	H S R F	**NPIRLP**	T T H E P A
Barley aleurain	F A D S	**NPIRPV**	T D R A A S
Potato 22 kDa protein	F T S E	**NPIVLP**	T T C H D D N
Potato cathepsin D inhibitor	F T S Q	**NLIDLP**	S

Figure 6.13 a Common motif in representative N-terminal propeptides.

Barley lectin	V̇ F Ȧ E Ȧ I Ȧ A N S T L̇ V̇ A E
Rice lectin	D G M̈ Ä Ȧ I L̇ A N N Q S V̇ S F E G ï E S V̇ Ȧ E L̇ V̇
Nicotiana tabacum chitinase	N G L̇ L̇ V̇ D Ṫ M̈
N. tabacum β-1,3-glucanase	V̇ S G G V̇ W D D S V̇ E T N Ȧ Ṫ Ȧ S L̇ V̇ S E M̈

Figure 6.13 b C-terminal extensions of vacuolar proteins; hydrophobic residues are indicated by dots (from Chrispeels and Raikhel 1992). With permission of Cell Press.

proCPY to the yeast vacuole (Fig. 6.12). A comparison of the sorting signals of procathepsin L, procathepsin D, CPY and proteinase A allows a consensus sequence motif 'serine-X-X-(+)-X-leucine' and a less-inclusive yeast sequence 'serine-X-X-(+)-proline-hydrophobic', where '+' indicates a basic or positively charged amino acid (McIntyre *et al.* 1994).

Like the yeast-soluble vacuolar proteins, plant vacuolar targeting information is located in N- or C- terminal propeptides or within the mature structure of the proteins. The vacuolar storage protein of tuberous roots of sweet potato sporamin and the vacuolar thiol protease aleurain from barley contain their targeting information within an N-terminal propeptide (NTPP) (Matsuoka and Nakamura 1991; Holwerda *et al.* 1992). A comparison of the amino acid sequences of these NTPPs, as well as those of potato 22kD protein and potato cathepsin D inhibitor, reveals that they share a short region of hydrophobic amino acids with a hydrophilic residue, Arg, in the centre: Asn-Pro-Ile-Arg-Leu-Pro (NPIRL/ P); Asn and Ile are conserved, Pro, Arg and Leu can be substituted (Fig. 6.13a). NPIRL/P in sporamin is critical for sorting, since a glycine substitution for the conserved isolucine or asparagine residues in the targeting sequence of sporamin results in the secretion of prosporamin from the cell.

On the other hand, the propeptides of C-terminus (CTPP) of a few vacuolar proteins contain vacuolar sorting signals. The CTPP of barley lectin is a hydrophobic 15-amino acid peptide which is necessary for proper sorting to the vacuole. Other Gramineae (Poaceae) lectins, such as wheat germ agglutinin and rice lectin, contain similar proteolytically processed CTPP and serve the same function (reviewed in Raikhel and Lerner 1991). Similarly, the heptapeptide CTPP vacuolar chitinase of tobacco is responsible for vacuolar sorting. When the C-terminal extension of barley lectin or tobacco chitinase A was engineered on to a secreted protein, cucumber chitinase, the resulting chimeric molecules were sorted to the vacuole, thereby demonstrating that these propeptides are sufficient for vacuolar targeting (Neuhaus *et al.* 1991). A similar heptapeptide in β-1, 3-glucanase may also contain sorting information (reviewed in Bol *et al.* 1990). A comparison of CTPP in a number of lectins and hydrolases (Fig. 6.13b) shows no consensus sequence, yet they are rich in hydrophobic residues and the sorting machinery may recognise this feature (Chrispeels and Raikhel 1992). Later analysis of barley lectin CTPP indicated that two short amino acid stretches—FAEAI and LVAE—are each sufficient for sorting of lectin to the vacuole (Bednarek and Raikhel 1992). Unger *et al.* (1994) recorded the amino acid sequences of the C-terminal extensions of vacuolar soluble isoenzyme I of β-fructofuranosidase (ßF) invertase, as Phe-Pro-Phe-Asp-Gln-Leu and that of isoenzyme II as His-Phe-Phe-Ala-Asp-Leu-Val-Ile. In analogy to the studies on barley lectin and tobacco chitinase, they suggested that the C-terminal extensions of the βF isoenzymes I and II may contain the information for vacuolar targeting.

Besides the cases of plant vacuolar protein sorted by NTPP or CTPP signals, some proteins contain targeting information within the mature molecule. In the field bean legumin, two segments of mature protein were shown to be effective in targeting to the vacuole: an N-terminal segment of 281 amino acids and a C-terminal segment of 76 amino acids. In bean phytohemagglutinin (PAA) an internal segment (amino acids 83–113) was both sufficient and necessary for directing yeast invertase to vacuoles in transgenic plants (Saalbach *et al.* 1991, Von Schaewen and Chrispeels 1993). However, these protein segments share no sequence identity. A 20-amino acid loop near the N-terminus of PAA includes the sequence Leu-Gln-Arg-Asp (LQRD) which is similar to the yeast CPY, Leu-Gln-Arg-Pro (LQRP). Targeting of 2S albumin of Brazil nut (*Bertholletia excelsa*) to the vacuole is mediated by CTPP of 4 amino acids and the adjacent 16 amino acids of the mature portion of the protein (Kirsch *et al.* 1996).

Collaborative research by K. Nakamura's group in Japan and N.V. Raikhel's group in the USA showed that NTPP and CTPP propeptides are interchangeable. Fusion proteins, in which NTPP is fused to the mature part of barley lectin or CTPP is fused to the sporamin, are localised in the vacuole of transgenic tobacco cells (Schroeder *et al.* 1993). Their experiments on double-labelling immunohistochemistry of thin sections show that both the proteins are targeted quantitatively to the same vacuole. Presumably, the same vacuolar protein-sorting machinery recognises both the proteins equally. However, Matsuoka *et al.* (1995) have shown that wortmannin, a specific inhibitor of mammalian phosphatidylinosital 3-kinase, completely inhibits CTPP-mediated vacuolar transport without any effect on NTPP-mediated transport. In the same system, Koide *et al.* (1997) have recently shown that the vacuolar-targeting function of the NPIRL sequence is not strictly dependent on its location at the N-terminus and that a C-terminallly located mutant NTPP can acquire some physicochemical properties of the C-terminal vacuole-targeting sequence. On the other hand, at least two different mechanisms may exist for vacuolar sorting in the same cell. Paris *et al.* (1996) observed that the same plant cell may contain two functionally distinct vacuolar compartments characterised by two different tonoplast intrinsic proteins alpha-TIP and TIP-Ma 27. Barley lectin is always present in alpha-TIP vacuoles but never in TIP-Ma 27 vacuoles, while aleurain, an acidophilic pratease, is present in TIP-Ma 27 vacuoles in the same cell, but never in alpha-TIP vacuoles in the early stages of development of the root tip.

At present, it is only possible to generalise that in certain cases short segments of NTPP or CTPP may be the signal system for vacuole targeting. In other cases, the internal amino acid sequences may play a vital role, or even a combination of signals may be necessary. It appears that plants, yeast and animals have a diversity of sorting signals. Two different soluble vacuolar lectins, seed lectin and DB58, from the bean *Dolichos biflorus* were expressed in yeast (Chao and Etzler 1994). Accumulation of lectins in the cytosol or their secretion and inability to be transported to the yeast vacuole suggest that plants and yeast utilise different signals for targeting soluble vacuolar proteins. The reports that the addition of a C-terminal HDEL motif causes an N-terminal deletion mutant of sporamin to be transported to the vacuole and that the human lysosomal enzyme β-glucuronidase is targeted to the vacuoles of plant cells, whereas a deletion mutant lacking 15 CTPP amino acids is secreted efficiently, may appear perplexing or may point to the conservation of peptide targeting motifs between plants and animals (vide review Chrispeels *et al.* 1995). Furthermore, plant aspartic proteinases contain an insert, which is not homologous to yeast and mammalian proteinases, but homologous to saposin, which is involved in a membrane associated mannose-6-phosphate-independent pathway for targeting proteins to lysosomes. The insert has been suggested to play a role as a vacuolar targeting determinant (Guruprasad et al. 1994). In yeast the transport of alkaline phosphate (ALP) to the vacuole depends on a clathrin adopter-like complex AP-3, but not on proteins necessary for transport through pre-vacuolar endosomes. Recently, Vowels and Payne (1998) identified leucine-valine-based sorting signals in ALP which direct the tonoplast enzyme to the vacuole in cells lacking clathrin function.

6.9.2 Direct pathway

In animal cells there may be pathways of protein transport which are not mannose-6-P-dependent. Direct uptake of proteins from the cytosol to lysosomes may occur selectively for proteins that bear a consensus KFERQ targeting signal (Dice 1990). In yeast, catabolite-sensitive cytosolic fructose-1, 6-biphosphatase may be directly taken up by the vacuole when the carbon source is switched from acetate to glucose (Chiang and Schekman 1991). Again, absence of any signal peptide sequence in vacuolar alpha-mannosidase indicates its independence of the normal secretory pathway. Höffte and Chrispeels (1992) have shown that the tonoplast protein (alpha-TIP) of tobacco cells is directed from the RER to the tonoplast by the C-terminal 48 amino acids which include the sixth membrane-spanning domain and the cytoplasmic tail. There are other examples of plant enzymes which directly enter the vacuole. The β-amylase in vegetative tissues of *Arabidopsis* (Monroe et al. 1991) and one lipoxygenase isozyme (vsp94) in the paraveinal mesophyll cells of soybean (Tranberger et al. 1991) have been localised in the vacuole, yet they probably do not have signal peptides. Vsp94 is possibly synthesised on free ribosomes and moves directly to the tonoplast without the involvement of membrane vesicles. It is intriguing to note that two other proteins—vsp24 and vsp27—which are co-localised in the same vacuoles, are transported through Golgi apparatus and Golgi vesicles (Franceschi and Giaquinta 1983).

6.9.3 Default mechanism

The default mechanism of delivery of a protein is defined by its delivery to a compartment, due to the absence of any specific signalling, mutation, or wrong signalling for another compartment. In mammalian cells, transport to the plasma membrane is thought to occur by default and delivery to or retention within other compartments of the secretory pathway, mediated by specific information contained in sorted proteins (Pfeffer and Rothman 1987). For example, the lysosomal membrane proteins first go to the plasma membrane and from there they are transported to the lysosomal membrane. This targeting requires information in the cytoplasmic tail as well as in the transmembrane domains (Peters et al. 1990).

6.9.3.1 *Absence of sorting signal*

By swapping domains of the Golgi membrane-bound enzyme dipeptidyl aminopeptidase A(DPAP-A) with tonoplast-bound related enzyme DPAP-B, Roberts et al. (1992) were able to demonstrate that in yeast the tonoplast is a default destination for integral membrane proteins. Both enzymes have signal membrane-spanning domains. No single domain of DPAP-B seems to be required for transport to the tonoplast, although the transmembrane domain is necessary to anchor it in the membrane. Mutations in the cytoplasmic tail of DPAP-A, on the other hand, cause the protein to proceed to the tonoplast and not to its original target, the plasma membrane. Roberts et al. proposed the vacuolar default model, which states that vacuolar membrane proteins do not require sorting information, because the default pathway for membrane proteins of the secretory pathway leads to the vacuole. Wilcox et al. (1992) reported that Kex2 protease, which processes pro-alpha-factor in the late Golgi compartment of yeast, contains a Golgi retention signal at the cytosolic tail. Mutation of one (Tyr713) of two tyrosine residues in the C-tail, or a nearby deletion, results in loss of Golgi localisation and the protease is transported to the vacuole by default. Chang and Fink (1995) reported that a mutant plasma membrane ATPase protein is not delivered to the plasma membrane, instead it is degraded in the vacuole. This agrees with the vacuolar default model. However, the information required for vacuolar localisation of the tonoplast protein alkaline phosphatase (ALP) of yeast resides in the N-terminal cytoplasmic tail and transmembrane regions of the protein (Klionsky and Emr 1990). Another vacuolar protein carboxypeptidase S (CPS) is synthesised at ER as type II integral membrane protein; the membrane-bound precursor transits through the Golgi cisternae and upon delivery to the vacuole is processed near the N-terminus to liberate the mature soluble vacuolar enzyme (Spormann et al. 1992). The default model does not apply to this case.

6.9.3.2   Garbage pathway

An alternative model suggests that abnormal, partially unfolded or defective proteins leaving the ER will be delivered to the vacuole via a 'garbage' pathway, although not much evidence exists for such a pathway. Many mislocalised membrane proteins of the secretory pathway in animal cells accumulate at the plasma membrane, which is considered to be their default destination (Williams and Fukuda 1990).

6.9.3.3   Molecular chaperons

Before leaving the ER for the cytosol, proteins may be subjected to strict quality control by other proteins known as molecular chaperons. Inside ER, the proteins attain correct tertiary and quaternary folding by a thermodynamically driven process aided by a binding protein (BiP) and disulfide bond formation by protein disulfide isomerase. In many cases, oligomerisation and glycosylation are also necessary. Molecular chaperons are a class of proteins that interact selectively with nascent polypeptides and are responsible for quality control by preventing export and degradation of the defective proteins (Klausner and Sitia 1990; Bergeron et al. 1994). The binding protein, BiP, originally known as immunoglobin heavy chain binding protein, is an example of a molecular chaperon. It interacts transiently with many normal secretory proteins and abnormally folded proteins in the ER in an ATP-dependent manner and is a member of the 70 KD heat shock protein (HSP70) family. BiP retains rice prolamines and corn zeins in the ER by facilitating their folding and assembly. In field beans, the vacuolar storage protein phaseolin, a glycoprotein, is made as monomers of 50kD that are assembled into trimers and move to the vacuoles via the Golgi apparatus. BiP binds to monomeric phaseolin but not to the trimeric state which is released by ER. Vitale et al. (1995) have shown that monomers of a C-deletion mutant of phaseolin are not transported to the vacuoles but are degraded in ER. Moreover, defective phaseolin is slowly degraded by endoglyosidase H and degradation does not involve transport along the secretory pathway (Pedrazzini et al. 1997). In other words, BiP is involved in the quality control process. When details of various mutations are considered, it is difficult to distinguish between the default and quality control models. But it is intriguing to note that BiP is a member of the 70kD heat shock protein (HSP-70) family and certain members of HSP-70 family are important for protein targeting to chloroplasts and mitochondria.

6.9.4   Autophagy

In the early part of the chapter the origin and development of vacuoles through the process of autophagy was discussed. The cytoplasm-to-vacuole (Cvt) pathway, described earlier, is also a form of autophagy. Moreover, existing vacuoles may undergo autophagy to cope with senescence or stress (e.g. Tuttle and Dunn 1995, Moriyasu and Ohsumi 1996) or with programmed differentiation (De 1992). Thus, there must be a constitutive mechanism by which non-specific entry of proteins occurs for routine degradation inside the vacuoles. Plant vacuoles may contain all the four major groups of endoproteinases (Canut et al. 1987; Callis 1995) with acid pH optima. Therefore, proteins destined to be degraded or even stored at acid conditions in the vacuole may be routinely targeted. The sorting signal for such molecular autophagy is not known. However, SNARE mechanisms for amminopeptidase I(API) and α-mannosidase have been discussed earlier. Chapter VII covers the process of autophagy in detail. Herman and Lamb (1992) have shown that in tobacco, arabinose-rich glycoproteins are internalised from the periplasmic space into the vacuoles by an autophagic-like process. In what may be called macro-autophagy, large masses of protein bodies may enter the vacuole as a regular process for storage. Levanony *et al.* (1992) used immuno-electron microscopy to show that wheat storage protein bodies (PB) are formed within the RER and that the Golgi complex and its associate vesicles are not involved in their transport to the vacuoles. Instead, the PBs are internalised by autophagy. After internalisation the PBs remain surrounded by pieces of membrane. This is not expected if the PB has entered by membrane fusion. The difference between vesicular transport from the Golgi and autophagy is that in the latter, the membrane surrounding the cargo protein ends in the vacuole and does not fuse with the tonoplast.

6.9.4.1 Concluding remarks

Four pathways for transporting proteins to vacuoles have been distinguished. Only the vesicular pathway is readily distinguishable. For example, in the case of mutant phaseolin, the default mechanism cannot be differentiated from the quality control model; and the direct pathway of α-mannosidase to the yeast vacuole is regarded as microautophagy.

Finally, as discussed above, there is no global mechanism of transport of proteins to vacuoles, since vacuoles themselves are of diverse origin, structure and function in different tissues of various groups of plants and animals. Similar views have been expressed by Bryant and Stevens (1998) in a recent review published since the completion of the present book.

6.10 Summary

- This chapter deals with all aspects of the origin and development of vacuoles in diverse cells. Although vacuoles generally develop from some pre-existing structure or another vacuole, their *de novo* origin as a special physico-chemical entity or liposome is a possibility in certain cases.
- Tonoplast is generally capable of very **fast growth**, and even of regeneration in physically evacuolated protoplasts, possibly by synthesis, direct transfer and subsequent anchorage of lipids and proteins into the membrane. Utilisation of reserve lipids by oxidation or lipolysis is also known to cause biogenesis and development of vacuoles in certain cases.
- The origin of **aleurone vacuoles** in storage parenchyma cells of many seeds occurs due to a large deposition of proteins in segments of ER or conversion of pre-existing vacuoles. In many cases, reserve substances and their respective hydrolases reside in the same structure and their subsequent utilisation results in a large, virtually empty vacuole.
- Different **lytic processes** by hydrolysing enzymes in ER, provacuoles, etc., may bring about formation of vacuoles. Fusion of provacuoles, localised engulfment of cytoplasmic mass by ER and its subsequent digestion may cause vacuolation. A Golgi apparatus-ER-lysosome (GERL) type origin of numerous provacuoles and programmed cellular autophagy may be routine in certain cells.
- With the **accumulation of biosynthates** or metabolites, certain membrane-bound structures may produce a vacuole. Either localised or general accumulation of protein in a segment of RER followed by fusion in certain cases, may lead to the formation of a vacuole. Similarly, SER may produce tannin-containing provacuoles which on fusion develop into tannin vacuoles. Provacuoles produced from SER/RER synthesising alkaloids give rise to alkaloid vacuoles. The development of a vacuole from ER is a reversible process in certain cases where vacuoles revert to ER.
- In certain cases, the **plasma membrane** may give rise to small vacuoles. In yeast, endocytosis of a number of substances produces a prevacuolar compartment or endosomes.
- The development of vacuoles from the **Golgi** in different cell types is well known. Numerous ultrastructural, electron cytochemical and biochemical studies have established that the

proteins synthesised in ER are transported and processed through the various Golgi cisternae. The Golgi vesicles containing the processed or semi-processed proteins are produced in the *trans* face and undergo fusion with the developing vacuole.
- Although **yeast** is not a typical plant cell, an enormous amount of information is available on the origin, development and division of vacuoles of this heterotroph which is highly amenable to molecular and genetic analysis.
  — Similar to the animal systems, intracellular transfer of proteins in yeast, as well as in plants, is mediated by three types of **coated vesicles**, viz. clathrin-coated vesicles (CCV), COPI-coated and COPII-coated vesicles, each with characteristic coating and modes of operation. In addition, certain adaptor protein complexes are associated with the vesicles. In essence, vesicle formation occurs when the specific coat proteins physically deform a patch of the membrane into a vesicle, sequester the membrane protein and package the cargo inside it. The vesicle then transports the membrane and the packaged protein to the target organelle. Disassembly of the coats allows the fusion of vesicle to the target membrane.
  — Whereas CCVs mediate transfer among the plasma membrane, endocytic and TGN compartments, COPI vesicles mediate intra-Golgi and ER-to-Golgi bidirectional transport and the COPII vesicles mediate anterograde transport from ER-to-Golgi only.
    The diversity of the cytoplasmically oriented integral membrane proteins of the vesicles and the adaptor complexes possibly indicates their role in cargo selection for the coated vesicles.
- Vesicular protein transport in numerous cell systems, including plants, involves several proteins for docking and fusion of transport vesicles with their target membranes. According to the **SNARE hypothesis**, complementary pairing between vesicle-SNARE and target-SNARE provides, in part, the specificity required for the targeting of the cargo to its appropriate destination. After the coupling of v- and t-SNAREs, N-ethylmaleimide-sensitive factor (NSF) and soluble NSF attachment proteins (SNAPs) catalyse the disassembly of SNARE complexes, allowing fusion of the transport vesicle with the target membrane.
  — The operation of the SNARE mechanism in yeast, as well as in plants, has been well established. The molecular mechanism, the various genes involved and the sequence of events have been elucidated and signal-recognition systems in a number of cases have been identified.
  — **Fusion** of homotypic or heterotypic membranes/vacuoles requires a number of molecular instruments including cognate v- and t-SNARE pins, alpha-SNAP and NSF. The **priming** of the steady state by NSF and LMA1 is followed by **tethering** of the membranes by a small GTP-binding protein. Membrane **docking** involves pairing of the cognate SNARE pins and bridging two membranes, which subsequently **triggers** Ca^{2+} flow and Ca^{2+}-calmodulin complex catalyses bilayer mixing and **fusion**.
  — At least in yeast cells, the **endosome** or prevacuolar compartment is a definite intermediate structure which mediates between Golgi vesicles and the vacuole, for both the membrane and lumenal proteins. In many plant cells what is called a **provacuole** performs the intermediary role.
  — Three distinct pathways exist through which both lumenal and membrane proteins may be transported to the vacuole of yeast as well as plants: (i) VPS-dependent or **CPY pathway** in which Golgi vesicles carry the protein to the prevacuolar compartment which finally passes them to the vacuole, (ii) alternative/**ALP pathway** in which the Golgi vesicles carrying certain proteins like alkaline phosphatase bypass the prevacuole and deliver them directly to the vacuole, and (iii) Cytoplasm-to-vacuole (**Cvt**) **pathway** in which certain proteins are wrapped in a double membrane carrier. The structure behaves like an autophagosome and is taken up by the vacuole as an autophagic process. Finally, all the multiple transport intermediates dock and/or fuse with the vacuole membrane due to functional and physical interaction between two proteins Vam7p and Vam3p as part of a t- SNARE complex on the tonoplast.

- Besides the various transport pathways which direct the proteins to their target, certain **sorting signals** are necessary in many cases. The sorting signals or vacuolar targeting information may be located at the N- or C-terminal or in an internal segment of the peptide. Such positive signals may not be present in many other proteins which are targeted to a vacuole either by direct uptake, by certain default mechanism or by autophagy.

In conclusion, although *de novo* origin of vacuoles is a distinct possibility in certain cases, vacuoles generally develop from pre-existing ones or from the endomembrane system by accumulation of lumenal material and growth of the vacuole membrane. The pattern of development of vacuoles varies widely in different cell systems. Depending on the program of cell development, very simple dilation of endomembranes may lead to the formation of a vacuole containing relatively simple inorganic or organic substances. In other cases, where discrete processing is required for proteins or glycoproteins, as luminal substance, a specific pathway involving Golgi vesicles is operative.

6.11 References

Adler, K. and Müntz, K., Origin and development of protein bodies in cotyledons of *Vicia faba*, *Planta*, **157**, 401, 1983.

Ahmadian-Tehrani, P., Application de la microscopic electronique a l'etude de la penetration du rouge neutre dans les cellules du meristeme radiculaire du Ble (*Triticum vulgare*). These Docteur Ingenieur, Marseille-Luminy, 1969.

Ahmed, S.U.L., Bar-Peled, M. and Raikhel, N.V., Cloning and sub-cellular location of an *Arabidopsis* receptor-like protein that shares common features with protein-sorting receptors in eukaryotic cells, *Plant Physiol.*, **114**, 325, 1997.

Amman, A., Wanner, G. and Zenk, M.H., Intracellular compartmentation of two enzymes of berberine biosynthesis in plant cell cultures, *Planta*, **167**, 310, 1986.

Baba, M., Osumi, M., Scott, S.V., Klionsky, D.J. and Ohsumi, Y., Two distinct pathways for targeting proteins from the cytoplasm to the vacuole/lysosome, *J. Cell Biol.*, **139**, 1687, 1997.

Bannykh, S.I. and Balch, W.E., Membrane dynamics at the endoplasmic reticulum-Golgi interface. *J. Cell Biol.*, **138**, 1, 1997.

Banta, L.M., Robinson, J.S., Klionsky, D.J. and Emr, S.D., Organelle assembly in yeast: Characterization of yeast mutants defective in vacuolar biogenesis and protein-sorting, *J. Cell Biol.*, **107**, 1369, 1988.

Barlowe, C. and Schekman, R., Sec12 encodes a guanine-nucleotide exchange factor essential for transport vesicle budding from the ER, *Nature*, **365**, 347, 1993.

Bassham, D.C. and Raikhel, N.V., An *Arabidopsis* VPS45p homologue implicated in protein transport to the vacuole, *Plant Physiol.*, **117**, 407, 1998.

Baur, P.S., and Walkinshaw, C.H., Fine structure of tannin accumulations in callus cultures of *Pinus elliotti*, *Can. J. Bot.* **52**, 615, 1974.

Becherer, K.A., Rieder, S.E., Emr, S.D. and Jones, E.W. A novel syntaxin homologue, Pep12p, required for the sorting of lumenal hydrolases to the lysosomal-like vacuole in yeast, *Mol. Biol. Cell.* **7**, 579, 1996.

Bednarek, S.Y. and Raikhel, N.V., Intracellular trafficking of secretory proteins, *Plant Mol. Biol.*, **20**, 133, 1992.

Bednarek, S.Y., Reynolds, T.L., Schroeder, M., Grabowski, R., Hengst, L., Gallwitz, D., and Raikhel, N.V., A small GTP-binding protein from *Arabidopsis thaliana* functionally complements the yeast YPT6 null mutant, *Plant Physiology*, **104**, 591, 1994.

Beevers, L., Clathrin-coated vesicles in plants, *Intl. Rev. Cytol.*, **167**, 1, 1996.

Beevers, L., and Raikhel, N.V., Transport to the vacuole: receptors and trans elements, *J. Exp. Bot.*, **49**, 1271, 1998.

Bensley, R.R., On the nature of canalicular apparatus of animal cells, *Biol. Bull.*, **19**, 179, 1910.

Bergeron, J.J.M., Brenner, M.B., Thomas, D.Y. and Williams, D.B., Calnexin: a membrane-bound chaperone of the ER, *Trends Biochem. Sci.*, **19**, 124, 1994.

Bergfield, R., Hong, Y-N., Kuhl, T. and Schopfer, P., Formation of oleosomes (storage lipid bodies) during embryogenesis and their breakdown during development in cotyledons of *Sinapsis alba* L. *Planta*, **143**, 297, 1978.

Berjak, P., Lysosomal compartmentation: Ultrastructural aspects of the origin, development and function of vacuoles in *Lepidium sativum*, *Ann. Bot.*, **36**, 73, 1972.

Berjak, P., and Villiers, T.A., Ageing in plant embryos. I. The establishment of the sequence of development and senescence in the root cap during germination, *New Phytol*, **69**, 929, 1970.

Bethke, P.C., Swanson, S.J., Hillmer, S. and Jones, R.L., From storage compartment to lytic organelle: the metamorphosis of the aleurone protein storage vacuole, *Ann. Bot.*, **82**, 399, 1998.

Bethke, P.C., Schuurink, R.C., and Jones, R.L., Hormonal signalling in cereal aleurone, *J. Exp. Bot.*, **48**, 1337, 1997.

Bol, J.F., Linthorst, H.J.M., and Cornelissen, B.J.C., Plant pathogenesis related proteins induced by virus infection, *Annu. Rev. Phytopathol.*, **28**, 113, 1990.

Bollini, R., Vitale A., Chrispeels, M.J., *In vivo* and *in vitro* processing of seed reserve protein in the endoplasmic reticulum: evidence for two glycosylation steps, *J. Cell Biol.*, **96**, 999, 1983.

Bowes, B.G., The origin and development of vacuoles in *Glechoma hederacea* L., *La Cellule*, **65**, 357, 1965.

Brown, J.W.S., Erland, D.R., Hall, T.C., Molecular aspects of storage protein synthesis during seed development, in 'The Physiology and Biochemistry of Seed Development and Germination' Kahn, A.A. Ed., Elsever Biomedical Press, Amsterdam, 3.

Bryant, N.J. and Stevens, T.H., Two separate signals act independently to localize a yeast late Golgi membrane protein through a combination of retrieval and retention, *J. Cell Biol.*, **136**, 287, 1997.

Bryant, N.J. and Stevens, T.H., Vacuole biogenesis in *Saccharomyces cerevisiae*: Protein transport pathways to the yeast vacuole, *Microbiol Mol. Biol. Revu.* **62**, 230, 1998.

Bryant, N.J., Piper, R.C., Weisman, L.S., and Stevens, T.H., Retrograde traffic out of the yeast vacuole to the TGN occurs via the prevacuolar/endosomal compartment, *J. Cell. Biol.*, **142**, 651, 1998.

Burd, C.G., Peterson, M., Cowles, C.R., Emr, S.D., A novel Sec-18p/NSF-dependent complex required for Golgi-to-endosome transport in yeast. *Mol. Bio. Cell*, **8**, 1089, 1997.

Burgoyne, R.D, and Clague, M.J., Annexins in the endocytic pathway, *Trends Biochem. Sci.*, **19**, 231, 1994.

Butler, J.M., Kirsch, T., Watson, B., Paris, N., Rogers, J.C. and Beevers, L., Interaction of the vacuolar targeting receptor, BP-80 with clathrin adaptors, *Plant Physiol.*, **114**, Supplement Abstract 1210, 1997.

Buvat, R., Recherches sur les infrastructures du cytoplasme, dans les cellules du meristeme apical, des ebauches foliaires et des feuilles developpees d'*Elodea canadensis*, *Ann. des. Sci.Nat. Bot.*, lle serie, 121, 1958.

Buvat, R., Diversite des vacuoles dans des cellules de la racine d'orge (*Hordeum sativum*). *Compt. Rend. Acad. Sci.* (Paris) **267**, 296, 1968.

Buvat, R., Origin and continuity of cell vacuoles, in 'Origin and Continuity of Cell Organelles', Reinert, J., and Ursprung, H., Eds, Springer Verlag, Berlin, 1971, 127.

Callis, J., Regulation of protein degradation, *Plant Cell*, **7**, 845, 1995.

Canut, H., Dupre, M., Carrasco, A. and Boudet, A.M., Proteases of *Melilotus alba* mesophyll protoplasts, *Planta*, **170**, 541, 1987.

Chafe, S.C., and Durzan, D.J., Tannin inclusions in cell suspension cultures of white spruce, *Planta*, **113**, 251, 1973.

Chang, A. and Fink, G.R., Targeting of the yeast plasma membrane [H^+]ATPase: A novel gene AST1 prevents mislocalization of mutant ATPase to the vacuole, *J. Cell Biol.*, **128**, 39, 1995.

Chao, Q. and Etzler, M.E., Incorrect targeting of plant vacuolar lectins in yeast, *J. Biol. Chem.*, **269**, 20866, 1994.

Cheon, C.I., Lee, N-G., Siddique, A-BM, Bal, A.K. and Verma, D.P.S., Roles of plant homologs of Rab1p and Rab7p in the biogenesis of peribacteroid membrane, a subcellular compartment formed *de novo* during root nodule symbiosis, *EMBO, J.* **12**, 4125, 1993.

Chiang, H.L., Schekman, R., Regulated import and degradation of a cytosolic protein in the yeast vacuole, *Nature*, **350**, 313, 1991.

Chrispeels, M.J., Biosynthesis, processing and transport of storage proteins and lectins in legume cotyledons, *Phil. Trans. Roy. Soc. London*, B **304**, 309, 1984.

Chrispeels, M.J. and Raikhel, N.V., Short peptide domains target proteins to plant vacuoles, *Cell*, **68**, 613, 1992.

Chrispeels, M.J., Green, P.J. and Nasrallah, J.B., Plant cell biology comes of age, *Plant Cell*, **7**, 237, 1995.

Clark, G.B. and Roux, S.J., Annexins of plant cells, *Plant Physiol.* **109**, 1133, 1995.

Colt, W.M., and Endo, R.M., Ultrastructural changes in *Pythium aphanidermatum* zoospores and cysts during encystment, germination and penetration of primary lettuce roots, *Phytopathology*, **62**, 751, 1972.

Conceicao, A. d-S., Marty-Mazans, D., Bassham, D.C., Sanderfoot, A.A., Marty, F. and Raikhel, N.V., The syntaxin homolog AtPEP12p resides on a late post-Golgi compartments in plants, *Plant Cell*, **9**, 571, 1997.

Conradt, B., Haas, A. and Wickner, W., Determination of four biochemically distinct sequential stages during vacuole inheritance *in vitro*, *J. Cell Biol.*, **126**, 99, 1994.

Conradt, B., Shaw, J., Vida, T., Emr, S. and Wickner, W., *In vitro* reactions of vacuole inheritance in *Sacchoromyces cerevisiae*, *J. Cell Biol.*, **119**, 1469, 1992.

Coulomb, P., and Coulomb, C., Processus d'autophagie cellulaire dans les cellules de meristemes radiculaires en etat d'anoxie, *Compt. Rend, Acad. Sci., Paris*, **274**, 214, 1972.

Cowles, C.R., Emr, S.D. and Horazdovsky, B.F., Mutations in the VPS45 gene, a SEC1 homologue, result in vacuolar protein sorting defects and accumulation of membrane vesicles, *J. Cell Sci.*, **107**, 3449, 1994.

Cowles, C.R., Odorizzi, G., Payne, G.S. and Emr, S.D., The AP-3 adaptor complex is essential for cargo selective transport to the yeast vacuole, *Cell*, **91**, 109, 1997b.

Cowles, C.R., Snyder, W.B., Burd, C.G. and Emr, S.D., An alternative Golgi-to-vacuole delivery pathway in yeast: identification of a sorting determinant and required transport component, *EMBO J.*, **16**, 2769, 1997a.

Craig, S., Goodchild, D.J. and Miller, C., Structural aspects of protein accumulation in developing pea cotyledons II. Three-dimensional reconstructions of vacuoles and protein bodies from serial sections, *Aust. J. Plant Physiol.*, **7**, 329, 1980.

Cresti, M., Pacini, E., and Sarfatti, G., Ultrastructural studies on the autophagic vacuoles in *Eranthis hiemalis* endosperm, *J. Submicr. Cytol.*, **4**, 33, 1972.

Creutz, C.E., The annexins and exocytosis, *Science*, **258**, 924, 1992.

Dahms, N.M., Lobel, P. and Kornfeld, S., Mannose-6-phosphate receptors and lysosomal enzyme trageting., *J. Biol. Chem*, 264, 12115, 1989.

Dainty, J., The structure and possible functions of the vacuole, in 'Plant Cell Organelle', Pridham, J.B., Ed., Academic Press, London, 1968, 40.

Darsow, T., Rieder, S.E. and Emr, S.D., A multispecificity syntaxin homologue, Vam3p, essential for autophagic and biosynthetic protein transport to the vacuole, *J. Cell Biol.*, **138**, 517, 1997.

Dawes, C.J. and Barilotti, C.D., Cytoplasmic organization and rhythmic streaming in growing blade of *Caulerpa prolifera*, *Amer. J. Bot.*, **56**, 8, 1969.

De, D.N., Lytic activity in plant cell differentiation, in 'Realm of Differentiation', Habibullah, M., Ed., Neelkanth, Pub. New Delhi, 1992, 21.

De Vries, H., Plasmalytische Studien über die Wand der Vakuolen, *Jahrb. Wiss. Bot.*, **16**, 465, 1885.

Deliu, C., Craciun, C., Craciun, V. and Tamas, M., Ultrastructural and biochemical study of *Berberis parviflora* root metristem and cell cultures, *Plant Science*, **96**, 143, 1994.

Dice, J.F., Peptide sequences that target cytosolic proteins for lysosomal proteolysis, *Trend. Biochem. Sci.*, **15**, 305, 1990.

Domozych, D.S., The endomembrane system and mechanism of membrane flow in the green alga, *Gloeomonas Kupfferi* (Volvocales, Chlorophyta). I. An ultrastructural analysis, *Protoplasma*, **149**, 95, 1989a.

Domozych, D.S. Ibid.II. A cytochemical analysis, *Protoplasma*, **149**, 108, 1989b.

Douce, R. and Joyard, J., Biochemistry and function of the plastid envelope, *Ann. Rev. Cell Biol.*, **6**, 173, 1990.

Fineran, B.A., Association between endoplasmic reticulum and vacuole in frozen-etched root tips, *J. Ultrastruct. Res.*, **43**, 75, 1973.

Franceschi, V.R. and Giaquinta, R.T., The paraveinal mesophyll of soybean leaves in relation to assimilate transfer and compartmentation. I. Ultrastructure and histochemistry during vegetative development, *Planta*, **157**, 411, 1983.

Frey-Wyssling, A., Submicroscopic Morphology of Protoplasm, 2nd. Ed. Elsevier Pub., Houston, 1953.

Frey-Wyssling, A., and Mühlethaler, K., Ultrastructural Plant Cytology, Elsevier, Amsterdam, 1965.

Gamalei, Y.V., Leaf Phloem, Nauka, Leningrad, (in Russian), 1990, 123.

Gay, J.L. and Greenwood, A.D., Structural aspects of zoospore production in *Saprolegnia ferax* with particular reference to the cell and vacuolar membranes, in 'The Fungus Spore', Proc. Symp. Calston Res. Soc., Madelin, M.F., Ed., Butterworth, London, **18**, 95, 1970.

Gay, J.L., Greenwood, A.D., and Heath, I.B., The formation and behaviour of vacuoles (vesicles) during oosphere development and zoospore germination in *Saprolegnia*, *J. Gen. Microbiol.*, **65**, 233, 1971.

Gonzalez, L. Jr. and Scheller, R.H. Regulation of membrane trafficking: Structural insights from a Rab/Effector complex. *Cell*, **96**, 755, 1999.

Griesbach, R.J. and Sink, K.C. Evacuolation of mesophyll protoplasts, *Plant Sci. Lett.*, **30**, 297, 1983.

Griffiths, G. and Simons, K., The Golgi network: sorting at the exit site of the Golgi complex, *Science*, **234**, 438, 1986.

Griffiths, G., Hoflack, B., Simons, K., Mellman, I. and Kornfeld, S., The mannose-6-phosphate receptor and the biogenesis of lysosomes, *Cell*, **52**, 329, 1988.

Grove, S.N., Fine structure of zoospore encystment and germination in *Pythium aphanidermatum*, *Amer. J. Bot.*, **57**, 745, 1970.

Guilliermond, A., The cytoplasm of the Plant cell, *Chronica Botanica*, Waltham, Mass, 1941, 146.

Guruprasad, K., Toermaekangas, K., Kervinen, J. and Blundell, T.L., Comparative modelling of barley-grain aspartic proteinase: a structural rationale for observed hydrolytic specificity. *FEBS Lett.* **352**, 131, 1994.

Haas, A., Conradt, B. and Wickner, W., G-protein ligands inhibit *in vitro* reactions of vacuole inheritance, *J. Cell Biol.*, **126**, 87, 1994.

Hanson, P.I., Roth, R., Morisaki, H., John, R. and Heuser, J.E., Structure and conformational changes in NSF and its membrane receptor complexes visualized by quick-freeze/deep-etch electron microscopy, *Cell*, **90**, 523, 1997.

Hara-Nishimura, I., Nishimura, M. and Akazawa,T., Biosynthesis and intracellular transport of 11S globulin in developing pumpkin cotyledons, *Plant Physiol*. **77**, 747, 1985.

Hara-Nishimura, I., Shimada, T., Hatano, K., Takeuchi, Y., Nishimura, M., Transport of storage proteins to protein storage vacuoles is mediated by large precursor-accumulating vesicles, *Plant Cell*, **10**, 825, 1998.

Harris, N. and Boulter, D., Protein body formation in cotyledons of developing cowpea (*Vigna unguiculata*) seeds, *Ann. Bot.*, **40**, 739, 1976.

Hasilik, A. and Tanner, W., Carbohydrate moiety of carboxypeptidase Y and perturbation of its synthesis, *Eur. J. Biochem.*, **91**, 567, 1978b.

Hauri, H.P. and Schweizer, A., The endoplasmic reticulum Golgi intermediate compartment. *Curr. Opin. Cell Biol.*, **4**, 600, 1992.

Hay, J.C., Chao, D.S., Kuo, C.S. and Scheller, R.H., Protein interactions regulating vesicle transport between the endoplasmic reticulum and Golgi appraratus in mammalian cells, *Cell*, **89**, 149, 1997.

Herman, E.M. and Lamb, C.J., Arabinogalactan-rich glycoproteins are localized on the cell surface and in intravacuolar multivesicular bodies, *Plant Physiol.*, **96**, 264, 1992.

Herman, E.M., Li, X., Su, R.T., Larsen, P., Hsu, H. and Sze, H., Vacuolar type H^+-ATPases are associated with the endoplasmic reticulum and provacuoles of root tip cells, *Plant Physiol*, **106**, 1313, 1994.

Hilling, B. and Amelunxen, F., On the development of the vacuole. II Further evidence for endoplasmic reticulum origin, *Eur. J. Cell Biol.* **18**, 195, 1985.

Hillt, K.J. and Stevens, T.H., Vam22p is a novel endoplasmic reticulum-associated protein required for assembly of the yeast vacuolar H^+-ATPase complex, *J. Biol. Chem.* **270**, 22329, 1995.

Hinkelmann, W., Licht-und elektronenmikroskopische Untersuchungen Über den Abbau der Reserveproteine in keimenden Erbsen. Phd Thesis, Technische Hochschule, Braunschweig, 1966.

Hoch, H.C., and Mitchell, J.E., The ultrastucture of zoospores of *Aphanomyces eutiches* and their encystment and subsequent germination, *Protoplasma*, **75**, 113, 1972.

Höffte, H. and Chrispeels, M.J., Protein sorting to the vacuolar membrane, *Plant Cell*, **4**, 995–1004, 1992.

Hoh, B., Hinz, G., Jeong, B-K. and Robinson, D.G., Protein storage vacuoles form *de novo* during pea cotyledon development, *J. Cell Sci.*, **108**, 299, 1995.

Holstein, S.E.H., Drucker, M. and Robinson, D.G., Identification of beta-type adaptin in plant clathrin-coated vesicles, *J. Cell Sci.*, **107**, 943, 1994.

Holtzman, E., Novikoff, A.B., and Villaverde, H., Lysosomes and GERL in normal and chromatolytic neurons of the rat ganglion nodosum, *J. Cell Biol.* **33**, 419, 1967.

Holwerda, B.C., Padgett, H.S. and Rogers, J.C., Proaleurain vacuolar targeting is mediated by short contiguous peptide interactions, *Plant Cell* **4**, 307, 1992.

Hörtensteiner, S., Martinoia, E. and Amrhein, N., Reappearance of hydrolytic activities and tonoplast proteins in the regenerated vacuole of evacuolated protoplasts, *Planta*, **187**, 113, 1992.

Hörtensteiner, S., Martinoia, E. and Amrhein, N., Factors affecting the re-formation of vacuoles in evacuolated protoplasts and the expression of the two vacuolar proton pumps, *Planta*, **192**, 395, 1994.

Huang, A. H. C., Oil bodies and oleosins in seeds, *Ann. Revu. Plant Physiol. Plant Mol. Biol.*, **43**, 177, 1992.

Huang, A.H.C., Oleosins and Oil bodies in seeds and other organs, *Plant Physiol.*, **110**, 1055, 1996.

Johnson, L.M., Bankaitis, V.A. and Emr S.D., Distinct sequence determinants direct intracellular sorting and modification of a yeast vacuolar protease, *Cell*, **48**, 875, 1987.

Jones, E.W., The synthesis and function of proteases in *Saccharomyces*: genetic approaches, *Ann.Rev. Genet*, **18**, 233, 1984.

Jones, R.L., and Price, J.M., Gibberellic acid and the fine structure of barley aleurone cells. III. Vacuolation of the aleurone cell during the phase of ribonuclease release, *Planta*, **94**, 1991, 1970.

Keen, J.H., Willingham, M.C. and Pastan, I.H., Clathrin-coated vesicles: isolation, dissociation and factor-dependent reassociation of clathrin baskets, *Cell*, **16**, 303, 1979.

Khoo, V. and Wolf, M.J., Origin and development of protein granules in maize endosperm, *Amer. J. Bot.*, **57**, 1042, 1970.

Kirsch, T., Paris, N., Butler, J.M., Beevers, L. and Rogers, J.C., Purification and initial characterization of a potential plant vacuolar targeting receptor, *Proc. Natl. Acad. Sci., USA*, **91**, 3403, 1994.

Kirsch, T., Saalbach, G., Raikhel, N.V. and Beevers, L., Interaction of a potential vacuolar targeting receptor with amino-and carboxy-terminal targeting determinants, *Plant Physiol.*, **111**, 469, 1996.

Klausner, R.D. and Sitia, R., Protein degradation in the endoplasmic reticulum, *Cell*, **62**, 611, 1990.

Klionsky, D.J., Nonclassical protein sorting to the yeast vacuole, *J.Biol. Chem.*, **273**, 10807, 1998.

Klionsky, D.J. and Emr, S.D., Membrane protein sorting: biosynthesis, transport and processing of yeast vacuolar alkaline phosphatase, *EMBO J.*, **8**, 2241, 1989.

Klionsky, D.J. and Emr, S.D., A new class of lysosomal/vacuolar protein sorting signals, *J. Biol. Chem.*, **265**, 5349, 1990.

Klionsky, D.J., Herman, P.K., and Emr, S.D., The fungal vacuole: composition, function and biogenesis, *Microbiol. Rev.*, **54**, 266, 1990.

Koide, Y., Hirano, H., Matsuoka, K. and Nakamura, K., The N-terminal propeptide of the precursor to sporamin acts as a vacuole-targeting signal even at the C-terminus of the mature part in tobacco cells, *Plant Physiol.*, **114**, 863, 1997.

Koncalova, M.N.,Vitalfärbung der Meristemzellenvakuolen bei Weizenpflanzen mit neutral rot. Licht- und elektronmikroskopische Untersuchungen, *Protoplasma*, **60**, 195, 1965.

Kornfeld, S., Structure and function of the mannose-6-phasphate/insulin-like growth factor II receptors, *Annu. Rev. Biochem.*, **61**, 307, 1992.

Kornfeld, S. and Mellman, I., The biogenesis of lysosomes, *Annu. Rev. Cell Biol.*, **5**, 483, 1989.

Krishnan, H.B., Franceschi, V.R. and Okita, T.W., Immunochemical studies on the role of Golgi complex in protein-body formation in rice seeds, *Planta*, **169**, 471, 1986.

Kuehn, M.J., Hermann, J.M. and Schekman, R., COPII-cargo interactions direct protein sorting into ER-derived transport vesicles, *Nature*, **391**, 187, 1998.

Larkins, B.A. and Hurkman, W.J., Synthesis and deposition of zein in protein bodies of maize endosperm, *Plant Physiol.* **62**, 256, 1978.

Larson, D.A., Fine-structural changes in the cytoplasm of germinating pollen, *Amer. J. Bot.*, **52**, 139, 1965.

Levanony, H., Rubin, R., Altschuler, Y. and Galili, G., Evidence for a novel route of wheat storage proteins to vacuoles, *J. Cell Biol.*, **119**, 1117, 1992.

Lin, H.B., Harley, S.M., Butler, J.M. and Beevers, L., Multiplicity of clathrin light chain-like polypeptides from developing pea (*Pisum sativum*) cotyledon, *J. Cell Sci*, **103**, 1127, 1992.

Lin, R.C. and Scheller, R.H., Structural organization of the synaptic exocytosis core complex, *Neuron*, **19**, 1087, 1997.

Lupashin, V.V. and Waters, M.G., t-SNARE activation through transient interaction with a Rab-like guanosine triphosphates, *Science*, **276**, 1255, 1997.

Maitra, S.C., and De, D.N., Ultrastructure of root cap cells: Formation and utilization of lipids, *Cytobios* (UK) **5**, 111, 1972.

Maniotis, A. and Schilwa, M., Microsurgical removal of centrosomes blocks cell reproduction and centriole generation in BSC-1 cells, *Cell*, **67**, 495, 1991.

Manton, I., Observations on stellate vacuoles in the meristem of *Anthoceros*, *J. Exp. Bot.*, **13**, 161, 1962.

Marinos, N.G.,Vacuolation in plant cells, *J. Ultrastruct. Res.*, **9**, 177, 1963.

Martinez-Menárguez, J.A., Geuze, H.J., Slot, J.W. and Klumperman, J., Vesicular tubular clusters between the ER and Golgi mediate concentration of soluble secretory proteins by exclusion from COP1-coated vesicles, *Cell*, **98**, 81, 1999.

Marty, F., Cytochemical studies on GERL, provacuoles and vacuoles in root meristematic cells of *Euphorbia*, *Proc. Nat. Acad. Sci., U.S.*, **75**, 852, 1978.

Marty, F., Branton, D., and Leigh, R.A., Plant vacuoles, in 'The Biochemistry of Plants: A Comprehensive Treatise', Tolbert, N.E. Ed., Academic Press, New York, 1980, 625.

Marty, P. Role du systeme mebranaire vacuolaire dans la differenciation des laticiferes d'*Euphorbia characias* L., *Comp. Rend. Acad. Sci., Paris*, **271**, 2301, 1970.

Matile, P., Aleurone vacuoles as lysosomes, *Z. Pflanzenphysiol*, **58**, 365, 1968a.

Matile, P., Lysosomes of root tip cells in corn seedling, *Planta*, **79**, 181, 1968b.

Matile, P., and Moor, H., Vacuolation: origin and development of the lysosomal apparatus in root tip cells, *Planta*, **80**, 159, 1968.

Matsuoka, K. and Nakamura, K., Propeptide of a precursor to a plant vacuolar protein required for vacuolar targeting, *Proc. Natl. Acad. Sci. USA*, **88**, 834, 1991.

Matsuoka, K., Bassham, D.C., Raikhel, N. and Nakamura, K., Different sensitivity to wortmannin of two vacuolar sorting signals indicates the presence of different sorting machineries in tobacco cells, *J. Cell Biol.*, **130**, 1307, 1995.

Matsuoka, K., Orci, L., Amherdt, M., Bednarek, S.Y., Hamamoto, S., Schekman, R. and Yeung, T., COPII-coated vesicle formation reconstituted with purified coat proteins and chemically defined liposomes, *Cell*, **93**, 263, 1998.

McBride, H.M., Rybin, V., Murphy, C., Giner, A., Teasdal, R. and Zerial, M. Oligomeric complexes link Rab5 effectors with NSF and drive membrane fusion via interactions between EEA1 and syntaxin 13, *Cell*, **98**, 377, 1999.

McIntyre, G.F., Godbold, G.D. and Erickson, A.R., The pH-dependent membrane association of procathepsin L is mediated by a 9-residue sequence within the propeptides, *J. Biol. Chem.*, **269**, 567, 1994.

Mellman, I., Fuchs, R. and Helenius, A., Acidification of the endocytic and exocytic pathways, *Ann. Rev. Biochem.*, **55**, 663, 1986.

Mesquita, J.F., Electron microscope study of the origin and development of the vacuoles in root tip cells of *Lupinus albus*, L., *J. Ultrastruct. Res.*, **26**, 242, 1969.

Mesquita, J.F., Ultrastructure de formations comparables aux vacuoles autophagiques dans les cellules des racines de l'*Allium cepa L.* et du *Lupinus albus L.*, *Cytologia*, **37**, 95, 1972.

Monroe, J.D., Salminen, M.D., and Preiss, J., Nucleotide sequence of cDNA clone encoding a β-amylase from *Arabidopsis thaliana*, *Plant Physiol.*, **97**, 1599, 1991.

Morano, K.A. and Klionsky, D.J., Differential effects of compartment deacidification on the targeting of membrane and soluble proteins to the vacuole in yeast, *J. Cell Sci.*, **107**, 2813, 1994.

Moriyasu, Y. and Ohsumi, Y., Autophagy in tobacco suspension-cultured cells in response to sucrose starvation, *Plant Physiol.*, **111**, 1233, 1996.

Moss, S.E., Annexins, *Trends Cell Biol.*, **7**, 87, 1997.

Munier-Lehmann, H., Mauxion, F. and Hoflack, B., Function of the two mannase-6-phosphate receptors in lysosomal enzyme transport, *Biochem. Soc. Trans.*, **24**, 133, 1996.

Nägeli, C., Pflanzenphysiologische Untersuchungen, Nägeli, C. and Cramer, C., Eds, Schulthess, Zürich, 1855.

Nakano, A. and Muramatsu, M., A novel GTP-binding protein, Sar1p, is involved in the transport from the endoplasmic reticulum to the Golgi apparatus, *J. Cell Biol.*, **109**, 2677, 1989.

Napier, J.A., Stobert, A.K., Shewry, P.R., The structure and biogenesis of plant oil bodies: The role of the ER membrane and the oleosin class of proteins, *Plant Mol. Biol.*, **31**, 945, 1996.

Neuhaus, J.M., Sticher, L., Meins, F.Jr. and Boller, T., A short C-terminal sequence is necessary and sufficient for the targeting of chitinases to the plant vacuoles, *Pro. Natl. Acad. Sci. USA*, **88**, 10362, 1991.

Newell, J.M., Leigh, R.A. and Hall, J.L., Vacuole development in cultured evacuolated oat mesophyll protoplasts, *J. Exp. Bot.*, **49**, 817, 1998.

Nichols, B.J., Ungermann, C., Pelham, H.R.B., Wickner, W.T. and Haas, A., Homotypic vacuolar fusion mediated by t-and v-SNAREs, *Nature*, **387**, 199, 1997.

Nothwehr, S.F., Conibear, E. and Stevens, T.H., Golgi and vacuolar membrane proteins reach the vacuole in vps1 mutant yeast cells via the plasmamembrane, *J.Cell Biol.*, **129**, 35, 1995.

Nothwehr, S.F., Roberts, C.I. and Stevens, T.H., Membrane protein retention in the yeast Golgi apparatus: dipeptidyl aminopeptidase A is retained by a cytoplasmic signal containing aromatic residues, *J. Cell Biol.* **121**, 1197, 1993.

Novick, P. and Brennwald, P., Friends and family: the role of the Rab GTPases in vesicular traffic, *Cell*, **75**, 597, 1993.

Novick, P. and Zerial, M., The diversity of Rab proteins in vesicle transport, *Curr. Opin. Cell Biol.*, **9**, 496, 1997.

Novikoff, A.B., Enzyme localization and ultrastructure of neurons, in 'The Neuron', Hyden, H., Ed., Elsevier Pub., Amsterdam, London, New York, 1967, 255.

Novikoff, A.B., Lysosomes: a personal account, in 'Lysosomes and Storage Diseases', Hers, H.G., and van Hoof, F., Eds, Academic Press, New York, 1973, Chap. 1.

Novikoff, A.B., The endoplasmic reticulum, a cytochemist's view (a review), *Proc. Nat. Acad Sci., US*, **73**, 2781, 1976.

Nunnari, J. and Walter, P., Regulation of organelle biogenesis, *Cell*, **84**, 389, 1996.

Oberbeck, K., Drucker, M. and Robinson, D.G., V-ATPase and pyrophosphatase in endomembranes of maize roots, *J. Exp. Botany*, **45**, 235, 1994.

Orci, L., Glick, B.S. and Rothman, J.E., A new type of coated vesicular carrier that appears not to contain clathrin: its possible role in protein transport within the Golgi stack, *Cell*, **46**, 171, 1986.

Orci, L., Palmer, D.J., Amherdt, M., and Rothman, J.E., Coated vesicle assembly in the Golgi requires only coatomer and ARF proteins, *Nature*, **364**, 732, 1993.

Orci, L., Stamnes, M., Ravazzola, M., Amherdt, M., Perrelet, A., Sollner, T.H. and Rothman, J.E., Bidirectional transport by distinct populations of COPI-coated vesicles, *Cell*, **90**, 335, 1997.

Ory, R.L., and Henningsen, K.W., Enzymes associated with protein bodies isolated from ungerminated barley seeds. *Plant Physiol.*, **44**, 1488, 1969.

Osafune, T., Tsuboi, S., Ehara, T., Satoh, Y. and Yamada, M., The occurrence of non-specific lipid transfer proteins in developing castor bean fruits, *Plant Sci.*, **113**, 125, 1996.

Parham, R.A. and Kaustinen, H.M., On the site of tannin synthesis of plant cells, *Bot. Gaz.*, **138**, 465, 1977.

Paris, N., Rogers, S., Jiang, L., Kirsch, T., Beevers, L., Phillips, T. and Rogers, J.C., Molecular cloning and further characterization of probable plant vacuolar sorting receptor, *Plant Physiol.* **115**, 29, 1997.

Paris, N., Stanley, C.M., Jones, R.L. and Rogers, J.C., Plant cells contain two functionally distinct vacuolar compartments, *Cell*, **85**, 563, 1996.

Pedrazzini, E., Giovinazzo, G., Bielli, A., de Virgilio, M., Frigerio, L., Pesca, M., Faoro, F., Bollini, R., Ceriotti, A. and Vitale, A., Protein quality control along the route to the plant vacuole, *Plant Cell*, **9**, 1869, 1997.

Pernollet, J.C. Protein bodies of seeds: ultrastructure bichemistry, biosynthesis and degradation, *Phytochemistry*, **17**, 1473, 1978.

Perroud, P.F., Crespi, P., Crèvecoeur, M., Fink, A.,Tacchini, P. and Greppin, H., Detection and characterization of GTP-binding proteins on tonoplast of *Spinacia oleracea*, *Plant Sci.*, **122**, 23, 1997.

Peters, C. and Mayer, A., Ca^{2+}/calmodulin signals completion of docking and triggers a late step of vacuole fusion, *Nature*, **396**, 575, 1998.

Peters, C., Braun, M., Weber, B., Wendland, M., Schmidt, B., Pohlman, R., Waheed, A. and Von Figura, K., Targeting of a lysosomal membrane protein: A tyrosine-containing endocytosis signal in the cytoplasmic tail of lysosomal acid phosphatase is necessary and sufficient for targeting to lysosomes, *EMBO J.* **9**, 3497, 1990.

Pfeffer, R.S. and Rothman, J.E., Biosynthetic protein transport and sorting by the endoplasmic reticulum and Golgi, *Ann. Rev. Biochem.*, **56**, 829, 1987.

Pfeffer, W., Zur Kenntnis der Plasmahaut und der Vakuolen, nebst Bemerkungen über den Aggregatzustand des Proptoplasmas und über osmotische Vorgänge, *Abb Math. Phys. Kgl. Sachs, Ges. Wiss.*, **16**, 185, 1890.

Pickett-Heaps, J.D., Further observations on the golgi apparatus and its functions in cells of the wheat seedling, *J. Ultrastruct. Res.*, **18**, 287, 1967a.

Pickett-Heaps, J.D., Ultrastructure and differentiation in *Chara* sp. I. Vegetative Cells, *Austral. J. Biol. Sci.*, **20**, 446, 1967b.

Pickett-Heaps, J.D., Green Algae, Sinauer Assoc., Sunderland, Mass. USA, 1975, 478.

Piper, R.C., Bryant, N.J. and Stevens, T.H., The membrane protein alkaline phosphatase is delivered to the vacuole by a route that is distinct from the VPS-dependent pathway, *J. Cell Biol.*, **138**, 531, 1997.

Pishvaee, B. and Payne, G.S., Clathrin-coats' threads laid bare, Minireveiw, *Cell*, **95**, 443, 1998.

Porter, K.R., and Machado, R.D., Form and distribution of ER during mitosis in cells of onion root tip, *J. Biophys. Biochem. Cytol.*, **7**, 167, 1960.

Poux, N., Nouvelles observations sur la nature el l'origine de la membrane vacuolaire des cellules vegetales, *J. Microscopie*, **1**, 55, 1962.

Poux, N., Localisation des phosphates et de la phosphatase acide dans les cellules des embryos de ble' (Tr. vulg. Vill) lors de la germination, *J. Microscopie*, **2**, 557, 1963.

Poux, N., Localisation de l'activite phosphatasique acide et de phosphates dans les grains d'aleurone. I. Grains d'aleurone referment á la fois golboides et cristalloides, *J. Microscopie*, **4**, 771, 1965.

Presley, J.M., Cole, N.B., Schroer, T.A., Hirschberg, K., Zaal, K.J.M. and Lippincott-Schwartz, J., ER-to-Golgi transport visualized in living cells, *Nature*, **389**, 81, 1997.

Raikhel, N.V. and Lerner, D.R., Expression and regulation of lectin genes in cereals and rice, *Dev. Genet.*, **12**, 255, 1991.

Raikhel, N.V., Ahmed, S.U., Bassham, D., Conceicao, A. d-S., Kirchhausen, T., Marty, F., Marty-Mazars, D., Rapoport, I., and Sanderfoot, A., Transport to the vacuole: Receptors and trans elements, *Abstract. IX Int. Cong. Plant Tiss Cell Cult.* IAPTC, Jerusalem, 1998, 11.

Raymond, C.K., Howald-Stevenson, I., Vater, C.A. and Stevens, T.H., Morphological classification of the yeast vacuolar protein sorting mutants: Evidence for a prevacuolar compartment in class E vps mutants, *Mol. Biol. Cell*, **3**, 1389, 1992.

Raymond, C.K., Howald-Stevenson, I., Vater, C.A. and Stevens, T.H., Morphological classification of the yeast vacuolar protein sorting mutants: evidence for a prevacuolar compartment in class E vps mutants. *Mol. Biol. Cell*, **3**, 1389, 1992.

Riezmann, H., Endocytosis in yeast: several of the yeast secretory mutants are defective in endocytosis, *Cell*, **40**, 1001, 1985.

Roberts, C.J., Nothwehr, S.F. and Stevens, T.H., Membrane protein sorting in the yeast secretory pathway: evidence that the vacuole may be the default compartment, *J. Cell Biol.* **119**, 69, 1992.

Rothman, J.E., Mechanisms of intracelluar transport, *Nature*, **372**, 55, 1994.

Rothman, J.E. and Orci, L., Molecular dissection of the secretory pathway, *Nature*, **355**, 409, 1992.

Rothman, J.E., Raymond, C.K., Gilbert, T., O'Hara, P.J. and Stevens, T.H., A putative GTP-binding protein homologous to interferon-inducible Mx proteins performs an essential function in yeast protein sorting, *Cell*, **61**, 1063, 1990.

Saalbach, G., Jung, R., Kunze, G., Saalbach, I., Adler, K., and Muentz, K., Different legumin protein domains act as vacuolar targeting signals. *Plant Cell*, **3**, 695, 1991.

Sato, M.H., Nakamura, M., Ohsumi, Y., Kouchi, H., Kondo, M., Hara-Nishimura, I., Nishimura, M. and Wada, Y., The AtVAM3 encodes a syntaxin-related molecule implicated in the vacuolar assembly in *Arabidopsis thaliana*, *J. Biol. Chem.*, **272**, 24530, 1997.

Sato, T.K., Darsow, T. and Emr, S.D., Vam7p, a SNAP-25-like molecule, and Vam3p, a syntaxin homologue, function together in yeast vacuolar protein trafficking, *Mol. Cell. Biol.* **18**, 5308, 1998.

Schekman, R., Ready... aim... fire!, *Nature,* **396**, 514, 1998.

Schekman, R. and Mellman, R., Does COPI go both ways? *Cell,* **90**, 197, 1997.

Schimmoeller, F. and Riezman, H., Involvement of Ypt7p, a small GTPase, in traffic from late endosome to the vacuole in yeast, *J. Cell Biol.* **106**, 823, 1993.

Schmid, S.L., Clathrin-coated vesicle formation and protein sorting: an integrated process. *Annu. Rev. Biochem.,* **66**, 511, 1997.

Schroeder, M.R., Borkhsenious, O.N., Matsuoka, K., Nakamura, K. and Raikhel, N., Colocalisation of barley lectin and sporamin in vacuoles of transgenic tobacco plants, *Plant Physiol.,* **101**, 451, 1993.

Scott, S.V., Baba, M., Ohsumi, Y. and Klionsky, D.J., Aminopeptidase I is targeted to the vacuole by a nonclassical vesicular mechanism, *J. Cell Biol.* **138**, 37, 1997.

Seals, D.F. and Randall, S.K., A vacuole-associated annexin protein, VCaB42, correlates with the expansion of tobacco cells, *Plant Physiol.,* **115**, 753, 1997.

Seeger, M. and Payne, G.S., Selective and immediate effects of clathrin heavy chain mutations on Golgi membrane protein retention in *Saccharomyces cerevisiae, J. Cell Biol,* **118**, 531, 1992.

Shimada, T., Kuroyanagi, M., Hara-Nishimura, I. and Nishimura, M., A potential sorting receptor for vacuolar proteins from dense vesicles of developing pumpkin cotyledons, *Plant Physiol.,* **114**, Supplement. Abstract 1206, 1997.

Simonsen, A., Lippe, R., Christoforidis, S., Gaulliers, J-M., Brech, A., Callaghan, J., Toh, B-H., Murphy, C., Zerial, M. and Stenmark, H., EEA1 links PI(3)K function to Rab5 regulation of endosome fusion, *Nature,* **394**, 494, 1998.

Söllner, T., Whiteheart, S.W., Brunner, M., Erdjument-Bromage, H., Geromanos, S., Tempst, P. and Rothman, J.E., SNAP receptors implicated in vesicle targeting and fusion, *Nature,* **362**, 318, 1993.

Spormann, D.O., Heim, J. and Wolf, D.H., Biogenesis of the yeast vacuole (lysosome). The precursor forms of the soluble hydrolase carboxypeptidase S are associated with the vacuolar membrane. *J. Biol. Chem.,* **267**, 8021, 1992.

Springer, S., Spang, A. and Scheckman, R., A primer on vesicle budding. *Cell,* **97**, 145, 1999.

St. Angelo, A.J., Ory. R.L., and Hansen, H.J., Localization of acid proteinase in hemp seed, *Phytochemistry,* **8**, 1135, 1969.

St. Angelo, A.J., Yatsu, L.Y., and Altschul, A.M., Isolation of edestin from aleurone grains of *Cannabis sativa, Arch. Biochem. Biophys.,* **124**, 199, 1968.

Staehlin, L.A. and Moore, I., The plant golgi: structure, functional organization and trafficking mechanism, *Annu. Rev. Plant Physiol. Plant Mol.Biol.,* **46**, 261, 1995.

Staswick, P.E., Novel regulation of vegetative storage protein genes, *Plant Cell,* **2**, 1, 1990.

Stevens, T., Esomon, B. and Schekman, R., Early stages in the yeast secretory pathway are required for transport of carboxypeptidase Y to the vacuole, *Cell,* **30**, 439, 1982.

Strasburger, E., Die pflanzlichen Zellhäute, *Jahrb. Bot.,* **31**, 511, 1898.

Swain, E. and Poulton, J.E., Utilization of amygdalin during seedling development of *Prunus serotina, Plant Physiol,* **106**, 437, 1994.

Thoma, S., Hecht, U., Kippers, A., Botella, J., de Vries, S. and Somerville, C., Tissue specific expression of a gene encoding a cell wall-localized lipid transfer protein from *Arabidopsis, Plant Physiol.* **105**, 35, 1994.

Tranberger, T.J., Franceschi, V.R., Hildebrand, D.F. and Grimes, H.D., The soybean 94-kilodalton vegetative storage protein is a lipoxygenase that is localized in paraveinal mesophyll cell vacuoles, *Plant Cell,* **2**, 973, 1991.

Treffry, T., Klein, S., and Abrahamsen, M., Studies of fine structural and biochemical changes in cotyledons of germinating soybeans, *Aust. J. Biol. Sci.* **20**, 859, 1967.

Tuttle, D.L. and Dunn, W.A., Divergent modes of autophagy in the methylotrophic yeast *Pichia pastoris*, *J. Cell Sci.*, **108**, 25, 1995.

Unger, C., Hardegger, M., Lienhard, S. and Sturm, A., cDNA cloning of carrot (*Daucus carota*) soluble acid ß-fructofuranosidases and comparison with the cell wall isoenzyme, *Plant Physiol.*, **104**, 4351, 1994.

Ungermann, C., Sato, K, and Wickner, W., Defining the functions of trans-SNARE pairs, *Nature*, **396**, 543, 1998.

Vaan der Eb, A.A., and Nieuwdorp, P.J., Electron microscopic structure of the aleurone cells of barley during germination, *Acta. Bot. Neerl.*, **15**, 690, 1967.

Valls, L.A., Winther, J.R. and Stevens, T.H., Yeast carboxypeptidase Y vacuolar targeting signal is defined by four propeptides aminoacids, *J.Cell Biol.*, **111**, 361, 1990.

Valls, L.A., Hunter, C.P., Rothman, J.H. and Stevens, T.H., Protein sorting in yeast: the localisation determinant of yeast vacuolar carboxypeptidase Y resides in the propeptide, *Cell*, **48**, 887, 1987.

Van der Wilden, W., Herman, E.M., and Chrispeels, M.J., Protein bodies of mung bean cotyledons as autophagic organelles, *Proc. Nat. Acad. Sci. US*, **77**, 428, 1980.

Verma, D.P.S., Cheon, C-I. and Hong, Z., Small GTP-binding proteins and membrane biogenesis in plants, *Plant Physiol.*, **106**, 1, 1994.

Vida, T.A., Huyer, G. and Emr, S.D., Yeast vacuolar proenzymes are sorted in the late golgi complex and transported to the vacuole via a prevacuolar endosome-like compartment, *J. Cell Biol.*, **121**, 1245, 1993.

Villiers, T.A., Cytolysosomes in long-dormant plant embryo cells, *Nature*, 214, 1356, 1967.

Villiers, T.A., Lysosomal activities of the vacuole in damaged and recovering plant cells, *Nature New Biol.*, **233**, 57. 1971.

Vitale, A., Bielli, A. and Ceriotti, A., The binding protein associates with monomeric phaseolin, *Plant Physiol.*, **107**, 1411, 1995.

Von Schaewen, A. and Chrispeels, N., Identification of vacuolar sorting information in phytohemagglutinin, an unprocessed vacuolar protein, *J. Exp. Bot. Suppl.*, **44**, 339, 1993.

Vowels, J.J. and Payne, G.S., A dilucine-like sorting signal directs transport into an AP-3-dependent, clathrin independent pathway to the yeast vacuole, *EMBO J.*, **47**, 2482, 1998.

Wada, Y., Ohsumi, Y. and Anraku, Y., Genes for directing vacuolar morphogenesis in *Saccharomyces cerevisiae*, *J. Biol. Chem.*, **267**, 18665, 1992.

Webb, G.C., Zhang, J., Garlow, S.J., Wesp, A., Riezman, H. and Jones, E.W. Pep7p provides a novel protein that functions in vesicle-mediated transport between the yeast Golgi and endosome, *Mol. Biol. Cell.*, **8**, 871, 1997.

Weber, T., Zemelman, B.V., McNew, J.A., Westermann, B., Gmachi, M., Parlati, F., Sollner, T.H. and Rothman, J.E., SNARE pins: Minimal machinery for membrane fusion, *Cell*, **92**, 759, 1998.

Weisman, L.S. and Wickner, W., Intervacuolar exchange in the yeast zygote: a new pathway in organelle communication, *Science* 241, 589, 1988.

Welters, P., Takegawa, K., Emr, S.D., Chrispeels, M.J., AtVPS34, a phosphatidyl 3-Kinase of *Arabidopsis thaliana*, is an essential protein with homology to a calcium-dependent lipid binding domain, *Proc. Nat L. Acad. Sci. USA*, **91**, 11398, 1994.

Went, F.A.F.C., Die Entstehung der Vakuolen in den Fortpflanzungszellen, *Jahrb. Wiss. Bot.*, **21**, 299, 1890.

Whaley, G.W., Mollenhauer, H.H., and Leech, J.H., The ultrastructure of the meristematic cells, *Amer. J. Botany*, **47**, 401, 1960.

Whaley, W.G., Kephart, J.E., and Mollenhauer, H.H., The dynamics of cytoplasmic membranes during development, in 'Cellular Membranes in Development', Locke, M., Ed., Academic Press, New York, 1964, 135.

Wilcox, C.A., Redding, K., Wright, R. and Fuller, R.S., Mutation of a tyrosine localization signal in the cytosolic tail of yeast Kex2 protease disrupts Golgi retention and results in default transport to the vacuole, *Mol. Biol. Cell*, **3**, 1353, 1992.

Williams, M.A. and Fukuda, M., Accumulation of membrane glycoproteins in lysosomes requires a tyrosine residue at a particular position in the cytoplasmic tail, *J. Cell Biol.*, **111**, 955, 1990.

Wilson, T.P., Canny, M. J., McCully, M. E. and Lefkovisch, L.P., Breakdown of cytoplasmic vacuoles: a model of endoplasmic membrane rearrangement, *Protoplasma*, **155**, 144, 1990.

Wirtz, K.W.A., Phospholipid transfer porteins, *Ann, Rev. Biochem.*, **60**, 73, 1991.

Yaklich, R.W. and Herman, E.M., Protein storage vacuoles of soybean aleurone cells accumulate a unique glycoprotein, *Plant Sci*, **107**, 57,1995.

Yoshihisa, T. and Anraku, Y., A novel pathway of import of alpha-mannosidase, a marker enzyme of vacuolar membrane *in Saccharomyces cerevisiae*, *J. Biol. Chem.* **265**, 22418, 1990.

Zorn, G.A., Lucas, J.J and Kates, K.R., Purification and characterization of regenerating mouse L929 karyoplasts, *Cell*, **18**, 659, 1979.

VII
Functions of vacuoles

7.1 Transport of molecules across membranes

Like all phospholipid bilayers, the tonoplast is permeable to small hydrophobic molecules such as O_2, CO_2, N_2, benzene and small uncharged polar molecules like H_2O, urea, ethanol and glycerol. However, transport of inorganic and organic ions, large uncharged polar molecules like glucose and sucrose, or large charged polar molecules — e.g. amino acids, ATP^{4-}, glucose-6-phosphate^{2-} — need special devices of the biomembranes. Molecules are transported across membranes by: (i) **passive transport** which needs no expenditure of energy and moves down the electrochemical or concentration gradient; and (ii) **active transport** which needs expenditure of energy against the electrochemical or concentration gradient.

Passive transport may take place unaided as **simple diffusion**, where molecules move from a more concentrated state to a less concentrated state in a spontaneous, thermodynamically favourable process. Small hydrophilic and uncharged molecules are capable of simple diffusion across the tonoplast. Another widespread mechanism is known as **facilitated diffusion**, where specific substances are transported across the tonoplast (like many other membranes) without any direct energy expenditure. This mechanism is conducted only by proteins which are integral membrane components and are specific for a given molecule or type of molecule. These are known as **transport proteins**. The two major classes of transport proteins are: (i) **carrier proteins** — so-called 'moving parts' shift the specific molecules passively across the membrane. They bind the specific solute and undergo a series of conformational changes to transfer the solute. A group of carrier proteins known as **permeases** or **transporters** are very specific for a substance and are competitively inhibited by structural analogues as has been indicated in amimo acid permeases and glucose permease. **ATP-binding cassette (ABC)** transporters utilise ATP as a direct energy source for organic solute transport. *In addition, certain **group translocators**, like those for glucose, belong to this class.* (ii) **Channel proteins** form a narrow hydrophilic core across the lipid bilayer allowing down-hill movement of small specific inorganic ions. The channel proteins may again be divided into three sub-groups: **porins, ion channels** and **ion transporters**. The porins are multipass transmembrane proteins that cross the lipid bilayer as a (beta) barrel configuration. They allow water or selected hydrophilic solutes of generally up to 600 daltons to diffuse across the membrane. Ion channels are distinguished from carriers by their several orders of magnitude higher flux rates and their mediation of the passive flux of ions across membranes. The electrochemical gradient across the membrane in which the channel resides determines whether, and in which direction, ions move. Carriers may move ions either with or against their substrate concentration gradients and may function as uniporters or co-transporters.

The channel proteins which are specifically concerned with the transport of inorganic ions are called **ion channels**. These channels are about 1000 times as fast as the transport rate mediated by carrier proteins. Ion channels transport through mebranes at the rate of 10^6 to 10^8 ions per second per channel protein. However, they are not coupled to an energy source and ion transport takes place down the electrochemical gradient across the lipid bilayer. In contrast to the porins, the ion channels show *ion selectivity*, permitting passage only of specific ions through a narrow pore. Moreover, the ion channels are not continuously open but fluctuate between 'open' and 'closed' states operated by 'gates' or in response to a specific stimulus. Thus the **voltage-gated channels** are stimulated by changes in voltage across the membrane, the **mechanically gated channels** are stimulated by mechanical stress like stretching and the **ligand-gated channels** are stimulated by chemical ligands including protein phosphorylation-dephosphorylation, nucleotides, etc. There are two types of voltage-gated channels: (i) slow-vacuolar type (SV-type) channels show characteristically slow activation time constant, a marked outward rectification at positive membrane potential and are widely distributed. (ii) Fast-vacuolar type (FV-type) show fast activation time constant, are voltage-independent or weakly voltage-dependent and active at both positive and negative membrane potential. The group of channel proteins called **ion transporters** should be considered less as a free pore and more like an enzyme-substrate system. Unlike the enzyme-substrate reactions, the transported ion is not covalently modified. However, the transport is indirectly aided by energy provided by other operating systems. Thus the ion co-transporters conduct what is called **secondary active transport**. The process, in contrast to simple diffusion, is not always proportional to the solute concentration, and is dependent on the saturation and binding constant of the channel protein.

7.2 UPTAKE OF WATER

The passage of various molecules across the cell membrane depends partly on the size and partly on the hydrophobic nature of the molecule, because the interior of the lipid bilayer is hydrophobic. Thus water, the smallest uncharged polar molecule, has the highest permeability coefficient across the membranes. The entry or exit of water through the membrane is due to either or both of the following: (i) the difference in the concentration of dissolved substances on the two sides of the membrane (since water moves by the passive process of osmosis from a lower solute concentration to a higher one); (ii) the difference in hydrostatic pressure between the two solutions.

7.2.1 Aquaporin

Sidel and Soloman (1957), working on human red cells, suggested that the bulk flow of water occurs through pores in the membrane. Over the decades, enough information has become available to indicate that many membranes contain water channels which permit bulk flow of water. The water-selective channels in plants are made of proteins and have been called aquaporin by Chrispeels and co-workers (Agre *et al.* 1993). The structure of tonoplast aquaporins and their homologues has been discussed in Chapter IV. The exclusivity of certain aquaporins for water has been demonstrated by the fact that no ion or metabolite has been found to be co-transported with water (Maurel *et al.* 1994). Aquaporins are localised both in the tonoplast (Höfte *et al.* 1992), as well as in the plasma membrane (Daniels *et al.* 1994). Recent experiments by Niemietz and Tyerman (1997) support the previous finding that tonoplast water channels are mercury sensitive, whereas some plasma membrane equivalents lack the mercury sensitive site (Daniels *et al.* 1994). While some aquaporins are constitutively produced, the aquaporin RD28 is induced by desiccation and is localised in the plasma membrane (Yamaguchi-Shinozaki *et al.* 1992; Daniels *et al.* 1994). Whether the channels for water movement are open or closed seems to be determined by the phosphorylation state of aquaporins (Maurel *et al.* 1995). This appears to provide control of flow rate and, further, to determine equilibrium with respect to chemical potential between extra- and intra-cellular water. Yet, under different water stress conditions aquaporin activity may change. Under salinity stress aquaporin mRNAs increase in *Arabidopsis* (Yamaguchi and Shinozaki *et al.* 1992), whereas the ice plant transiently down-regulates aquaporin mRNA under similar stress (Yamada *et al.* 1995). Shiratake *et al.* (1997b) have reported that during the active cell

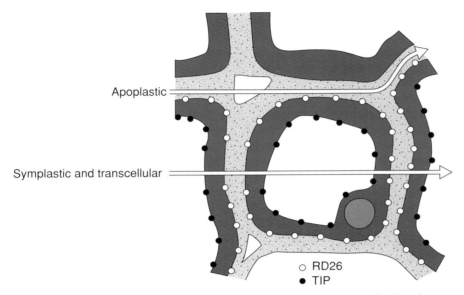

Figure 7.1 Possible routes for hydraulic water flow through a living tissue: apoplastic and symplastic/transcellular. TIP, shown in the tonoplast, is constitutively expressed, whereas RD28, shown in the plasma membrane, is expressed only as a result of desiccation (from Chrispeels and Maurel 1994). With permission of the American Society of Plant Physiologists.

expansion stage of the development of the young pear fruit, the level of the water channel protein VM23 is especially high. Thus VM23 may play an important role in the rapid expansion of cells as a vacuolar water channel.

After the entry to the cytoplasm from the surrounding bathing solution, water, along with dissolved substances, can take one of two routes:

i) they may be transported across the tonoplast to the large vacuole of mature plant cells, or
ii) they may move through plasmodesmata to the cytoplasm of the neighbouring cell.

Very often, vacuoles may not be involved at all in different types of transport activities. To understand medium distance and long distance transport, Münch (1930) differentiated two types of spaces in the plant body: (i) *apoplast* includes all spaces outside the plasmalemma barrier of cells plus intercellular spaces, lumen of xylem tracheids and vessels, etc; (ii) *symplast* is the space inside the plasmalemma. When the protoplasts of different cells are interconnected by plasmodesmata, they form a symplastic system. As far as water transport is concerned, it is now generally agreed that water moves from cell to cell either through the symplast or through cell wall–plasmalemma barriers, and local equilibrium should occur between the protoplast and the surrounding cell wall. This does not exclude apoplast water movement which probably also occurs, but the symplast appears to be the dominant path (Boyer 1985). Various studies indicate that the preferred route of water transport may depend on the tissue, organ, and physiological status of the plant, as well as on the osmotic and/or hydrostatic pressure (Steudle 1992). Figure 7.1 shows the possible route for hydraulic water flow through a tissue.

7.2.1.1 Histological considerations

Dual uptake

An investigation by Epstein and Hagen (1952) on the effect of external cation concentration on the rate of uptake revealed that, in general, ion uptake by plant tissue with increasing external concentration approaches a saturation or maximal rate. From the discovery by Friend and

Noggl (1958) that two systems participate in ion uptake, showing saturation in two different concentration ranges, it is now generally agreed that for a given ion there are two distinct mechanisms of absorption.

i) System 1, which operates at low concentration with high affinity for the ions, and with a low Michaelis constant.

ii) System 2, which operates only at concentrations higher than those giving essentially the maximal rate of absorption vice mechanism 1 with a low affinity for the ions and a high Michaelis constant.

According to the Torii-Laties' (1966) hypothesis, the two systems of ion uptake occur on two different membranes in the cell, namely the plasmalemma and the tonoplast, and they are in series with each other. This means that System 1, dealing with ion uptake at low concentration, is located at the outermost barrier of the cell, i.e. the plasmalemma; and System 2, operating the ion uptake at high concentration, is located within the cell, i.e. the tonoplast. Support for this dual uptake mechanism comes from a wide range of experimental evidence (Lüttge and Higinbotham 1974). In a computer simulation, Pitman (1969) took into account a considerable number of mutually interdependent parameters, i.e. the fluxes at the plasmalemma and the tonoplast in opposite directions and the concentrations in the external medium, the cytoplasm and the vacuole. He concluded that tonoplast plays a role in anion transport and that the active transport mechanisms at the plasmalemma and tonoplast must be coupled with each other. This can be done by a common source of energy or, in the simplest case, by dependence of both pumps on the ion concentration in the cytoplasm.

Feedback inhibition

The negative correlation between uptake and **internal ion concentration** is well known (Cram 1976). Since influx to the vacuole in general is more strongly reduced than plasmalemma influx by ion accumulation, negative feedback between vacuolar ion concentration and fluxes to the vacuole have been proposed (Cram and Laties 1971). Reductions in plasmalemma influx are also associated with ion accumulation. The cytoplasmic ion concentration is maintained within rather narrow limits, in most eucaryotic cells. Hence, the most likely targets for this feedback would be tonoplast and plasmalemma transporters. Under conditions of nutrient deprivation, it is reasonable to presume that cytoplasmic concentration is maintained at the expense of vacuolar concentrations. Considering relative vacuolar and cytoplasmic volumes, this could be achieved with only small changes of vacuolar concentration.

Whole body transport

In general, the medium distance and long distance transport systems for ions through the symplast may not involve the vacuole, except for peripheral influence or storage. This is very clearly explained in a scheme of transport and accumulation in a leaf with salt bladders. Apart from transport across the tonoplast in the bladder cell, vacuoles are not really involved (Lüttge 1969) (Fig. 7.2). Thus symplast transport bypasses the vacuole. Similarly in the case of the rector recreation model outlined by Lüttge and Schnepf (1976), both symplastic transport and metabolically controlled membrane transport do not involve vacuole or tonoplast.

The general pattern of ion transport in a typical plant suggested by Lüttge and Higinbotham (1979) can be summarised as follows:

From the external medium or soil, the roots take up ions in their apoplastic free space. The influx and efflux at the plasmalemma of the root hairs, epidermal and cortical cells control ion uptake and ion release in the cytoplasmic phase. On the other hand, the influx and efflux at the tonoplast control the amount of accumulation or rate of mobilisation of ions in the vacuole.

At the level of the whole organ, the distribution of ions is regulated by the symplastic transport system, which also regulates ions across a barrier in the apoplastic space which is marked by the

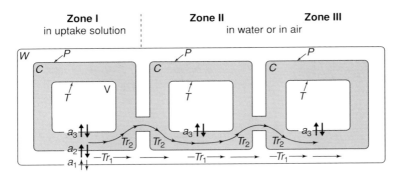

Figure 7.2   Simplified scheme of symplastic transport. *Zone I*, uptake region; *Zones II and III*, regions into which transport occurs; a_1, a_2 and a_3, influx and efflux of the compartments apoplastic free space, cytoplasm and vacuole. *Thick arrows*, membrane-controlled transport, Tr_1, apoplastic transport in the free space; Tr_2, symplastic transport; *W*, cell wall; *P*, plasmalemma; *C*, cytoplasm; *T*, tonoplast; *V*, vacuole (from Lüttge 1969). With permission of Springer Verlag.

casparian strip. For long distance transport through the central cylinder or stele, the efflux from the stellar parenchyma cells loads the tracheids and vessels.

7.3   Uptake of ions

7.3.1   General considerations

Electrolytes may appear in different concentrations on either side of a plant membrane and in many plant cells the total ionic flux is mainly due to movements of K^+, Na^+ and Cl^-, although H^+ and OH^- fluxes may also be considerable. The fluxes of these ions across the membrane are caused by **gradients in the chemical potential** which create **electrical potential differences** across the membrane. This difference is termed **diffusion potential** and arises as a result of the differential mobilities of the ions involved and the condition of electrical neutrality which is maintained. To determine the direction in which the diffusion of an electrolyte will occur, two gradients need to be considered: the chemical gradient, which depends upon the concentration difference of the substance between the two compartments, and the electrical difference, which depends upon the difference in charge. Together these differences are referred to as the **electrochemical gradient**.

Donnan equilibrium/Donnan phase

Cells contain, as well as mobile ions, a whole range of immobile ions or charges. Biological structures, surfaces, etc., within cells possess large charged molecules like proteins, pectin and organic phosphate esters. The region containing such immobile charged particles is referred to as the Donnan phase. The immobile carboxyl group of pectin and the proteins of membranes, including those inside the vacuole or tonoplast, may present such a phase with multivalent anions at physiological pH. In any system, the total number of negatively charged particles must equal the number of postitively charged, i.e. the number of equivalents of anions on one side of the membrane must equal the number of equivalents of cations on the other side. This results in an important phenomenon, the Donnan equilibrium, which requires **passive diffusion** of ions across the membrane.

The classification and definition of each of the mechanisms of transport of small molecules and ions across a biomembrane has been stated above. In a living cell none of these mechanisms operates in isolation, and in the tonoplast, as in all other membranes, various mechanisms may operate with or against one another.

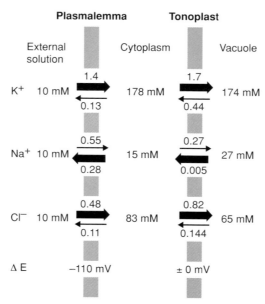

Figure 7.3 Summary of electrochemical data on oat (*Avena*) coleoptile cells: K^+, Na^+, and Cl^- compartmentation at 10 mM external concentration. Membrane potentials and flux sizes (arrows) in pmol s^{-1} cm^{-2}. *Heavy arrows*, active fluxes; *thin arrows*, passive fluxes (from Pierce and Higinbotham 1970). With permission of the American Society of Plant Physiologists.

7.3.2 Accumulation of ions

It is well known that plant cells can develop a high level of internal concentration of ions from a low external concentration. A large proportion of ions are accumulated in the central vacuole which may, in certain cases, occupy 90% or more of the cell volume. Quantitative data for the active transport at the plasmalemma and tonoplast may be obtained from the measurement of membrane potentials, ion fluxes and ion concentrations. A summary of electrochemical data on *Avena* coleoptile cells is given by Pierce and Higinbotham (1970) (Fig. 7.3). It is widely accepted that active influxes for K^+, Na^+ and Cl^- exist at the tonoplast in most land plants. At the tonoplast of the fresh water alga *Nitella*, for example, there is an active influx of Na^+ and Cl^-, whereas in sea water alga it is Na^+ and K^+ (Fig. 7.4).

The mechanism of active transport originally proposed for animal cell membrane by Skou (1974) involves an enzyme (Na^+-K^+)-ATPase with Mg^+ as co-factor, which causes the hydrolysis of ATP and results in the transport of Na^+ and K^+ in opposite directions. The enzyme system is located within the membrane and is directly related to the activity of the pump which moves Na^+ out of and K^+ into the cell. Both the pump and the enzyme require Na^+ and K^+ and are inhibited specifically by ouabain (stropanthin G). The experiments dealing with the kinetics of K^+ and Rb^+ uptake and ATPase activity in relation to KCl and RbCl concentration in the medium have provided strong evidence for an involvement of ATPase in ion transport (Fisher et al. 1970; Hodges 1976).

Lin et al. (1977) reported that isolated *Tulipa* petal vacuoles contain 148 mM K^+, 50 mM Na^+, 33 mM Mg^{2+}, 8mM Ca^{2+} and 32mM Cl^-; the same concentrations found in the whole protoplasts. They found evidence for an ATP-dependent H^+ transport in isolated vacuoles and ion-stimulated ATPases in the tonoplast. Kylin and Hansson (1971) reported diverse types of alkali transporting ATPases of the plasmalemma and tonoplast in four cultivars of sugar beet.

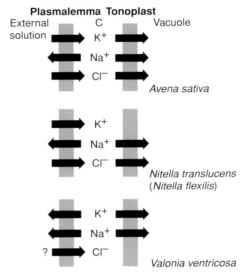

Figure 7.4 Active ion fluxes in cells of a higher plant (*Avena*) in a fresh water alga (*Nitella*) and in a sea water alga (*Valonia*) (from Pierce and Higinbotham 1970). With permission of the American Society of Plant Physiologists.

7.3.2.1 Primary active transport

It is often found that the concentration of the two most important cations Na^+ and K^+ across the plasma membrane or tonoplast is not what is expected for Donnan electrochemical equilibrium. (vide Table 5.2). The ability of a cell to develop and maintain such an extreme gradient across the membrane can only occur by active transport, which is the process by which an ion is moved against a gradient of electrochemical potential. This process is formally endergonic and involves intervention of metabolic energy and often carriers, enzymes and co-factors. Very often it involves consuming chemical energy from ATP, NADH or other energy-rich compounds.

The primary transport activity against the electrochemical gradient is operated by two inwardly directed electrogenic proton pumps, one energised by ATP and the other by PPi (inorganic pyrophosphate). The proton concentration gradient and the electrical gradient generated by these two pumps provide the driving force for the transport of solutes across the membrane. The presence of the two proton pumps, H^+-ATPase and PPi-energised H^+-PPase on the tonoplast catalyses inward electrogenic H^+-translocation from cytosol to vacuole lumen to establish inside positive trans-tonoplast H^+-electrochemical potential difference for the energisation of H^+-coupled solute transport. Simultaneous or successive measurements of both ATP hydrolysis and ATP-generated H^+-pumping indicate that the tonoplast H^+-ATPase is an anion sensitive ATPase stimulated by Cl^- and inhibited by NO^{3-}. Except in a few cases, tonoplast ATPase is insensitive to vanadate, a potent inhibitor of the plasma membrane ATPase. Chapter IV describes the molecular characteristics of both the enzymes and V_oV_1-ATPase structure and sub-unit composition.

Vacuolar-type ATPase

Using micro-capillary glass electrodes, electrical potential between the vacuole and the external medium can be measured. It is usually about 100mV. This electrical potential arises largely as a result of two processes, diffusion and active electrogenic transport. The possible presence of electrogenically active ATPase was first shown in the lutoid membrane of *Hevea* (d'Auzac 1977).

Using the anthocyanin pigments within the vacuoles as pH indicators, Wagner and Lin (1982) have shown that Mg-ATPase can cause acidification of vacuolar sap. On the basis of transport studies on isolated tonoplast vesicles, Sze (1984, 1985) detected the presence of electrogenic H^+-pumping ATPase in the tonoplast. The tonoplast ATPase is relatively abundant in oat cells and is estimated to represent 4–8% of the total tonoplast protein. The 'tonoplast-type' enzyme acidifies the vacuolar sap and generates a proton motive force. Mandala and Taiz (1985) reported similar MgATP-dependent acidification of vacuolar sap of maize coleoptile protoplasts. A number of workers (Miller et al. 1984; Thom and Komor 1984, 1985) studied the effect of ATP on the electrical membrane potential difference of isolated vacuoles. ATP was found to induce a change in electrical gradient to more positive values, indicating that the ATP dependent proton pump is electrogenic. Studies by Joachem et al. (1984) also indicate the presence of an *inwardly directed* electrogenic proton pump on the tonoplast. The shift of membrane potential difference and pH induced by ATP establishes that an electrochemical proton gradient or proton motive force (PMF) energises the tonoplast. The PMF can persist in the absence of an exogenous energy source and the vacuoles may maintain the energised state. This can also be the driving force for a secondary transport system (vide review by Nelson 1988). Using electron-cytochemical techniques, Balsamo and Uribe (1988) demonstrated the presence of ATPases on the tonoplast of CAM plants. Now ATPase is recognised as a ubiquitous tonoplast enzyme and its catalytic site is exposed to the cytoplasm (Bowman and Bowman 1985). V-ATPase, the nitrate-sensitive ATPase (ns-ATPase), is also found in the Golgi complex, ER, plasma membrane and in clathrin-coated vesicles. Lately it has been demonstrated by Matsuoka et al. (1997) that all V-ATPases are not identical in tobacco cells and the V-ATPases in the Golgi complex and the tonoplast are distinguishable in terms of their biochemical properties.

Vacuolar-type PPase

In tonoplast preparations phosphohydrolases have also been detected. Mg^{2+}-dependent pyrophosphatase has been found in the membranes of isolated vacuoles of red beetroot (Leigh and Walker 1980) and tulip petals (Wagner and Mulready 1983). Rea and Poole (1985) demonstrated that the pyrophosphatase (PPase) in a tonoplast-type membrane fraction from red beet microsomes is a proton pump. Additional reports (Chanson et al. 1985; Wang et al. 1986) confirmed the existence of a tonoplast bound cation-sensitive pyrophosphatase activity in establishing a proton-electrochemical gradient.

Using plasma membrane-permeabilised cells of *Chara*, Shimmen and MacRobbie (1987) demonstrated that both ATP- and PPi-dependent H^+-activities coexist in the tonoplast. Using the neutral red accumulation in the vacuole as an index of H^+-transport activity, they showed that ATP- and PPi-dependent neutral red accumulations have different sensitivities to monovalent ions. In view of the finding that H^+-translocating PPase activity is twice as high as H^+-translocating ATPase activity in isolated vacuolar membranes, Takeshige and Hager(1988) contend that H^+-translocating PPase can play an important role in regulating the cytoplasmic pH or in generating an electrochemical H^+ gradient across the vacuolar membrane. The V-PPase shows an absolute dependence on K^+ for maximum activity. Again, V-PPase may be involved in K^+ transport into the vacuole (Davies et al. 1992; Obermeyer et al. 1996; Gordon-Weeks et al. 1997). Hand in hand with the detection of a H^+-ATPase in typical vacuoles, the presence of PPase in lutoids of *Hevea* latex has been reported. The structure of H^+-ATPase and PPase has been described in Chapter IV.

7.3.2.2 Secondary active transport

For an ion to be transferred across a membrane, both a driving force and a pathway are required. The two primary H^+ transport systems serve to build a proton motive force (PMF) which has two components: pH gradient and electrical gradient. Either or both of these components may be used to energise various transporter-aided solute transport across the tonoplast. To demonstrate the presence of such porters in the tonoplast, accompanying proton transport

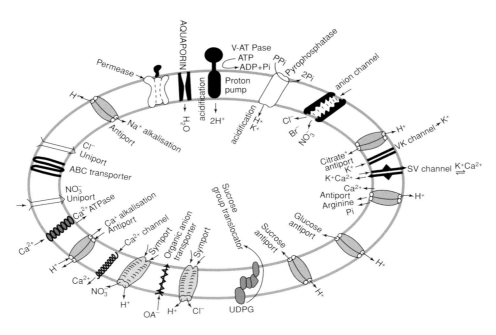

Figure 7.5 Diagram representing the transport processes through the tonoplast.

should be demonstrated, as well as solute transport. Such ion fluxes should also be reversible in the presence of an appropriate ionophore (small hydrophobic molecules).

There are three types of secondary transport mechanism (Fig. 7.5):

i) *Uniport*: When the transport of an ion or small molecule, against a concentration gradient, takes place uncoupled to the movement of any other molecule or ion. These are systems carrying a net charge driven by electrical gradient. For example, Cl^- and NO_3^- transport into the vacuole may take place by such uniports (Blumwald and Poole 1985b; Kaestner and Sze 1987).

ii) *Antiport*: When the transport of an ion or small molecule is achieved by coupling its movement to that of another ion or molecule, in the opposite direction. This system carries both H^+ and substrate, and is driven by pH gradient. Active transport of cations through antiport mechanism across the tonoplast operates by coupling the flux of cations to the opposite flux of H^+. The isolation and identification of membrane ion transfer of Ca^{2+}/H^+ (Schumaker and Sze 1986; Joyce et al. 1988), Na^+/H^+ (Blumwald et al. 1987; Blumwald and Poole 1985), sucrose/H^+ (Brisken et al. 1985), as well as Ca^+/Na^+, K^+/Na^+, Na^+/Ca^{2+}, Ca^{2+}/H^+ and Mg^{2+}/H^+ have all been indicated. The occurrence of a Na^+/H^+ antiporter in tonoplasts (Garbarino and DuPont 1988, 1989) has removed all doubts.

iii) *Symport*: When an ion or molecule is co-transported with another ion or molecule in the same direction. These are systems carrying both H^+ and substrates and are driven by pH gradient. The coupling of opposite charges takes place. The operation of a NO_3^{3-}/H^+ symport has been suggested for the retrieval of NO_3^- from the vacuole (Blumwald and Poole 1985), although certain doubt about this symport has been raised (Pope and Leigh 1988). In addition, glucose/H^+, lactose/H^+, amino acid/Na^+, glucose/Na^+ have also been suggested/demonstrated.

The operation of secondary active transport of anions is suggested by the elevated concentration of anions in vacuoles, an extreme example of which is the sulfuric acid-accumulating vacuoles of the marine brown alga *Desmarestia*. In addition to primary active H^+-transport, a powerful sulfate accumulation mechanism must be operative to bring about a concentration of 0.2–0.4 N

H_2SO_4 and a pH of 0.5 to 0.8 inside these vacuoles (McClintock et al. 1982). Similar ion transport operation is also expected in the case of membrane-bound statocysts of *Chara* rhizoids containing solid barium sulfate (Sievers and Schmitz 1982). Again, Cl^- may move into the vacuoles as a consequence of primary active H^+-transport, and inorganic phosphate may be taken up by an exchange for Cl^- (Hager and Helme 1981). The existence of anion carriers is also demonstrated by the MgATP-driven anion uptake in isolated vacuoles and tonoplast vesicles, e.g. malate in mesophyll vacuoles of a C_3 (Martinoia et al. 1985) and a CAM plant (Nishida and Tominaga 1987) and Cl^- in barley mesophyll vacuole (Martinoia et al. 1986). It has been shown in the case of lutoids that the electrochemical gradient generated by proton pumping ATPases governs the accumulation of various ions such as citrate, basic amino acid and Ca^{2+}.

7.3.2.3 Uptake of anions

Depending on the anion concentration in the cytoplasm and vacuole and the demands of cellular metabolism, these transport systems may serve to mobilise Cl^- and NO_3^- in the vacuole for use in the cytoplasm (symporters) or to drive these anions into the vacuole (antiporters). The pH component of the PMF generated by the tonoplast H^+-ATPase or H^+-PPase could provide the energy for the movement. Schumaker and Sze (1987) demonstrated that H^+-coupled Cl^- (H^+/Cl^-) and H^+-coupled NO_3^- (H^+/NO_3^-) transport systems operate in tonoplast. This coupled transport refers to an apparent interdependence of H^+ and anion movements which occur at the same time via proteinaceous porters.

Using a chloride-sensitive fluorescent probe on isolated tonoplast vesicles, Pope and Leigh (1990) concluded that Cl^- is transported via a uniport which, although unaffected by SO_4^{2-}, is completely inhibited by NO_3^-, indicating that NO_3^- and Cl^- are transported by the same porter. Again, the results obtained by Miller and Smith (1992), using nitrate-selective microelectrodes, indicate that nitrate transport into the vacuole is mediated by an H^+/NO_3^- symport mechanism. On the other hand, Chodera and Briskin (1990) reported that ClO_3^- or NO_3^- influx at the tonoplast is mediated by a uniport which responds to the positive interior membrane potential. Thus the mechanism of development of an ionic osmoticum in the vacuole is being gradually understood. By the activity of the independent H^+-pumps, *viz* H^+-ATPase and pyrophosphatase, associated with the tonoplast, a pH gradient (inside acid) is generated. This gradient is used to drive **cation transport** via a cation/H^+ antiport, whereas *anion uptake* is thought to result from electrophoretic movement of anions in response to the positive membrane potential. In other words, anions move through *anion channels* in the tonoplast in response to positive membrane potential generated by ATPase, and increasing the pH, as expected from the chemi-osmotic hypothesis (Mitchell 1976). Pope and Leigh (1987) and others have shown that Cl^-, Br^- and NO_3^- move into the tonoplast vesicles in response to the positive membrane potential generated by a PPi-dependent H^+-pump. Since SO_4^{2-} and malate do not cause stimulation of pH gradient there may be a certain degree of specificity of the anion channels. Moreover, a possibility remains that an anion channel may or may not be physically associated with the proton-pumping enzyme. In other words, the question of physical structure, association of anion channels and the pump is still open.

In most cells a metabolically active pool of inorganic phosphate (Pi) can be distinguished from an inactive Pi pool and Pi is exchanged between the pools for metabolic necessities (Bieleski 1973). Various studies have revealed that whereas the cytoplasmic Pi content is maintained at a certain degree of constancy, the vacuolar amount can vary enormously. When Pi is available in abundance it is stored in the vacuole and released to the cytoplasm on demand. Thus in most cases the vacuole maintains a cytoplasmic homeostasis of Pi (Schröppel-Meier and Kaiser 1988; Mimura et al. 1990). In their studies on vacuoles of mesophyll cells of barley leaves, Mimura et al. (1990) have shown that cytoplasmic Pi homeostasis can be maintained under Pi starvation by export from the vacuole into the cytoplasm. A similar phenomenon has also been recorded for the NO_3^- pool in barley root cells. Using double-barrelled nitrate-selective micoelectrodes, Van der Leij et al.

(1998) measured the remobilisation of vacuolar nitrate. They concluded that during the first 24 h of nitrate withdrawal, vacuolar nitrate can be readily mobilised to supply the nitrogen demands of the seedling and to maintain the cytosolic nitrate concentration. Such export is directed against a concentration gradient and presumably also against an electrical potential. An active proton pump produces an electrical potential across the tonoplast which may drive anion uptake. Again, the high cytosolic concentration of proteins which carry a net negative charge produces a diffusion potential even in the absence of proton pumping activity. More information on Pi homeostasis has been obtained from recent studies by Booth and Guidotti (1997) on yeast vacuoles which provide a major storage conpartment for phosphate. Isolated intact vacuoles take up large amounts of added (^{32}P) phosphate by counter flow exchange with phosphate present in the vacuoles at the time of their isolation. Exchange mediated by the newly identified bidirectional phosphate transporter is faster than the unidirectional efflux of phosphate from vacuoles. The transporter is highly selective and strongly pH-dependent with increasing activity at lower pH. Of the various organic anions in plants, malate and citrate play central roles as intermediates in the TCA cycle and CO_2 fixation, and when accumulated at elevated levels these acids are stored in vacuoles. Their uptake is stimulated by MgATP and inhibited by inhibitors of ATPase. Malate and citrate reveal similar transport specificities and succinate appears to use the same transporter at least in *Heavea* lutoids. Recently the transporter from *Hevea* lutoid known as citrate-binding protein has been critically analysed and is found to represent a member of a novel class of proteins (Rentsch *et al.* 1995).

7.3.2.4 Role of ion channels

It is now evident that besides the primary and secondary active mechanism of transport of ions, ion channels are increasingly recognised for their role in a range of physiological processes (vide review by Schroeder *et al.* 1994; Ward *et al* 1995; De Boer and Wegner 1997). Of the various ion channels, the K^+ channel is the most studied (vide review by Maathius *et al.* 1997). K^+ channels are known to regulate membrane voltage and provide inportant pathways for long-term physiological K^+ uptake and release. They are also involved in stomatal and leaf movement, cell expansion and growth, loading/unloading into the xylem, cation nutrition, intracellular solute redistribution and cytosolic volume control. Turgor and water relations are also influenced by K^+ channels. Using the patch-clamp technique, Hedrich and co-workers (Hedrich and Neher 1987; Hedrich *et al.* 1988) first described the existence of both slow-vacuolar (SV) and fast vacuolar (FV) channels in tonoplast. The SV channels are active at negative potential and are involved in anion release for vacuoles. Thus the SV-type are activated by increases in free Ca^{2+} and cause anion release. In contrast, the FV-type, which is activated by both positive and negative potentials, is activated by decrease in free Ca^{2+} and may be involved in anion accumulation within the vacuole (Schroeder and Hedrich 1989). A third type of K^+ channel, which is highly selective for K^+, was identified by Ward and Schroeder (1994) in the tonoplast of *Vicia faba*. These vacuolar K^+ (VK) channels are voltage independent and activate at cytosolic Ca^{2+} elevations in the physiological range. Recent studies on osmoregulation in stomatal guard cells, to be discussed later, have thrown more light on the gated-channel-Ca^{2+}-calmodulin system. Various transport processes across the tonoplast are schematically presented in Fig. 7.5.

7.4 pH REGULATION BY VACUOLES

Depending on the physiological state of the cell the vacuolar pH (pHv) in plants may show considerable variation. The change in pHv during development is best exemplified during the fruit maturation of a lemon when the pHv can drop from 6.2 to as low as 2.2, whereas in the epicotyl cells of a lemon, the pHv is about 5.5 (Müller *et al.* 1996). To study the mechanism of hyperacidification, the kinetics of ATP-driven protein pumping by tonoplast vesicles from lemon fruits and epicotyls were compared and found to differ in respect to inhibitors, oxidation parameters, etc. These results suggested the possible presence of two H^+-ATPases in the fruit preparation: a V-ATPase and a vanadate-sensitive H^+-ATPase. Subsequent studies on the sub-unit composition showed that the fruit V-ATPase is enriched in two polypeptides of 33/34 and 16 kD (Müller et al. 1997). The electron micrographs of negatively stained juice sacs showed that the stalks of the V-ATPases are thicker than those of epicotyl V-ATPases. In photosynthetic cells, the pHv can flucuate due to the intensity of light. In plants with Crassulaceanacid metabolism, the pHv can vary diurnally from a value of 3 when organic acids are accumulated at night to 6 during the day when the organic acids are metabolised (Lüttge 1987). Stresses such as anoxia, osmotic shock and temperature shock can also affect pHv. Swanson and Jones (1996) have recently shown that the pHv of freshly isolated aleurone cells is 6.6, but after incubation in gibberellic acid (GA_3), the pH fell to 5.8. In contrast, abscisic acid (ABA) caused little or no acidification. It is now clear that the proton-ATPases control the acidification of the vacuolar sap and formation of a positive membrane potential inside. Since the stoichiometry of the proton pump is probably 2 H^+/ATP, at equilibrium, a pH gradient of more than three units should be formed. The over-acidification of the sap may be controlled by ATPase inhibitors, phospholipids and loosely coupled mechanism of proton pumping modulated by factors such as anions. The proton-pumping activity of vacuolar ATPase is enhanced by Cl^- ions and inhibited by NO_3^-.

In addition to H^+-ATPase, the possibility that other proton and cation exchange systems may be involved in vacuolar pH regulation has been examined by Guern and co-workers (Mathieu et al. 1989; Guern et al. 1989). With the aid of a ^{31}P NMR technique these workers studied the vacuolar pH and the trans-tonoplast pH modifications induced by the activity of the two proton pumps and by the proton exchanges catalysed by the Na^+/H^+ and Ca^{2+}/H^+ antiports in isolated intact vacuoles prepared from *Catharanthus roseus* cells. They recorded intravacuolar acidification by hydrolytic and vectorial activities of H^+-PPase comparable to that of H^+-ATPase. The vacuoles incubated with Na^+ (10mM) or Ca^{2+} (1mM) alkalised the intravacuolar sap considerably. Thus, in maintaining the H^+-ion concentration inside the vacuole the role of H^+-transporting systems is much more than the intrinsic buffering capacity of the vacuolar sap. The relative role of two proton pumps H^+-ATPase and H^+-PPase in maintaining vacuolar acidity has been critically investigated. Work by Rea and Poole (1993) and Brauer et al. (1992, 1995) suggest that in the absence or suppression of activity of H^+-ATPase vacuolar acidity is maintained by H^+-PPase. Recently, Brauer et al. (1997) have shown that in conditions which lead to the accumulation of ADP, like anoxia or KCN, PPase activity may be able to maintain the vacuolar acidity in maize root hair cells. Shiratake et al. (1997b) observed that during the development of pear fruit, V-PPase is the major H-pump of the vacuolar membranes of young fruit, while the contribution of V-ATPase increases with fruit development until finally it becomes the major H^+-pump during the later stages of development. Similar developmental changes in V-PPase activity have been reported by Nakanishi and Maeshima (1998) in the growing hypocotyls of mung bean. Whereas V-PPase activity is high in the elongating region of the 3-day-old hypocotyl, it is extremely low in the 5-day-old mature region. Again the relative activity of the two protein pumps in the tonoplast may be different in different tissues. Fänha et al. (1998) demonstrated that H^+-ATPase was more active than H^+-PPase in coleoptile tonoplast vesicles, whereas in the vesicles H^+-PPase was clearly dominant.

In certain cases, anion exchange systems (Blumwald and Poole 1985; Kaestner and Sze 1987) and malate exchange (Chang and Roberts 1969; Kurkdjian et al. 1985) are also likely to be involved in pH regulation. Besides programmed pH changes, the vacuole may act as a buffer for additional or sudden changes in the pH of the cytoplasm. A direct study of the changes in the pH of vacuoles (pHv) and cytoplasm was carried out by vacuolar perfusion technique in the giant cells of *Chara australis* by Moriyasu et al. (1984a). In these giant cells the pHv is about 5 and they found that pHv changed by only 0.5 pH units 24 hours after a change of external pH (pH_O) from 5.5 to 10.5. When the pHv was raised to pH 6 by vacuolar perfusion, it gradually returned to the original value. This process was inhibited by dicylohexyl carbodiimide (DCCD), suggesting that an

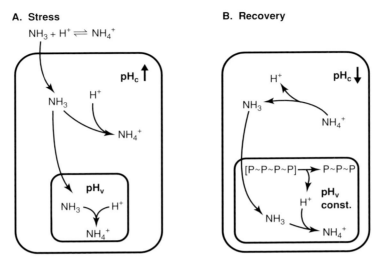

Figure 7.6 Proposed mechanism of the role of intravacular polyphosphate hydrolysis in amine compartmentation and pH homeostasis (after Pick and Weiss 1991). With permission of the American Society of Plant Physiologists.

H^+-pump is involved in acidification of the vacuole. On the other hand, pHv quickly recovered its original value when acidified to pH 4.2 after an initial overshoot. This process was DCCD insensitive. Thus the balance between the passive leakage of H^+ from vacuole to cytoplasm and active H^+ secretions from cytoplasm to vacuole seems to play a key role in pHv regulation. The role of H^+-ATPase in *Chara* tonoplast is supported by the fact that the pH-sensitivity of the tonoplast potential was suppressed by DCCD. Also, isolated vacuolar vesicles accumulated H^+ only in the presence of MgATP (Moriyasu *et al.* 1984b). Since then a number of studies, for example, Reid *et al.* (1989) and Fox and Ratcliffe (1990), have shown that both the cytoplasmic and vacuolar compartments are highly buffered and are resistant to internal pH changes arising from external pH offence. Horn *et al.* (1992) demonstrated by ^{31}P-NMR and fluorescence spectroscopy of cultured soybean cells that extracellular pH changes are rapidly communicated to the vacuole without significantly perturbing the cytoplasmic pH.

On the other hand, a change in the cytoplasmic pH can be counteracted by the vacuole. This has been clearly demonstrated in the case of mesophyll cells of leaves exposed to acidifying gases. In mesophyll cells vacuoles occupy about 80% of cellular volume and the vacuole's capacity for proton storage greatly exceeds that of the cytoplasm (Pfanz and Heber 1986). Obviously, damage by proton accumulation in the cytoplasm can be prevented, or at least postponed, if protons are transported from the cytoplasm into the vacuole. In a computer analysis, Laisk *et al.* (1988) have shown that in the case of SO_2 fluxes in leaves, sulfate and H^+ transfer into the vacuole can preserve cytoplasmic pH values at the expense of the vacuolar pH as long as the sulfate and H^+ transport system of the tonoplast can cope with protons and the sulfate produced in the cytoplasm during the influx of SO_2.

Another aspect of the buffering capacity of vacuoles is exhibited by amine accumulation in the cells of halotolerant alga *Dunaliella salina*, as reported by Pick *et al.* (1990). They have shown that influx of ammonia in these cells induces a rapid alkalinisation of the cytoplasm, followed by recovery of the cytoplasmic pH. The amines are compartmentalised in the acidic vacuoles which serve as a high-capacity buffering system and as a safeguard against cytoplasmic alkalinisation and uncoupling of photosynthesis. Moreover, this work has demonstrated that the cells' capacity to accumulate amines depends on hydrolysis of polyphosphates in these vacuoles (Pick and Weiss 1991). The researchers propose that compartmentation and protonation of the amine inside the acidic vacuoles would relieve the cytoplasmic pH stress and provide an effective buffering system (Fig. 7.6).

In other cases, a non-proton pump factor, possibly Donnan equilibrium, is involved in maintaining the acid pH of the vacuoles, parallel to what has been demonstrated in the case of acidic lysosomes from mice liver (Moriyama et al. 1992). Plant vacuoles are known to contain membrane-impermeable forms of negatively-charged molecules which could form a Donnan potential and maintain the acidic internal pH by preventing free diffusion of protons. Moriyama et al. (1992) proposed that protons are first transported into lysosomes by H^+-ATPase; then, negatively charged molecules in the lysosomes reach Donnan-equilibrium with the protons and maintain a stable acidic pH. This acidic pH could be maintained unless monovalent cations are added externally.

7.4.1 Role of oxido-reductases

This account of transport across the tonoplast would not be complete without recording the possible existence of redox enzymes. An antimycin-insensitive NADH-cytochrome C oxidoreductase, capable of functioning as a proton pump in the presence of exogenous cytochrome C was found in lutoid membrane (d'Auzac et al. 1982). According to Marin et al. (1985), while the ATPase proton pump operates on the lutoidic tonoplast and acidifies the vacuole contents, an NADH cytochrome C-reductase electrogenic proton pump handles the efflux of protons from inside the lutoids to the cytosol. These two pumps have a slightly different optimum pH than the physiological pH of latex which varies between 6.6 and 7.3. This leads to the suggestion that these two tonoplast enzymes operate simultaneously as a biophysical and biochemical pH-stat using a very active phosphoenolpyruvate carboxylase to control the cytoplasmic pH and lutoid sap pH. In addition, d'Auzac et al. (1987) reported the presence of a non-specific NAD(P)H-quinone reductase producing toxic form of oxygen. They propose that the two reductases are linked as a possible control mechanism of electrons and dissolved oxygen. Therefore, the regulation of vacuolar pH is a complex dynamic process, in which various tonoplast features, as well as external and internal factors, are involved.

7.5 SALT TOLERANCE

A plant's salt tolerance is derived from a special physiological feature, i.e. its ability to accumulate sufficient ions to maintain growth while avoiding either a water deficit or an excess of ions. If the water potential of the protoplast remains above that of the external medium, or if ions accumulate to such a high concentration in the cell wall to cause dehydration of protoplast, then a water deficit may arise. Again, if the concentration of ions exceeds the limits compatible with normal enzymatic activity, then ion toxicity will result.

The salt tolerance of halophytes is another aspect of vacuolar function. Many plants are tolerant to a saline environment by accumulating high intracellular concentrations of Na^+ and Cl^- and it is generally accepted that these ions are sequestered in the vacuole, where compatible solutes, such as sugars, proline and glycinebetaine function to balance the osmotic pressure of the cytoplasm. Flowers (1975) concluded that halophytes' ability to accumulate a large concentration of ions seems to be correlated with the ability to retain these ions in vacuoles. The available data indicate that the maximum Na^+ concentration in the cytoplasm is perhaps 150 mM whereas the vacuolar concentration in most halophytes is in excess of 500 mM. If the concentration in the cytoplasm and vacuole were 100 and 500 mM respectively, a potential difference across the tonoplast of some 40 mv would be required for the ions to be in passive equilibrium. Hazibagheri and Flowers (1989) grew a halophyte Sueda maritima in 200 mM NaCl and studied the Na and Cl concentrations by X-ray microanalysis of freeze substituted thin sections. The concentration of Na and Cl in the vacuoles was about four times that of the cytoplasm or cell walls, whereas K was more concentrated in the cell walls and cytoplasm than in the vacuoles. The vacuolar Na concentration was about 12 times higher than that of K.

Binzel et al. (1988) studied the intracellular compartmentation of Na^+ and Cl^- ions in salt-adapted tobacco cells. Utilising steady state efflux kinetic analysis and scanning electron energy dispersive x-ray microanalysis, they observed that Na^+ and Cl^- were compartmentalised in vacuoles, at concentration of 780 and 620 mM respectively, while cytoplasmic concentrations were maintained at 96 mM. The ability to establish and maintain such steep concentration gradients must involve a high proton motive force-generating capacity across the plasma membrane and the tonoplast. H^+-ATPase in the tonoplast creates a proton gradient that may serve as energy sources for Na^+ extrusion from cytoplasm via Na^+/H^+ antiports. Evidence for Na^+/H^+ exchange across both the tonoplast and plasma membrane of salt-tolerant plants has been reported. Niemitz and Willenbrink (1985) found evidence for Na^+/H^+ exchange across the tonoplast of intact vacuoles from red beetroots. Although Na^+ did not inhibit ATP hydrolysis, it inhibited formation of a pH gradient. Blumwald and Poole (1985) reported that addition of Na^+ to tonoplast vesicles from red beetroots caused a preformed pH gradient to dissipate. They proposed that there was an electrically neutral exchange of Na^+ for H^+, via a Na^+/H^+ antiport. Garbarino and DuPont (1989) demonstrated that NaCl induces Na^+/H^+ exchange due to activation of an existing protein in the tonoplast of barley roots. Along with V-ATPase, H^+-PPase is involved in energising the tonoplast for salt accumulation (Rea and Sanders 1987). Thus, the two proton pumps create a proton gradient that may serve as an energy source for the vacuolar accumulation of Na via Na^+/H^+ antiports in the roots of both halophytic and glycophytic plants. Several studies have shown that Na^+/H^+ antiport activity is exceedingly important for Na uptake. Na^+/H^+ antiport activity increased in sugar beet cell suspension when grown in the presence of NaCl (Blumwald and Poole 1987). It is also rapidly induced in barley root cells. Additional evidence on the role of the antiport comes from Staal et al. (1991), who reported that tonoplast Na^+/H^+ antiport activity could be detected only in the tonoplast vesicles of the salt-tolerant Plantago maritima but not in those of the salt-sensitive Plantago media. On the basis of their studies on Na^+/H^+ antiporter protein, Fukuda et al. (1998) concluded that the amount of antiporter in vacuolar membrane may be one of the most important factors determining salt tolerance. Other mechanisms besides Na^+/H^+ transporter may also contribute. On the basis of transfer capacity of the vacuolar sulfate uptake system, Kaiser et al. (1989) concluded that halotolerance of barley mesophyll cells depends on the capacity of the tonoplast sulfate transporter.

It is now accepted that membrane phospholipid composition is important in the modulation of membrane protein activity. Kasamo and Nouchi (1987) have shown that a specific phospholipid environment is required for optimal ATPase activity. Several other reports (Norberg and Liljenberg 1991; Douglas 1985; Mansour et al. 1994) indicate that changes in phospholipids and free sterols of the cell membranes may contribute to salt tolerance. Burgos and Donaire (1996) reported that although the relative distribution of free sterol molecular species was stable in isolated tonoplast vesicles from normal and NaCl-treated jojoba roots, the total phospholipid fatty acids and free sterols were increased in salt-stressed roots showing increased H^+ATPase activity.

7.6 OSMOREGULATION

Two physical components of osmoregulation are osmotic pressure and turgor pressure. Osmotic pressure is the hydrostatic pressure needed to prevent water movement through a semi-permeable membrane (tonoplast/plasma membrane) between the solution and pure water. Turgor pressure is the outwardly directed pressure exerted by the contents of the cell on the cell wall. The process of osmoregulation enables cells to survive and grow under a wide range of osmotic pressure which is exerted by neighbouring cells, apoplast or the environment. One of the basic roles of vacuoles in a plant cell is the maintenance and modulation of the turgor pressure of the cell. A change in the balance between water uptake and water loss to the apoplast or environment can change the cell turgor without changing the cell content of osmotically effective substances. However, in most physiological conditions, especially those involving movement, change of the osmotic properties by uptake or loss of solutes is the primary process.

7.6.1 Turgor for growth

Under favourable conditions, plant cells can elongate extremely fast at rates ranging from 20 to 75 µm per hour. The cell growth is necessarily accompanied by expansion and /or elongation which occurs when the somewhat elastic cell wall stretches under pressure created by water taken into the vacuole. The role of turgor on growth rate has been demonstrated in the internodal cells of *Nitella*. Green *et al.* (1971) showed that cell elongation takes place only if the turgor exceeds the yielding threshold of the wall. This contention is supported indirectly by studies on auxin-induced bursting of naked-protoplast of *Avena* coleoptiles. The dependency of bursting response upon the concentration of auxin closely follows the concentration dependency of extension growth in the respective organs. Apparently turgor pressure under hormonal regulation is one of the prerequisites for growth (Hall and Cocking 1974).

7.6.2 Tonoplast permeability

Water permeability of the tonoplast is extremely high (Kiyosawa and Tazawa 1977) and the cytoplasm is always in osmotic equilibrium. When the osmotic pressure is increased or decreased by a transcellular process, vacuolar osmotic pressure immediately follows suit. The standard procedure for obtaining intact vacuoles by osmotic lysis of the protoplast demonstrates another aspect of tonoplast permeability. From 0.6 M mannitol osmoticum, a drop to 0.06 M osmoticum causes bursting of the plasma membrane in 5 sec, but it does not affect the tonoplast. Although the procedure is very rapid (about 5 sec), the vacuole which was previously equilibrated presumably at 0.6 M, withstands the 0.06 M after protoplast lysis. This is indicative of the tonoplast's high tensile strength. It can be stretched to a great extent without disruption or release of sap (Wagner 1983, 1985).

After the discovery of the aquaporin water channel, tonoplast permeability became understood in terms of water channels. Recently, Maurel *et al.* (1997) compared the osmotic water permeability coefficients (pf) of tonoplast (TP) and plasma membrane (PM) vesicles from tobacco suspension cells. Their results suggest that water transport across the PM occurs mostly by diffusion across the lipid matrix. In contrast, water transport through the TP with a 100-fold higher Pf occurs through water channels. A high TP Pf suggests a role for the vacuole in buffering osmotic fluctuations occurring in the cytoplasm. Thus, the differential water permeabilities and water channel activities observed in the tobacco TP and PM point to an original osmoregulatory function for water channels in relation to the typical compartmentation of plant cells.

Turgor pressure leads to the rigidity of plant cells and consequently of tissues and also provides a large contact surface with the environment. Together with the cell wall, turgor pressure provides mechanical support for maintaining cell and tissue shape. The maintenance of aerial hyphae of many fungi is an ignored example of turgor pressure. Wiebe (1968) suggested that plants can achieve relatively large volumes by filling a large proportion of the cell volume with 'cheap' vacuolar solutes, thereby saving 'expensive' nitrogen-rich cytoplasm and also obtaining a large surface between the thin peripheral layer of cytoplasm and the environment. Turgor pressure also provides the mechanism to maintain proper water relations of terrestrial plants.

7.6.3 The mechanism of osmoregulation

7.6.3.1 Osmoregulation by sugars

Vacuoles are used by the cells both as a long-term storage house and as a temporary repository of solutes that may be used later for metabolism or to maintain cytoplasmic solute concentration (Leigh and Wyn Jones 1986). Accumulation of these solutes in large quantity, in addition to the inorganic salts, will make a large contribution to sap osmotic pressure and hence turgor pressure. In the case of export of the solute there would necessarily be a drop in osmotic pressure. The vacuole, in certain cases, has a mechanism to counter such a drop. One of the mechanisms is controlled breakdown and re-synthesis of osmotically inactive polymers like starch, polyphosphates, etc. The interconversion of sugar–starch or malate–starch is a widespread osmoregulation process. Similar interconversion of inorganic phosphate–polyphosphate in certain microorganisms is also such a process. Bacon et al. (1965) have demonstrated in the cells of Beta storage roots that when sugars are removed by washing, an activity of the acid invertase is induced leading to the hydrolysis of sucrose to glucose and fructose. Subject to the external osmotic pressure, glucose and fructose also accumulate to increase internal osmotic pressure. Moreover, when necessary, fructose may be preferentially exported, maintaining the concentrations of Na^+, K^+ Cl^- and betanin, unaffected. This has been shown in comparable experiments by Perry et al. (1987) in red beet cells. Thus, both internal osmotic pressure and turgor pressure are regulated during the mobilisation of sucrose in storage tissue.

Perry et al. (1987) suggest that the use of hexoses for osmotic purposes clearly limits the extent to which sucrose-derived carbon is released from the vacuole and made available to the rest of the cell. So, when hexoses are the only solutes available for turgor-pressure regulation this function is given priority over the sugars' metabolic requirements. For complete utilisation of all the carbon available in stored sucrose, alternative vacuolar solutes must be made available to replace the hexoses that have been accumulated for osmotic purposes. Inorganic salts are the convenient alternative osmotica. If salts are provided when the reducing sugars have replaced the sucrose, they are absorbed and the concentration of the reducing sugar declines. This indicates that salts replace sugars in the vacuole, releasing the sugars for metabolism. However, changes in salt and sugar concentration are not equal because osmotic pressure and turgor pressure increase.

Osmoregulation by ion transport across the tonoplast

Although organic solutes which are directly dependent on cellular metabolism are involved in osmoregulation, the most important mechanism involves the transport of inorganic ions across the tonoplast for:
i) maintenance of optimum turgor
ii) control of cell volume
iii) cellular metabolism and growth, and
iv) movement of organs.

Because the cytoplasm has a strict requirement for organic and inorganic solutes, their concentrations must be maintained at a reasonably constant level. The vacuole necessarily shows a wide variation in solute composition, although sufficiently low osmotic potential must be maintained to ensure both an adequate turgor and water potential equilibrium or H_2O homeostasis across the tonoplast. Since the water permeability of the tonoplast is extremely high, the cytoplasm is always in osmotic equilibrium with the vacuole. When the cytoplasmic osmotic pressure is increased or decreased by transcellular osmosis, the vacuolar osmotic pressure follows suit. Any change in cytoplasmic volume would be adjusted by vacuolar inorganic ions or by changes in organic solutes.

Besides the uncharged organic solutes, viz. sucrose, hexose, amino acids, quarternary ammonium compounds and secondary products, various inorganic ions are involved in the process. All these substances are very unequally distributed between the vacuole and the cytoplasm (including cytosol and organelles). The inorganic ions consist mainly of common nutrient ions Na^+, K^+, NO_3^- and Pi. In the cytoplasm, the main site for biochemical activity, inorganic nutrients like NH_4^+, NO_3^- and Pi participate as substrates and others, such as K^+ and Mg^{2+} as co-factors. K^+ and Mg^{2+} are maintained in a narrow concentration range but the nutrient ions exist in the vacuole for both short-term and long-term storage and turgor generation and their amounts are variable (vide review by Leigh and Wyn Jones 1986).

With the aid of transcellular osmosis, Kamiya and Kuroda (1956) concluded that *Nitella* cells regulate osmotic pressure rather than turgor pressure. Since regulation of inside osmotic pressure (n_i) in *Nitella* is achieved mainly by K^+, the question arises whether the cells regulate the osmotic pressure (n_i) or K^+ concentration. Nakagawa *et al.* (1974) have shown that *Nitella* cells are relatively insensitive to the ion species and ionic concentrations in the vacuole, as long as the osmotic pressure is maintained at the normal level. On the other hand, that the cells can regulate turgor pressure has been demonstrated in euryhaline brackish Characeae. Bisson and Kirst (1980) showed that the cells of *Lamprothamnium papulosum* can keep the turgor pressure almost constant for changes in the outside osmotic pressure (n_o) between 550 and 1350 mOsm. Similarly, another species, *L. succintum*, can regulate its turgor for outside osmotic pressure (n_o) ranging from zero to two times diluted artificial sea water (0.54 Osm).

Typically, the osmolarity of the vacuole is about 400 to 600 mOsm/L and inorganic salts provide a large part of this. Of the various ions, K^+ and Cl^- are the most important. Vacuolar concentrations range from 40 to 205 mM for K^+ and 38 to 170 mM for Cl^- for most non-halophytes (Lüttge and Higinbotham 1979). These two ions are also most responsive to the outside osmotic pressure. The total osmotica in the vacuole is generally restricted, but not the individual solute constituents. Flowers (1988) has shown that the accumulation of Cl^- within the vacuole is important both for the generation of turgor and for the control of cytoplasmic Cl^- concentration. Of the total leaf K^+ content in a K^+-sufficient leaf, 90% is sequestered in the vacuole. Walker *et al.* (1996) used triple-barrelled microelectrodes to obtain the first fully quantitative measurements of the changes in K^+ activity in the vacuole and cytosol of barley root cells grown in different K^+ concentrations. The measurements revealed that vacuolar K^+ activity declined linearly with decreases in tissue K^+ concentration, while cytosolic K^+ activity initially remained constant in both epidermal and cortical cells but then declined at different rates in each cell type. Again, since the major anion balancing K^+ is cytoplasmic phosphate, the Pi concentration of the vacuole is variable to a great extent. Starvation of Pi causing a large decrease in the total Pi content of the tissue is due entirely to the depletion of the vacuolar pool. In contrast to glycophytes, halophytes have to deal with Na^+. In order to balance external salinity, halophytes accumulate Na^+ in the vacuole. Active Na^+/H^+ antiport activity at the tonoplast sequesters Na^+ in the vacuole. Thus vacuolar Na^+ serves as an osmoticum necessary for H_2O homeostasis. Wyn Jones (1981) has suggested that halophytes, while retaining the cytoplasmic requirement for K^+, can accumulate NaCl in the vacuole to provide a low osmotic potential. Jeschke (1980) reviewed the evidence that the tonoplast fluxes favour Na^+ over K^+ in transport **to** the vacuole and K^+ over Na^+ in transport **from** the vacuole. Again, replacement of one anion by another in the vacuole is significant. The high NO_3^- levels in certain halophytes may exchange for Cl^- or may compete with Cl^- for uptake. Pope and Leigh (1990) have shown that Cl^- and NO_3^- may be transported by the same anion porter; Cl^- is preferentially stored in the vacuoles of many halophytes.

7.6.3.2 Turgor movements

A number of movements in plant cells and tissues, known as turgor movements, are chiefly due to turgor changes within the cell. In most cells these movements are conducted by the large vacuole which is responsible for the cell turgor. Changes in turgor result in changes in cell dimension and shape. Stomatal opening and pulvinar movement of leguminous leaves are the most prominent examples of turgor movement. In darkness when the guard cells close, the cell contains numerous small vacuoles. Upon induction of opening, these vacuoles inflate and coalesce into a single large vacuole. In *Albizzia*, turgor-mediated volume changes of the motor cells are reflected by a similar structural organisation of the vacuole. During cell shrinking, the large central vacuole splits up into several smaller vacuoles, while during swelling of the cells the vacuoles fuse and increase in volume.

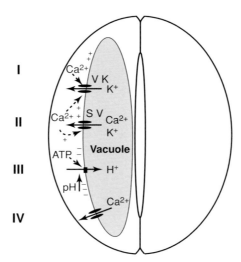

Figure 7.7 Model for ion channel-mediated K$^+$ and Ca^{2+} release from guard cell vacuoles during stomatal closure. **I** Increases in cytoplasmic Ca^{2+} activate selective vacuolar K$^+$ (VK) channels in the vacuolar membrane. The resulting K$^+$ influx into the cytoplasm causes a positive shift in vacuolar membrane potential. **II** The activation of slow vacuolar (SV) non-selective cation channels leads to the release of K$^+$ and Ca^{2+} from the vacuole. Increased cytoplasmic Ca^{2+} provides positive feedback to activate further both SV and VK ion channels. **III** Vacuolar proton pump activity is required to drive long-term K$^+$ efflux through VK channels and provides a mechanism for cytoplasmic alkalinisation observed in response to ABA. **IV** Ca^{2+} release channels activated by alkalinisation of the vacuolar lumen allow Ca^{2+} release from vacuoles at negative potentials on the cytosolic side of the membrane (from Ward *et al.* 1995). With permission of the American Society of Plant Physiologists.

Turgor movement of guard cells

The opening and closing of the stomata depend on the osmoregulation process in which swelling or shrinking of the guard cells takes place. The volume and turgor changes during reversible cell expansion of the guard cells require coordination of ion fluxes, and carbohydrate and organic acid biosynthesis between the tonoplast and the plasma membrane. As discussed earlier, the vacuole plays the primary role in turgor regulation, in a typical multicompartmental plant cell where it is the largest compartment containing most of the cellular water. The studies on ion fluxes in guard cells show the operation of electrogenic pumps, K$^+$ channels, and anion and stretch-regulated channels at the plasma membrane. The channels are activated by membrane depolarisation, hyperpolarisation or stretch. In particular, a detailed analysis of depolarisation-activated anion channels at the plasma membrane of broad bean guard cells has suggested a role for these channels in the salt efflux that mediates stomatal closing (Hedrich and Becker 1994; Ward *et al.* 1995). As well as two electrogenic pumps and the ubiquitous slow-activation vacuolar-type channels which are somewhat permeable to anions, highly selective anion channels have been identified in the tonoplast (Pantoja *et al.* 1992; Amodeo *et al.* 1994; Cerana *et al.* 1995).

During this decade, a network of ion channels have been identified in both the plasma membrane and tonoplast, which are deeply involved in perception of stimulus leading to an intracellular cascade of events. In a simplified scheme, Ward *et al.* (1995) have suggested that, in general, cell stimulation may cause anion channel activation. Anion efflux then results in plasma membrane depolarisation, which in turn triggers the activation of voltage-dependent Ca^{2+} channels that mediate Ca^{2+} flux. In the case of guard cells, the reduction of turgor is caused by an efflux of large amounts of K$^+$ and anions and a parallel conversion of malate to starch, leading to stomatal

182 Plant cell vacuoles: an introduction

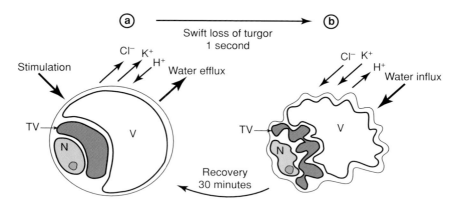

Figure 7.8 Schematic view of the variation in motor cell shape. The cell remains swollen if pulvini movement is prevented by diethylether (a), whereas it becomes shrunken after pulvini response (b). Motor cells contain two vacuole types, one tannin-rich (TV) located near the nucleus (N), the other aqueous and central (V), corresponding to the vacuole in most mesophyll cells. During shrinkage both vacuoles change their shape: the tannin vacuole forms many tubules that are connected, the aqueous vacuole develops invaginations in its external region (from Fleurat-Lessard et al. 1997). With permission of the American Society of Plant Physiologists.

closing. Ward and Schroeder (1994) have identified a voltage-independent K^+ channel (VK^+ channel) on the tonoplast. They have provided evidence for a direct link in the signalling pathway between increases in cytoplasmic Ca^{2+} resulting from stimuli, such as abscisic acid (ABA) and the initiation of K^+ release at the tonoplast. Consolidating the available data, Ward et al. (1995) proposed a model for ion-channel-mediated K^+ and Ca^{2+} release from guard cell vacuoles during stomatal closure (Fig. 7.7). In brief, ABA-induced cytosolic release of Ca^{2+} leads to the activation of the VK^+ channel. VK^+ channel activation provides positive shifts in tonoplast membrane potential (required for SV channel activation) which are permeable to both Ca^{2+} and K^+. The positive feedback leads to further release of Ca^{2+} causing loss of turgor in the guard cells.

Turgor movement of pulvinar motor cells

The most dramatic turgor movement is the seismonastic movements of *Mimosa pudica*, in which any stimulus like touch, impact, electrical impulses or heat may induce an extremely fast response. The response may start a mere 0.1 sec after stimulation and is completed in a few seconds. If the pinnules are stimulated first they will close first, then the leaflets and finally the petiole. The response is noted in the neighbouring leaves in succession and response spreads up and down the plant very fast, depending on the intensity of the original stimulus at rates up to 40–50 cm/sec. The response is brought about by the sudden loss of turgor of the cells of the pulvinus. The anatomy and ultrastructure of the pulvinus has been described in detail by Fleurat-Lessard (1981). The feature that is important to the present discussion is that the vacuolar compartment of the motor cells of the pulvinus contains two kinds of vacuoles: tannin and non-tannin or colloidal vacuoles. Whereas non-tannin vacuoles are electron-translucent, tannin vacuoles contain highly electron dense, amorphous material (Fig. 7.8). The size of the vacuoles, number per cell and the ratio between number of tannin and non-tannin vacuoles per cell vary considerably. Toriyama and Jaffe (1972) have observed changes in both size and shape of tannin vacuoles during seismonastic movement of the main pulvinus of *Mimosa*. The turgor changes in the pulvinar motor cells involved in the movements have been reviewed by Satter and Galston (1981). The changes in turgor occur simultaneously with fluxes of K^+, Cl^-, Ca^{2+} and other ions together with water fluxes. An X-ray microprobe analysis by Moysset and Simon (1991) of *Robinia pseudoacacia* motor cells has shown that, besides phenolic and polysaccharide compounds, tannin vacuoles contain appreciable amounts of Na^+, Ca^{2+} and K^+. Campbell et al. (1979) also suggested that tannin vacuoles may be involved in storage and release of Ca^{2+} and K^+.

In *Mimosa*, mechanical or other stimulation is thought to result in release of Ca^{2+} from tannin vesicles adjacent to the vacuole. Actin microfilaments are activated by Ca^{2+} to react with myosin. The actin–myosin interaction results in conformation changes of membrane proteins in the tonoplast, thus increasing its permeability, leading to passive efflux of Cl^- and K^+ and loss of turgor. It is interesting to note that calcium oxalate crystals have been found in plants with slower or imperceptible seismonastic movements, e.g. in *Albizzia* (Schrempf et al. 1976) and *Samanea* (Morse and Satter 1979), but have not been reported in *Mimosa* pulvini (Fleurat-Lessard 1981). On the basis of studies on *Robinia*, Moysset and Simon (1991) suggested that calcium oxalate crystals are characteristic of slow pulvinus and tannin vacuoles are typical of rapid pulvinus.

The similarity of movement between guard cells and motor cells of *Mimosa* cannot be taken too far even though both cases involve water and ion release from the vacuole. The collapse of turgor is fast (about 0.1 sec) in the case of *Mimosa* motor cells and relatively slow (2 secs or more) turgor loss mediated by ABA-voltage-gated -Ca^{2+} release in guard cells or auxin-ligand gated (IP3)-Ca^{2+} release in other cells.

It is likely that the stimulus in *Mimosa* hits the tonoplast directly, and changes in membrane thickness of tannin vacuoles due to compression or stretching directly influence ion transport via the ion channels. Although wind- and touch-induced rapid increase in the transcription of calmodulin would be too slow a mechanism (as shown in *Arabidopsis* (Bramm 1992)) for *Mimosa* sensitiveness, it would be interesting to explore the presence of a mechano-sensitive or stretch-induced Ca^{2+} channel in the tonoplast of *Mimosa* motor cells. Such a process has been suggested by Cosgrove and Hedrich (1991) in the guard cells of *Vicia faba*, for entry of Ca^{2+} to the cytosol across the plasma membrane. Recently Fleurat-Lessard et al. (1997) reported that non-tannin vacuoles in *Mimosa pudica* motor cells, in contrast to tannin vacuoles, were characterised by exclusive localisation of γ-TIP aquaporin and most of the V-ATPase in their tonoplast. The development of the pulvinus into a motor organ was accompanied by a more than three-fold increase per unit length of membrane in the abundance of both aquaporin and V-ATPase. In view of the fact that aqueous vacuoles have a volume about 27 times that of tannin vacuoles, most of the water and solutes are localised in the aqueous vacuole. However, the aqueous vacuole has 10 times more aquaporins per unit length of tonoplast than the tannin vacuole. Fleurat-Lessard et al. (1997) concluded that the aqueous vacuole is chiefly responsible for water and ionic fluxes. Therefore, it may be hypothesised that tannin vacuoles play a secondary role by releasing water as well as Ca^{2+} to bring about actin–myosin interaction with the tonoplast. However, the precise mechanism of the intracellular *superfast cascade* of events remains unclear.

7.7 ROLE OF VACUOLES IN CALCIUM HOMEOSTASIS

The two divalent cations Mg^{2+} and Ca^{2+} play a major role in metabolism. Although Mg^{2+} plays a key role in activity of many enzymes and other cellular functions, information on its mechanism of uptake and compartmentalisation is very scanty, especially with reference to the vacuole and tonoplast. Recently, Shaul et al. (1998) identified the Mg^{2+}/H^+ exchanger protein in the tonoplast. The transporter mediates Mg^{2+} uptake into the vacuole and possibly maintains the Mg^{2+} homeostasis in the various compartments of the cell.

It is universally accepted that Ca^{2+} is the pivotal ion species which actively participates and/or controls numerous physiological processes in the cytoplasm, although its concentration is maintained within a narrow range from a steady state level of 30 to 200mM in unstimulated resting cells. In contrast, the vacuolar concentration of Ca^{2+} is generally maintained at levels 3 to 4 orders of magnitude higher; levels which are potentially toxic. The cell uses this large concentration gradient to regulate various cytosolic processes by judicious regulation of Ca^{2+} fluxes. Ca^{2+} moiety may affect/control many physiological activities by being both a 'stabilisation' component and a 'message' component. The cytoplasmic concentration of Ca^{2+} may regulate various transport processes of ions in plant cells (Hepler and Wayne 1985; Kauss 1987; Mimura and Tazawa 1983). In *Chara* an increase in Ca^{2+} triggers a sequence of events leading to turgor regulation via activation of K^+ and Cl^- channels in the plasma membrane and tonoplast (Williamson and Ashley 1982). The presence of a few mM of external Ca^{2+}, essential for the induction of hypotonic turgor regulation in *Lamprothamnion succinctum*, suggests its role in activation of ion channels (Okazaki and Tazawa 1986), especially when the perfusion experiments are viewed together with the electrical characteristics of plasma membrane and tonoplast. Kikuyama and Tazawa (1987) reported that when Cl^- concentration in the vacuole of *Nitella* was drastically reduced by vacuolar perfusion, the direction of the tonoplast action potential was reversed. The tonoplast action potential can only be generated when the plasma membrane is excited. Thus excitation of tonoplast is likely to be coupled with the excitation of plasma membrane. In view of the increase of Ca^{2+} upon excitation, Ca^{2+} is thought to be the coupling factor. Lunevsky *et al.* (1983) suggested that an increase in cytoplasmic Ca^{2+} activates the Cl^- channel in *Nitellopsis*. Similar Ca^{2+}-dependent anion channel activation has also been demonstrated in a water mould, *Blastocladiella* (Caldwell *et al.* 1986). Direct evidence has been presented by Hedrich and Neher (1987) that changes in cytoplasmic Ca^{2+} regulate the action of two distinct types of voltage-dependent ion channels in the tonoplast. The slow vacuolar type (SV) channel appearing at Ca^{2+} levels above 10^{-7} M may be involved in the movement of organic and inorganic ions across the vacuolar membrane. When the Ca^{2+} concentration is low (10^{-8} M) a fast vacuolar type (FV) channel may operate. The FV channel may provide a pathway for anions which accumulate at positive vacuole potential and allow the equilibration of K^+. Elevation of cytoplasmic Ca^{2+}, following diverse physiological stimuli and reduction of the pump-generated positive potential, would open the SV type channel for the release of anions. Thus the two channel types modulated by physiological changes of cytoplasmic Ca^{2+} provide for versatile regulation of ion and metabolic fluxes between the cytoplasm and the vacuole. On the other hand, the tonoplast together with the plasma membrane is the regulator of cytoplasmic free calcium.

As well as ionic balance and control of ion channels, Ca^{2+} is involved in a number of physiological processes. The fluctuation of Ca^{2+} has been correlated with phototropism and geotropism (Gehring *et al.* 1990), response to hormone action (Felle 1988; McAinsh *et al.* 1990), touch and cold (Knight *et al.* 1991), regulation of protein kinase (Blackshear *et al.* 1988), chilling effect (Minorsky 1985), stress (Rengel 1992) dark-induced inhibition of sucrose synthesis (Brauer *et al.* 1990) and stomatal closure (Gelroy *et al.* 1990). Thus, the description by Rasmussen *et al.* (1990) of the cytosolic Ca^{2+} as a kind of cellular fire contained in a vacuole and retrieved for control of various physiological activities, is very appropriate.

7.7.1 Ca^{2+} transporters and channels

Two Ca^{2+} transporters maintain the high concentration of Ca^{2+} in the vacuole against a cytosolic gradient. A calcium proton exchanger Ca^{2+}/H^+ antiport, indirectly driven by the proton ATPase, catalyses Ca^{2+} accumulation across the tonoplast for high capacity intracellular storage inside the vacuole under low cytoplasmic Ca^{2+} concentration (Schumaker and Sze 1986). The other transporter, Ca^{2+}-ATPase, an ATP-dependent primary energised high-affinity Ca^{2+} pump, plays a significant role in modulating cytosolic Ca^{2+} levels in the physiological range (Pfeiffer and Hager 1993). Recently, Lommel and Felle (1997) reported that the uptake of Ca^{2+} from the medium by intact vacuoles of *Chenopodium album* is mediated predominantly by Ca^{2+}-ATPase. In addition to the above-mentioned transporters which are involved in uptake of Ca^{2+} inside the vacuole, the tonoplast has a number of voltage-gated and ligand-gated Ca^{2+} channels which are responsible for release of Ca^{2+} from the vacuole. A second type of Ca^{2+}-selective voltage-dependent and IP_3-insensitive channel has been reported in excited patches of tonoplast by Johannes *et al.* (1992). This channel is sensitive to gadolinium (Gd^{3+}) and not affected by changes in cytoplasmic Ca^{2+} concentrations (from nM to mM). A related type of Ca^{2+} permeable ion channel that is inhibited rather than activated by micromolar levels of Ca^{2+} has been identified by Gelli and Blumwald (1993). Recently, two more inward-rectifying voltage-gated Ca^{2+} release channels which co-reside in the guard cells of tonoplast have been reported by Allen and Sanders (1994). The channels are gated open by cytosol-negative transmembrane voltages, increases in vacuolar Ca^{2+} concentration and increases in vacuolar pH. They are inhibited by Gd^{3+} and nifedipine. The two channels differ in voltage dependence, gating characteristics and relative selectivity for Ca^{2+} and K^+.

Of the ligand-gated Ca^{2+} channels, an inositol 1,4,5-triphosphate (IP_3)-responsive Ca^{2+} channel is most important (Alexandre et al. 1990). Indeed, the vacuole may be the largest store of IP_3-releasable calcium (Canut et al. 1993; Lommel and Felle 1997). IP_3 is able to transfer external stimuli to the vacuole by triggering a transient release of Ca^{2+} across the tonoplast so that the cytosolic concentration is changed to a desired level. A third type of outward rectifying Ca^{2+} channels has been identified with non-physiological Ca^{2+} concentration in the mM range (Pontoja et al. 1992). These channels are thought to conduct Ca^{2+} from the cytosol into the vacuole following a rise in cytoplasmic Ca^{2+} in response to IP_3-induced Ca^{2+} release. In addition to the above mentioned channels, Allen et al. (1995) identified a cyclic ADP-ribose-activated channel in sugar beet vacuoles, which may interact with SV channels.

7.7.1.1 Ca^{2+} as second messenger

It is obvious from the above list of transporters and channels that Ca^{2+} must be an extraordinary cation whose concentration and mobility in various cell compartments is governed by a complex system of signal transduction. It has been suggested that Ca^{2+} acts as an intracellular secondary messenger conveying the stimulus or stress to which the cell is subjected, to the target structure or receptor molecule (Poovaiah and Reddy 1993). Ca^{2+} may directly communicate the signal to the effector or bind to certain proteins which carry the necessary message. In other words, increased levels of Ca^{2+} may induce formation of calcium-bound proteins, which in turn regulate the effector proteins. A number of such effector proteins or calcium-binding proteins (CaB) has been identified in plants. Moreover, Ca^{2+}-dependent protein kinases, phosphatases and Ca^{2+}-stimulated phospholipases have also been detected. Bush (1993) recognised three basic patterns of changes in cytosolic Ca^{2+}: i) mechanical stimulation, hypotonic shock and elicitors can cause a transient and large increase; ii) GA, cytokinin and light can induce a modest and steady-state increase/decrease; and iii) auxin and ABA can induce oscillatory changes with regular or irregular periods.

7.7.1.2 Ca^{2+}- binding proteins

Of the various CaBs, calmodulin (CaM) is the most prevalent and most studied. CaM utilises Ca^{2+} to modulate cellular functions. A group of Ca^{2+}-sensing proteins binds Ca^{2+} with high affinity and becomes activated or subsequently activates specific enzymes. CaM is considered to be part of the transduction chain for environmental signals like wind or touch which may cause a rapid increase in transcription of the CaM homologue in *Arabidopsis* (Bramm 1992). Weiser et al. (1991) have shown that cytosolic CaM mediates the Ca^{2+}-dependence of the SV-channel in the *Chenopodium* tonoplast. Askerlund (1997) on the other hand, has detected the presence of 111-kD Ca^{2+}-ATPase in the vacuolar membrane of cauliflower and ATP-dependent Ca^{2+} uptake by the intact vacuoles was found to be CaM stimulated and partly protonophore insensitive. Bethke and Jones (1994) reported that following exposure to GA, barley aleurone cells exhibited an increase in the level of CaM. They suggested that Ca^{2+} and CaM act as signal transduction elements mediating hormone-induced changes in SV-channel activity. Schulz-Lessdorf and Hedrich (1995) draw similar conclusions, i.e. that guard cell SV-type channels which might be responsible for the release of K^+, Cl^- and, to a lesser extent, Ca^{2+} during stomatal closure, could serve as intracellular sensors for changes in cytosolic Ca^{2+} (Ca^{2+}– CaM) and pH. A large (>10^4-fold) physiological gradient of Ca^{2+} from the vacuole to the cytoplasm could lead to release of Ca^{2+} even when net currents are directed into the vacuole (Schneggenburger et al. 1993). Moreover, on the basis of their studies on SV channels of tonoplast of guard cells, Ward and Schroeder (1994) concluded that Ca^{2+}-activated SV channels are Ca^{2+}-permeable. This ubiquity of SV channels, Ca^{2+}-permeation and Ca^{2+}- activation of SV channels suggest that these ion channels may provide an important mechanism for vacuolar Ca^{2+}-induced Ca^{2+}-release which may be important for the transduction of a variety of signals.

Another major group of CaBs is Ca^{2+}-dependent membrane binding proteins, the annexins. On the basis of amino acid sequence homologies, a number of annexins have been identified in various plants. For example, Seals et al. (1994) identified a 42kD Ca^{2+}-dependent membrane-binding protein (VCaB42) associated with the tonoplast. This annexin-like protein has a high amino acid homology with other annexins and is possibly associated with the cytosolic surface of the tonoplast. Seals et al. suggest that it resides in the cytosol and, as a signal transducer, monitors cytosolic Ca^{2+} levels, transmiting the information to the vacuole by binding with the tonoplast. Recent experiments (Seals and Randall 1997) indicate that VCaB42 may be involved in the vacuolation process of expanding cells as well as age-associated and hormonally induced changes in cell volume in tobacco suspension cultures. Thus the protein may be involved in the early event of vacuole biogenesis since annexins may mediate vesicle fusion.

7.8 METABOLIC FUNCTIONS

7.8.1 Crassulacean acid metabolism

A diurnal rhythm of malic acid content and pH in the vacuole is characteristic of CAM plants. They take up CO_2 through stomata and accumulate malic acid in vacuoles during the night, to be dicarboxylated and utilised photosynthetically the following day. This involves transport of a large quantity of malic acid across the tonoplast. The existence of a malic acid carrier has been suggested from a study of the tonoplast of *Bryophyllum daigremontianum* by Buser-Suter et al. (1982) and it is now known that H^+-translocating ATPase is responsible for the process (Aoki and Nishida 1984; Joachem et al. 1984; Martinoia et al. 1985).

In their studies on the CAM plant *Graptopetalum paraguayense*, Iwasaki et al. (1988) recorded a clear diurnal rhythm between pH and total malic acid content but the amount of negative charges due to unprotonated carboxyl groups of malic acid remained approximately constant. Negative charges are balanced by the positive charges of cations which are also constant throughout the diurnal CAM rhythm. The results provide evidence for the electroneutrality of the translocation of malic acid and protons across the tonoplast membrane.

In essence, malate transport is driven by the maintenance of an inside-positive electrochemical membrane potential gradient across the tonoplast energised by tonoplast V_oV_1-H^+-ATPase and H^+-inorganic orthophosphatase activity (Lüttge et al. 1995). Mal^{2-} is transported across the tonoplast via an ion channel or a malate transporter into the vacuole where it becomes protonated (Iwasaki et al. 1992). Recently, more information has been obtained on the transporter of malate and citrate. It has been shown that both acids are transported by a novel citrate-binding protein (Rentsch et al. 1995). It is interesting to note that the induction of CAM from C_3 photosynthesis is accompanied by *de novo* synthesis of tonoplast proteins, and increase in relative amounts of some distinct polypeptides including subunits of V-ATPase. NaCl-treatment induced a different polypeptide (Bremberger and Lüttge 1992).

As far as efflux of malate is concerned, Lüttge et al. (1975) showed that malate is exported from vacuoles by passive diffusion. This may also take place by a second channel (Iwasaki et al. 1992). Lüttge and Smith (1984) found that malic acid efflux from the vacuole increases exponentially with the malic acid content of the tissue. Although temperature affects CAM by more than one single general mechanism, its effect on tonoplast was first emphasised by Wilkins (1983). He showed from temperature-related phase shifts of the endogenous CO_2-exchange rhythms in CAM plants that alterations in temperature can influence the permeability of the tonoplast, thus controlling malic acid transport across this membrane. Friemert et al. (1988) concluded that in acidified tissue the increase of temperature at the begining of the light period increases the efflux of malate from the vacuole by changing the properties of the tonoplast. On the other hand, lowering the temperature at the end of the light period would facilitate loading the vacuole with malic acid. Recent studies indicate that the tonoplast is sensitive to various stress conditions. In *Mesembryanthemum crystallinum*, when CAM is induced by drought or salinity stress, it is accompanied by an increase in tonoplast ATPase activity by *de novo* enzyme synthesis as well as structural changes in the enzyme (Bremberger et al. 1988).

With the aid of the patch-clamp technique, malate movement through ion channels has been investigated by Hedrich et al. (1986). In the tonoplast of barley leaf, in addition to the proton pump, a non-selective ion channel has been detected. The large ion channel (60–80pS) shows a strong voltage dependence. Both K^+ and malate ions can be driven through this channel by their concentration gradient and the electrical gradient established by the proton pump. If electrogenic pumping depolarises (inside more positive) the tonoplast beyond the malate equilibrium potential, malate could accumulate in the vacuole through non-selective channels. Similarly, a reduction of pump activity could lead to a loss of vacuolar malate.

7.8.2   Homeostasis of amino acid levels

As stated in Chapter V, plant cells can store amino acids in the vacuole. It has also been noted that most of the free amino acids in the vacuoles of yeast and other fungi are metabolically inactive and not readily available for protein synthesis. Studies on purified vacuoles of yeast indicated that they contain a large part of the intracellular amino acid pool and most of the intracellular arginine (Wiemken and Dürr 1974). The vacuolar localisation of most of the free arginine was corroborated by differential extraction of the soluble pools from the cytosol and vacuoles (Hüber-Wälchle and Wiemken 1979).

The retention, loss and utilisation of amino acids reveal the role played by vacuole in cellular homeostasis. In yeast vacuoles, arginine and phosphate residues in polyphosphate which cannot pass membranes are present in approximately equimolar concentrations (Urech et al. 1978). When the cells are starved of phosphate, vacuolar polyphosphate is utilised without affecting the vacuolar arginine content (Dürr et al. 1979). Similar results were obtained in *Neurospora crassa* which also contains equimolar concentrations of arginine and polyphosphate-phosphate. When starved of phosphate, although most of the polyphosphate was mobilised, arginine still remained sequestered (Cramer et al. 1980). The storage of amino acids, possibly to guard against nitrogen deficiency by heterotrophic cells, is not so obvious in autotrophic green cells. The distribution of amino acids in the cytoplasm and vacuole of *Chara* internodal cells provides an interesting feature of compartmentation of this essential metabolite. With a volume ratio of vacuole to cytoplasm of 95:5, 90% of free amino acid is localised in the cytoplasm. When the characean cells are maintained in continuous light or continuous darkness, the amino acid content of the vacuole increases considerably, and cytoplasmic content remains relatively constant. This indicates that the vacuole acts as a reservoir in the homeostasis of cytoplasmic amino acid levels. When the vacuole was loaded with large amounts of various amino acids by perfusion technique they gradually disappeared at different rates. After about one day, most of the alanine introduced had disappeared from the vacuole and aspartic acid, glutamine, serine and glycine was also metabolised by the cytoplasm in various degrees (Sakano and Tazawa 1985).

In higher plant cells the vacuolar pool of amino acids appears to be maintained more critically. Many acids, especially the neutral ones, are able to permeate lipid bilayers by diffusion (Wilson and Wheeler 1973). The tonoplast is reported to contain a number of amino acid-permeases, for example, a malic acid-permease in the tonoplast of a CAM plant *Bryophyllum*. Yeast tonoplast has also been attributed with the presence of methionine-specific permeases (Schwencke and Robichon-Szulmajster 1976). A great deal of work on the presence of arginine-specific permease has been conducted by Boller (1977). Hanower *et al.* (1977) note that saturation kinetics, competitive inhibition and other features suggest transport of arginine and lysine in the lütoid vacuoles of *Hevea* is mediated by a permease. However, Boller and Wiemken (1986) note that arginine uptake in their experiments is due to exchange transport and not permease activity. Moreover, uptake of ATP-dependent arginine and other amino acids has been demonstrated in yeast (Sato *et al.* 1984). Pistocchi *et al.* (1988) have shown that exogenously applied polyamines can be recovered from carrot vacuoles. On the basis of saturation kinetics and the narrow pH optimum observed for spermidine uptake, the authors believe that polyamine uptake takes place by a carrier-mediated process. Of course, the possibility of faciliated diffusion or active transport still exists.

A number of specific transport systems for uptake of various amino acids have also been demonstrated. The existence of a transport system for uptake of phenylalanine (Homeyer and Schultz 1988), alanine, glutamine, leucine and methionine (Dietz *et al.* 1990), arginine and aspartic acid (Martinoia *et al.* 1991), and glycine (Goerlach and Williams-Hoff 1992) has been strongly indicated in barley mesophyll vacuoles. Repeated observations by a number of workers show convincingly that the vacuolar concentration of amino acids is extremely low compared to the cytosolic concentration. For example, Winter *et al.* (1994) recorded that in mesophyll cells of spinach leaf, cytosolic/vacuolar ratios of 41, 38 and 26 respectively, were found for concentrations of glutamine, glutamate and aspartate, which are the most abundant amino acids in leaves. Similar results were also obtained by Winter *et al.* (1992) in barley leaf mesophyll cells, indicating that amino acids are very efficiently extruded from the vacuoles. Even without a high rate of passive permeation one would not expect the concentrations of amino acids to be so highly imbalanced. This imbalance is possibly due to active extrusion of amino acids from vacuoles, with help from several amino acid translocators (Winter *et al.* 1993). Roos *et al.* (1997) has studied the regulation of amino acid transport between the vacuole and the cytoplasm in the hyphal cells of *Penicillium cyclopium in situ*. The accumulation of amino acids in the vacuole appeared to be the result of a dynamic equilibrium of active ATP-dependent uptake and energy-independent efflux. The latter was strongly accelerated after the vacuolar amino acid content had surpassed a threshold level. The efflux of the vacuolar amino acids was specifically controlled by cytosolic adenylates, which consist mainly of ATP.

7.8.3 Sucrose accumulation and transport

The presence of sucrose in vacuoles is well known. That it is actually accumulated in the vacuole is evident by the fact that in sugar beet tap root tissue, cytosolic sugar concentration is only 76mM compared to a vacuolar concentration of 514mM (Saftner *et al.* 1983). However, in photosynthesising protoplasts, the pattern of compartmentation and transport may be different. In barley protoplasts $H^{14}CO_3$ incorporation studies indicate that sucrose, along with other photosynthates, rapidly enters the vacuole (Kaiser *et al.* 1982). But in *Melilotus*, vacuoles appeared to keep out the newly made sucrose with only 10mM sucrose in the vacuole, and about 100 mM in the cystosol (Boller and Alibert 1983).

The mechanism of sucrose transport in vacuoles isolated from leaf tissue has been studied for barley plants by Kaiser and Heber (1984). They indicate that this transport is carrier-mediated but does not require metabolic energy. De Lean *et al.* (1988) have been able to reproduce most of their results and agree with their proposal of faciliated diffusion of sucrose across the tonoplast of the vacuole.

During the normal photosynthetic period, sucrose entry into the vacuole may be passive or active, with the vacuole acting as a storage pool. But its retrieval from the vacuole in darkness, or when the supply of triose phosphate from the chloroplast is limiting and sucrose synthesis in the cytosol is less than required for translocation, may involve active transport (Wilson and Lucas 1987). Bacterial cells use transport systems where the nutrient is chemically **modified during transport**. The modified substance accumulates in the cytoplasm and cannot pass across the plasma membrane into the medium in a process known as **group translocation**. It involves a cluster of enzymes which span the membrane and operate in sequence to use the product on the other side. The presence of a hexose group translocator at the tonoplast of grape pericarp cells was first proposed by Brown and Coombe (1984). Thom and Maretzki (1985) proposed that in sugarcane cells a series of tonoplast-bound enzymes, or enzymes closely associated with the tonoplast, provide a vectorial group translocation system for compartmentalisation of sucrose in excess of metabolic requirements. They showed (Thom et al. 1986; Delrot et al. 1986) that tonoplast vesicles convert externally supplied uridine diphosphate glucose (UDPGlc) to internally accumulated sucrose. The products of UDPGlc uptake in the vesicles were sucrose and sucrose phosphate which, on hydrolysis with alkaline phosphatase and invertase, showed that both hexose moieties are derived from UDPGlc. In this system no sucrose transport via an ATP-dependent system could be detected. However, the validity of the identification of the translocated product inside the vacuoles has been questioned. Preisser and Komor (1988) showed that when the vacuole was incubated with labelled UDP–Glc, what was received from inside the vacuoles was not sucrose but another disaccharide, laminaribiose. This was corroborated by Maretzki and Thom (1988).

In photosynthetic cells sucrose is synthesised from UDPGlc and fructose-6-P in a sequence of two reactions catalysed by sucrose–phosphate–synthase (SPS) and sucrose–phosphate–phosphase (SPP). SPS activity is localised in the cytosol while SPP activity is detected in the vacuoles of storage tissue of red beet, sugar beet and sugar cane stems (Hawker et al. 1987). This suggests that sucrose-P may be transported across the tonoplast to be dephosphorylated, as proposed by Brown and Coombe (1984). On the other hand, Echeverria and Salvucci (1991) reported that sucrose-P was not transported across the tonoplast of red beet storage cells. Hence SPS is not involved in sucrose synthesis and accumulation in storage cells. Voss and Weidner (1988) indicated that induced group translocation of sucrose takes place in isolated tonoplast vesicles of red beet. They identified all reaction intermediates and the end-product sucrose, by two-dimensional high-performance thin layer chromatography and autoradiography for vectorial synthesis of sucrose. They also agreed with Willenbrink's (1987) suggestion that a multienzyme complex consisting of five individual enzymes governs both the formation and the translocation of sucrose into vacuoles. They contend that in the tonoplast of red beet, direct sucrose uptake energised by tonoplast ATPase and coupled to an ATP-dependent H^+-gradient via a sucrose/proton antiport, as first proposed by Thom and Komor (1984a), operates parallel to the UDP-dependent group translocator. However, whether both mechanisms act simultaneously or competitively is not clear. Preisser and Komor (1991) did not obtain evidence for a Suc/H^+ antiport system in the isolated vacuoles of sugar cane suspension cells. Strict dependence on medium pH and inhibition of sucrose transport by p-chloromercuriphenylsulfonic acid (PCMBS) indicated that sucrose uptake into sugar cane vacuoles is a passive, carrier-mediated process. Moreover, the equal distribution of sucrose between cytosol and vacuole in all phases of growth cycle in the suspension culture cells of sugar cane supports the passive uptake model (Preisser et al. 1991). Getz (1991) showed that in intact red beet vacuoles sucrose is transported *in vitro* against a 200-fold concentration gradient in the presence of MgATP. Comparing sucrose uptake rates with corresponding H^+-export rates, a stoichiometry of approximately one proton per transported sucrose was estimated by Getz and Klein (1995). Thus the available data support a coupled Suc/H^+ antiport model as the mechanism of sucrose transport across the vacuole membrane. The stimulating effects of K^+ and Na^+ salts of Cl^- on sucrose (Getz et al. 1993) indicate that symport of negatively charged ions such as OH^- and Cl^- cannot be excluded. Along with sucrose, Keller (1992) and Greutert and Keller (1993) studied tetrasaccharide stachyose uptake in both intact vacuoles and isolated tonoplast vesicles of Japanese artichoke (*Stachys sieboldii*). They reported that stachyose and sucrose uptake was inhibited by fructose and raffinose, and reciprocally, by sucrose and stachyose, but not by glucose or galactose. The main structural feature common to all sugars recognised by the uptake systems seems to be a terminal fructosyl residue. It is interesting to note that a tetrasaccharide may use a similar, if not the same, porter for uptake as a disaccharide. The uptake of stachyose-like sucrose was stimulated by Mg-ATP and inorganic pyrophosphate, indicating a Suc/H$^+$ antiport system. Getz et al. (1993) provided immunological evidence for the existence of a carrier protein for sucrose transport in the tonoplast vesicles from red beet storage tonoplast.

The apparent confusion over the mechanism of sucrose uptake can be resolved only by accepting the fact that tonoplast of different cells of diverse plants and metabolic pattern may operate different methods of uptake in vacuoles. Similarly, the accumulation and uptake of hexose appears to take place by diverse mechanisms. Active glucose transport in tonoplast vesicles was first demonstrated by Thom and Komor (1984b). This was supported by demonstration of a direct coupling of 3-O-methly-D-glucose (OMG) transport to H$^+$-ATPase dependent proton transport (Rausch et al. 1987). Martinoia et al. (1987), on the other hand, could not find any stimulating effect of ATP on glucose uptake in vacuoles of barley protoplasts. Using heterotrophic suspension cells of sugar cane, Preisser et al. (1991) found an accumulation of hexose, but not of sucrose, in these vacuoles. On the basis of non-aqueous fractionation of tobacco leaves, Heineke et al. (1994) demonstrated that the vacuolar membrane is able to maintain a high concentration gradient of hexose in leaf cells. They concluded that leaf vacuoles contain transporters for the active uptake of glucose and fructose. Recently, Shiratake et al. (1997a, 1998b) studied the uptake of sugars by tonoplast vesicles in pear fruit. The uptake was highest for glucose and, in decreasing order, for fructose, sucrose and sorbitol. It was not stimulated by the addition of ATP and was significantly inhibited by PCMBS. They concluded that the PCMBS- sensitive uptake was mediated by the transporter of facilitated diffusion. Since OMG inhibited both glucose and fructose uptake, the same transporter may mediate both glucose and fructose at low concentrations, but this hexose transport system differs from the sucrose and sorbitol transport systems. In many storage cells the high content of hexoses results from breakdown of disaccharides, as well as from direct uptake of monosaccharides. In the root cells of carrot tap root storage cells (Sturm et al. 1995) and grape berries (Davies and Robinson 1996) it has been amply demonstrated that sucrose is downloaded from phloem, transported to the vacuole and then cleaved by vacuolar acid invertase (β-fructofuranosidase) to fructose and glucose. The rate of UDPGlc uptake by red beet vacuoles from the vascular bundle regions is greater than those isolated from storage tissue away from the conducting tissue (Getz 1987). These observations suggest UDPGlc may be the main sugar taken up by the vacuoles in cells adjacent to vascular tissue. As these cells may use ATP for glucose uptake from the *apoplast*, glucose may then be taken up by vacuoles by UDPGlc -dependent group translocators at the tonoplast. On the other hand, storage cells away from the conducting tissue mainly import sucrose via the **symplastic** route and this imported sucrose will be taken up by vacuoles by means of the Suc/H$^+$ antiport, as more ATP is available in the cytoplasm (Willenbrink 1987).

Attention is being given to gene cloning and immunochemical localisation of the sugar transport proteins from various organisms. Several plant genes have been cloned that encode members of the sugar transport subgroup of the **major facilitator superfamily** of transporters (Marger and Saier 1993). Chiou and Bush (1996) have reported the cloning, expression and localisation of sucrose transporter from the tonoplast of sugar beet. This protein, which mediates sugar partitioning between the vacuole and cytoplasmic compartments, has an estimated mass of 54kD — very close to the mass of the sucrose carrier protein of (55–66kD) identified by Getz et al. (1993) in red beet tonoplast.

Although a great deal of information is available on the molecular genetics of a number of hexose carriers of the plasma membrane, e.g. glucose carriers in *Chlorella kessleri* (Sauer and Tanner 1989), a glucose transporter from *Arabidopsis thaliana* (Sauer et al. 1990) and sink-specific H$^+$/monosaccharide co-transporter from *Nicotiana tabaccum* (Sauer and Stadler 1993), a specific hexose carrier of the tonoplast is yet to be isolated and cloned. Like the sugar transport proteins, the monosaccharide transporters belong to the major facilitator superfamily of genes.

7.8.4 Biosynthesis inside vacuoles

As well as the transport of various metabolites to the vacuole, the biosynthesis of a number of substances inside the vacuole is gradually being understood. In some vacuoles the tonoplast may undertake the conversion process, while in others the enzymes of the sap will carry out the metabolism. In diverse biosynthetic systems in cells, a vacuole may take part as a compartment in which:
  i)   the final metabolites are admitted and stored,
  ii) the intermediate metabolites are admitted and after a few steps of metabolism the final metabolite is stored, or
  iii) the intermediates are admitted and after one or more steps are transported out to the cytosol or the apoplastic space.

7.8.4.1 Processing of sugars

The simplest of the various processing systems that occur in the vacuole is the formation of hexoses from sucrose. The accumulation of sugar in the form of glucose and fructose in roughly equal amounts is one of the main features of the ripening process of grape berries (Kliewer 1965). The enzyme invertase (ß-fructosidase) inside vacuoles of the parenchyma cells of the berries catalyses the hydrolysis (Leigh *et al.* 1979). There is some evidence that this also happens in the developing tomato fruit (Daman *et al.* 1988).

7.8.4.2 Biosynthesis of glycosides

Fructans

The storage and metabolism of fructans in the vacuoles of Jerusalem artichoke (*Helianthus tuberosus* L.) demonstrate the use of vacuoles for compartmentation of a metabolic pathway. In the tubers a huge amount of reserve food is stored in the form of fructan (polyfructosylsucrose), a water-soluble non-structural carbohydrate. The specific type of fructan in Jerusalem artichoke is inulin which has a low degree of polymerisation by (2 - 1)-β-D-fructofuranosidic linkage. Edelman and Jefford (1968) found four enzymes involved in inulin metabolism: SST (sucrose-sucrose-1 fructosyl transferase) forms the trisaccharide isoketose from sucrose; FFT (fructan-fructan-1-fructosyl transferase) transfers single terminal fructosyl residues from inulin to inulin or sucrose, and two FEHS (fructan exohydrolase) hydrolyse single terminal fructosyl residues off the chain ends of inulin. Using specific vacuole markers, Frehner *et al.* (1984) demonstrated that all the inulin enzymes and low polymers of inulin saccharides were vacuolar. They suggested that the key steps of the metabolism of inulin take place in the large central vacuole. Similar sub-cellular compartmentation of enzymes of inulin metabolism has also been demonstrated in barley leaves by Wagner *et al.* (1983). Glucosylation similar to that in artichokes may occur in the vacuole of leaf cells of *Hevea brasiliensis*. Gruhnert *et al.* (1994) have shown that the cyanogenic glucoside linamarin is localised exclusively in the central vacuole. The last step in the biosynthesis of cyanogenic glucosides is the glucosylation of the unstable hydroxynitrile, catalysed by a UDPGlc-dependent glucosyl-transferase. These researchers argued that glucosylation takes place in the vacuole.

7.8.4.3 Anthocyanins

The role of tonoplast and vacuole sap in transport, modification and storage of anthocyanin is becoming clear. Fritsch and Griesbach (1975) suggested that some enzymes involved in anthocyanin biosynthesis may be tonoplast-bound. Certain transferase activities have also been found to be associated with the tonoplast (Löffelhardt and Kopp 1981). A high amount of chalcone synthase in the vacuole has been demonstrated (Hopp *et al.* 1985). Hopp and Seitz (1987) demonstrated that anthocyanin is transported across the tonoplast only on acylation with sinapic acid and that deacylated glycosides were not taken up by isolated vacuoles. Moreover, acylated anthocyanin is transported by a selective carrier and might be trapped by a pH-dependent conformational change (ion trap) of the molecule inside the acid vacuolar sap. Matern et al. (1986) investigated the transport of apigenin 7-O-(6-O-malonylglucoside) inside the vacuole. They concluded that uptake by the isolated vacuoles depended on the malonylglucose moiety and that the malonylated flavonoid is trapped in the vacuoles by a switch in their conformation caused by the acidic vacuolar conditions. Thus acylation of glucosides may be a common phenomenon for tonoplast transport. However, anthocyanin may also be taken up by a method that is more like a xenobiotic (see later in this chapter). Marrs et al. (1995) and Marrs (1996) recorded that in maize, due to the expression of the bz2 gene, anthocyanin is conjugated with glutathione in a reaction catalysed by glutathione S-transferase and that this conjugate is then transported to the vacuoles by a glutathione pump in the tonoplast. In barley, the C-glycosyl flavone isovitexine is transported into vacuoles by a glutathione-independent mechanism requiring a vacuolar H^+-ATPase (Klein et al. 1996).

In many cell types dense anthocyanin-containing globular structures known as cyanoplasts are found in vacuoles (Politis 1959). Such cyanoplasts are also produced in vacuoles of suspension culture cells of sweet potato when they are subjected to continuous illumination (Nozue and Yasuda 1985). Recently, Nozue et al. (1995, 1997) have shown that a 24kD protein (VP24) accumulates in the anthocyanin-containing vacuoles and VP24 is involved in trapping large amounts of anthocyanin to form the cyanoplasts.

Another category of compartmentation is typically presented by flavonoid metabolism in mesophyll cells of rye leaves. Luteolin is converted to luteolin 7-0-monoglucuronide in the cytosol of the mesophyll cells, converted to the corresponding diglucuronide and transported readily to the vacuole. Anhalt and Weissenböck (1992) showed that luteolin-7-0-diglucuronide is converted to its corresponding triglucuronide (luteolin-7-0-diglucuronyl-4'-O-glucuronide) by the action of the corresponding glucuronosyl transferase inside the vacuole. Finally the triglucuronide is transported to the cytosol and the apoplastic space. According to Grotewald et al. (1998), in suspension cells of maize, under the influence of specific transcriptional activators, anthocyanin is accumulated in the lumen of SER. These become dilated and filled with granular material and eventually become anthocyanoplasts. In subsequent stages of development these anthocyanoplasts coalesce into a single vacuole. Different patterns of uptake (Pecket and Small 1980; Nozzolillo and Ishikura 1988) of various anthocyanins and related compounds by the vacuole appear to operate in different cell systems.

7.8.4.4 Ethylene metabolism

Many secondary metabolites, particularly phenolic compounds, frequently accumulate in plant cells as glycosides. Glycoside formation increases solubility and mobility and has been suggested to facilitate transport and storage processes. Sweet clover (*Melilotus alba*) leaves contain large amounts (up to 8% dry wt) of O-coumaric acid glucoside, which is exclusively located in vacuoles. The glucosyltransferase glucosylating the phenolic O-coumaric acid is extravacuolar (Alibert et al. 1982). The reported presence of glucosyltransferases in the tonoplast (Fritsch and Griesbach 1975; Löffelhardt and Kopp 1981) has led to the suggestion that the same may be the case for O-coumaric acid. However, the presence of these enzymes in the tonoplast has been questioned (Hrazdina and Wagner 1985).

Guy and Kende (1984) have demonstrated that ethylene is converted from 1-aminocyclopropane-1-carboxylic acid (ACC) in the vacuoles of *Pisum sativum*. Isolated vacuoles contained most of the precursor and converted it into ethylene. Sharma and Strack (1985) and Strack and Sharma (1985) studied the possible involvment of vacuole in the metabolism of hydroxycinnamic acid conjugates in the cotyledons of radish. They showed that the vacuoles contain 1-sinapolyglucose and sinapolymalate as well as the enzyme catalysing their transconjugation. The vacuolar pH was found to be favourable for transconjugation.

7.8.4.5 Processing of storage proteins

It has been gradually accepted that the proteins of dicotyledonous plants are stored in protein bodies which are vacuolar in nature (Akazawa and Hara-Nishimura 1985). The processing of storage proteins in the developing cotyledons of pumpkin is an excellent example of metabolic and synthetic activity of vacuoles. In these cells the trimeric proglobulin molecules are synthesised in RER and transported by dense vesicles to the vacuole. An endoproteolytic processing enzyme, cysteine protease, present in the soluble vacuolar matrix, processes the trimeric proglobulin to the mature hexameric globulin. The same enzyme is involved in the processing of protrypsin inhibitor (Hara-Nishimura et al. 1985).

The vacuolar processing activities that convert proglycinin to glycinin in maturing soybean seeds (Scott et al. 1992) and proricin to ricin in maturing castor bean seeds (Harley and Lord 1985) have also been studied. The 11S globulin-vacuolar processing enzyme (VPE), which is a 37 kD cysteine protease, is responsible for converting various proprotein precursors with a broad range of molecular structure into their respective mature forms (Hara-Nishimura et al. 1993). The unique enzyme VPE cleaves peptide bonds on the C-terminal side of exposed asparagine residues present on molecular surfaces of various proproteins. The detection of low levels of VPE in different non-storage tissues of castor bean, including roots and mature laeves (Hiraiwa et al. 1993) and the expression of VPE genes in vegetative organs of *Arabidopsis* (Kinoshita et al. 1995) suggest that VPE is widely distributed and operative in many plant organs and involved in many physiological functions of the vacuole. Lately, Hiraiwa et al. (1997) showed that the aspartic endopeptidase is localised in the matrix of the protein-storage vacuoles of castor bean, but the enzyme is unable to convert any of three endosperm proproteins, pro 2S albumin, proglobulin and proricin into their mature forms, while the purified VPE could convert all three proproteins. Their studies indicated that aspartic endopeptidase may play a role in trimming the C- terminal propeptides from the sub-units produced by the action of VPE.

Several seed proteins, such as lectins which are composed of two distinct polypeptide chains linked together by disulfide bonds are also known to be synthesised as a single polypeptide precursor and the protein bodies, i.e. the vacuoles, are the site of endoproteolytic cleavage of the precursor molecules. Similar processing of Concanavalin A has been indicated by Faye and Chrispeels (1987) in the vacuoles of jack bean (*Canavalia ensiformis*) storage parenchyma cells. Con A is first synthesised as pro-Con A, a glyco-protein precursor processed in the vacuole where mature protein is generated through the combined action of endopeptidase and carboxypeptidase activity. Cleavage on the carboxyterminal side of an asparagine residue is a widespread feature of proteolytic processing. The processing of Con A involves at least two endoproteolytic events to excise the glycopeptide from the centre of pro Con A, followed by a polypeptide ligation resulting in a transposition of the two large polypeptides generated by the excision of the glycopeptide. Another pro-protein in jack bean undergoes change to produce two polypeptides of α-mannosidase and this processing step occurs in the protein storage vacuoles (Faye et al. 1988). In their study of the maturation and intracelluar transport of one of the pathogenesis related proteins β-1,3-glycanase in cultured tobacco cells, Sticher et al. (1992) showed that the enzyme with an N-glycosylated C-terminal extension is processed to its mature form inside the vacuole by the removal of a C-terminal extension.

Runeberg-Roos et al. (1994) detected aspartic proteinase from barley grains as well as from the vacuoles of barley leaf and root cells. They demonstrated that the proteinase is co-localised with barley lectin (Fig. 7.9) and cleaves off 13 amino acid residues at the c-terminus. Lately, Romero et al. (1997) have shown that in barley leaves, thionins are synthesised as precursors with a signal peptide and a long C-terminal acidic peptide that is processed inside the vacuole by a vacuolar 70 kD proteinase. A number of similar vacuolar processing systems are known in yeast. For example, Wolff et al. (1996) reported that in yeast the vacuolar aspartyl protease

194 Plant cell vacuoles: an introduction

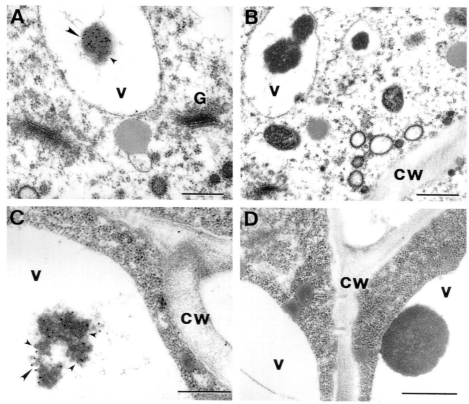

Figure 7.9 Immunohistochemical co-localisation of aspartic proteinase (HvAP) and barley lectin (BL) in root tip cells (A and B) and embryonic roots (C and D) of barley. Sections in A and C were probed with antibody to mature BL, followed by secondary antibody linked to 10 nm gold particles, and then were probed with antibody to HvAP, followed by secondary antibody linked to 15 nm gold particles. Sections B and D were probed with control antibodies. The presence of both small (10 nm) and large (15-nm) gold particles (arrows) on the electron dense structures inside vacuole (A and C) indicates co-localisation of HvAP and BL. Bar = 0.5 μm (from Runeberg-Ross et al. 1994. Courtesy: Natasha V. Raikhel.) With permission of the American Society of Plant Physiologists.

proteinase A is encoded as a preproenzyme. The active mature enzyme is generated by a specific proteolysis of the preproenzyme by the vacuolar endoprotease proteinase B.

7.8.4.6 Biosynthesis of alkaloids

The vacuole has special significance as a site for biosysthesis and storage of a number of highly priced drugs for treatment of hypertension and circulatory diseases and others with potent antitumour properties. Various types of alkaloids including indoles, isoquinolines and nicotines are synthesised and/or stored in the vacuoles of many species, especially Apocynaceae, Solanaceae and Papaveraceae. Perhaps the most investigated of all the medicinal plants, *Catharanthus roseus* (L) G. Don. (syn. *Vinca rosea*) of Apocynaceae, commonly known as Madagascar periwinkle or *Noyantara*, can yield more than 100 alkaloids from different parts of the plant, many of them possessing remarkable pharmacological activity. The most important of these are the antileukemic alkaloids vinblastine and vincristine and the antihypertensive alkaloids, ajmalicine and serpentine, which have a sedative effect (vide review by Moreno *et al.* 1995). In order to understand the formation and storage of alkaloids, their localisation has been investigated in cell suspensions

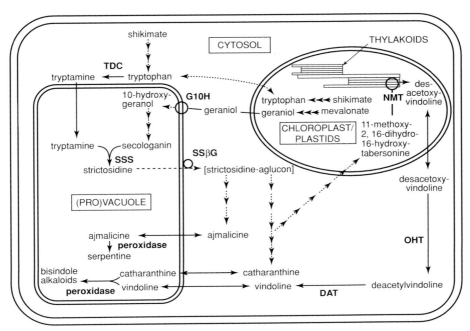

Figure 7.10 Compartmentalisation of terpenoid indole alkaloid biosynthesis in *Catharanthus roseus*. The localisation of enzymatic steps indicated with dashed arrows is hypothetical. Circles indicate membrane-associated enzymes (from Meijer *et al.* 1993b). With permission of the Botanical Society of Japan.

of *C. roseus* by Neuman *et al.* (1983). They noticed that alkaloids are stored within the vacuoles of particular cells which possess a vacuolar pH of 3 in contrast to 'normal' cells which have a vacuolar pH of 5. The mechanisms involved in the accumulation of alkaloids are not clear. It has been suggested that they may be transported by energy requiring specific carriers and possibly by a proton-antiport carrier system (Deus-Neumann and Zenk 1986; Mende and Wink 1987; Yamamdo *et al.* 1989). It is also possible that neutral alkaloids diffuse across the membranes towards an equilibrium and since there is a pH gradient with the lowest pH inside the vacuole the alkaloids will be accumulated by an ion trap mechanism. The studies on uptake and accumulation of ajmalicine into isolated vacuoles of cultured cells of *C. roseus* support the ion-trap hypothesis (Blom *et al.* 1991). In addition to an ion-trap mechanism the alkaloid may accumulate by binding to complexes located in the vacuole (Renaudin 1989). Blom *et al.* (1991) have shown that the basic isoenzymes of *peroxiodases* within the vacuole oxidise ajmalicine to serpentine. By the transformation of ajmalicine into the charged serpentine a trap is created to retain the alkaloids more efficienctly in the vacuoles. The synthesis of a number of alkaloids in *C. roseus* starts with secologanin, a monoterpene formed within the vacuole through several enzymatic steps initiated after the hydroxylation of geraniol by the enzyme geraniol-10-hydroylase (G10H). G10H is a membrane-bound cytochrome P-450 monooxygenase, consisting of two parts, a cytochrome P-450 and a NADPH: cytochrome P-450 reductase. The fact that G10H has been obtained from provacuolar membranes suggests that geraniol available in the cytosol may pass through the tonoplast to be converted to 10-hydroxygeraniol (Meijer *et al.* 1993a). Tryptamine, which is transported from the cytosol, is coupled with secologanin by strictosidine synthase (SSS) to form a gluco-alkaloid 3-alpha (S) strictosidine, the universal precursor for all monoterpenoid indole alkaloids. The first step in alkaloid biosynthesis after the formation of strictosidine is the removal of the sugar moiety to form an unstable aglycan by strictosidine-β-glucosidase (SG). Whereas SSS has been located inside the vacuole, SG is bound to the outside of the tonoplast

(Stevens *et al*. 1993). Thus, strictosidine formed inside the vacuoles is transported to the cytoplasm where the glucose moiety will be removed leading to the biosynthesis of other alkaloids. Bisindole alkaloids are also synthesised inside the vacuole by the coupling of the two precursors catharanthine and vindoline, which are transported to the vacuole from the cytosol. Smith *et al.* (1988) indicated that a peroxidase mediates the dimerisation of cantharanthine and vindoline to anhydrovinblastine, which in turn may be converted to the natural dimeric alkaloids vinblastine, catharine and leurosine. A hypothetical compartmentation of alkaloid biosynthesis and the role of the vacuole is proposed by Meijer *et al.* (1993b) (Fig. 7.10).

7.9 Lytic functions

A major function of the vacuole is its lytic function which is discussed as:
i) Presence of hydrolytic enzymes and their action inside the vacuole,
ii) Occurrence of autophagy under normal and stress conditions,
iii) Occurrence of heterophagy, and
iv) Presence and accumulation of breakdown products in the vacuole.

7.9.1 Hydrolytic activity

In animal cells, the lytic enzymes are compartmentalised in discrete structures known as lysosomes and their various derivatives. Whether typical lysosomes are present in plant cells has long been debated. However, it is accepted that in many plant cells distinct bodies surrounded by a single membrane and containing one or more hydrolytic enzymes are present. The presence of lytic enzymes in the vacuoles has been demonstrated in a wide variety of species by techniques of enzyme cytochemistry at both light microscopic and electron microscopic levels. Various hydrolytic enzymes have been identified after biochemical extraction from isolated cells, as well as from isolated vacuoles. A large number of publications on vacuoles indicate the presence of one or more of the various hydrolases in numerous different species ranging from algae, fungi and ferns to gymnosperm, monocotyledons and dicotyledons. All the three main groups of hydrolases, viz. peptidases, esterases and glycosidases, have been demonstrated. Both types of peptidases, viz. endopeptidases and exopeptidases, are found in abundance. Three esterases, viz. phosphomonesterases, phosphodiesterases and acetylesterases, are also prevalent. The glycosidases are also widespread, especially in the storage vacuoles. In addition to the hydrolytic enzymes and non-hydrolytic cleaving enzymes, lyases have also been found in vacuolar sap. The presence of a wide variety of degradative enzymes has been discussed in detail and presented in Table 5.3.

Thus, the dangerous degrading enzymes are compartmentalised in the vacuole. Moreover, the lysosomal nature of the plant vacuole is established by demonstrating the latent activity of hydrolases, i.e. the substrate should not be attached by the hydrolases unless the membrane of the vacuole is disrupted.

7.9.2 Proteolytic activity

A significant aspect of vacuolar compartmentation has been highlighted by the discovery that endopeptidases are exclusively vacuolar and that aminopeptidases are widely distributed throughout the cytoplasm. Canut *et al.* (1987) studied the effectiveness of degradation of cytosolic and vacuolar enzymes in the mesophyll protoplast of *Melilotus alba*. When a typical cytosolic enzyme in the presence of purified amino- and endopeptidases was considered, only vacuolar endopeptidases were effective in its degradation. Vacuolar hydrolytic enzymes were very stable during digestion and very resistant to proteolytic attack. Thus, the vacuole is a very effective compartment for proteolytic activity acting on cytosolic enzymes. The breakdown of the cytosolic proteins or enzymes during their intracellular turnover must involve the entry of these proteins into the vacuole. Evidence of such transfer has been reported (Canut *et al.* 1985a, b). Similarly, exit of vacuolar proteases for cytosolic protein degradation has also been suggested. Special attributes, like a distinct half-life, presence or absence of co-factors, etc., may make the cytosolic enzyme distinctly susceptible, in contrast to hydrolytic enzymes which are markedly resistant to proteolysis.

7.9.2.1 Degradation of extraneous proteins

Another line of evidence on compartmentation of proteolytic activity in vacuoles comes from the vacuolar perfusion studies of Moriyasu and Tazawa (1986a) on characean plant cells. They **partially replaced** vacuolar sap with artificial cell sap containing bovine serum albumin (BSA) and incubated it in artificial pond water. After a while the cell sap was isolated by vacuolar perfusion and analysed for BSA content. No BSA was detected after 16 hours of incubation. A similar result was obtained with casein. If the vacuolar sap was **completely replaced** by perfusion no proteolytic activity was noted. They also reported that in *Chara* the haemoglobin-digesting activity was largely (85%) localised in the vacuole (Moriyasu and Tazawa 1986b). Whereas all of the carboxypeptidase activity could be detected in the vacuolar sap, aminopeptidase activity with alkaline pH optimum could be detected in both cytoplasm and vacuole.

The undamaged state of the vacuole may be an important aspect of protein degradation. Canut *et al.* (1985a) noted that protein degradation occurred in isolated vacuoles, but not in sonicated and broken vacuoles. Similarly, when Moriyasu and Tazawa (1988b) tested whether BSA can be degraded if it is incubated with the isolated vacuolar sap, they noted that BSA was not proteolysed in the incubation mixture, although it could be proteolysed *in situ* in the vacuole.

7.9.2.2 Proteolysis of mutant proteins

A number of critical experiments point to very specific proteolytic activity of the vacuole. In common bean (*Phaseolus vulgaris*) phaseolin is synthesised on the ER and accumulated in protein storage vacuoles. Because normal phaseolin is deficient in sulfur amino acids, a mutant HiMet was constructed to test the prospects of enhancing the methionine content of phaseolin by inserting a sequence that encodes a 15-amino acid helical peptide containing 6 methionines in the protein (Hoffman *et al.* 1988). Recently, Pueyo *et al.* (1995) have shown that HiMet proteins are selectively degraded by vacuolar proteases. A mutant protein noted by Chang and Fink (1995) suffered a similar fate. They studied a class of *pma1* mutants of *Saccharomyces cerevisiae* whose growth is arrested at *the restrictive temperature* because newly synthesised ATPase fails to be targeted to the cell surface. Although the mutant ATPase is not substantially defective for catalytic activity, it is defective for delivery to the plasma membrane and is instead, degraded in the vacuole. Noda *et al.* (1995) developed a novel system for monitoring autophagy by constructing cells in which modified vacuolar alkaline phosphatase is expressed as an inactive precursor form in the cytosol. Under starvation conditions, the precursor is gradually processed to the mature form which is finally located in the vacuole. Another example of vacuolar proteolytic activity is exemplified by the mating partners of yeast which secrete peptide pheromone **a**-factor and (**alpha**) factors. These pheromones bind receptors on the cell surface of the mating partner, stimulate transcription of genes involved in conjugation and coordinate the cells for cell and nuclear fusion. After binding to the receptor, alpha-factor is internalised, enters the vacuole via the endosome pathway and is degraded by a vacuolar hydrolase (Fig. 6.9) (Chvatchko *et al.* 1986; Shimmoeler and Riezman 1993).

7.9.3 Nucleolytic activity

In constrast to proteolytic activity, nucleolytic activity appears to play a minor role in the degradation of nucleic acids. However, Abel and Glund (1986) reported that in suspension cultured tomato cells, the bulk (60–80%) of intracellular RNA-degrading activity is localised in the vacuole. These researchers isolated the enzyme and characterised it as RNase I (Abel *et al.* 1989). The occurrence of RNA-oligonucleotides shorter than 80 nucleotides and their characterisation in cultured tomato vacuoles led Abel *et al.* (1990) to conclude that nucleotides may be products of vacuolar RNase I action and that plant vacuoles are involved in cellular nucleolytic processes. Although occasional reports of the occurrence of acid DNases are published about heterotrophic organisms and some higher plants, vacuoles are not considered to have a role in routine DNA-degrading activity.

7.9.4 Autophagy

7.9.4.1 Organelle breakdown inside vacuoles

The first detection of autophagy in plant cells at the ultrastructural level was made by Poux (1963) in the vacuoles of apical meristem and the young leaf of maize. The existence of cytoplasm together with organelles like mitochondria and Golgi bodies in the process of degeneration was observed in the vacuoles with intact tonoplast. Similar observations in plant cells were made by Sievers (1966) in the apical cells of rhizoids of the alga *Chara foetida*, and by Villiers (1967) in the embryos of *Fraxinus* in which Gomori-positive vacuoles contain cytoplasmic organelles. Gifford and Stewart (1968) reported that in the shoot apex cells of *Bryophyllum*, a CAM plant, proplastids with lipid inclusions are occasionally discharged into the vacuoles. In a study of differentiation of lignified collenchyma of *Eryngium*, Wardrop (1968) reported that vacuolar structures containing cytoplasmic matter are discharged into the central vacuole and at terminal stages of differentiation the remains of vacuolar bodies could be seen in the large central vacuole.

7.9.4.2 Membranous enclosure of cytoplasm

In the fungus *Phycomyces blakesleeanus*, Thornton (1968) studied the vacuolation process in differentiating sporangiophores. Sporangiophores contain vesicles with either a single membrane or a pair of concentric membranes. The occurrence of organelles or their debris in the vesicles suggests that the segments of ER systematically isolate and degrade parts of its own cytoplasm and discard the residues in the vacuole. Similarly, in the root meristems of the gourd *Cucurbita pepo* Coulomb and Buvat (1968) reported the presence of degenerating cytoplasmic particles in vacuole-like structures. Some of the electron micrographs by Buvat (1971) dealing with autophagy are very revealing. In barley roots, cupuliform vacuoles surrounding a protrusion of intact, degenerated or totally digested cytoplasm can be noticed. The best demonstration of the development of the vacuole as a result of autophagy was given by Marty (1978) in the meristematic cells of *Euphorbia*, where lysosomal tubes wrap themselves around portions of cytoplasm, merge laterally and digest the contents.

During the growing season, in the apical initials and the central mother cells of the apical meristem of Douglas fir, signs of autophagy have been recorded by Krasowski and Owens (1990). Vacuoles contained partially or entirely engulfed portions of cytoplasm and coarse fragments of cytoplasm-like or membranous material. The ER appeared to be associated with the formation of autophagic vacuoles. Similar autophagy was also noticed in rib meristem cells in mid winter.

7.9.4.3 Invagination of tonoplast

So, under normal conditions, the vacuolar membrane could be expected to show invagination and other features to indicate that cytoplasmic fragments and organelles have been engulfed by or discharged into the vacuole. Numerous electron micrographs obtained after ultrathin sectioning and freeze-etching by various workers substantiate the process of autophagy in plant cells. Invaginating tonoplast of vacuoles in root meristem cells of *Zea mays* has been shown by Matile and Moor (1968). Fineran (1970, 1971) has shown that local growth of the tonoplast permits the cytoplasmic protuberances to extend into the lumen of the vacuole. After invagination is complete, intravacuolar vesicles are pinched off from the tonoplast. In cases where autophagy is induced, similar invagination of the tonoplast has been recorded by Gupta and De (1988). The protruding mass of the cytosol and organelles is pinched off from the tonoplast, released inside the vacuole and finally degraded.

Thus a survey of literature shows that at the ultrastructural level, autophagy may take place by a number of processes. In the process where it leads to the formation of autophagic vacuoles, certain membranes or smooth tubules of ER gradually enclose a part of the cytoplasm. By gradual fusion of the fenestrated network, the tubules or membranes sequester a part of the cytoplasm. The hydrolases present in between the concentric membranes act on the enwrapped cytoplasm to cause lysis. As the inner membrane dissolves a vacuole results with a single outlining membrane (Marty et al. 1980). In another process, autophagy takes place by invagination of the tonoplast and a cytoplasmic protrusion enters the existing vacuole. Either by discharge or strangulation, the contents of the protrusion are released in the vacuolar cavity where they are degraded by various digestive enzymes. In this way a mass of the cytoplasm or selected organelles may be degraded. The recognition of these processes and their appearance at the ultrastructural level has led to many reports on autophagy in diverse plant tissues. The enclosure of cytoplasmic mass by ER, concentric rings of membranes and partially digested structures in small or young vacuoles has almost always been interpreted as autophagic phenomena.

Depending on the physiology and type of tissue, autophagy may be diverse in appearance at the ultrastructural level. The normal route of wheat storage proteins to vacuoles is a case in point. Wheat prolamine is not glycosylated and does not pass through the Golgi cisternae. Hence, from ER it is transported directly to the storage vacuoles via intermediate structures called protein bodies (PB). Levanony et al. (1992) have shown that PB enters the vacuoles by an autophagic-like process. The surfaces of PB inside the vacuole and the adjacent regions are highly enriched with loosely attached membranous material. It seems likely that during internalisation to the vacuoles, the PB became surrounded by the tonoplast membrane which was subsequently detached and degraded during further fusion inside the vacuole. The difference between vesicular transport from the Golgi and autophagy is that after autophagy the membrane surrounding the cargo protein ends up in the vacuole and does not fuse with the tonoplast.

When metabolic conditions are disturbed, cells may undergo certain autophagic processes. In a series of publications, Ohsumi and co-workers (Baba et al. 1997) recorded the autophagic process caused by starvation in *Saccharomyces cerevisiae*. Under nutrient-deficient conditions, the yeast cell sequesters its own cytoplasmic components into vacuoles in the form of 'autophagic bodies' (Takashige et al. 1992). With the aid of two cytosolic enzyme markers they observed that starvation induces formation of autophagosomes, which consist of a double membrane enclosing a portion of the cytosol non-selectively. The outer membrane of the autophagosome fuses with the vacuolar membrane and the resulting inner single membrane structure, the autophagic body, is delivered to the vacuole, where it is rapidly disintegrated (Fig. 7.11) (Baba et al. 1994; Noda et al. 1995). It is intriguing to note that this process of macroautophagy is topologically the same as the biosynthetic pathway of targeting aminopeptidase I from cytoplasm to vacuole as a spherical complex (Cvt) of double membrane-bound structure of ~150 nm diam. (Baba et al. 1997; vide review by Klionsky 1998). Such multilamellar enodosome-like structures are more abundant in vacuolar protein sorting mutant vps 28 (Rieder et al. 1996). It is remarkable that fine genetic analyses by Scott et al. (1996) have shown that cytoplasm-to-vacuole targeting and autophagy employ the same machinery to deliver proteins to the yeast vacuole. Complementation analysis of yeast mutants defective for cytoplasm-to-vacuole targeting (cvt) aminopeptidase I (API) and autophagy (apg) revealed seven overlapping complementation groups between these two sets of mutants. In addition, all 14 *apg* complementation groups are defective in the delivery of API to the vacuole. Similarly, the majority of non-overlapping *cot* complementation groups appear to be at least partially defective in autophagy (vide review by Bryant and Stevens 1998). Tuttle and Dunn (1995) have shown that in methanol utilising yeast *Pichia pastoris* the sequestering of peroxisomes to the vacuole takes place when major alterations of cellular metabolism are initiated upon a shift in carbon source to ethanol or glucose. In the ethanol-induced system the individual peroxisomes are sequestered into autophagosomes by wrapping membranes, which then fuse with the vacuole.

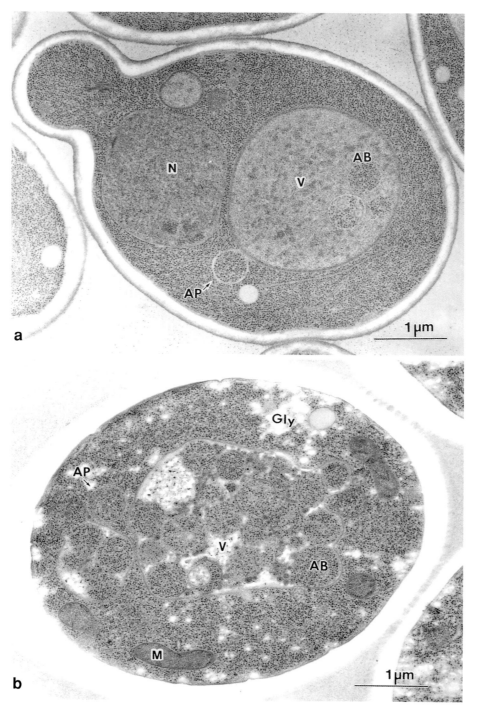

Figure 7.11 Ultrastructure of yeast cell (BJ926) under starvation conditions. **a** A spherical body outlined by a double membrane (arrow) appears in the cytosol after carbon-starvation for 3h. **b** Under nitrogen starvation condition for 2h, the vacuole contains many autophagic bodies. (Courtesy: Y. Ohsumi, from Baba et al., *J. Cell Biol.* **124**, 903, 1994.) With permission of Rockefeller University Press.

This process is known as macroautophagy, in contrast to microautophagy, which is induced by glucose where finger-like protrusions of the tonoplast engulf clusters of peroxisomes. When starvation experiments were conducted on plant cells in culture, formation of autolysosomes in the cytoplasm was not detected.

Moriyasu and Ohsumi (1996) showed that when the tobacco cells in suspension culture in MS medium containing sucrose are transferred to MS medium without sucrose, many small spherical bodies accumulated in the perinuclear region of the cytosol only in the presence of cysteine protease inhibitor. Cytochemical enzyme analysis using 1-naphthylphosphate and beta-glycerophosphate as substrates showed that these vesicles contained acid phosphatases. The researchers suggest that the degradation is accomplished through an autaphagic process via the formation of spherical vesicles known as autolysosomes.

7.9.5 Heterophagy

The ingestion of non-nutrient foreign substances in plant cells is broadly known as heterophagy. Vacuoles are not as directly involved in heterophagy compared with their role in autophagy. The reason is that the cell wall is a physical barrier against particulate foreign substances. From the determination of the pore size of cell walls of living plant cells by Carpita *et al.* (1979), it is clear that particles larger than 4 nm are not likely to cross the cell wall. The transport of cationic ferritin as shown by Tanchak *et al.* (1984) and lectin-gold conjugates as shown by Hillmer *et al.* (1986) may be the limit for particulate substances passing through the cell wall.

7.9.5.1 Types of heterophagy

After crossing the cell wall, foreign substances have to encounter the plasma membrane and may be internalised by i) Simple diffusion; ii) Membrane transport; iii) Fluid-phase endocytosis; or iv) Plasmalemmasome formation. A number of small molecules can cross the plasma membrane by simple diffusion. The plasma membrane transport activity for foreign molecules is best exemplified by the vital dyes. For example, Swain and De (1997) indicated that uptake of two vital dyes, methylene blue and toluidine blue by living protoplasts, takes place because the plasma membrane can reduce the impermeable cation of the basic dye to its leucobase form, which being non-polar can easily enter the living protoplasts. After entering the cell, the leuco dye reionises in the cell sap as well as in the vacuole and imparts its colour to the living colourless protoplast. The phenomenon of fluid-phase endocytosis, by which extracellular fluid is internalised by formation of plasma membrane vesicles, is best studied by transport of a membrane-impermeant probe Lucifer Yellow CH (LYCH). The highly fluorescent LYCH is found to enter the cytosol and also the vacuole by fluid-phase endocytosis (reviewed by Robinson and Hillmer 1990). It has been demonstrated by Cole *et al.* (1991) in maize and carrot cell suspension cultures and by Oparka *et al.* (1991) in onion epidermal tissue that LYCH enters the cytoplasm and is rapidly sequestered in the vacuole, the latter process being inhibited by probenicid, an organic anion inhibitor. However, Wright *et al.* (1992), working on the transport system in vacuolating oat aleurone protoplasts, negated the possibility of fluid-phase endocytosis of LYCH. They showed that the dye was completely excluded from non-vacuolate protoplast, and, as culturing induced vacuolation, the uptake of the dye increased exponentially. They suggest that the uptake of LYCH is due to the development of highly coordinated membrane transport systems on both plasmalemma and tonoplast.

7.9.5.2 Plasmalemmasomes

In the typical process of endocytosis in which particulate matter is taken up by animal cells, the plasma membrane invaginates inside the cytoplasm and pinches off to form an intracellular vesicle containing the ingested particle. Although endocytosis is not a regular phenomenon in plant cells, similar concave-shaped invaginations of the plasma membrane, consisting of single or multiple vesicles (plasmalemmasomes) have been noted in various cells and tissue types.

In plant cells of high solute fluxes, invaginations of the plasma membrane to form finger-like processes, multivesicular bodies or plasmalemmasomes are frequently observed. The idea of plasmalemmasomes being involved in internalisation of substances from the plasma membrane to the vacuole was proposed by Merchant and Robards (1968) and Robards and Kidwai (1969). That ferritin molecules are taken into protoplasts by coated pits and vesicles and are delivered to the vacuole was first reported by Fowke (1986). In a typical animal endocytosis-like process soybean suspension culture cell protoplasts have been shown to take up suface-bound cationic ferritin into multivesicular bodies (Record and Griffing 1988). Herman and Lamb (1992) advocated a plasmalemmasome-mediated endocytosis pathway for internalisation of extracellular glycoprotein. With the aid of immunogold-electron microscopy they have shown that arabinogalactan-rich glycoprotein is distributed on the external face of the plasma membrane and plasmalemmasomes which finally enter the vacuole for degradation.

7.9.5.3 Accumulation of breakdown products

The presence and accumulation of various substances in the vacuole has been documented since the early days of research (vide Chapter V). However, precise examples of accumulation breakdown products is limited. Gupta and De (1983a, b, 1985b, 1988) have shown that acridine orange induces autophagy in various types of cells. The damaged cellular organelles and cytoplasmic mass enter the vacuole by tonoplast invagination (Fig. 7.12). After intravacuolar digestion, the abiological bodies in the vacuole were identified as masses of acridine orange. Düggelin *et al.* (1988) reported that two fluorescent lipofuscin-like compounds formed due to breakdown of chlorophyll in the senescent primary leaves are located exclusively in the vacuole.

In senescent cotyledons of rape, *Brassica napus*, the vacuoles contain three non-fluorescent chlorophyll catabolites which account for practically all the chlorophyll broken down. As leaves yellow, the porphyrin moiety of chlorophyll is cleaved into colourless linear tetrapyrrolic catabolites which are eventually deposited in the central vacuoles of mesophyll cells. Recently, Hinder *et al.* (1996), in studies with vacuoles isolated from barley mesophyll cells, demonstrated that the tonoplast is equipped with a directly energised carrier that transports tetrapyrrolic catabolites of chlorophyll into the vacuole.

A number of workers have reported the presence of myelin structures in the vacuole. These structures are spherical bodies made of concentric membranes which are known to occur naturally in diverse plants and fungi (Thomas and Issac 1967; Bowes 1969). Gupta (1985a) noted their occurrence in the root tip cells of *Vicia* treated with acridine orange. Sangwan *et al.* (1989) recorded their normal occurrence in the vacuoles of microspores of *Datura*. They are, apparently, undigested membranes. The occurrence of breakdown products of cytoplasmic material may be taken as a proof for the autophagic and lytic functions of vacuoles.

7.9.5.4 Host–pathogen interactions

Internalisation of pathogens

Phagocytosis is an unusual phenomenon in plant cells. In cases of invading pathogens, vacuoles rarely exhibit heterophagy. In most cases of virus infection the virions are surrounded by a membrane, even if it is in the vacuole. Milicic (1963) detected virus particles in vacuoles soon after infection with tobacco-mosaic and turnip-mosaic viruses. Bundles of paracrystalline needles occurred in vacuoles of systemically infected plants and in those of localised lesions. In TMV infections, the cells with needles in the vacuoles also showed typical hexagonal crystals in the cytoplasm. These crystals were agitated by Brownian movement in the vacuoles. Electron microscope studies by Esau and Cronshaw (1967) have shown that the particles of TMV in vacuoles are delimited by tonoplast and placed between the plasma membrane and the cell wall. When the mycelium of a pathogenic fungus infects a plant host, the mycelium is not directly exposed to the vacuolar sap and is protected from vacuolar enzymes. In endobiotic, host-symbiotic cases the parasite-

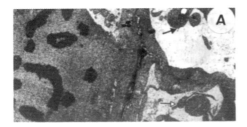
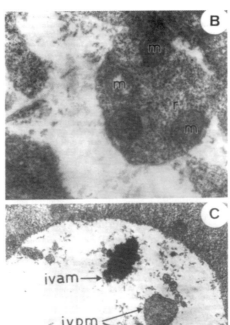

Figure 7.12 Electron micrographs of acridine orange (AO) — treated root tip cells of *Vicia faba*. Fixed in 1.5% formaldehyde-glutaraldehyde, post-fixation in osmium tetroxide. Post-stained in uranyl acetate and lead citrate. **a** Cells treated with 1 ppm AO for 4 days. Tonoplast invaginations are apparent (solid arrow). Vacuoles show intravacuolar bodies (open arrow). Presence of metaphase chromosomes indicates healthy and actively dividing cells × 7500.
b–c Ultrastructure of cells treated with 50 ppm AO for 6 h and 12 h respectively.
b Part of a vacuole containing protoplasmic masses ranging from 0.3 to 3μm in diameter. The invagination is seen engulfing cytoplasmic organelles along with cytosol. Four mitochondria (M) and cytosol containing ribosomes (r) form the intravacuolar mass × 40 000. **c** Vacuoles displaying two types of intravacuolar bodies (i) ivpm – intravacuolar protoplasmic mass with structural features similar to cytosol (ii) ivam – intravacuolar abiological matter, presumably the aggregates of AO × 16 000 (from Gupta and De 1988). With permission of Urban and Fischer Verlag.

symbiont is generally separated by a definite interface. In the case of haustorium of *Erysiphae graminis* var. *tritici*, the haustorial lobes are bounded by a dark line comprising the invaginated host plasmalemma, a thin layer of host cytoplasm and the host tonoplast. Similar instances are found in cases of mycorrhizal associations (Scannerini and Bonfante-Fasola 1983) and rust diseases (Harder and Chong 1984). Bacterial pathogens in higher plants are usually located in the intercellular spaces between the host cells or within non-living cells of vascular tissue. Observations by Drews *et al.* (1988) suggest that multivesicular bodies or plasmalemmasomes may proliferate as a consequence of infection of soybean leaves by *Pseudomonas*. In other words, true phagocytosis may be unknown in the cells of higher plants.

A phenomenon like heterophagy is seen when *Rhizobium* is released in the early stages of development of root nodules in leguminous plants. Prasad and De (1971) have shown in *Phaseolus mungo*, that once the tip of the infection thread is dissolved, rhizobia are released, either in vacuoles or in the cytoplasm close to the vacuole. Although bacteria are initially free-floating in the vacuole or cytoplasm, they are soon enclosed by membranes derived from the outer membrane of the nuclear envelope. On the other hand, Truchet and Coulomb (1973) and Bal and co-workers (Verma *et al.* 1978) have shown that in soybean root nodules the bacteroids are enveloped by a membrane derived from the plasma membrane. Although some of the invading rhizobia may be degraded by vacuolar enzymes, most of them survive till they are enclosed by the membrane envelope. Apart from other functions involving nitrogen fixation, the enveloping membrane protects the bacteria against hydrolases, although during degeneration of the nodules, a large number of bacteroids would also be digested. On the other hand, the vesicles which contain the bacteroids may be considered as a kind of heterophagosome, because the infection of the host cells involves a phagocytosis-like uptake of bacteria from the infection thread (Truchet and Coulomb 1973). An interesting aspect of lytic activity in vacuoles has been demonstrated by Bal (1985). Soon after release of bacteria, vacuoles of many infected cells show debris of infection threads, whereas vacuoles of uninfected cells are free of such debris (Fig. 7.13). Bal suggests that the infection thread degrades rapidly and thus the process cannot be seen in all samples.

Defense response

The defence response of plants against invading pathogens involves a wide variety of biochemical changes in synthetic and metabolic pathways, as reviewed by Bowles (1990). It may include callose formation, biosynthesis of lignin-like material, phytoalexin synthesis, release of H_2O_2, production of hydrolytic enzymes, synthesis of protease inhibitors and hypersensitive cell death. Most of these processes directly or indirectly involve the vacuole and provide another aspect of vacuole function.

Almost all pathogenic fungi, including necrotrophic fungi with varying degrees of parasitism and biotrophic species exhibiting obligate parasitism, secrete various cell wall-lytic enzymes including cellulases, hemicellulases, pectic enzymes and proteinases (Wood 1967). As a reaction to the pathogens, the host may produce/release enzymes which may again be counter-attacked by an inhibitor protein secreted by the pathogen (Albersheim *et al.* 1969; Albersheim and Valent 1974). This battle of enzymes is expected to be revealed in the cytoplasmic structure of the host and/or the pathogen. Pitt and Coombes (1968) demonstrated that in potato tuber cells, subcellular particles rich in hydrolases (lysosomes) undergo swelling and disruption when the tuber is infected by *Phytophthora erythroseptica*. A similar phenomenon has been reported by Wilson *et al.* (1970) in the case of mycoparasitism of *Piptocephalis virginiana* on *Mycotypha microspora*. Within the parasitised host cell, lysosome-like particulate bodies are swollen and often disrupted. Histochemical studies revealed a heavy concentration of acid phosphatase, acid DNase and aryl sulfatase in the surface of the haustoria of *P. virginiana*. Pitt and Galpin (1973) showed that infection of potato leaves by *Phytophthora infestans* resulted in reduced particulate activity of acid phosphatase accompanied by increased activity in the soluble phase of the cell. Again, an increase in lysosomal activity has been noticed in the cells of the protocorm of orchid *Dactylorhiza purpurella* after the infection by the endophytic fungus (Williamson 1973). Thus, in the battle of enzymes, lysosomal structures are directly involved.

Pathogenesis-related proteins

A group of host-encoded proteins, called pathogenesis-related (PR) proteins, are often produced by the host as a reaction to pathogens. The *de novo* expression of one or more proteins is induced by diverse biotic or abiotic substances called elicitors. A given pathogen may induce as many as 10 PR proteins as reported in the case of systemic infection of tomato plants by citrus exocortis viroid (CEVd) (Granell *et al.* 1987). Based on their structure and function, the PR proteins have been classified into five families PR-1 to PR-5 (Bowles, 1990; Linthorst 1991). Most families can be subdivided into two or more classes by structural differences, as well as properties. While class I proteins are antifungal, vacuolar and basic, class II proteins are not antifungal, extracelluar and acid isoforms of some class I proteins. Examples of class I vacuolar proteins are basic chitinase, basic beta-1,3-glucanase, osmotin (AP-24, fungus-induced, tobacco, 24kD) and thaumatin (P-23

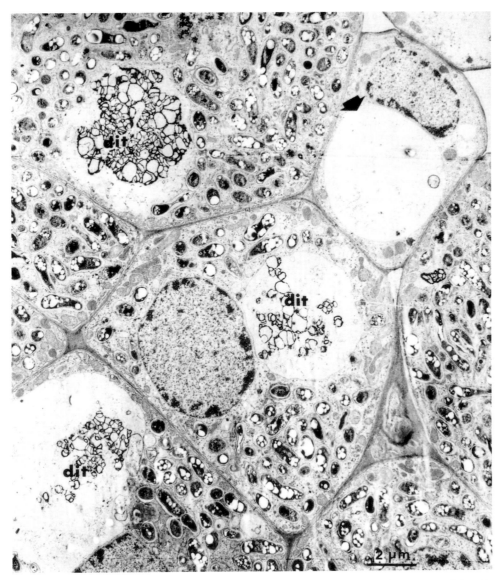

Figure 7.13 A low power electron micrograph showing several infected cells of a young nodule with vacuoles containing degrading infection threads (dit). Note the vacuole of the uninfected cell (marked by a thick arrow) is devoid of any infection thread debris. × 5500 (from Bal 1985). With permission of Faculty Press.

viroid-induced, tomato, 23 kD). As well as existing in soluble form, PR proteins may be localised in the inclusion bodies of the vacuole. The vacuoles of tobacco cell culture subjected to osmotic stress contain inclusion bodies in which a PR protein osmotin accumulates preferentially (Singh et al. 1987). Similarly, the CEVd-induced P-23 tomato protein is selectively localised in the inclusion bodies (Rodrigo et al. 1993). Stimuli of different sorts induce the expression of very similar genes and target the protein to the vacuoles. The vacuolar 23kD PR protein (P-23)-coding DNA clones were isolated from both viroid-induced and ethylene-induced libraries of tomato (Rodrigo et al. 1993). Similarly, a chitin-binding antifungal vacuolar protein CBP-20 can be induced by both tobacco mosaic virus and wound (Ponstein *et al.* 1994). It is also interesting to note that, at least in tobacco, whereas the class I proteins are strongly induced upon wounding, class II proteins are not (Bredrode *et al.* 1991). In tomato, the salt-induced protein osmotin is very similar to the protein induced by viroid infection and both are targeted to the vacuole (King *et al.* 1988; Ruiz-Medrano *et al.* 1992). The homologies between the two proteins imply that both of them may have antifungal activity and osmotic adaptation. Kononowicz *et al.* (1992) proposed that osmotin could have a primary role in low water potential situations by protecting plasma membranes as part of a mechanism causing permeabilisation and a secondary role as a vacuole-associated permeabilising agent acting against the plasma membrane of pathogens.

Elicitor-enzyme-vacuole-discharge model
Cell wall-degrading enzymes, chitinase and β-1,3-glucanase act on the wall of fungal pathogens (Boller 1985). In many plants these two enzymes accumulate rapidly following a pathogen attack after elicitor treatment, as well as in response to the plant stress hormone ethylene. With the aid of immuno-cytochemical and biochemical fractionation techniques, Mauch and Staehelin (1989) demonstrated that all the chitinase and most of the β-1,3-glucanase accumulate in the vacuoles of ethylene-treated bean leaves. Immunogold cytochemical techniques could detect both the enzymes in the large vacuolar protein aggregates which are absent in control leaves. A small amount of β-1,3-glucanase is also accumulated in the middle lamella of the cell wall. Mauch and Staehelin (1989) proposed a model for the defence mechanism of plant cells against fungal pathogens (Fig. 7.14). When fungal pathogens come in contact with the β-1,3-glucanase molecules localised in the middle lamella, oligosaccharide fragments are released from the β-1,3-glucan-containing cell wall. These oligosaccharides act as (exo-)elicitors. Through a signal transduction pathway, an increase in the transcription of the genes for chitinase and β-1,3-glucanase takes place. Then both the enzymes are processed through the Golgi apparatus and transported to the vacuole or the extracellular space. According to the model, when the same host cells are lysed during pathogenesis, the lethal amount of enzymes stored in the vacuole suddenly floods the pathogens and kills them. Application of plant hormone ethylene to the leaves of *Phaseolus vulgaris* induces the accumulation of the same enzymes in large electron-dense aggregates in the vacuoles (Mauch *et al.* 1992).

Parallel to the studies on ethylene-treated bean leaves, Spanu *et al.* (1989) studied chitinase activity in roots of *Allium porrum* during development of a vesicular-arbuscular mycorrhizal symbiosis with *Glomus versiforme*. The penetration of the fungus in the root causes an early defence reaction by the plant and an increase in chitinase activity which subsides later. An antiserum raised against bean-leaf chitinase was used for the immunocytochemical localisation of the enzyme with fluorescent and gold-labelled probes. Chitinase was localised in the vacuole as well as in the intercellular spaces, but the enzyme was never observed to be actually bound to the fungal wall. Since the chitin in the fungal wall does not come in contact with plant chitinase, the researchers concluded that it is not involved in the development of the mycorrhizal fungus. So far, these observations agree with Mauch and Staehelin's model, where β-1,3-glucanase has been implicated in the recognition process.

The hypersensitive reaction
One of the mechanisms by which a host plant or a non-host plant may be resistant to a pathogen is hypersensitivity. Plants showing a hypersensitive reaction (HR) are extremely sensitive to the pathogen and the tissue surrounding the point of infection undergoes very rapid necrosis, mostly within hours of infection. Thus the pathogen is deprived of nutrition, the progress of the infection is inhibited and the pathogen is killed.

The general symptoms of HR caused by viruses, bacteria and fungi are similar and include change in the permeability of the cell membrane, loss of electrolytes and water, and granulation

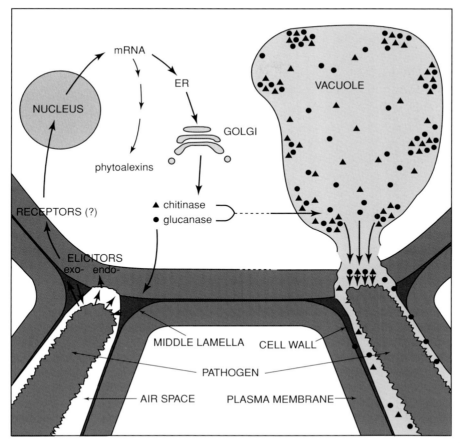

Figure 7.14 Model outlining the roles of chitinase and β-1,3-glucanase in a bean plant's defence against pathogen attacks (from Mauch and Staehelin 1989). With permission of the American Society of Plant Physiologists.

and browning of the cytoplasm (Nicholson *et al.* 1977; Goodman and Plurad 1971). Thus, HR response must be attributed to the alteration in membrane structure and function leading to decompartmentalisation and uncontrolled mixing of enzymes and substrates within cells.

On the basis of present literature indicating i) severe disorganisation of the cytoplasm of the penetrated cell; ii) discontinuity of the lysosomal membrane; iii) swelling and disruption of lysosomes (Pitt and Coombes 1968); iv) release of hydrolases in the cell sap (Pitt and Coombes 1968, Pitt and Galpin 1973); and v) host cell necrosis preceding the death of the pathogen (Maclean *et al.* 1974; Shimony and Friend 1975), Gupta (1985) proposed that the HR is triggered by release of hydrolytic enzymes which are normally compartmentalised inside plant lysosomes or vacuoles. Initially invasion by the pathogen leads to a change in cell membrane permeability which is accompanied by a change in the permeability of tonoplast or lysosome membranes. Due to altered permeability, the tonoplast releases hydrolytic enzymes from the vacuoles to the cytoplasm causing its degradation and leading to granulation and browning of the cytoplasm. Ultimately these hydrolytic enzymes digest away a few host cells surrounding the point of infection and may also act on the pathogen through the labilisation of their lysosomic membrane. This scheme is compatible with the model of Mauch and Staehelin (1989).

However, recent research shows that the key feature of HR is the oxidative burst, characterised by a dramatic increase in the level of reactive oxygen intermediates produced by elicited plant cells

(Apostol *et al.* 1989). It consists predominantly of the superoxide anion, hydrogen peroxide and the hydroxyl radical (Mehdy 1994), which are also known as active oxygen species (AOS). Since the oxidative burst has been shown to precede the HR in all systems tested, it is likely that the release of hydrolytic enzymes due to altered permeability of the tonoplast is the intermediate stage of HR before degradation of cytoplasm. Indeed, it is known that two particularly harmful and early effects of AOS production are membrane lipid peroxidation in which fatty acids are converted to 4-hydroxyalkenals and oxidative damage to DNA (Kappus 1985; Berhane *et al.* 1994).

Phytoalexins

Another defence arsenal used by host plants is the production of a group of low mass antibacterial compounds, often phenolic or terpenoids, collectively known as phytoalexins (Ersek and Kiraly 1986; Ebal and Griesbach 1988). For example, pterocarpan phytoalexins, phaseollin or glyceollin are produced when soybean is challenged by pathogens or abiotic stress and causes disruption of membrane function (Giannini *et al.* 1990). The authors also demonstrated that glyceollin, a mixture of at least three isomers, increased the proton and Ca^{2+} conductance of *Phytophthora* plasma membrane, in addition to electrolyte leakage and membrane disorganisation of the plant tissue. Giannini *et al.* (1991) have further shown that very low concentrations of glyceollin inhibit ATP-dependent H^+-transport across the soybean tonoplast. Thus the collapse of the proton gradient across the tonoplast leads to leakage of components from the vacuole and eventual cell death. The rupture of the plant cell thus exposes the invading organisms to phytoalexins, vacuolar enzymes and possibly other induced defence products.

In many plants flavonoids act as phytoalexins. For example, in sorghum, 3-deoxyanthocyanidin is a potent vacuolar phytoalexin. In several leguminous species including chickpea, soybean and alfalfa, isoflavonoid phytoalexin is stored in vacuoles (Mackenbrock *et al.* 1993). Medicarpin, a major isoflavonoid phytoalexin of alfalfa and several other leguminous species, undergoes glutathionation by glutathione S-transferase to form medicarpin-GS which is then transported across the tonoplast by GS-X pumps for vacuolar storage (Li *et al.* 1997). Elicitation infection may then cause release of isoflavones from the vacuole (Park *et al.* 1995).

7.9.6 Lytic activity in differentiation

As discussed above, meristematic cells show distinct lytic activity to produce vacuoles and/or existing vacuoles undergo autophagy. If it is accepted that a meristematic cell is devoid of a typical vacuole, then vacuolation itself should be considered as the first step to differentiation in the vectorial development of the meristematic tissue. In this connection, the experiments on evacuolation of protoplast are interesting. Evacuolation is the process of removal of vacuoles from isolated protoplasts by subjecting the cells to centrifugation through a Percoll gradient (Griesbach and Sink 1983). Burgess and Lawrence (1985) have shown that evacuolated tobacco mesophyll protoplasts show signs of ER activity within 6 hr of recovery. By ring-like configurations ER may enclose a cytoplasmic mass or cellular organelles, including mitochondria, suggesting autophagic function. ER was also observed as short fragments of cisternae with distended ends. Within 12 hrs of recovery a large central vacuole was formed, possibly by autophagy.

7.9.6.1 Autophagy in vascular differentiation

Autophagic activity may cause degradation of the cytoplasm or may be selective for a certain organelle or part of the cytoplasm. Gahan and Maple (1966) noted that on the way to differentiation, β-glycerophosphatase activity was detected first in the lysosome-like particles and was later found to be diffuse throughout the cell, including the vacuole. In the early stages of development of sieve tubes in roots of broad bean and shoots of tomato, lysosome-like particles exhibiting positive reactions for acid β-glycerophosphatase, napthal ASBI phosphatase, and esterase activities, were found to be evenly distributed throughout the cytoplasm of the cell (Gahan and McLean

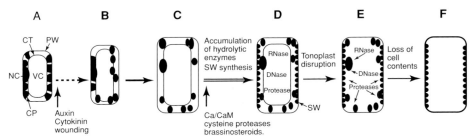

Figure 7.15 Cellular events that occur during transdifferentiation of Zinnia Mesophyll cells into TEs. **a** isolated mesophyll cell; **b** de-differentiated cell; **c** TE precursor cell; **d** immature TE; **e** maturing TE; **f** mature TE. Transdifferentiation is induced by wounding and a combination of auxin and cytokinin. Broken, single and double arrows show the progress of stages I, II, and III, respectively. The transition from stage II to stage III appears to be regulated by calcium (Ca)/CaM, cysteine proteases, and endogenous brassinosteroids. At the start of stage III, genes that are involved in both secondary wall synthesis and autolysis are expressed. Hydrolytic enzymes, such as DNases, RNases, and proteases, may accumulate in the vacuole. The disruption of the tonoplast causes these enzymes to invade the cytoplasm and attack various organelles, resulting in the formation of a mature TE that has lost its cell contents. CP, chloroplast; CT, cytoplasm; NC, nucleus; PW, primary wall; SW, secondary wall; VC, vacuole (from Fukuda 1997). With permission of the American Society of Plant Physiologists.

1969). Apart from the particulate occurrence of the enzymes, various hydrolytic enzymes in the sieve elements have been detected using histochemical and cytochemical methods (Wardlaw 1974). The presence of acid phosphatase containing digestive vacuoles with membranous contents in the companion cells has been recorded by Esau and Charvat (1975).

It has long been known that meristematic cells destined to be primary xylem cells lose their cell contents by general breakdown of the cytoplasm (Wodzicki and Brown 1973). During the last decade, a number of laboratories have been engaged in research on the differentiation of tracheary elements (TE) in suspension cultures of Zinnia cells and a lot of information on the ultrastructure, biochemistry and molecular genetics is available (vide review by Fukuda 1996). Whereas *in planta* the procambial cells differentiate into a primary xylem, *in vitro* the mesophyll cells of Zinnia in suspension can be induced directly to form TE (Fukuda and Komamine 1980) (Fig. 7.15). On the basis of cytological and biochemical processes and identification of numerous genetic markers, the trans-differentiation is divided into three stages (Fukuda 1997). Stage I involves the functional 'de-differentiation' process including disarray of chloroplast, cessation of photosynthesis and initiation of calmodulin. In stage II, along with preparation for secondary wall (SW) synthesis, the process of programmed cell death (PCD) is initiated with the synthesis of hydrolytic enzymes like cysteine protease (Minami and Fukuda 1995), serine protease (Beers and Freeman 1997) and a number of endonucleases. Several Zn^{2+}-activated DNases and RNases are also closely associated with TE differentiation. The signal peptides of the N-termini of the proteases and the endonucleases suggest that their precursors are probably transported into the vacuole where they are processed to the activated form. It is interesting that the optimal pH of both cysteine protease and DNase is 5.5 which corresponds to the pH in the vacuole. Groover et al. (1997) noted the presence of small vacuoles which cause partial loss of density of the cytoplasm at this stage. Recently, Groover and Jones (1999) reported a 40 kD serine protease secreted during TE differentiation and concomitant SW synthesis, and specific proteolysis of the extracellular matrix is necessary and sufficient to trigger Ca^{2+} influx.

Stage III involves autolysis and secondary wall formation. The influx of Ca^{2+} is followed by rapid collapse of the large hydrolytic vacuole and cessation of cytoplasmic streaming. After the disruption of the tonoplast, organelles with single membranes such as Golgi bodies and ER

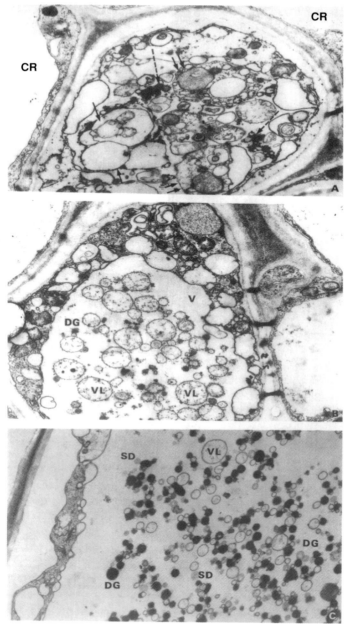

Figure 7.16 Stages in the development of laticifer in *Calotropis gigantea*. **a** A portion of a laticifer at its middle stage of development showing the process of extensive degeneration of cellular organelles and vacuolation. Autophagy is evident in a few membrane-bound cytoplasmic masses (double arrow) in which numerous membrane-bound vesicles are undergoing degeneration. Other vacuoles or vesicles (single arrow) show an advanced state of degeneration. Electron-dense particulate matter (broken arrow) starts to develop (× 8600). **b** An advanced state of development of a laticifer in which a large central vacuole (V) contains numerous vesicles (VL) containing some granular matter. Electron-dense globules (DG) are all more or less spherical discrete structures (× 8600). **c** The central vacuole of a mature laticifer shows three types of structures: membrane-bound vesicles (Vl), electron-dense globules (DG) and semidense particles (SD), (×12500) (A.T. Roy and D.N. De, *Ann. Bot.*, **70**, 443, 1992). With permission of Academic Press.

become swollen and then rupture. Subsequently, organelles with double membranes are degraded (as well as the nucleus). The TE dramatically lose most of their organelles within a few hours after disruption of the tonoplast and the entire contents of the cells disappear within 6 hrs after the first visible evidence of SW thickening.

7.9.6.2 Autophagy in laticifer development

Autophagic activity has also been recorded in the formation of laticifers of many plants. In cells of the shoot apex of *Euphorbia characias*, Marty (1970, 1971) demonstrated extensive autophagic activity which led to the degradation of a large amount of the cytoplasm in the central region of the cell, leaving a thin layer of cytoplasm along the plasma membrane. Esau and Kosaki (1975) reported partial autolysis of cytoplasmic components in vacuoles of articulated laticifers of *Nelumbo nucifera*. The laticifer formation in *Calotropis gigantea* has been studied by Roy and De (1992). The transformation of the parenchymatous cells starts with the enlargement of the small vacuoles and formation of new vacuoles, concomitant with the degeneration of cell organelles. This leads to formation of various cytoplasmic islands with diverse membranous components. At a later stage, the central region is occupied by a large vacuole housing vesicles of various dimensions and small spheres of slightly electron dense latex particles (Fig. 7.16).

As well as general and random autolysis of the cytoplasm, selective autophagy has been noted in various cases of cellular differentiation. Gifford and Stewart (1968) noted in the shoot-apical cells of *Bryophyllum* and *Kalanchoe* that proplastids containing certain inclusions are transferred to vacuoles by tonoplast invagination. Villiers (1971) observed that, in the cells of *Fraxinus* embryos subjected to prolonged dormancy, vacuoles engulfed specifically proplastids and mitochondria, which were presumably damaged.

7.9.6.3 Autophagy in cotyledonary development

One of the best examples of lytic activity is shown in the protein bodies in the cells of seed cotyledons. For example, in the 17-day-old developing cotyledonary cells of carrot, a few lipid bodies invaginated in the vacuole and distinct amoeboid-like lipid bodies crossing the tonoplast were observed (Dutta *et al.* 1991). In leguminous seeds, storage protein together with substances like phytin and lectin is accumulated in structures which are equivalent to vacuoles. During seedling growth the reserve proteins are mobilised, the protein bodies appear empty and their outlining single membranes fuse to form the central vacuole (vide Chapter VI). Van der Wilden *et al.* (1980) have demonstrated the presence of hydrolases α-mannosidase, N-acetyl-β-glucosaminidase, ribonuclease, acid phosphatase, phosphodiesterase and phospholipase D. Their ultrastructural studies indicate that invagination of the protein body membrane causes portions of the cytoplasm to be internalised within the protein bodies, resulting in the formation of autophagic vesicles.

7.9.6.4 Autophagy in gametogenesis

The process of gametogenesis is the most important cell differentiation in the life cycle of a plant. Sangwan *et al.* (1989) have shown that during microsporogenesis of *Datura*, autophagic vacuoles are formed in the cytoplasm of the early meiotic cells and during pollen development cytoplasmic organelles pass into the large central vacuole and undergo degradation. They suggest that these two processes are related to the overall turnover of cytoplasmic constituents and are probably related to the sporophytic to gametophytic transition. Finally, autophagy is also involved in the release of sperm. The sperm cell is a very precisely programmed cell destined for fusion with the egg nucleus. The cytoplasm of the sperm cell is not only expendable, but its dissolution is essential for liberating the sperm nucleus prior to fusion. Ultrastructure study of isolated sperm cells of *Zea mays* by Wagner and Dumas (1989) has revealed that the cytoplasm of the sperm cell undergoes dissolution by both types of autophagy, i.e. by concentric stacking of ER and by direct degradation of the cytoplasmic mass in the pre-existing vacuole.

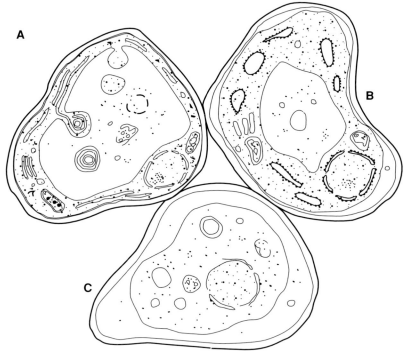

Figure 7.17 Schematic diagrams of the changes in the ultrastructure of mesophyll cells during senescence of the corolla of *Ipomoea purpurea* (Morning Glory). **a** Autophagic activity of the vacuole; invaginations of the tonoplast result in sequestration of cytoplasmic material into the lysosomal compartment (i.e. the central vacuole according to the interpretation of Matile). **b** Shrinkage of the vacuole, dilution of the cytoplasm and inflation of cytoplasmic membrane systems. **c** Autolysis is initiated by ultimate breakdown of the tonoplast (from Matile and Winkenbach 1971). With permission of Oxford University Press.

7.9.6.5 Autophagy in senescence

Senescence is the ultimate state of differentiation leading to programmed cell death. The petals of mature flowers provide a striking example of rapid senescence. By means of biochemical and ultrastructural techniques Matile and Winkenbach (1971) followed the changes in the activity of the acid hydrolases and breakdown of cytoplasmic organisation of the wilting corolla of Morning Glory (*Ipomoea purpurea*). As the corolla undergoes senescence the hydrolytic enzymes, viz. DNase, RNase, protease and β-glucosidase, increase markedly, and the various organelles undergo progressive degeneration with increasing vacuolar activity. At first the invagination of the tonoplast draws the cytoplasmic constituents within the lumen of the vacuole which acts as the lytic compartment for the cytoplasm and gradually the organelles are degraded. Further activity of the hydrolases leads to the bursting of the tonoplast and digestion of the nucleus (Fig. 7.17). The senescing petals of carnation (*Dianthus caryophyllus*) exhibited ultrastructural changes in mesophyll vacuoles from the pre-senescent stage where occasional rupture of the tonoplast was noticed. In the pre-climacteric petals, the vacuoles of some mesophyll cells were seen to contain numerous small vesicles and some mitochondria and in some other cells tonoplast appeared to burst, resulting in a general mixing of the cell contents. As the petals passed from the climacteric to the post-climacteric stage, rupture of the tonoplast led to large-scale ultrastructural disorganization (Smith *et al.* 1992). Thus, in programmed cell death autophagy of cytosolic constituents by the vacuole and/or rupture of tonoplast causing release of hydrolases leads to cellular disintegration. Smart (1994) recorded that the activity of several hydrolytic enzymes, most notably cysteine proteases localised in the vacuole, increases dramatically during leaf senescence. This catabolism frees carbons and nitrogen for mobilisation to the shoots and roots. The senescing cell and its organelles are disassembled in an ordered fashion: chloroplasts diassemble before other organelles by autolysis rather than by autophogy.

Matile (1992) stated that mesophyll chloroplasts are converted into 'gerontoplast' with degenerating internal ultrastrure during leaf senescence and yellowing, accompanied by the disassembly of thylakoid pigment-protein complexes and breakdown of chlorophyll into phytol and water-soluble porphyrin derivatives. The water-soluble porphyrins are exported from the gerontoplast and accumulated ultimately in the vacuoles of senescing cells where they remain until the end of the senescence period (Matile et al. 1988). This intracellular excretion storage of linear tetrapyrroles is conducted by ABC transporters in the tonoplast and serves to protect the cytosol from photo-oxidative damage which accompanies production of free radicals (Hinder et al. 1996). Finally however, the tonoplast bursts and autolysis is complete.

7.10 Detoxification of SO_2

As sulfur is an essential element for plant nutrition and sulfates are the principal form of uptake of the element, detoxification of SO_2 is only an issue when there is excess sulfate in the soil or SO_2 is present in the air. As has been discussed earlier, an excess of a substance in the cytoplasm may be sequestered in the vacuole. In some form or other, excess sulfur is finally detoxified by the vacuole. When barley seedlings are grown hydroponically in the presence of high concentrations of potassium sulfate, the sulfate concentration in the mesophyll vacuoles increases by a factor of about 10 or more. Similar results were obtained with experiments on isolated vacuoles (Kaiser et al. 1989). Although uptake of sulfate ions cannot be considered in isolation, sulfate tolerance of barley mesophyll cells depends on the capacity of the sulfate transporter of the tonoplast.

On the other hand, gaseous uptake of SO_2 into the leaves occurs by passive diffusion. Inside the cytoplasm SO_2 is hydrated, and the hydration products HSO_3^-, SO_3^{2-} and H^+ are trapped in the slightly alkaline cytoplasm, and not in the acidic vacuole (Pfanz et al. 1987).

In most plants, excess SO_4 is stored in vacuoles, where it remains metabolically inactive. Accumulation of $BaSO_4$ in *Chara* and sulfuric acid in *Desmarestia* are examples of extreme cases. Irrespective of the form of available sulfur, i.e. whether it is sulfur dioxide, hydrogen sulfide or cysteine, it is stored as SO_4. When there is excess sulfur in the reduced form it may accumulate as glutathione, which appears to be the storage form of reduced sulfur in higher plants; but the storage glutathione in the vacuole is not yet clear. (Only a part of glutathione is found in the vacuole.)

In *Lemna minor*, as well as in carrot cells, sulfate is actively transported at both plasmalemma and tonoplast (Thoiron et al. 1981; Cram 1983a, b). However when excess sulfate is made available, the tonoplast regulates the influx by feedback inhibition (Cram 1983a). Rennenberg (1984) outlined a model for the influx of sulfate into cellular compartments. With increasing sulfate concentration in the cytoplasm, vacuolar sulfate concentration will increase. This will cause a decrease in the active influx of sulfate and the efflux by facilitated diffusion will increase until both fluxes are equal. Negative feedback control by sulfate may prevent the vacuolar sulfate concentration from increasing indefinitely with increasing external supplies of sulfate. Kaiser et al. (1989) have shown that sulfate accumulation is powered by a tonoplast ATPase. The main cation accompanying sulfate during net uptake into vacuoles is K^+ by the K^+/H^+ antiport mechanism. The role of vacuoles in the absorption of atmospheric SO_2 has also been emphasised by the work of Kaiser et al. The sulfate-pumping capacity of tonoplast of barley mesophyll cells is more than sufficient to cope with transport requirements for sulfate when leaves are exposed to SO_2 pollution within permissible limits (140 µg SO_2 m^{-3}). Moreover, vacuolar detoxification capacity appears to be limited by the vacuole's capacity for proton storage, not by that for sulfate storage. The phytotoxic sulfite and bisulfite anions may be detoxified by either reduction or oxidation. The primary product of reduction of H_2S may be emitted and the main secondary products are cysteine and organic compounds. However, it appears that the major detoxification mechanism is oxidation to sulfate, which is actively sequestered in the vacuole by a symport mechanism in which Ca^{2+}, K^+ and Mg^{2+} serve as dominating cations for sulfate in coniferous trees (Hüve et al. 1995).

7.11 Detoxification of xenobiotics

Non-nutrient, foreign soluble or particulate substances which may enter the plant cell have been considered in the section on heterophagy. But, xenobiotic chemicals which are cytotoxic must be detoxified to the process of detoxification by plant cells. There is no sharp line of demarcation between heterophagy and detoxification, except that detoxification deals with xenobiotics and herbicides. Xenobiotics are diverse synthetic chemicals and herbicides are lipophilic electrophiles that penetrate the barriers surrounding the symplast, i.e. the extracelluar matrix and plasma membrane as well as the other endomembrane systems. Plant cells are able to metabolise and detoxify many xenobiotics by a variety of enzymatic reactions. Ishikawa (1992) described the metabolism and detoxification of xenobiotics in three phases. In phase I (activation), the compound is oxidized, reduced, or hydrolysed to expose or introduce a functional group of the appropriate reactivity for phase II enzymes. In phase II (conjugation), the activated derivative is conjugated with hydrophilic substances, such as glutathione, glucuronic acid or glucose. In phase III (elimination), the conjugate is either excreted to the extracellular medium or sequestered in an intracellular compartment. Kreuz et al. (1996) and Rea et al. (1998) suggest phase IV in which the transported conjugates are further substituted and degraded to yield transport inactive derivatives. After phase II the resulting conjugates are i) generally inactive toward the initial target site; ii) more hydrophilic and less mobile in the plant than the parent substance; or iii) susceptible to further processing which may include secondary conjugation, degradation and compartmentation (Kreuz et al. 1996). The phase I process of oxidation, reduction or hydroxylation may be carried out by a number of enzymes, of which cytP450-dependent mono-oxygenases play a pivotal role. There are a number of CytP450 forms which may mediate hydroxylation of aromatic rings or alkyl groups, or hetero-atom release (Barret 1995). After oxidation or hydrolysis, which usually introduces a hydroxy, amino or carboxylic acid function, the herbicide is subjected to rapid glycosylation by a glucosyltransferase utilising UDP-GLc as the usual sugar donor. The glucosyltransferases which are needed for biosynthesis of secondary metabolites are present in the plant cytosol either as soluble or membrane-bound forms. On the other hand, numerous studies have established the existence of glutathione-S-transferase (GST) families comprising isozymes possessing varying degrees of substrate specificity for particular herbicides (Kreuz 1993).GST conjugates glutathione (GSH) to the electrophilic substrate (X–Z) with concomitant displacement of a nucleophil (Z, eg. halogen, phenolate or alkyl sulfoxide):

$$X - Z + GSH \xrightarrow{GST} X - GS + HZ$$

The GSH conjugates of herbicides in plants may undergo extensive processing to Cys or N-malonylcysteine and other conjugates. Since the two classes of herbicide complexes, viz. glucosylated and glulathione-conjugated, may be harmful to the cytosol, they are either processed further or sequestered in a cellular compartment. No information is available on the excretion of the detoxified xenobiotic to the extracellular matrix or apoplast across the plasma membrane. However, a number of workers have investigated the mechanism of vacuolar transport. Vacuoles are known to accumulate a wide range of unrelated compounds varying in molecular weights, steric configurations and charges, many of which may be due to the presence of a unidirectional transporter in the tonoplast. Oparka et al. (1996) suggested that a probenecid-sensitive uptake mechanism on the tonoplast may function as a part of a detoxification system for the removal of xenobiotics from the cytosol. Martinoia et al. (1993) studied the uptake of a number of glutath-

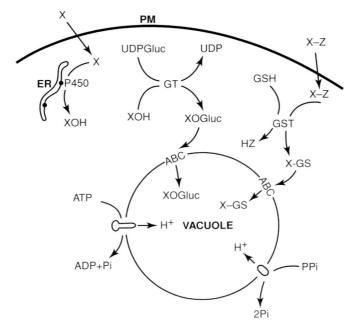

Figure 7.18 Schematic representation of the role of ATP-binding cassette (ABC) transporters in detoxification of xenobiotic X and X–Z. Z, the nucleophile moiety displaced in the GST-catalysed reaction. PM, plasma membrane; P450, CytP450 monooxygenase; GT, glucosyl transferase, GSH, glutathione; GST, glutathione S-transferase

ione conjugates by isolated barley vacuoles and concluded that they accumulated in the vacuole even when the vacuolar H^+-ATPase was inhibited and the pH gradient between the cytosol and the vacuole was abolished, indicating the presence of a specific transport ATPase. Moreover, it is possible that only a single transporter recognises all the different forms of the glutathione conjugates (Tommasini et al. 1993).

Li et al. (1995a, b) have shown that, unlike most other characterised organic solute transport in plants, uptake of the model compound S-2,4-dinitrophenyl glutathione (DNP-GS) and glutathione-S-conjugates by *Vigna* tonoplast vesicles is directly energised by MgATP. The detection of glutathione-S-conjugate transporter satisfied the minimum requirement of an element involved in the detoxification of xenobiotics, viz. enhancement of its activity in response to exposure of the intact organism to the target compound. The degradation of GSH conjugates to the Cys conjugate has often been observed for herbicides and may represent a further detoxification step. Uptake of herbicide glucosides increases in the presence of Mg-ATP and unlike glutathione conjugates, their uptake is inhibited by other glucosides (Gaillard et al. 1994). The ATP-binding-cassette (ABC) family of transporters present in the tonoplast may function to sequester xenobiotics, as well as endogenous compounds into the vacuole (Fig. 7.18). Xenobiotics, like other compounds in the cytoplasm, will require glutathione canjugation to be transported across the tonoplast via the ABC transporter (Swanson and Jones 1996). Recently, Lu et al. (1998) identified a gene, *AtMRP2* from *Arabidopsis*, that encodes a multispecific ABC-transporter competent in the transport of both GS-conjugates as well as chlorophyll catabolites. It is intriguing that the same gene is responsible for both and it is now accepted (vide review by Rea et al. 1998) that the members of the ATP-binding cassette (ABC) transport superfamily, the largest protein family known, are capable of a multitude of transport functions. They are directly energised by Mg^{2+}-ATP rather than by the electrogenic pumps, V-ATPase and PPase. A major subclass of the superfamily, MRPs (or multidrug resistance associated proteins as named by their animal prototypes), are represented by GS-X pumps (glutathione-conjugate or multispecific organic anion Mg^{2+}-ATPase), which participate in the transport of exogenous and endogenous amphipathic anions and glutathionated compounds from the cytosol into the vacuole. Several other such pumps have been detected and implicated in the transport of i) large amphipathic organic ions like chlorophyll catabolites (Alfenito et al. 1998), bile acid-like substances e.g. digitonin, sterol or lipid derivatives and glucuronate conjugates; ii) glucose conjugates; and iii) phytochelatins. Thus they are pivotal in herbicide detoxification, cell pigmentation, alleviation of oxidative damage, and storage of antimicrobial compounds. They may also be involved in channel activity and/or heavy metal chelates (discussed in the next section).

For Phase IV, or transformation, some information is available on storage and excretion of the end product. Wolf et al. (1996) recorded that exposure of barley leaves to alachlor results in massive vacuolar accumulation of alachlor-GS. Colemen et al. (1997) demonstrated *in situ* glutathionation of monochlorobimane to bimane-GS and accumulation of the latter in vacuoles of cultured cells. Very high accumulative capacity vacuolar GS-conjugate pumps have been demonstrated in studies by Li et al. (1995) and Wolf et al. (1996). Measurements of DNP-GS(S-(2,4-dinitrophenyl)-glutathione) by vacuolar membrane vesicles and of the distribution of the glutathionated herbicide, alachlor-GS, between the vacuolar and non-vacuolar compartments of barley leaves yield and accumulation ratio above 50. The detoxified substance may not always be stored in the vacuole. The transport of bound residues to the extracellular matrix has been noted in certain cases (Sandermann 1992).

7.12 Phytoremediation of heavy metals

The removal of pollutants from the environment by green plants, a process known as phytoremediation (Cunningham and Berti 1993; Raskin et al. 1994) is well known. Plants have a unique sorption potential for a large array of metallic cations from the environment. The uptake of toxic metals could be either through adsorption and/or absorption and is possible through some physiological adaptiveness and homoeostasis. The metals are transported to various compartments, especially to apparent intercellular or intracellular free space, vacuoles and other such bodies, to sequester the metals away from the main metabolic pathways. Formation of immobilised heavy metal containing crystals have been described for Cd, Co, Fe, Pb, Sr and an increase in Ca-oxalate crystals observed under heavy exposure to the metal. In general, the metals may form metal-organic acid, metal-sulfide, metal-phytate, or metal-peptide complexes in plants.

Zinc

When exposed to Zn salts, intravacuolar bodies in root meristematic cells of the grass *Festuca rubra* were found to contain Zn (Davies et al. 1991). Exposure to zinc salts actually induces vacuolation to various degrees. An exposure to a subtoxic concentration of Zn can induce a 2.93 fold and 6.78 fold increase in total vacuolar volume fraction in rice and wheat root meristematic cells, whereas rye roots exhibit no increase in vacuolation (Davies et al. 1992). In the case of Zn-tolerance Mathys (1977) suggested that Zn was chelated in the cytosol by malate, and this complex was delivered to the vacuole where it was sequestered as Zn-oxalate. A high malate content can be correlated with high Zn tolerance. However, this model has been criticised for a lack of critical experimental evidence. Thurman and Rankin (1982) have shown that the metal may indeed be bound to organic acids in the vacuole. Godbold et al. (1984) have demonstrated a high correlation between citrate content and Zn-accumulation in the root cortical cells of *Deschampsia caespitosa* and argued that citrate may have a role in Zn-tolerance. In the same species, Van Steveninck et al. (1987) showed that Zn is complexed with phytate in small vacuoles in root cortical cells. This observation is especially interesting in view of the fact that Zn^{2+} can bind tightly at the multiple phosphate groups of phytate (inositol hexophosphate).

X-ray microanalysis of Zn accumulation in *Thlaspi caerulescens*, a known hyperaccumulator of metals, by Vázquez et al. (1994), showed that Zn was compartmentalised in vacuoles of roots as well as leaves. When exposed to high concentrations of Zn, the plant exhibited a higher concentration of the metal in leaves than in roots. Zn storage is achieved by the formation of Zn-rich crystals in vacuoles of epidermal and subepidermal cells of leaves. Both the Zn/P element ratios found in the crystals and the absence of Mg indicate that, in contrast to other plant species, phytate is not the main storage form for Zn in *I. caerulescens*. On the other hand, vacuolar citrate may be the central mechanism for sequestration of Zn in plants exposed to either low or high levels of the metal (Wang *et al.* 1992). They used computer modeling of chemical equilibria to predict the metal-ligand species in the vacuoles of suspension culture cells of tobacco.

Cadmium

Like Zn, cadmium can also make organic acid complexes and be retained in the vacuole. Computer simulation also showed that in moderate concentrations the metal may form stable complexes as Cd-citrate despite an abundance of malate (Wang *et al.* 1991). On the other hand, a number of reports showed that Cd-peptides are chiefly located in vacuoles of leaf protoplasts of plants or cultured cells grown under Cd-pollution conditions. (Vögeli-Lange and Wagner, 1990; Krotz *et al.* 1989). Electron probe microanalyses by Heuillet *et al.* (1986) and Van Stevenick *et al.* (1990) provide conclusive evidence that vacuoles are the site of deposition of Cd. A class of cysteine-rich, metal-binding proteins, known as phytochelatins, may have an important role in cellular tolerance to toxic metals (Tomsett and Thurman 1988; Rauser 1990). Phytochelatins consist of repeating units of γ-glutamylcysteine followed by a C-terminal glycine [poly-(γ-Glu-Cys)$_n$-Gly peptides]. Although certain studies have shown that Zn is not generally bound to phytochelatin in vacuoles (Wang *et al.* 1992), on exposure to Cd a significant proportion of intracellular Cd is bound as a heterogenous Cd-phytochelatin sulfide complex, production of which is necessary for Cd-detoxification. The complex is localised in the vacuole (Steffens 1990; Wagner 1994). Salt and Wagner (1993) identified a Cd^{2+}/H^+ antiport on the tonoplast, which may be responsible for Cd-accumulation in the vacuole. Salt and Rauser (1995) have reported that when the oat roots were exposed to moderate to high levels of Cd, the metal was found to be primarily associated with the phytochelatin. The authors have shown that Cd-complexes of the phytochelatin are transported across the tonoplast by a MgATP-dependent, vanadate-sensitive, possibly ATP-binding cassette type (ABC) transporter which utilises an ATPase-proximate energy source for active organic solute transport. On the basis of studies on Cd-sensitive mutants of yeast *Schizosaccharomyces pombe*, Oritz *et al.* (1994) have shown that an ABC-transporter associated with the tonoplast is responsible for transport of both phytochelatin and Cd-phytochelatin complexes. It may be concluded that the vacuole is intimately associated with the process of detoxification of metals. Whereas a moderate amount of the metal can be complexed with organic acid inside the vacuole, high doses of the metal may cause, at least in a number of cases, the induction of phytochelatin synthesis and subsequent transport of the metal-phytochelatin complex to the vacuole. On the other hand, a novel gene *YCF1* or cadmium factor (a close homologue of *MRP1* of the ABC family), which confers resistance to cadmium salts, was isolated from yeast cells. Recently, Li *et al.* (1997) studied the substrate requirements, kinetics and Cd^{2+}/glutathione stoichiometry of cadmium uptake and the molecular weight of the transport-active complex which demonstrated that YCF1 selectively catalyses the transport of bis(glutathionato) cadmium. Recently, Ghosh *et al.* (1999) have shown that in yeast cells the same gene is responsible for removal of arsenite (As III) from the cytosol by sequestering it in the vacuole as the glutathione conjugate. Hence, the cadmium-factor gene encodes a MgATP-energised glutathione-S-conjugate transporter responsible for vacuolar sequestration of organic compounds after their S-conjugation with glutathione.

Nickel

Although the vacuole has long been suspected of being a site for accumulation of nickel in plant roots, no direct evidence is yet available. Lately Gries and Wagner (1998) have made a critical attempt to study Ni transport along with Ca and Cd into the tonoplast vesicles of oat roots. Although

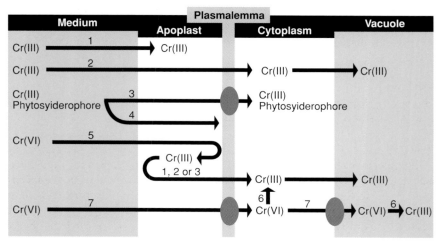

Figure 7.19 Chromium uptake model based on current hypothesis of micronutrient uptake. (1) retention of Cr(III) in the apoplast; (2) Diffusion of Cr(III) through plasmalemma and tonoplast; (3) Binding of Cr(III) by phytosiderophores and transport of the complex through plasmalemma (in monocots); (4) binding of the Cr(III) phytosiderophore complex to plasmalemma; (5) reduction of Cr(VI) to Cr(III) by root reductase; (6) reduction of Cr(VI) in cytoplasm or vacuole (7) transport of Cr(VI) by anion carriers (from Barceló and Poschenrieder 1997). Courtesy J. Barceló. With permission of Franco-Angeli.

Ni was associated with the vesicles, its relative rate of accumulation/association is very low compared to that of Ca and Cd. Protonophores and the potential Ni ligands citrate and histidine, nucleoside triphosphates or PPi did not stimulate Ni association with vesicles. Comparison of Ni versus Ca and Cd associated with vesicles, using various membrane perturbants, indicate that while Ca and Cd are rapidly and principally antiported to the vesicle sap, Ni is only associated with the membrane in a hardly dissociable condition. Hence, it has been concluded that no Ni^+/H^+ antiport exists in oat tonoplast and the vacuole is not a major compartment for Ni accumulation.

Chromium

Because chromium does not cause widespread environmental problems and phytotoxicity due to Cr under field conditions is unknown, Cr toxicity has not been widely investigated. Of the common oxidation states of Cr in inorganic compounds, +II, +III and +VI, the Cr(III) and Cr(VI) are biologically significant. On the basis of diverse mutagenic and carcinogenic effects of Cr(VI) on various organisms, Gebhart (1984) regarded it as a highly cytotoxic substance. Vázquez et al. (1987) studied the ultrastructural changes in the cells of *Phaseolus vulgaris* which were treated with Na_2CrO_4, that is, Cr in hexavalent form, in the nutrient solution. They noted that besides other ultrastructural damages, the membranes, especially of the vacuole in root and stem cells, were damaged and certain precipitates were also detected in the vacuoles. Membrane damage by Cr(VI) has also been described in the fronds of two fresh water plants, *Lemna minor* and *Pistia stratiotis* (Bassi et al. 1990). In view of i) the similar effective ionic radius of Cr(III) and Fe(III); ii) understanding of the mechanism of absorption of Fe as a micronutrient (Welch 1995); iii) binding of Cr(III) by phytosiderophores (plant homologues of animal non-heme Fe-transport protein); and iv) certain other similarities in uptake of the two metals, Barceló and Poschenrieder (1997) proposed a model of Cr uptake (Fig. 7.19). According to this model, what is finally accumulated in the vacuole is the trivalent form of Cr. However, the mechanism of transport of Cr(III) across the tonoplast is not known. Since Cr(III) tightly binds to carboxyl groups of free amino acids as well as to those in proteins, it may form metaloprotein complexes and also some low molecular weight complexes containing nicotinic acid, glycine, glutamic acid and cysteine (Toepfer *et al.* 1977). The formation of Cr-phytochelatin remains to be confirmed. As discussed earlier, it is possible that certain ABC transporters could be involved in sequestration of Cr-conjugates in the vacuole.

Aluminium

Unlike chromium, aluminium may cause toxicity problems in acid soils (pH <5.0) for crop productivity, which can be ameliorated by application of lime or gypsum. Al accumulates mainly in the apoplast and crosses the plasma membrane slowly, so toxicity occurs mainly in the apoplast (vide review by Kochian 1995). Aluminium was not detectable by specific staining in root tip vacuoles of either tolerant or sensitive wheat exposed for 48 hr to Al (Tice *et al.* 1992). Electron probe X-ray microanalysis by several authors (Marienfeld *et al.* 1995; Lazof *et al.* 1997) failed to detect any Al deposit inside the cytoplasm or vacuoles of roots of plants treated with Al. However, Barceló *et al.* (1996) detected Al in electron-dense deposits in the root tip vacuoles of maize plants treated with Al for 24 to 120 hr. Recently, Vázquez *et al.* (1999) have reported that, in the roots of a tolerant maize variety, no Al was detected in cell walls after 24 hr, but an increase in vacuolar Al was observed after 4 hr of exposure. After 24 hr, higher amounts of Al, P and Si were observed in the vacuolar deposits. Moreover, the mineral contents of P, Ca, Zn and Mg in these deposits are similar to those reported for phytate deposits from maize (Mikus *et al.* 1992). Thus formation of phytate inside the vacuole may be responsible for Al tolerance.

It appears that the vacuole plays a distinct role as the terminal point in phytoremediation of a number of metals. A toxic metal may be conjugated to low molecular organic acids, glutathione peptides, phytochelatin or phytate. Certain metals may be transported by proton antiporters and others may be conjugated in the cytoplasm to be sequestered in the vacuole by specific ABC transporters. Thus, when a cell is subjected to an undesirable metal or even an essential metal in high dosage, it uses certain effective metabolic pathways to conjugate it and finally store it in the safe custody of a vacuole in an inert form. The tolerance or resistance of a plant depends, at least partly, on this capacity for detoxification and sequestration in the vacuole.

7.13 SCAVENGING OF ACTIVE OXYGEN SPECIES

Various stress conditions like wounding, attack of pathogens, and exposure to xenobitics, heavy metals, UV and ionising radiation (reviewed by Inz and Van Montagu 1995), induce an oxidative system with formation of molecular species of active oxygen: superoxide (O_2), hydrogen peroxide (H_2O_2), hydroxyl radical (OH), singlet oxygen ($^1O_2$) or peroxyl radical. These active oxygen species (AOS) are normally involved in major aerobic biochemical reactions like respiratory or photosynthetic electron transport and are produced in copious quantities by several enzyme systems. While endogenous AOS are readily utilised by the cell metabolism, exogenous AOS have to be rapidly processed or scavenged to prevent oxidative damage. AOS are strongly nucleophilic and react with biomolecules to generate reactive electrophiles. In this respect glutathione is extremely important because it acts both as a reducing agent protecting the cell from AOS and as a nucleophile protecting the cell from the electrophilic products of AOS action. Although a direct connection between oxidative stress and the GS-X pump on the tonoplast has not been established, it is known that a glutathione S-conjugate pump on the tonoplast transports oxidised glutathione (GSSG) into the lumen (Li *et al.* 1995) and many of the oxidative stresses are known to induce specific glutathione S-transferases (vide review by Rea *et al.* 1998). In addition to glutathione, a number of efficient antioxidants like ascorbate, carotenoids and flavonoids are found in plant cells. Many studies have indicated that flavonoids, which are the most common secondary metabolite in vascular plants, can directly scavenge AOS. Flavonoids are largely localised in vacuoles and it is unlikely that various active radicals will pass through

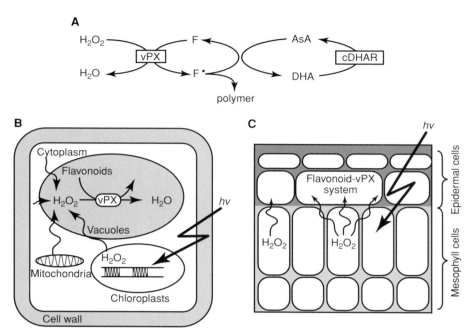

Figure 7.20 A proposed diagram for the protective function of flavonoids during stress and growth. **A** Scheme of the H_2O_2-scavenging mechanism by flavonoids. vPX, Vacuolar peroxidase; F, flavonoid; F˙ flavonoid radical; AsA, ascorbic acid; DHA, dehydroascorbic acid; *hv*, light energy; and cDHAR, cytosolic dehydroascorbic acid reductase. The diffusive nature of H_2O_2 enables vPX to scavenge it in vacuoles, even if the generating site is other than a vacuole (B). This concept can be expanded to the cell–cell interaction. The photoproduced H_2O_2 may leak out from mesophyll cells and be scavenged in epidermal cells that have a high flavonoid content (C) (from Yamasaki *et al.* 1997). With permission of the American Society of Plant Physiologists.

the tonoplast, since degradation of lipids of the tonoplst would be their first hit. However, unlike other AOS, H_2O_2 is stable and able to diffuse across membranes. On the other hand, flavonoids are electron donors for peroxidation and Takahama (1989) has shown that the flavonol-peroxidase reaction may act as an H_2O_2-scavenging system *in vivo* as well as *in vitro*. Yamasaki *et al.* (1997) reported that the flavonoids in the leaves of a tropical tree *Schefflera arboricola* have the potential to act as electron donors to peroxidase and that vacuolar peroxidase participates in H_2O_2-induced oxidation of flavonoids. Thus, the flavonoid-peroxidase reaction can function as a mechanism for H_2O_2-scavenging in plants (Fig. 7.20). Although the scheme is attractive, further investigations are necessary to determine its validity and efficiency in other cell systems.

7.14 Summary

Gross function

The gross function of the vacuole is **storage of water** along with **dissolved** or **undissolved** substances in its lumen. The transport of molecules across the tonoplast takes place either by **passive transport** which is independent of energy, or by **active transport** which requires energy.
- Passive transport may take place by **simple diffusion** or **facilitated diffusion** aided by certain **transport proteins** which come in two classes: i) **carrier proteins** which bind transiently with solute, and include **transporters** and **group translocators**; and ii) **channel proteins** which consist of **porins, ion channels** and **ion transporters**.

- The ion channels are not coupled to energy, yet provide a very fast rate and ion specificity. They may be open or closed due to a stimulus of voltage across the membrane, mechanical stress or chemical ligands. They are indirectly aided by energy and conduct secondary active transport.
- Bulk flow of water across the tonoplast takes place through a porin, known as **aquaporin**, which is partly controlled by state of phosphorylation, salinity, etc.

Uptake of ions

The large accumulation of ions in the cell, especially in the vacuole, occurs chiefly through the two transport processes.

- **Primary active transport** is the process by which an ion is moved against a gradient of electro chemical potential energised by either of the two pumps, vacuolar type H^+-ATPase (V-ATPase) and H^+-pyrophosphatase (H^+-PPase). Both pumps on the tonoplast catalyse inward electrogenic H^+-tranlocation from cytosol to vacuolar lumen to establish inside positive trans-tonoplast H^+-electrochemical potential difference for the energisation of H^+-coupled solute transport. Both V-ATPase and H^+-PPase activities may coexist in the tonoplast but V-ATPase is the predominant proton pump and in many cases H^+-PPase takes over only when V-ATPase fails or is inhibited.
- **Secondary active transport** is the process in which the transporter is aided by pH gradient and/or electrical gradient. Three types of secondary active mechanisms are i) *Uniport*, which transports an ion or a small molecule against a concentration gradient, uncoupled to the movement of any other molecule; ii) *Antiport*, which transports an ion or small molecule by coupling its movement to that of another ion or molecule in the opposite direction, and iii) *Symport*, which co-transports an ion or molecule with another ion or molecule in the same direction.
- **Uptake of anions** generally takes place through symport or even uniports which are *anion channels*. The anions move through the anion channel in the tonoplast in response to positive membrane potential generated by V-ATPase. In addition, there may be specific transports in the tonoplast for various organic or inorganic anions.

Vacuolar pH regulation

The pH of the vacuole may vary considerably from the pH of the cytoplasm. In general, H^+-ATPase is chiefly responsible for maintaining the vacuolar pH. However, depending on the cell system and its metabolic activity, H^+-PPase may control vacuolar pH. Besides programmed pH changes, the vacuole may act as a buffer for additional or sudden changes in the pH of the cytoplasm. In many cases, the balance between the passive leakage of H^+ from vacuole to cytoplasm and active H^+ transport from cytoplasm to vacuole seems to play a key role in vacuolar pH regulation.

Salt tolerance

The salt tolerance of the cell or plant is derived from its capacity to tolerate sufficient ions to maintain growth while avoiding water deficit or excess of ions. Compared to the cytoplasm, vacuoles compartmentalise large amounts of ions and the tonoplast proton pumps and ion transporters are responsible for such accumulation. Membrane phospholipid may play a role in the process.

Osmoregulation

Osmoregulation has two components: turgor pressure and osmotic pressure. One of the vacuole's basic functions is the maintenance and modulation of turgor pressure by controlled uptake and loss of water. Adequate turgor pressure is essential to maintain the plant cell's rigidity as well as its growth. Due to the tonoplast's high water permeability, the vacuole is always in osmotic equilibrium with the cytoplasm. Salts and sugars are the main solutes which affect osmotic pressure and hence turgor pressure. Thus, osmoregulation is brought about by one or both the following processes: i) interconversion of organic solutes to soluble and insoluble components; and ii) ion transport across the tonoplast.

Turgor movements

Turgor movements in various plant cells, tissues and organs are due to changes in turgor pressure of the vacuoles of participating cells. The collapse and recovery of the vacuole responsible for guard cell movement is caused by operation of electrogenic pumps as well as K^+ channels, Ca^{++} channels of the tonoplast, in response to cytosolic changes in anions and Ca^{2+}. The turgor movement of leaf pulvinar motor cells also involves fluxes of water, K^+, Cl^- and Ca^{2+} from the vacuole. Tannin-bound Ca^{2+} and Ca^{2+}-oxalate play a secondary role.

Ca^{2+} Homeostasis

Ca^{2+} is the pivotal ion species which actively controls numerous physiological processes in the cytoplasm. Its concentration in the cytoplasm is maintained within a narrow range by influx or efflux of Ca^{2+} from the vacuole which may maintain the ion at a level of 3 to 4 orders of magnitude of the cytoplasm. The multiplicity of transporters and channels for Ca^{2+} suggests that the most prevalent divalent cation acts as an intracellular second messenger. A number of Ca^{2+}-binding proteins, including the most studied calmodulin, modulate many vacuolar fluxes of Ca^{2+} in response to diverse stimuli.

Metabolic functions

Vacuoles control a large number of metabolic functions of the cell.
- During **Crassulacean acid metabolism**, malic acid produced by dark assimilation of CO_2 is transported by a malate transporter in the tonoplast into the vacuole. During the day, malate in the vacuole is exported either by passive diffusion or mediated by a non-selective ion channel.
- In the **homeostasis of amino acids**, vacuoles play a significant role in yeast and other fungi where they maintain a high concentration of certain amino acids. In higher plants, specific transport systems for uptake of various amino acids by the vacuole have been demonstrated. However, the amino acids are maintained at an extremely low critical concentration and any excess is efficiently exported from the vacuole.
- Depending on the metabolic state, **sucrose accumulation** by the vacuoles of different cell types may take place by diffusion mediated by carrier protein or by sucrose/H^+ antiporter. In the storage cells of sugar cane and red beet, a multienzyme complex in the tonoplast may govern the formation and translocation of sucrose into vacuoles.
- **Biosynthesis inside vacuoles** may be conducted by certain enzymatic processes in the lumen. One simple example is that of conversion of sucrose to glucose and fructase by invertase in the vacuoles of ripening fruit cells. Again, in the vacuoles of Jerusalem artichoke, sucrose is converted to inulin by the action of a series of enzymes. A number of enzymes for uptake of anthocyanin and subsequent processing have been detected in the tonoplast and lumen. Anthocyanins may be trapped by conformational change or conjugated with protein for retention inside the vacuole. Similarly, ethylene is formed from its precursor.
- The best example of **vacuolar processing** is noted in the developing protein bodies of seeds. In the storage bodies, the vacuolar processing enzyme, cysteine protease, converts trimeric proglobulin to hexameric globulin, proglycinin to glycinin, or proricin to ricin in different plant cotyledons. Concanavaline A is first synthesised as Pro-Con A which undergoes combined action of endopeptidase and carboxypeptidase. Other vacuolar enzymes like proteinases undertake processing activity in different cell systems including yeast.
- Perhaps **biosynthesis of alkaloids** in the cells of the medicinal plant *Catharanthus roseus* is the best illustration of vacuolar compartmentalisation of specific metabolic activity. For example,

the precursor substances tryptamine and geraniol, produced in the cytosol, are transported to the vacuole and conjugated there to form strictosidine. This unstable gluco-alkaloid is readily passed out of the vacuole for removal of glucose moiety and subsequently converted to ajmalicine or cantharidine which may again enter the vacuole for peroxidation/dimerisation to form stable akalolids for storage.

Lytic functions
- The **lytic functions** of the vacuole are evident from the presence of both hydrolytic and proteolytic enzymes in its lumen. That the enzymes are present and are released by the alteration of the permeability or disruption of the tonoplast has been amply demonstrated. During intracellular turnover, programmed or induced differentiation, or death, vacuolar endopeptidases are effective in degradation of the cytoplasmic proteins and enzymes, but not the vacuolar hydrolases. Extraneous proteins, if introduced inside a vacuole, can be readily degraded. Mutant or defective proteins are also known to be taken up by the vacuole and digested. The presence of RNA-degrading enzymes has been reported in certain vacuoles.
- Another aspect of lytic functions is **autophagy**, which involves degradation of intracellular structures and substances after their ingestion by the vacuole. Besides organelle breakdown inside the vacuole, autophagy may be initiated by RER or some endomembrane enclosing and isolating a certain mass of the cytoplasm. The membrane-bound sac may produce lytic enzymes and gradually degrade its contents to form an autophagic vacuole. In other cases, cytoplasmic fragments and organelles invaginate the tonoplast. The intruding mass of the cytosol together with the organelles is pinched off from the tonoplast and finally digested. Thus, autophagy is a routine process in many differentiating and developmental systems. An example is provided by wheat storage proteins which are synthesised by ER to form protein bodies. These enter the vacuole by typical autophagy. On the other hand, perturbation in nutrient requirement may induce autophagy. In yeast, starvation may induce formation of autophagosome which fuses later with the vacuole. If the methanol-utilising yeast *Pichia* cells are shifted to ethanol or glucose as the carbon source, the constitutive peroxisomes become redundant and they are soon sequestered into autophagosomes for subsequent fusion with the vacuole.
- **Heterophagy**, which denotes ingestion of foreign non-nutrient substances, is not a normal process in plant cells. Heterophagy of dissolved substances may take place by various mechanisms. In certain cases, internalisation may involve fluid-phase endocytosis. Typical animal-like vesicle-mediated endocytosis is undertaken for ferritin-like substances.
- In plants, **host-pathogen** interaction does not involve typical animal-like phagocytosis. However, the well-studied *Rhizobium* infection of leguminous root cells exhibits some semblance of phagocytosis. The infection is initiated with the release of the bacteria close to the vacuole and the bacteria are protected against lytic enzymes by a peribacteroid membrane. These bacteria-housing vesicles may be considered as heterophagosome. The remnants of the infection thread can be noted in vacuoles and are readily digested.

  Certain substances produced by pathogens called **elicitors** may induce production of host-encoded proteins called **pathogenesis-related** (**PR**) proteins. Elicitors may induce transcription of genes for fungal wall-degrading enzymes like chitinase and glucanase which are stored in vacuoles. When the host cell is lysed, the vacuolar enzymes are released to destroy the pathogen.

  Hypersensitive reaction (**HR**) is indicated by necrosis of cells surrounding the point of infection. HR is caused by release of the vacuolar degrading enzymes which may be preceded by formation of active oxygen species (oxidative burst) in the elicited host cell. Another defence arsenal is a group of anti-microbial compounds called **phytoalexins** which are constitutively produced or elicited by pathogens. Phytoalexins are generally stored in vacuoles and are active when needed.

- As the first step towards **cellular differentiation**, autophagy in the form of ingestion of cytoplasmic organelles inside the vacuole or rupture of the vacuole is necessary.

    Cytochemical and biochemical studies or differentiation of sieve tubes and tracheary elements from the precursor cells have indicated that differentiation is invariably associated with the production and release of vacuolar enzymes. Extensive or partial autophagy in a series of cells is necessary for laticifer development in many plants.

    Autophagy has also been extensively recorded in the development of cotyledonary leaves and is a well-known process during embryo sac development. During microgametogenesis autophagy is evident in the development of pollen tubes from pollen grains as well as in the dissolution of the cytoplasm proximal to the sperm nuclei.

    Senescence of cells and tissues is always associated with uptake of cytoplasmic organelles or their breakdown substance by the vacuole. Ultimately the tonoplast is ruptured and the vacuolar enzymes dissolve most of the cell structure.

Detoxification of SO_2

Generally, the excess amount of sulfate which is absorbed by the roots is sequestered in the vacuole. When SO_2 is in excess in the atmosphere the sulfate-pumping activity of the tonoplast of the leaf mesophyll cells is generally sufficient to cope with excess sulfate formed in the cytoplasm.

Detoxification of xenobiotics

Xenobiotic or harmful compounds in the environment are detoxified by plant cells in four phases: i) activation, by which the xenobiotic is oxidised, reduced or hydrolysed; ii) conjugation by which the activated substance is conjugated with a hydrophilic substance; iii) elimination, in which the conjugate is secreted or sequestered in the vacuole; and iv) termination, in which the conjugate is converted to inert derivatives. A large number of xenobiotics are conjugated with glutathione in the cytosol and subsequently the glutathione-conjugate is transported into the vacuole by glutathione-conjugate pump, a member of the family of ATP-binding cassette transporters.

Phytoremediation of heavy metals

Phytoremediation involves uptake of heavy metals by plants from the environment. If absorbed in excess, Zn is generally conjugated with organic acids like oxalic acid or malic acid in the cytoplasm and later compartmentalised as Zn-deposits in the vacuole. However, Cd, in addition to conjugation with organic acids, is chiefly conjugated with a class of cysteine-rich metal-binding proteins, called phytochelatin and transported to the vacuole. Cd-glutathione conjugates may also be sequestered in the vacuole. Cr is usually compartmentalised in its trivalent (Cr (III)) form in the vacuole as a Cr-organic acid complex or even as a Cr-phytochelatin complex. In cases of excess Al, the metal is detected as Al-phytate in the vacuole.

Scavenging of active oxygen species

Scavenging of active oxygen species (AOS) is the latest in a long list of beneficial roles played by vacuoles in the life of a cell or a plant. Various stress conditions produce AOS, viz. superoxide, hydrogen peroxide, hydroxyl radical, singlet oxygen and peroxyl radical. Glutathione (GS) is the most important molecule which scavenges AOS in the cytoplasm and forms oxidised glutathione (GSSG) which is readily transported to the vacuole. When H_2O_2, a stable AOS, diffuses across the tonoplast, the flavonoids in the vacuole may act as a H_2O_2-scavenging system by flavonoid-peroxidase reaction.

7.15 REFERENCES

Abel, S. and Glund, K., Localization of RNA-degrading enzyme activity within vacuoles of cultured tomato cells, *Physiol. Plant.*, **66**, 79, 1986.

Abel, S., Blume, B. and Glund, K., Evidence for RNA-oligonucleotides in plant vacuoles isolated from cultured tomato cells, *Plant Physiol.*, **94**, 1163, 1990.

Abel, S., Krauss, G.J. and Glund, K., Ribonuclease in tomato vacuoles: high performance liquid chromatagraphic analysis of ribonucleolytic activities and base specificity, *Biochem. Biophys. Act.*, **998**, 145, 1989.

Agre, P., Sasaki, S. and Chrispeels, M.J., Aquaporins - a family of water channel proteins, *Am.J. Physiol.*, **265**, F461, 1993.

Akazawa, T. and Hara-Nishimura, I., Topographic aspects of biosynthesis, extracellular secretion, and intracellular storage of proteins in plant cells, *Ann. Rev. Plant Physiol*, **36**, 441, 1985.

Albersheim, P. and Valent, B.S., Host-pathogen interactions. VII. Plant pathogens secrete proteins which inhibit enzymes of the host capable of attacking the pathogen, *Plant Physiol.*, **53**, 684, 1974.

Albersheim, P., Jones, T.M. and English, P.D., Biochemistry of the cell wall in relation to infective processes, *Ann. Rev. Phytopathol.*, **7**, 171, 1969.

Alexandre, J., Lassalles, J.P., and Kado, R.T., Opening of Ca^{2+} channels in isolated red beet root vacuole membrane by inositol-1,4,5 triphosphate, *Nature*, **343**, 567, 1990.

Alfenito, M.R., Souer, E., Goodman, C.D., Buell, R., Mol, J., Koes, R., and Walbot, V., Functional complementation of anthocyanin sequestration in the vacuole by widely divergent glutathione S-transferases, *Plant Cell*, **10**, 1135, 1998.

Alibert, G., Boudet, A.M. and Rataboul, P., Transport of O-coumaric acid glucoside in isolated vacuoles of sweet clover, in Plasmalemma and Tonoplast: Their functions in the plant cell, Marme, D., Marre, E. and Hartel, R., Eds, Elsevier Biomedical Press, Amsterdam, 1982, 193.

Allen, G.J. and Sanders, D., Two voltage-gated cadmium release channels coreside in the vacuolar membrane of broad bean guard cells, *Plant Cell*, **6**. 685, 1994.

Allen, G.J., Muir, S.R. and Sanders, D., Release of Ca^{2+} from individual plant vacuoles by both InsP3 and cyclic ADP-ribose, *Science*, **268**, 735, 1995.

Amodeo, G., Escobar, A. and Zeiger, E., A cationic channel in the guard cell tonoplast of *Allium cepa*, *Plant Physiol.*, **105**, 999, 1994.

Anhalt, S. and Weissenböck, G, Subcellular localization of luteolin glucuronides and related enzymes in rye mesophyll, *Planta*, **187**, 83, 1992.

Aoki, K. and Nishida, K., ATPase activity associated with vacuoles and tonoplast vesicles isolated from the CAM plant, *Kalanchoe daigremontiana*. *Physiol. Plant*, **60**, 21, 1984.

Apostol, I., Heinstein, P. and Low, P., Rapid stimulation of an oxidative burst during elicitation of cultured plant cells: role in defense and signal transduction, *Plant Physiol.*, **90**, 109, 1989.

Askerlund, P., Calmodulin-stimulated Ca^{2+}-ATPases in the vacuolar and plasma membranes in cauliflower, *Plant Physiol.*, **114**, 999, 1997.

Baba, M., Ohsumi, M., Scott, S.V., Klionsky, D.J. and Ohsumi, Y. Two distinct pathways for targeting proteins from cytoplasm to vacuole/lysosome, *J. Cell. Biol.*, **139**, 1687, 1997.

Baba, M., Takashige, K., Baba, N. and Ohsumi, Y., Ultrastructural analysis of the autophagic process in yeast: Detection of autophagosomes and their characterization, *J. Cell Biol.*, **124**, 903, 1994.

Bacon, J.S.D., MacDonald, I.R., and Knight, A.H., The development of invertase activity in slices of root of *Beta vulgaris* L. washed under aseptic conditions, *Biochem. J.*, **94**, 175, 1965.

Bal, A.K., Vacuolation and infection thread in root nodules of soybean, *Cytobios*, **42**, 41, 1985.

Balsamo, R.A., and Uribe, E.G., Plasmalemma and tonoplast-ATPase activity in mesophyll protoplasts, vacuoles and microsomes of the Crassulacian-acid-metabolism plant *Kalanchoe daigremontiana*, *Planta*, **173**, 190, 1988.

Barceló, J., and Poschenrieder, Ch., Chromium in plants, in 'Chromium Environmental Issues', Canali, S., Tittarelli, F., and Sequi, P., Eds, Franco Angeli, Milan, 1997, 102.

Barceló, J., Poschenrieder, Ch., Vázquez, M.D. and Gunse, B., Aluminium phytotoxicity, *Fertilizer Res.*, **43**, 217, 1996.

Barrett, M., Metabolism of herbicides by cytochrome P450 in corn, *Drug Metab. Durg Interact.*, **12**, 299, 1995.

Bassi, M., Corradi, M.G., and Realini, M., Effects of chromium (VI) on two fresh water plants, *Lemna minor* and *Pistia stratiotes*. I. Morphological observations, *Cytobios*, **62**, 27, 1990.

Beers, E.P., and Freeman, T.B., Protease activity during tracheary element differentiation in *Zinnia* mesophyll cultures, *Plant Physiol.*, **113**, 873, 1997.

Berhane, K., Wiedersten, M., Engstron, A., Kozarich, J.W. and Mannervik, B., Detoxification of base propenals and other Alpha unsaturated aldehyde products of radical reactions and lipid peroxidation by human glutathione S-transferases, *Proc. Natl. Acad. Sci. USA*, **91**, 1480, 1994.

Bethke, P.C. and Jones, R.L., Ca^{2+}-calmodulin modulates ion channel activity in storage protein vacuoles of barley aleurone cells, *Plant Cell*, **6**, 277, 1994.

Bieleski, R.L., Phosphate pools, phosphate transport and phosphate availability, *Annu. Rev. Plant Physiol.*, **24**, 225, 1973.

Binzel, M.L., Hess, F.D., Bressan, R.A., and Hasegawa, P.M., Intracellular compartmentation of ions in salt adapted tobacco cells, *Plant Physiol.*, **86**, 607, 1988.

Bisson, M.A. and Kirst, G.O., *Lamprothamnium*, a euryhaline charophyte. I. Osmotic relations and membrane potential at steady state, *J. Exp. Bot.*, **31**, 1223, 1980.

Blackshear, P.J., Naim, A.C. and Kuo, J.F., Protein kinases 1988: a current perspective, *FASEB J.* **2**, 2957, 1988.

Blom, T.J.M., Sierra, M., Van Vilet, T.B., Franke-van Dijk, M.E.I., deKoning, P., van Iren, F., Verpoorte, R. and Libbenga, K.R., Uptake and accumulation of ajmalicine into isolated vacuoles of cultured cells of *Catharanthus roseus* (L.) G. Don. and its conversion into serpentine, *Planta*, **183**, 170, 1991.

Blumwald, E., and Poole, R.J., Na^+/H^+ antiport in isolated tonoplast vesicles from storage tissue of *Beta vulgaris*, *Plant Physiol.*, **78**, 163, 1985a.

Blumwald, E., and Poole, R.J., Nitrate storage and retrieval in *Beta vulgaris*: effect of nitrate and chloride on proton gradients in tonoplast vesicles, *Proc. Nat. Acad. Sci. US*, **82**, 3683, 1985b.

Blumwald, E., Cragoe, E.J., and Poole, R.J., Inhibition of Na^+/H^+ antiport activity in sugar beet tonoplasts by analogs of amiloride, *Plant Physiol.*, **85**, 30, 1987.

Boller, T., Die Arginin-Permease der Hefe vacuole, Ph.D.Thesis No. 5928, Swiss Fed. Inst. Technol., Zürich, 1977.

Boller, T., Induction of hydrolases as a defense reaction against pathogens, in 'Cellular and Molecular Biology of Plant Stress', Key, J.L. and Kosuge, T., Eds, Alan R. Liss, New York, 1985, 247.

Boller, T. and Alibert, G., Photosynthesis in protoplast from *Melilotus alba*: distribution of products beween vacuole and cytosol, *Z. Pflanzenphysiol.*, **110**, 231, 1983.

Boller, T., and Wiemken, A., Dynamics of vacuolar compartmentation, *Ann. Rev. Plant Physiol.*, **37**, 137, 1986.

Booth, J.W. and Guidotti, G., Phosphate transport in yeast vacuoles, *J.Biol. Chem.*, **272**, 20408, 1997.

Bowes, B.G., Electron microscopic observations on myelin like bodies and related membranous elements in *Glechoma hederacea*, L., *Z. Pflphysiol.*, **60**, 414, 1969.

Bowles, D.J., Defense related proteins in higher plants, *Annu. Rev. Biochem.*, **58**, 873, 1990.

Bowman, E.J. and Bowman, B.J., The H⁺-translocating ATPase in vacuolar membranes of *Neurospora crassa*, in 'Biochemistry and Function of Vacuolar Adenosin-Triphosphatase in Fungi and Plants', Marin, B.P. Ed., Springer Verlag, Berlin, 1985, 131.

Boyer, I.S., Water Transport, *Ann. Rev. Plant Physiol.*, **36**, 473, 1985.

Bramm, J., Regulation of expression of calmodulin and calmodulin-related genes by environmental stimuli in plants, *Cell Calcium* **13**, 457, 1992.

Brauer, D., Conner, D. and Tu, S., Effects of pH on proton transport by vacuolar pumps from maize roots, *Physiol plant.*, **86**, 63, 1992.

Brauer, D., Otto, J. and Tu, S-I, Nucleotide binding is insufficient to induce cold inactivation of the vacuolar type-ATPase from maize roots, *Plant Physiol. Biochem.*, **33**, 555, 1995.

Brauer, D., Unkalis, J., Triana, R., Sachar-Hill, Y. and Tu, S-I., Effects of bafilomycin A1 and metabolic inhibitors on the maintenance of vacuolar acidity in maize root hair cells, *Plant Physiol.*, **113**, 809, 1997.

Brauer, M., Sanders, D. and Still, M., Regulation of photosynthetic sucrose synthesis: a role for calcium, *Planta*, **182**, 236, 1990.

Braun, Y., Hassidim, M., Lerner, H.R., and Reinhold, L., Evidence for a Na⁺/H⁺ antiporter in membrane vesicles isolated from roots of the halophyte *Atriplex nummularia*, *Plant Physiol*, **87**, 104, 1988.

Bredrode, F.F., Linthorst, H.J.M. and Bol, J.F., Differential induction of acquired resistance and PR gene expression in tobacco by virus infection, ethephon treatment, UV light and wounding, *Plant Mol. Biol.*, **17**, 1117, 1991.

Bremberger, C. and Lüttge, U., Dynamics of tonoplast proton pumps and other tonoplast proteins of *Mesembryanthemum crystallinum* L. during the induction of Crassulacean acid metabolism, *Planta*, **188**, 575, 1992.

Bremberger, C., Haschke, H., and Lüttge, U., Separation and purification of the tonoplast ATPase and pyrophosphatase from plants with constitutive and inducible crassulacean acid metabolism, *Planta*, **175**, 465, 1988.

Brisken, D.P., Thornby, W.R., and Wyse, R.E., Membrane transport in isolated vesicles from sugar beet taproot: Evidence for sucrose/H⁺ antiport, *Plant Physiol*, **78**, 871, 1985.

Brown, S.G. and Coombe, B.G., Proposal for hexose group transport at the tonoplast of grape pericarp cell, *Physiol. Veg.*, **22**, 231, 1984.

Bryant, N.J. and Stevens, T.H., Vacuole biogenesis in *Saccharomyces cerevisiae*: Protein transport pathways to the yeast vacuole, *Microbiobiol. Mol. Biol. Rev.*, **62**, 230, 1998.

Burgess, J. and Lawrence, W., Studies of the recovery of tobacco mesophyll protoplasts from an evacuolation treatment, *Protoplasma*, **126**, 140, 1985.

Burgos, P.A. and Donaire, J.P., H⁺-ATPase activities of tonoplast-enriched vesicles from non-treated and NaCl-treated jojoba roots, *Plant Sci.*, **118**, 167, 1996.

Buser-Suter, C., Wiemken, A., and Matile, P., A malic acid permease in isolated vacuoles of a crassulacean acid metabolism plant, *Plant Physiol.*, **69**, 456, 1982.

Bush, D.S., Regulation of cytosolic calcium in plants, *Plant Physiol.*, **103**, 7. 1993.

Buvat, R., Origin and continuity of cell vacuoles, in 'Origin and Continuity of Cell Organelles', Reinert, J. and Ursprung, H., Eds, Springer Verlag, Berlin, 1971, 127.

Caldwell, J.H., Brunt, J.V. and Harold, F.M., Calcium-dependent anion channel in the water mold *Blastocladiella emersonii*, *J. Membr. Biol.*, **86**, 85, 1986.

Campbell, N.A., and Thomson, W.W., Effects of lanthanum and ethylenediamine tetraacetate on leaf movements of *Mimosa*, *Plant Physiol.*, **60**, 635, 1977.

Canut, H., Alibert, G. and Boudet, M., Hydrolysis of intracellular proteins in vacuoles isolated from *Acer pseudoplatanus* L. cells, *Plant Physiol.*, **79**, 1090, 1985.

Canut, H., Alibert, G., Carrasco, A. and Boudet, A.M., Rapid degradation of abnormal proteins in vacuoles from *Acer pseudoplatanus* L. cells. *Plant Physiol.*, **81**, 460, 1986.

Canut, H., Carrasco, A., Rossignol, M., and Ranjeva, R., Is vacuole the richest store of IP_3-mobilizable calcium in plant cells? *Plant Sci.*, **90**, 135, 1993.

Canut, H., Dupre, M., Carrasco, A. and Boudet, A.M., Proteases of *Melilotus alba* mesophyll protoplasts. II General properties and effectiveness in degradation of cytosolic and vacuolar enzymes, *Planta*, **170**, 541, 1987.

Carpita, N., Sabularse, D., Montezinos, D. and Delmer, D.P., Determination of the pore size of cell walls of living plant cells, *Science*, **205**, 1144, 1979.

Cerana, R. Giromini, L. and Colombo, R., Malate-regulated channels permeable to anions in vacuoles of *Arabidopsis thaliana, Aust. J. Plant Physiol.*, **22**, 115, 1995.

Chang, A. and Fink, G.R., Targeting of the yeast plasma membrane [H^+] ATPase: A novel gene ASTI prevents mislocalisation of mutant ATPase to the vacuole, *J. Cell Biol.*, **128**, 39, 1995.

Chang, K. and Roberts, J.K.M., Observation of cytoplasmic and vacuolar malate in maize root tips by ^{13}C NMR spectroscopy, *Plant Physiol.*, **89**, 197, 1989.

Chanson, A., Fichmann, J., Spear, D., and Taiz, L., Pyrophosphate-driven proton transport by microsomal membranes of corn coleoptiles, *Plant Physiol.*, **79**, 159, 1985.

Chiou, T-J. and Bush, D.R., Molecular cloning, immunochemical localisation to the vacuole, and expression in transgenic yeast and tobacco of a putative sugar transporter from sugar beet, *Plant Physiol.* **110**, 511, 1996.

Chodera, A.J. and Briskin, D.P., Chlorate transport in isolated tonoplast vesicles from red beet (*Beta vulgaris* L.) storage tissue, *Plant Sci.*, **67**, 151, 1990.

Chrispeels, M.J. and Maurel, C., Aquaproins: The molecular basis of facilitated water movement through living plant cells, *Plant Physiol.*, **105**, 9, 1994.

Chvatchko, Y., Howald, I. and Riezman, H., Two yeast mutants defective in endocytosis are defective in pheromone response, *Cell*, **46**, 355, 1986.

Cole, L., Coleman, J., Kearns, A., Morgan, G. and Hawes, C., The organic anion transport inhibitor, probenecid, inhibits the transport of Lucifer Yellow at the plasma-membrane and the tonoplast in suspension-cultured plant cells, *J. Cell Sci.* **99**, 545, 1991.

Coleman, J.O.D., Randall, R., and Blake-Klaff, M.A.A., Detoxification of xenobiotics in plant cells by glutathione conjugation and vacuolar compartmentalization: a fluorescent assay using monochlorobimane, *Plant Cell Environ*, **20**, 449, 1997.

Cosgrove, D. J. and Hedrich, R., Stretch-activated chloride, potassium and calcium channels coexisting in plasma membranes of guard cells of *Vicia faba* L., *Planta*, **186**, 143, 1991.

Coulomb, C., and Buvat, R., Processus de degenerescence cytoplasmique partielle dans les cellules de jeunes racines de *Cucurbita pepo, Compt. Rend. Acad. Sci.*, Paris, **267**, 843, 1968.

Cram, W.J., Negative feedback regulation of transport in cells: the maintenance of turgor, volume and nutrient supply, in 'Encyclopedia of Plant Physiology', New Ser. 2A, Lüttge, U. and Pitman, M.G., Eds, Springer Verlag, Berlin, 1976, 284.

Cram, W.J., Characteristics of sulfate transport accross plasmalemma and tonoplast of carrot root cells, *Plant Physiol.*, **72**, 204, 1983a.

Cram, W.J., Sulfate accumulation is regulated at the tonoplast, *Plant Sci. Lett.*, **31**, 329, 1983b.

Cram, W.J., and Laties, G.G., The use of short term and quasi-steady influx in estimating plasmalemma tonoplast influx in barley root cells at various external and internal salt concentrations, *Aust. J. Biol. Sci.*, **24**, 633, 1971.

Cramer, C.L., Vaugh, L.E. and Davis, R.H., Basic amino acids and inorganic polyphosphates in *Neurospora crassa*: independent regulation of vacuolar pools, *J. Bactriol.*, **142**, 945, 1980.

Cunningham, S.D., and Berti, W.R., Remediation of contaminated soils with green plants: an overview. *In Vitro Cell. Dev. Biol.* **29**, 207, 1993.

d'Auzac, J., ATPase membranaire de vacuoles lysomales: les lutoids du latex d'*Hevea brasiliensis*, *Phytochemistry*, **16**, 1881, 1977.

d'Auzac, J., Crestin, H., Marin, B., and Lioret, C., A plant vacuolar system: the lutoids from *Hevea brasiliensis* latex. *Physiol. Veg.*, **20**, 311, 1982.

d'Auzac, J., Chrestin, H., and Marin, B., Biochemical and enzymatic components of a vacuolar membrane: Tonoplasts of lutoids from *Hevea latex*, *Methods Enzymol.*, **148**, 87, 1987.

Daman, S., Hewitt, J. Nieder, M. and Bennet, A.B., Sink metabolism in tomato fruit. II. Phloem unloading and sugar uptake, *Plant Physiol.*, **87**, 731, 1988.

Daniels, M.J., Mirkov, T.E., and Chrispeels, M.J., The plasma membrane of *Arabidopsis thaliana* contains a mercury-insensitive aquaporin that is a homolog of the tonoplast water channel protein TIP, *Plant Physiol.*, **106**, 1325, 1994.

Davies, C. and Robinson, S.P., Sugar accumulation in grape berries: Cloning of two putative vacuolar invertase cDNA and their expression in grapevine tissues, *Plant Physiol.*, **111**, 275, 1996.

Davies, J.M., Poole, R.J., Rea, P.A. and Sanders, D., Potassium transport into vacuoles energized directly by a proton-pumping inorganic pyrophosphatase, *Proc. Natl. Acad. Sci. USA*, **89**, 11701, 1992.

Davies, K.L., Davies, M.S. and Francis, D., Zinc-induced vacuolation in root meristematic cells of *Festuca rubra*, *Plant Cell Environ*, **14**, 399, 1991.

Davies, K.L., Davies, M.S. and Francis, D., Zinc-induced vacuolation in root meristematic cells of cereals, *Ann. Bot.*, **69**, 21, 1992.

De Boer, A.H., and Wegner, L.H., Regulatory mechanisms of ion channels in xylem parenchyma cells, *J. Exp. Bot.*, **48**, 411, 1997.

De Leon, J.L.D., Daie, J. and Wyse, R., Tonoplast stability and survival of isolated vacuoles in different buffers, *Plant Physiol.*, **88**, 251, 1988.

Delrot, S., Thom, M. and Maretzki, A., Evidence for a uridine-5'-di-phosphate-glucose protected p-chloromercuribenzene sulfonic acid-binding site in sugarcane vacuoles, *Planta*, **169**, 64, 1986.

Deus-Neuman, B. and Zenk, M.H., Accumulation of alkaloids in plant vacuoles does not involve an ion trap mechanism, *Planta*, **167**, 44, 1986.

Dietz, K-J., Jäger, R., Kaiser, G. and Martinoia, Amino acid transport across the tonoplast of vacuoles isolated from barley mesophyll protoplasts, *Plant Physiol.*, **92**, 123, 1990.

Douglas, T.J., NaCl effects on 4-desmethylsterol composition of plasma membrane-enriched preparations from citrus roots. *Plant Cell Environ.*, **8**, 687, 1985.

Drews, G., Ziser, K., Shrock-Vietor, U. and Golecki, J.R., Cellular responses of soybean to virulent and avirulent strains of *Pseudomonas syringae* pv glycinea, *Eur. J. Cell Biol.*, **46**, 369, 1988.

Düggelin, T., Schellenberg, M., Borttik, K. and Matile, P., Vacuolar location of lipofuscin- and proline-like compounds in senescent barley leaves, *J. Plant Physiol.*, **133**, 492, 1988.

Dürr, M., Urech, K., Boller, T., Wiemken, A., Schwenke, J. and Nagy, M., Sequestration of arginine by polyphosphate in vacuoles of yeast (*Saccharomyces cerevisiae*), *Arch. Microbiol.*, **121**, 169, 1979.

Dutta, P.C., Appleqvist, L.A., Gunnarson, S. and von Hofsten, A., Lipid bodies in tissue culture, somatic and zygotic embryo of *Daucus carota* L., *Plant Sci.*, **78**, 259, 1991.

Ebal, J. and Grisebach, H., Defense strategies of soybean against the fungus *Phytophthora megasperma* f. sp.glycinea: a molecular analysis. *Trend. Biochem. Sci.*, **13**, 23, 1988.

Echeverria, E. and Salvucci, M.S., Sucrose phosphate is not transported into vacuoles or tonoplast vesicles from red beet (*Beta vulgaris*) hypocotyl, *Plant Physiol.* **96**, 1014, 1991.

Edelman, J. and Jefford, T.G., The mechanism of fructan metabolism in higher plants as exemplified in *Helianthus tuberosus*, *New Phytol.*, **67**, 517, 1968.

Epstein, E., and Hagen, C.E., A kinetic study of the absorption of alkali cations by barley roots, *Plant Physiol.*, **27**, 457, 1952.

Ersek, T. and Kiraly, Z., Phytoalexins: warding off compounds in plants?, *Physiol. Planta*, **68**, 343, 1986.

Esau, K. and Charvat, I.D., An ultrastructural study of acid phosphatase localization in cells of *Phaseolus vulgaris* phloem by the use of azo dye method, *Tissue Cell*, **4**, 619, 1975.

Esau, K. and Cronshaw, J., Relation of tobacco mosaic virus to the host cells, *J. Cell Biol.*, **33**, 665, 1967.

Esau, K. and Kosaki, H., Laticifers in *Nelumbo nucifera* Gaertn.: distribution and structure. *Ann. Bot.* **39**, 713, 1975.

Evans, D.E., Regulation of cytoplasmic free calcium by plant cell membranes, *Cell Biol. Int. Report*, **12**, 383, 1988.

Faye, L. and Chrispeels, M.J., Transport and processing of the glycosylated precursor of Concanavalin A in jackbean, *Planta*, **170**, 217, 1987.

Faye, L., Greenwood, J.S., Herman, E.M., Stürm, A. and Chrispeels, M.J., Transport and posttransitional processing of the vacuolar enzyme alpha-mannosidase in jackbean cotyledons, *Planta*, **174**, 271, 1988.

Felle, H., Auxin oscillations of cytosolic free calcium and pH in *Zea mays* coleoptiles, *Planta*, **174**, 495, 1988.

Fineran, B.A., Organization of the tonoplast in frozen-etched root tips, *J. Ultrastruct. Res.*, **33**, 574, 1970.

Fineran, B.A., Ultrastructure of vacuolar inclusions in root tips, *Protoplasma*, **72**, 1, 1971.

Fisher, D.B., Hansen, D., and Hodges, T.K., Correlation between ion fluxes and ion-stimulated adenosine triphosphatase activity of plant roots, *Plant Physiol.*, **46**, 812, 1970.

Fleurat-Lessard, P. Ultrastructural features of the starch sheath cells of the primary pulvinus after gravistimulation of the sensitive plant (*Mimosa pudica* L.), *Protoplasma*, **105**, 177, 1981.

Fleurat-Lessard, P., and Millet, B.J., Ultrastructural features of cortical parenchyma cells, motor cells in stamen filaments of *Berberis canadensis* and tertiary pulvini of *Mimosa pudica*, *J. Exp. Bot.*, **35**, 1332, 1984.

Fleurat-Lessard, P. Frangne, N., Maeshima, M., Ratajczak, R., Bonnemain, J.L., and Martinoia, E., Increased expression of vacuolar aquaporin and H^+-ATPase related to motor all function in *Mimosa pudica* L., *Plant Physiol*. **114**, 827, 1997.

Flowers, T. J., Halophytes, in 'Ion Transport in Plant Cells and Tissues', Baker, D.A. and Hall, J.L. Eds, North-Holland, Amsterdam, 1975, Chap. 10.

Flowers, T.J., Chloride as a nutrient and as an osmoticum, Advances in Plant Nutrition, Tinker, P.B., Läuchli, A., Eds, Praeger, New York, 1988, Vol 3, 55.

Fowke, L.C. Investigations of cell structure using cultured plant cells and protoplasts, *Int. Congr. Plant Tissue Cell Cult.* **6** Mett, 19 (Abstract), 1986.

Fox, G.G. and Ratcliffe, R.G., ^{31}P-NMR observations on the effect of the external pH on the intracellular pH values in plant cell suspension cultures, *Plant Physiol.*, **93**, 512, 1990.

Frehner, M., Keller, F. and Wiemken, A., Localization of fructan metabolism in the vacuoles isolated from protoplasts of Jerusalem artichoke tubers (*Helianthus tuberosus* L.), *J. Plant Physiol.*, **116**, 197, 1984.

Fried, M., and Noggle, J.C., Multiple site uptake of individual ions by roots as affected by hydrogen ion, *Plant Physiol.*, **33**, 139, 1958.

Friemert, V., Heiniger, D., Kluge, M., and Ziegler, H., Temperature effects on malic acid efflux from the vacuoles and on the carboxylation pathways in Crassulacian-acid-metabolism plants, *Planta*, **174**, 453, 1988.

Fritsch, H. and Griesbach, H., Biosynthesis of cyanidin in cell cultures of *Haplopappus gracilis*, *Phytochemistry*, **14**, 2437, 1975.

Fukuda, H., Xylogenesis : Initiation, progression and cell death, *Ann Rev. Plant Physiol. Plant Mol. Biol.*, **47**, 299, 1996.

Fukuda, H., Tracheary element differentiation, *Plant Cell.*, **9**, 1147, 1997.

Fukuda, H., and Komamine, A., Establishment of an experimental system for the tracheary element differentiation from single cells isolated from the mesophyll of *Zinnia elegans*, *Plant Physiol.*, **65**, 57, 1980.

Fukuda, A., Yazaki, Y., Ishikawa, T., Koike, S. and Tanaka, Y., Na^+/H^+ antiporter in tonoplast vesicles from rice roots, *Plant Cell Physiol.*, **39**, 196, 1998.

Gahan, P.B. and Maple, A.J., The behaviour of lysosome-like particles during cell differentiation, *J. Exp. Botany*, **17**, 151, 1966.

Gahan, P.B. and McLean, J., Subcellular localization and possible functions of acid beta-glycerophosphatases and naphthal esterases in plant cells, *Planta*, **89**, 126, 1969.

Gaillard, C. Dufaud, A., Tommasini, R., Kreuz, K., Amrhein, N. and Martinoia, E., A herbicide antidote (safener) induces the activity of both the herbicide detoxifying enzyme and of a vacuolar transporter for the detoxified herbicide, *FEBS Lett.* **352**, 219, 1994.

Garbarino, J., and DuPont, F.M., NaCl induces a Na^+/H^+ antiport in tonoplast vesicles from barley roots, *Plant Physiol.*, **86**, 231, 1988.

Garbarino, J. and Du Pont, F.M., Rapid induction of Na^+/H^+ exchange activity in barley root tonoplast, *Plant Physiol.*, **89**, 1, 1989.

Gebhart, E., Mutagenität, Karzinogenität, Teratogenität, in 'Metalle in der Unwelt', Merian, E., Ed., Verlag Chemie, Weinheim, 1984, 237.

Gehring, C.A., Williams, D.A., Cody, S.H., and Parish, R.W., Phototropism and geotropism in maize coleoptile are spatially correlated with increases in cytosolic free calcium, *Nature*, **345**, 528, 1990.

Gelli, A. and Blumwald, E., Calcium retrieval from vacuolar pools: characterization of a vacuolar calcium channel, *Plant Physiol.*, **102**, 1139, 1993.

Getz, H.P., Accumulation of sucrose in vacuoles released from isolated beet root protoplasts by both direct sucrose uptake and UDP-glucose-dependent translocation, *Plant Physiol. Biochem.*, **25**, 573, 1987.

Getz, H.P. Sucrose transport in tonoplast vesicles of red beet roots is linked to ATP hydrolysis, *Planta*, **185**, 261, 1991.

Getz, H.P. and Klein, M., Characteristics of sucrose transport on the tonoplast of red beet (*Beta vulgaris* I.) storage tissue, *Plant Physiol.*, **107**, 459, 1995.

Getz, H.P., Grosclaude, J., Kurkdjian, A., Lelievre, F., Maretzki, A. and Guern, J. Immunological evidence for the existence of carrier protein for sucrose transport in tonoplast vesicles from red beet (*Beta vulgaris* L.) root storage tissue, *Plant Physiol.* **102**, 751, 1993.

Ghosh, M., Shen, J. and Rosen, B.P., Pathways of As(III) detoxification in *Saecharomyces cerevisiae*, *Proc. Natl. Acad. Sci. USA*, **96**, 5001, 1999.

Giannini, J.L., Holt, J.S. and Briskin, D.P., The effect of glyceollin on proton leakage in *Phytophthora megasperma* f. sp. glycinea plasma membrane and red beet tonoplast vesicles, *Plant Sci.*, **68**, 39, 1990.

Giannini, J.L., Holt, J.S. and Briskin, D.P., The effect of glyceollin on soybean (*Glycine max* L.) tonoplast and plasma membrane vesicles, *Plant Sci.*, **74**, 203, 1991.

Gifford, E.M., and Stewart, K.D., Inclusions of proplastids and vacuoles in the shoot apices of *Bryophyllum* and *Kalanchoe*, *Am. J. Bot.*, **53**, 269, 1968.

Gilroy, S., Read, N.D. and Trewavas, Elevation of cytoplasmic calcium by caged calcium or caged inositol triphosphate initiates stomatal closure, *Nature*, **346**, 769, 1990.

Godbold, D.L., Horst, W.J., Collins, J.C., Thurman, D.A. and Marschner, H., Accumulation of zinc and organic acids in roots of zinc tolerant and non-tolerant ecotypes of *Deschampsia caespitosa*, *J. Plant Physiol*, **116**, 59, 1984.

Goerlach, J. and Willims-Hoff, D., Glycine uptake into barley mesophyll vacuoles is regulated but not energized by ATP, *Plant Physiol.*, **99**, 134, 1992.

Goodman, R.N. and Plurad, S.B., Ultrastructural changes in tobacco undergoing the hypersensitive reaction caused by plant pathogenic bacteria, *Physiol. Plant Pathol.*, **1**, 11, 1971.

Gordon-Weeks, R., Koren'kov, V.D., Steele, S.H. and Leigh, R.A., Tris is a competitive inhibitor of K^+ activation of the vacuolar H^+-pumping pyrophosphatase, *Plant Physiol.*, **114**, 901, 1997.

Granell, A., Belles, J.M. and Conejero,V., Induction of pathogenesis-related proteins in tomato citrus exocortis viroid, silver ions and ethophon, *Physiol. Mol. Plant Pathol*, **31**, 83, 1987.

Green, P.B., Erickson, R.O., and Buggy, J., Metabolic and physical control of cell elongation rate, *Plant Physiol.*, **47**, 423, 1971.

Greutert, H. and Keller, F., Further evidence for stachyose and sucrose/H^+ antiporters on the tonoplast of Japanese artichoke (*Stachys siebolodi*) tubers, *Plant Physiol.*, **101**, 1317, 1993.

Gries, G.E. and Wagner, G.J., Association of nickel versus transport of cadmium and calcium in tonoplast vesicles of oat roots, *Planta*, **204**, 390, 1998.

Griesbach, R. and Sink, K., Evacuolation of mesophyll protoplasts, *Plant Sci. Lett.*, **30**, 297, 1983.

Groover, A., and Jones, A.M., Tracheary element differentiation uses a novel mechanism coordinating programmed cell death and secondary cell wall synthesis, *Plant Physiol.*, **119**, 375, 1999.

Groover, A., DeWitt, N., Heidel, A., and Jones, A., Programmed cell death of plant tracheary elements differentiating *in vitro*, *Protoplasma*, **196**, 197, 1997.

Grotewald, E., Chamberlin, M., Snook, M., Siame, B., Butter, L., Swenson, J., Moddock, S., St. Clair, G. and Bowen, B., Engineering secondary metabolism in maize cells by ectopic expression of transcription factors, *Plant Cell*, **10**, 721, 1998.

Gruhnert, C., Biehl, B. and Selmer, D., Compartmentation of cyanogenic glucosides and their degrading enzymes, *Planta*, **195**, 36, 1994.

Guern, J., Mathieu, Y., Kurkdjian, A., Manigault, P., Manigault, J., Gillet, B., Beloeil, J.C. Lallemand, J.Y., Regulation of vacuolar pH of plant cells. II. A ^{31}PNMR study of the modifications of vacuolar pH in isolated vacuoles induced by proton pumping and cation/H^+ exchanges, *Plant Physiol*, **89**, 27, 1989.

Gupta, H.S. and De, D.N., Functional analogy of plant vacuoles with animal lysosomes, *Curr. Sci.*, **52**, 680, 1983a.

Gupta, H.S. and De, D.N., Uptake and accumulation of acridine orange by plant cells, *Proc. Ind. Nat. Sci. Acad., India* **B 49**, 553, 1983b.

Gupta, H.S., Acridine orange-induced formation of myelin-like structures in plant cell vacuoles, *Indian J. Genet*, **45**, 133, 1985a.

Gupta, H.S., Plant lysosomes: aspects and prospects, *Curr. Sci.*, **54**, 554, 1985b.

Gupta, H.S. and De, D.N., Mechanism of drug detoxification by plant cells as studied by fluorescence microscopy, *J. Assam. Sci. Soc.*, **28**, 1, 1985c.

Gupta, H.S. and De, D.N., Acridine-orange induced vacuolar uptake of cytoplasmic organelles in plant cells: an ultrastructural study, *J. Plant Physiol.*, **132**, 254, 1988.

Guy, M. and Kende, H., Conversion of 1-aminocyclopropane-1-carboxylic acid to ethylene by isolated vacuoles of *Pisum sativum* L., *Planta*, **160**, 281, 1984.

Hager, A., and Helme, M., Properties of an ATP-fueled, Cl-dependent proton pump localised in membranes of microsomal vesicles from maize coleoptiles, *Z. Naturforsch.*, **36c**, 997, 1981.

Hajibagheri, M.A. and Flowers, T.J., X-ray micro analysis of ion distribution within root cortical cells of the halophyte *Suaeda maritima* (L.) Dum., *Planta*, **177**, 131, 1989.

Hall, M.D., and Cocking, E.C., The response of isolated *Avena coleoptile* protoplasts to indole-3-acetic acid, *Protoplasma*, **79**, 225, 1974.

Hanower, P., Brzozowska, I., and Niamien N'Goran, M., Absorption des acides amines par les lutoides du latex d'*Hevea brasiliensis*, *Physiol. Plant*, **39**, 299, 1977.

Hara-Nishimura, I., Nishimura, M. and Akazawa, T., Biosynthesis and intracellular transport of 11S globulin in developing pumpkin cotyledons, *Plant Physiol.*, **77**, 747, 1985.

Hara-Nishimura, I., Takeuchi, Y. and Nishmura, M., Molecular characterization of a vacuolar processing enzyme related to a putative cysteine proteinase of *Schistosoma mansoni*, *Plant Cell*, **5**, 1651, 1993.

Harder, D.E., and Chong, J., Structure and physiology of haustoria, in 'The Cereal Rusts', Bushnell, W.R., and Roelfs, A.P., Eds, Academic Press, New York, 1984, 431.

Harley, S.M. and Lord, J.M., *In vitro* endoproteolytic cleavage of castorbean lectin precursors, *Plant Sci.*, **41**, 111, 1985.

Haupt, W., Physiology of movement, *Prog. Botany*, **44**, 222, 1982.

Hawker, J.S., Smith, G.M., Phillips, M. and Wiskich, J.T., Sucrose phosphatase associated with vacuole preparations from red beet, sugar beet and immature sugarcane stem, *Plant Physiol.*, **84**, 1281, 1987.

Hedrich, R. and Becker, D., Green circuits: the potential of plant specific ion channels, *Plant Mol. Biol.*, **26**, 1637, 1994.

Hedrich, R. and Neher, E., Cytoplasmic calcium regulates voltage dependent ion channels in plant vacuoles, *Nature*, **329**, 833, 1987.

Hedrich, R., Flügge, V.I., and Fernandez, J.M., Patch-clamp studies of ion transport in isolated plant vacuoles, *FEBS. Letters*, **204**, 228, 1986.

Hedrich, R., Barbier-Brygoo, H., Felle, H., Flügge, U.I., Lüttge, U. Maathuis, F.J.M., Mark, S., Prins, H.B.A., Raschke, K., Schnabl, H., Schroeder, J.I., Struve, I., Taiz, L., and Zeigler, P., General mchanisms for solute transport across the tonoplast of plant vacuoles: a patch-clamp survey of ion channels and proton pumps, *Bot. Acta*, **101**, 7, 1988.

Heineke, D., Wildenberger, K., Sonnewald, U., Willmitzer, L., and Heldt, H.W., Accumulations of hexoses in leaf vacuoles: Studies with transgenic tobacco plants expressing yeast-derived invertase in the cytosol, vacuole or apoplasm, *Planta*, **194**, 29, 1994.

Hensel, W., Movement of pulvinated leaves, *Prog. Botany*, **49**, 171, 1987.

Hepler, P.K., and Wayne, R.O., Calcium and plant development, *Ann. Rev. Plant Physiol.*, **36**, 397, 1985, 439.

Herman, E.M. and Lamb, C.J., Arabinogalactan-rich glucoproteins are localised on the cell surface and in intravacuolar multivesicular bodies, *Plant Physiol.*, **98**, 264, 1992.

Heuillet, E., Moreau, A., Halpern, S., Jeanne, N. and Puiseux-Dao, S., Cadmium binding to a thiol-molecule in vacuoles of *Dunaliella bioculata* contaminated with $CdCl_2$: electron probe microanalysis. *Biol.Cell*, **58**, 79, 1986.

Higinbotham, N., Electropotential of plant cells, *Ann. Rev. Plant Physiol.*, **24**, 25, 1973.

Hillmer, S., Depta, H. and Robinson, D.G., Confirmation of endocytosis in higher plant protoplasts using lectin-gold conjugates. *Eur. J. Cell Biol.* **41**, 142, 1986.

Hinder, B., Schellenberg, M., Rodoni, S., Ginsburg, S., Vogt, E., Martinoia, E., Matile, P. and Hörtensteiner, S., How plants dispose of chlorophyll catabolites: Directly energized uptake of tetrapyrrolic breakdown praducts into isolated vacuoles, *J. Biol. Chem.*, **271**, 27233, 1996.

Hiraiwa, N., Takeuchi, Y., Nishimura, M. and Hara-Nishimura, I., A vacuolar processing enzyme in maturing and germinating seeds: its distribution and associated changes during development, *Plant Cell Phyiol.* **34**, 1197, 1993.

Hiraiwa, N., Kondo, M., Nishimura, M. and Hara-Nishmura, I., An aspartic endopeptidase is involved in the breakdown of propeptides of storage proteins in protein storage vacuoles of plants, *Eur. J. Biochem.*, **246**, 133, 1997.

Hodges, T.K., ATPase, associated with membranes of plant cells, *Encl. Plant Physiol.*, **2A**, 260, 1976.

Hoffman, L.M., Donaldson, D.D., and Herman, E.M., A modified storage protein is synthesised, processed and degraded in the seeds of transgenic plants, *Plant Mol. Biol.*, **11**, 717, 1988.

Höfte, H., Hubbard, L., Reizer, J., Ludevid, D., Herman, E.M. and Chrispeels, M.J., Vegetative and seed-specific isoforms of a putative solute transporter in the tonoplast of *Arabidopsis thaliana*, *Plant Physiol.*, **99**, 561, 1992.

Homeyer, U. and Schultz, G., Transport of phenylalanine into vacuoles isolated from barley mesophyll protoplasts, *Planta*, **176**, 378, 1988.

Hopp, W. and Seitz, H.U., The uptake of acylated anthocyanin into isolated vacuoles from a cell suspension culture of *Daucus carota*, *Planta*, **170**, 74, 1987.

Hopp, W., Hinderer, W., Petersen, M. and Seitz, H.U., Anthocyanin containing vacuoles isolated from the protoplasts of *Daucus carota* cell cultures, in 'The Physiological Properties of Plant Protoplast', Pilet, P.E. Ed., Springer Verlag, Berlin, 1985, 122.

Horn, M.A., Meadows, R.P., Apastol, I., Jones, C.R., Gorenstein, D.G., Heinstein, P.F. and Low, P.S., Effect of elicitation and changes in extracellular pH on the cytoplasmic and vacuolar pH of suspension-cultured soybean cells, *Plant Physiol.*, **98**, 680, 1992.

Hrazdina, G. and Wagner, G.J., Compartmentation of plant phenolic compounds: sites of synthesis and accumulation, in 'Ann. Proc. Phytochem. Soc. Europe', Vol. 25, Sumere, C.P., Lea, P.J., Eds, Oxford Uni Press, Oxford, 1985, 119.

Hüber-Wälchle, V. and Wiemken, A., Differential extraction of soluble pools from the cytosol and the vacuole of yeast (*Candida utilis*) using DEAE-dextran, *Arch. Microbiol.*, **120**, 141, 1979.

Hüve, K., Dittrich, A., Kinderman, G., Slovik, S. and Heber, U., Detoxification of SO_2 in conifers differing in SO_2-tolerance: A comparison of *Picea abies*, *Picea pungens* and *Pinus sylvestris*, *Planta*, **195**, 578, 1995.

Inz, D. and Van Montagu, M., Oxidative stress in plants, *Curr. Opin. Biotechnol.*, **6**, 153, 1995.

Ishikawa, T., The ATP-dependent glutathione S-conjugate export pump, *Trends Biochem. Sci.*, **17**, 463, 1992.

Iwasaki, I., Arata, H. and Nishimura, M., Ionic balance during malic acid accumulation in vacuoles of a CAM plant *Graptopetalum paraguayense*, *Plant Cell Physiol.*, **29**, 643, 1988.

Iwasaki, I., Arata, H., Kijima, H. and Nishimura, M., Two types of channels involved in the malate ion transport across the tonoplast of a Crassulacean acid metabolism plant, *Plant Physiol.*, **98**, 1494, 1992.

Jeschke, W.D., Roots : Cation selectivity and compartmentation, involvement of protons and regulation, in 'Plant Membrane Transport: Current Conceptual Issues', Spanswick, R.M., Lucas, W.J. and Dainty, J., Eds, Elsevier/North Holland, Amsterdam, 1980, 17.

Joachem, P., Rona, J.P., Smith, J.A.C. and Lüttge, U., Anion-sensitive ATPase activity and proton transport in isolated vacuoles of species of the CAM genus Kalanchoe, *Physiol. Plant*, **62**, 410, 1984.

Johannes, E., Brosnan, J.M. and Sanders, D., Parallel pathways for intracellular Ca^{2+} release from the vacuole of higher plants, *Plant J.*, **2**, 97, 1992.

Joyce, D.C., Cramer, G.R., Reid, M.S. and Bennett, A.B., Transport properties of the tomato fruit tonoplast, III. Temperature dependence of calcium transport, *Plant Physiol.*, **88**, 1097, 1988.

Kaestner, K.H., and Sze, H., Potential-dependent anion transport in tonoplast vesicles from oat roots, *Plant Physiol.*, **83**, 483, 1987.

Kaiser, G. and Heber, U., Sucrose transport into vacuoles isolated from barley mesophyll protoplasts, *Planta*, **161**, 562, 1984.

Kaiser, G., Martinoia, E. and Wiemken, A., Rapid appearance of photosynthetic products in the vacuoles isolated from barley mesophyll protoplasts by a new fast method, *Z. Pflanzenphysiol.*, **107**, 103, 1982.

Kaiser, G., Martinoia, E., Schröppel-Meier, G. and Heber, U., Active transport of sulfate into the vacuole of plant cells provides halotolerance and can detoxify SO_2, *J. Plant Physiol.*, **133**, 756, 1989.

Kamiya, N. and Kuroda, K., Artificial modification of the osmotic pressure of the plant cell, *Protoplasma*, **46**, 423, 1956.

Kappus, H., Lipid peroxidation: mechanisms, analysis, enzymology and biological relevance, in 'Oxidative Stree', Ed. Sies, H., Acad. Press. London, 1985, 273.

Kasamo, K. and Nouchi, I., The role of phospholipid in plasma membrane ATPase activity in *Vigna radiata* L. (Mung bean) roots and hypocotyls, *Plant Physiol.*, **83**, 823, 1987.

Kauss, H., Some aspects of calcium-dependent regulation in plant metabolism, *Ann. Rev. Plant Physiol.*, **38**, 47, 1987.

Keller, F., Transport of stachyose and sucrose by vacuoles of Japanese artichoke (*Stachys sieboldi*) tubers, *Plant Physiol.*, **98**, 442, 1992.

Kikuyama, M., and Tazawa, M., Tonoplast action potential in *Nitella* in relation to vacuolar chloride concentration, *J. Membr. Biol.*, **92**, 95, 1987.

King, G.J., Turner, V.A., Hussey, C.E., Wurtele, E.S. and Lee, M., Isolation and characterization of a tomato cDNA clone which codes for a salt-induced protein, *Plant Mol. Biol.*, **10**, 401, 1988.

Kinoshita, T., Nishmura, M. and Hara-Nishimura, I., Homologues of a vacuolar processing enzyme that are expressed in different organs in *Arabidopsis thaliana*, *Plant Mol. Biol.* **29**, 81, 1995.

Kiyosawa, K., and Tazawa, M., Hydraulic conductivity of tonoplast-free *Chara* cells. *J. Membr. Biol.*, **37**, 157, 1977.

Klein, N., Wissenboeck, G., Dufaud, A., Gaillard, C., Kreuz, K., and Martinoia, E., Different energization mechanisms drive the vacuolar uptake of a flavonoid glycoside and a herbicide glucoside, *J. Biol. Chem.*, **271**, 29666, 1996.

Kliewer, W.M. Changes in the concentration of glucose, fructose and total soluble solids in flowers and berries of *Vitis vinifera*. *Am. J. End. Vitic.* **16**, 101, 1965.

Klionsky, D.J., Nonclassical protein sorting to the yeast vacuole, *J. Biol. Chem.*, **273**, 10807, 1998.

Knight, M.R., Campbell, A.K., Smith, S.M. and Trewavas, A.J., Transgenic plant aequorin reports the effects of touch and cold shock and elicitors on cytoplasmic calcium, *Nature*, **353**, 524, 1991.

Kochian, L.V., Cellular mechanisms of aluminium toxicity and resistance in plant, *Annu. Rev. Plant Physiol. Plant Mol. Biol.*, **46**, 237, 1995.

Kononowicz, A. K., Nelson, D. E., Singh, N. K., Hasegawa, P. M. and Bressan, R.A., Regulation of the osmotin promoter, *Plant Cell*, **4**, 513, 1992.

Krasowski, M.J. and Owens, J.N., Seasonal changes with apical zonations and ultrastructure of coastal Douglas-fir seedlings (*Pseudotsuga manziesii*), *Amer J. Bot*, **77**, 245, 1990.

Kreuz, K., Herbicide safeners: recent advances and biochemical aspects of their mode of action, in 'Proc. Brighton Crop Protect. Conf. Weeds', Brighton, UK, 1993, 1249.

Kreuz, K., Tommasini, R. and Martinoia, E., Old enzymes for a new job: Herbicide detoxification in plants, *Plant Physiol.*, **111**, 349, 1996.

Krotz, R.M., Evangelou, B.P. and Wagner, G.J., Relationship between cadmium, zinc, Cd-peptide and organic acid in tobacco suspension cells, *Plant Physiol.* **91**, 780, 1989.

Kurkdjian, A., Quiquampoix, H., Barbier-Brygoo, H., Pean, M., Manigault, P., and Guern, J. Critical evaluation of methods for estimating the vacuolar pH of plant cells, in 'Biochemistry and Function of Vacuolar Adenosine Triphosphatase in Fungi and Plants, Marin, B.P., Ed., Springer Verlag, Berlin, Heidelberg, New York, Tokyo, 1985, 98.

Kylin, A., and Hansson, G., Transport of sodium and potassium, and properties of (sodium + potassium) activated adenosine triphosphatases: possible connection with salt tolerance in plants, in 'Proc. 8th Colloq. Intern. Potash. Inst., Bern, Int. Potash Inst., 1971, 64.

Laisk, A., Pfanz, H., and Heber, U., Sulfur-dioxide fluxes into different cellular compartments of leaves photosynthesizing in a polluted atmosphere, *Planta*, **173**, 241, 1988.

Lazof, D.B., Goldsmith, J.G. and Linton, R.W., The *in situ* analysis of intracellular aluminium in plants, in 'Progress in Botany', **58**, 112, 1997.

Leigh, R.A. and Walker, R.R., ATPase and acid-phosphatase activities associated with vacuoles isolated from storage roots of red beet (*Beta vulgaris* L.), *Planta*, **150**, 222, 1980.

Leigh, R.A., and Wyn Jones, R.G., Cellular compartmentation in plant nutrition: the selective cytoplasm and the promiscuous vacuole, in 'Advances in Plant Nutrition', Tinker, P.B. and Läuchli, A., Eds, Praeger, New York, 1986, 249.

Leigh, R.A., Rees, T., Fuller, W.A. and Banfield, J., The location of acid invertase activity and sucrose in the vacuoles of storage roots of beet root (*Beta vulgaris*), *Biochem. J.*, **178**, 539, 1979.

Levanony, H., Rubin, R., Altschuler, Y. and Galili, G., Evidence for a novel route of a wheat storage proteins to vacuoles, *J. Cell Biol*, **119**, 1117, 1992.

Lew, R.R., and Spanswick, R.M., Characterization of anion effects on nitrate sensitive ATP-dependent proton pumping activity of soyabean (*Glycine max.* L.) seedling root microsomes, *Plant Physiol.*, **77**, 352, 1985.

Li, Z-S, Alfenito, M., Rea, P., Walbot, V. and Dixon, R.A., Vacuolar uptake of the phytoalexin medicarpin by glutathione-conjugate pump, *Phytochemistry*, **45**, 689, 1997.

Li, Z-S, Zhen, R.G. and Rea, P.A., 1-Chloro-2,4-Dinitrobenzene-elicited increase in vacuolar glutathione S-conjugate transport activity, *Plant Physiol.*, 109, 177, 1995.

Li, Z-S., Lu, Y-P., Zhen, R-G., Szczypka, M., Thiele. D.J. and Rea, P.A., A new pathway for vacuolar cadmium sequestration in *Saccharomyces cerevisiae*: YCF 1- catalyzed transport of bis (aglutathionoto) cadmium, *Proc. Natl. Acad. Sci. USA*, **94**, 42, 1997.

Li, Z-S, Zhao, Y. and Rea, P.A., Magnesium adenosine 5'-triphosphate-energized transport of glutathione S-conjugates by plant vacuolar membrane vesicles, *Plant Physiol.*, **107**, 1257, 1995.

Lin, W., Wagner, G.J., Siegelman, W., and Hind, G., Membrane-bound ATPase of intact vacuoles and tonoplasts isolated from mature plant tissue, *Biochem. Biophys. Acta*, **465**, 110, 1977.

Linthorst, H.J.M., Pathogenesis related proteins of plants, *Crit. Rev. Plant Sci.* **10**, 123, 1991.

Löffelhardt, W. and Kopp, B., Subcellular localization of glucosyltransferases involved in cardiac glycoside glucosylation in leaves of *Convallaria majales*, *Phytochemistry*, **20**, 1219, 1981.

Lommel, C. and Felle, H., Transport of Ca^{2+} across the tonoplast of intact vacuoles from *Chenopodium album* L. suspension cells: ATP-dependent import and inositol-1, 4,5-triphosphate induced release, *Planta*, **201**, 477, 1997.

Lu, Y.P., Lia, Z-S., Drozdowicza, Y.M., Hörtensteiner, S.H., Martinoia, E. and Rea, P.A., ATMRP2, an *Arabidopsis* ATP binding cassette transporter able to transport glutathione S-conjugates and chlorophyll catabolites: Functional comparisons with ATMRP1, *Plant Cell*, **10**, 1, 1998.

Lunevsky, V.Z., Zherelova, O.M., Vostrikov, I.Y. and Berestovsky, G.N., Excitation of characeal cell membranes as a result of activation of calcium and chloride channels, *J. Membr. Biol.*, **72**, 43, 1983.

Lüttge, U., Carbon dioxide and water demand: Crassulacean acid metabolism (CAM), a versatile ecological adaptation exemplifying the need for integration in ecophysiological work, *New Phytol*, **106**, 593, 1987.

Lüttge, U. and Ball, E., Electrochemical investigation of active malic acid transport at the tonoplast into the vacuoles of the CAM plant *Kalanchoe daigremontiana*, *J. Membr. Biol.*, **47**, 401, 1979.

Lüttge, U., and Higinbotham, N., Transport in Plants, Springer-Verlag, New York, 1979, Chap. 12.

Lüttge, U., Smith, J.A.C. and Marigo, G., Membrane transport, osmoregulation, and the control of CAM, in 'Crassulacean Acid Metabolism', Ting, I.P. and Gibbs, M., Eds, Waverly Press, Baltimore, 1982, 69.

Lüttge, U., and Schnepf, E., Organic Substances, *Enc. Plant. Pysiol.*, **2B**, 244, 1976.

Lüttge, U., and Smith, J.A.C., Mechanisms of passive malic acid efflux from vacuoles of the CAM plant *Kalanchoe daigremontiana*, *J. Membr. Biol.*, **81**, 149, 1984.

Lüttge, U., Ball, E., and Tromballa, H.W., Potassium independence of osmoregulated oscillations of malate-levels in the cells of CAM leaves, *Biochem. Physiol. Pflanz.*, **167**, 267, 1975.

Lüttge, U., Fischer-Schliebs, E., Ratajczak, R., Kramer, D., Berndt, E. and Kluge, M., Functioning of the tonoplast in vacuolar C-storage and remobilization in crassulacean acid metabolism, *J.Exp. Bot.* **46**, 1377, 1995.

Mackenbrock, V., Gunia, W. and Barz, W., Accumulation and metabolism, of medicarpin and maackinian malonylglucosides in elicited chickpea (*Cicer arietinum* L.) cell suspension cultures, *J. Plant Physiol.*, **142**, 385, 1993.

Macklon, A.E.S., Calcium fluxes at plasmalemma and tonoplast, *Plant Cell Environ*, **7**, 407, 1984.

Maclean, D.J., Sargent, J.A., Tommerup, I.C. and Ingram, D.S., Hypersensitivity as the primary event in resistance to fungal parasites, *Nature*, **249**, 186, 1974.

Mandala, S. and Taiz, L., Proton transport in isolated vacuoles from corn coleoptile, *Plant Physiol.*, **78**, 104, 1985.

Mansour, M.M., Van Hasselt, P.R. and Kuiper, P.J.C., Plasma membrane lipid alterations induced by NaCl in winter wheat roots, *Physiol. Plant*, **92**, 473, 1994.

Maretzki, A. and Thom, M., UDP-glucose-dependent sucrose translocation in tonoplast vesicles from stalk tissue of sugar cane, *Plant Physiol.*, **83**, 235, 1987.

Maretzki, A. and Thom, M., High performance liquid chromatography-based re-evaluation of disaccharides produced upon incubation of sugarcane vacuoles with UDP-glucose, *Plant Physiol.*, **88**, 266, 1988.

Marger, M.D. and Saier, J.M.H., A major superfamily of transmembrane facilitators that catalyse uniport, symport and antiport. *Trends Biochem Sci.*, **18**, 13, 1993.

Marienfeld, S., Lehmann, H. and Stelzer, R., Ultrastructural investigations and EDX-analyses of Al-treated oat (*Avena sativa*) roots, *Plant Soil*, **171**, 167, 1995.

Marin, B.P., Evidence for an electrogenic adenosine triphosphatase in *Hevea* tonoplast vesicles, *Planta*, **157**, 324, 1983.

Marin, B.P., and Komor, E., Isolation, purification and subunit structure of H^+-translocating ATPase from *Hevea* latex, *Plant Physiol.*, **75**, 163, 1984.

Marin, B., Preisser, J., and Komor, E., Solubilization and purification of the ATPase from the tonoplast of *Hevea*, *Eur. J. Biochem.*, **151**, 131, 1985.

Marrs, K.A., The functions and regulation of glutathione S-transferase in plants, *Annu. Rev. Plant. Physiol. Plant Mol. Biol.*, **47**, 127, 1996.

Marrs, K.A., Alfenito, M.R., Lloyd, A.M., and Walbot, V., A glutathione S-transferase involved in vacuolar transfer encoded by the maize gene Bronze-2, *Nature*, **375**, 397, 1995.

Martinoia, E., Kaiser, G., Schramm, M.J. and Heber, U., Sugar transport across the plasmalemma and the tonoplast of barley mesophyll protoplasts: evidence for different transport systems., *J. Plant Physiol.*, **131**, 467, 1987.

Martinoia, E., Flügge, U.I., Kaiser, G., Heber, G., and Heldt, H.W., Energy-dependent uptake of malate into vacuoles isolated from barley mesophyll protoplast, *Biochem. Biophys. Acta*, **806**, 311, 1985.

Martinoia, E., Grill, E., Tommasini, R., Kreuz, K. and Amrhein, N., An ATP-dependent glutathione S-conjugate 'export' pump in the vacuolar membrane of plants, *Nature*, **364**, 247, 1993.

Martinoia, E., Schramm, M.J., Kaiser, G., Kaiser, W.M., and Heber, U., Transport of anions in isolated barley vacuoles I. Permeability to anions and evidence for a Cl^- uptake system, *Plant Physiol.*, **80**, 895, 1986.

Martinoia, E., Thume, M., Vogt, E., Rentsch, D. and Deitz, K-J., Transport of arginine and aspartic acids into isolated barley mesophyll vacuoles, *Plant Physiol.*, **97**, 664, 1991.

Marty, F., Role du systeme membranaire vacuolaire dans la differenciation des laticiferes d'*Euphorbia characias* L., *Comp. Rend. Acad. Sci.* (Paris), **271**, 2301, 1970.

Marty, F., Differenciation des plastes dans les laticiferes d'*Euphorbia characias*, L., *Comp. Rend. Acad. Sci.*, (Paris) **272**, 223, 1971.

Marty, F., Cytochemical studies on GERL, provacuoles, and vacuoles in root meristematic cells of *Euphorbia*, *Proc. Nat. Acad. Sci., USA*, **75**, 852, 1978.

Marty, F., Branton, D. and Leigh, R.A., Plant vacuoles, in 'The Biochemistry of Plants: A Comprehensive Treatise', Tolbert, N.E., Ed., Academic Press, New York, 1980, 625.

Matern, V., Reichenbach, C. and Heller, W., Efficient uptake of flavonoids into parsley (*Petroselinum hortense*) vacuoles requires acylated glucosides, *Planta*, **167**, 183, 1986.

Mathieu, Y., Guern, I., Kurkdjian, A., Manigault, P., Manigault J., Zielinska, T., Gillet, B., Beloeil, J.C. and Lallemand, J.Y., Regulation of vacuolar pH of plant cells. I. Isolation and properties of vacuoles suitable for ^{31}P-NMR studies, *Plant Physiol*, **89**, 19, 1989.

Mathys, W., The role of malate oxalate and mustard oil glucosides in the evolution of zinc resistance in herbage plants, *Plant Phyiol.*, **40**, 130, 1977.

Matile, P., 'Crop Photosynthesis: Spatial and Temporal Determinants', Eds Baker, N.R., and Thomas, H., Elsevier, Amsterdam, 1992, 413.

Matile, P., and Moor, H., Vacuolation: origin and development of the lysosomal apparatus in root tip cells, *Planta*, **80**, 159, 1968.

Matile, P. and Winkenbach, F., Function of lysosomes and lysosomal enzymes in the senescing corolla of the morning glory *Ipomoea purpurea*. *J. Exp. Bot.*, **22**, 759, 1971.

Matile, P., Ginsburg, S., Schellenberg, M., and Thomas, H., Catabolites in senescing barley leaves are localised in the vacuole of mesophyll cells, *Proc. Natl. Acad. Sci, USA*, **85**, 9529, 1988.

Matsuoka, K., Higuchi, T., Maeshima, M. and Nakamura, K., A vacuolar-type H^+ATPase in a non-vacuolar organelle is required for the sorting of soluble vacuolar protein precursors in tobacco cells, *Plant Cell*, **9**, 533, 1997.

Mauch, F. and Staehelin, L.A., Functional implications of the subcellular localization of ethylene-induced chitinase and beta-1,3-glucanase in bean leaves, *Plant Cell*, **1**, 447, 1989.

Mauch, F., Meehl, J.B. and Staehlin, A.B., Ethylene-induced chitinase and beta-1,3-glucanase accumulate specifically in the lower epidermis and along vascular strands of bean leaves, *Planta*, **186**, 367, 1992.

Maurel, C., Kado, R.T., Guern, J. and Chrispeels, M.J., Phosphorylation regulates the water channel activity of the seed-specific aquaporin γ-TIP, *EMBO. J.*, **14**, 3028, 1995.

Maurel, C., Reizer, J., Schroeder, J.I., Chrispeels, M.J. and Saier, M.H. Jr., Functional characterization of *Escherichia coli* glycerol facilitator GlpF in *Xenopus occytes*, *J. Biol. Chem.*, **269**, 11869, 1994.

Maurel, C., Tacnet, F., Josette, G., Guern, J. and Ripoche, P, Purified vesicles of tobacco cell vacuolar and plasma membranes exhibit dramatically different water permeability and water channel activity, *Proc. Natl. Acad. Sci. USA*, **94**, 7103, 1997.

McAinsh, M.R., Brownlee, C. and Hetherington, A.M., Abscisic acid-induced elevation of guard cell cytosolic Ca^{2+} precedes stomatal closure, *Nature*, **343**, 186, 1990.

McClintock, M., Higinbotham, N., Uribe, E.G. and Cleland, R.E., Active, irreversible accumulation of extreme levels of H_2SO_4 in the brown alga, *Desmarestia*, *Plant Physiol*, **70**, 771, 1982.

Mehdy, M., Active oxgen species in plant defense against pathogens, *Plant Physiol*, **105**, 467, 1994.

Meijer, A.H., de Waal, A. and Verpoorte, R., Purification of the cytochrome P-450 enzyme geraniol-10 hydroxylase from cell cultures of *Catharanthus roseus.*, *J. Chromatogr.*, **635**, 237, 1993a.

Meijer, A.H., Verpoorte, R. and Hoge, J.H.C., Regulation of enzymes and genes involved in terpenoid indole alkaloid biosynthesis in *Cathananthus roseus*, *J. Plant Res.*, **3**, 145, 1993b.

Mende, P. and Wink, M., Uptake of quinolizidine alkaloid lupanine by protoplasts and isolated vacuoles of suspension cultured *Lupinus polyphyllus* cells. Diffusion of carrier-mediated transport?, *J. Plant Physiol.* **129**, 229, 1987.

Merchant, R. and Robards, A.W., Membrane systems associated with the plasmalemma of plant cells, *Ann. Bot.*, **32**, 457, 1968.

Mikus, M., Boba'k, M. and Lux, A., Structure of protein bodies and elemental composition of phytin from dry germ of maize (*Zea mays* L.) *Bot. Acta.*, **105**, 26, 1992.

Milicic, D., Viruskörper in Zellsafte, *Protoplasma*, **57**, 602, 1963.

Miller, A.J. and Smith, S.J., The mechanism of nitrate transport across the tonoplast of barley root cells, *Planta*, **187**, 554, 1992.

Miller, A.J., Brimelow, J.J. and John, P., Membrane potential changes in vacuoles isolated from storage roots of red beet (*Beta vulgaris* L.), *Planta*, **160**, 59, 1984.

Mimura, T., and Tazawa, M., Effect of intracellular Ca^{2+} on membrane potential and membrane resistance in tonoplast free cells of *Nitellopsis obtusa*, *Protoplasma*, **118**, 49, 1983.

Mimura, T., Dietz, K.J., Kaiser, W., Schramm, M.J., Kaiser, G. and Heber, U., Phosphate transport across biomembranes and cytosolic phosphate homeostasis in barley leaves, *Planta*, **180**, 139, 1990.

Minami, A., and Fukuda, H., Transient and specific expression of cysteine endopeptidase during autolysis in differentiation tracheary elements from *Zinnia* mesophyll cells, *Plant Cell Physiol.*, **36**, 1599, 1995.

Minorsky, P.V., A heuristic model of chilling injury in plants: A role for calcium as the primary physiological transducer of injury, *Plant Cell Environ.*, **8**, 75, 1985.

Mitchell, P., Vectorial chemistry and the molecular mechanics of chemiosmotic coupling: power transmission by protocity, *Biochem. Soc. Trans.*, **4**, 399, 1976.

Moreno, P.R.H., Van der Heijden and Verpoorte, R., Cell and Tissue cultures of *Catharanthus roseus*: A literature survey II Updating from 1988 to 1993, *Plant Cell Tiss. Org. Cult.*, **42**, 1, 1995.

Moriyama, Y., Maeda, M. and Futai, M., Involvement of a non-proton pump factor (possibly Donnan-type equilibrium) in maintenance of an acidic pH in lysosomes, *FEBS Lett.* **302**, 18, 1992.

Moriyasu, Y. and Ohsumi, Y., Autophagy in tobacco suspension-cultured cells in response to sucrose starvation, *Plant Physiol.*, **111**, 1233, 1996.

Moriyasu, Y. and Tazawa, M., Distribution of several proteases inside and outside the central vacuole of *Chara australis*, Cell Struct. Funct., **11**, 81, 1986a.

Moriyasu, Y. and Tazawa, M., Plant vacuole degrades exogenous proteins, *Protoplasma*, **130**, 214, 1986b.

Moriyasu, Y., and Tazawa, M., Degradation of proteins artificially introduced into vacuoles of *Chara australis*, *Plant Physiol.*, **88**, 1092, 1988.

Moriyasu, Y., Shimmen, T., and Tazawa, M., Vacuolar pH regulation in *Chara australis*, Cell Struct. Funct., **9**, 225, 1984a.

Moriyasu, Y., Shimmen, T., and Tazawa, M., Electrical characteristics of the vacuolar membrane of *Chara* in relation to pHv regulation, *Cell Struct. Funct.*, **9**, 235, 1984b.

Morse, M.J. and Satter, R.L., Relationships between motor cell ultrastructure and leaf movements in *Samanea saman*, *Physiol. Planta*, **46**, 338, 1979.

Moysset, L. and Simon, E., Secondary pulvinus of *Robinia pseudoacacia* (Leguminosae): structural and ultrastructural features, *Am. J. Bot.*, **78**, 1467, 1991.

Müller, M.L., Irkens-Kiesecker, V., Kromer, D. and Taiz, L., Purification and reconstitution of the vacuolar H^+-ATPases from lemon fruits and epicotyls, *J. Biol. Chem.*, **272**, 12762, 1997.

Müller, M.L., Irkens-Kiesecker, V., Rubinstein, B. and Taiz, L., On the mechanism of hyperacidification in lemon - comparison of the vacuolar H^+-ATPase activities of fruits and epicotyls, *J. Biol. Chem.*, **271**, 1916, 1996.

Münch, E., Die Staffbewegungen in der Pflanze, Gustav Fisher, Jena, 1930.

Nakagawa, S., Kataoka, H. and Tazawa, M., Osmotic and ion regulation in *Nitella*, *Plant Cell Physiol.*, **15**, 457, 1974.

Nakanishi, Y. and Maeshima, M., Molecular cloning of vacuolar H^+-pyrophosphatase and its developmental expression in growing hypocotyl of mung bean. *Plant Physiol.* **116**, 589, 1998.

Nelson, N., Structure, function and evolution of proton-ATPases, *Plant Physiol.*, **86**, 1, 1988.

Neuman, D., Kraus, G., Heike, M. and Gröger, D., Indole alkaloid formation and storage in cell suspension cultures of *C. roseus*, *Planta Med.* **48**, 20, 1983.

Nicholson, R.L., van Scoyoc, S., Williams, E.B. and Kuc, J., Host-pathogen interaction preceeding the hypersensitive reaction of *Malus* sp. to *Venturia*, *Phytopathology*, **67**, 108, 1977.

Niemietz, C.M. and Tyerman, S.D., Characterization of water channels in wheat root membrane vesicles, *Plant Physiol.*, **115**, 561, 1997.

Niemietz, C., and Willenbrink, J., The function of tonoplast ATPase in intact vacuoles of red beet is governed by direct and indirect ion effects, *Planta*, **166**, 545, 1985.

Nishida, K., and Tominaga, O., Energy-dependent uptake of malate into vacuoles isolated from CAM plant, *Kalanchoe daigremontiana*, *J. Plant Physiol.*, **127**, 385, 1987.

Noda, T., Matsuura, A., Wada, Y. and Ohsumi, Y., Novel system for monitoring autophagy in the yeast *Saccharomyces cerevisiae*, *Biochem. Biophys. Res. Comm.*, **210**, 126, 1995.

Norberg, P. and Liljenberg, P., Lipids of plasma membranes prepared from oat root cells: Effect of induced water deficit tolerance. *Plant Physiol.*, **96**, 1136, 1991.

Nozue, M. and Yasuda, H., Occurrence of anthocyanoplasts in suspension culture of sweet potato, *Plant Cell Rep.*, **4**, 252, 1985.

Nozue, M., Kubo, H., Nishimura, M. and Yasuda, H., Detection and characterization of a vacuolar protein (VP24) in anthocyanin producing cells of sweet potato in suspension culture, *Plant Cell Physiol.*, **36**, 883, 1995.

Nozue, M., Yamada, K., Nakamura, T., Kubo, H., Kondo, M. and Nishimura, M., Expression of vacuolar protein (VP24) in anthocyanin-producing cells of sweet potato in suspension culture, *Plant Physiol.*, **445**, 1065, 1997.

Nozzolillo, C. and Ishikura, N., An investigation of the intracellular site of anthocyanoplasts using isolated protoplasts and vacuoles, *Plant Cell Rep.*, **7**, 389, 1988.

Obermeyer, G., Sommer, A. and Bentrup, F.W., Potassium and voltage dependence of the inorganic pyrophosphatase of intact vacuoles from *Chenopodium rubrum*, *Biochim. Biophys. Acta*, **1284**, 203, 1996.

Okazaki, Y., and Tazawa, M., Involvement of calcium ion in turgor regulation upon hypotonic treatment in *Lamprothamnium succintum*, *Plant Cell Environ*, **9**, 185, 1986.

Oparka, K. J., Murant, E.A., Wright, K. M., Prior, D.A.M. and Harris, N., The drug probenecid inhibits the vacuolar accumulation of fluorescent anions in onion epidermal cells, *J. Cell Sci.*, **99**, 557, 1991.

Oritz, D.F., McCue, K.F. and Ow, D.W., *In vitro* transport of phytochelatins by the fission yeast ABC-type vacuolar membrane protein (abstract no.25), *Plant Physiol.*, **105**, S-17, 1994.

Pantoja, O., Gelli, A. and Blumwald, E., Characterization of vacuolar malate and K^+ channels under physiological conditions, *Plant Physiol.*, **100**, 1137, 1992a.

Pantoja, O., Gelli, A. and Blumwald, E., Voltage-dependent calcium channels in plant vacuoles, *Science*, **225**, 1567, 1992b.

Park, H.M., Hakamatsuka, T., Sankawa, U. and Ebizuka, Y., Rapid metabolism of isoflavonoids in elicitor-treated cell suspension cutlures of *Pueraria lobata*, *Phytochemistry*, **38**, 373, 1995.

Pecket, C.R. and Small, C.J., Occurrence, location and development of anthocyanoplasts, *Phytochemistry*, **19**, 2571, 1980.

Perry, C.A., Leigh, R.A., Tomos, A.D., Wyse, R.E., and Hall, J.L., The regulation of turgor pressure during sucrose mobilisation and salt accumulation by excised storage-root tissue of red beet, *Planta*, **170**, 353, 1987.

Pfanz, H., and Heber, U., Buffer capacities of leaves, leaf cells and leaf cell organelles in relation to fluxes of potentially acidic gases, *Plant Physiol.*, **81**, 597, 1986.

Pfanz, H., Martinoia, E., Lange, O.L. and Heber, U., Flux of SO_2 into leaf cells and cellular acidification by SO_2., *Plant Physiol.*, **85**, 928, 1987.

Pfeiffer, W. and Hager, A. Ca^{2+}-ATPase and a Mg^{2+}/H^+-antiporter present on tonoplast membranes from roots of *Zea mays* L., *Planta*, **191**, 377, 1993.

Pick, U. and Weiss, M., Polyphosphate hydrolysis within acidic vacuoles in response to amine-induced alkaline stress in the halotolerant alga *Dunaliella salina*, *Plant Physiol.*, **97**, 1234, 1991.

Pick, U., Bental, M., Chitlaru, E. and Weiss, M., Polyphosphate hydrolysis - a protective mechanism against alkaline stress? *FEBS Lett.* **274**, 15, 1990.

Pierce, W.S. and Higinbotham, N., Compartments and fluxes of K$^+$, Na$^+$ and Cl$^-$ in *Avena* coleoptile cells, *Plant Physiol.*, **46**, 666, 1970.

Pistocchi, P., Keller, F., Bagni, N., and Matile, P., Transport and subcellular localization of polyamines in carrot protoplasts and vacuoles, *Plant Physiol.*, **87**, 514, 1988.

Pitman, M.G., Simulation of Cl$^-$ uptake by low salt barley roots as a test of models of salt uptake, *Plant Physiol.*, **44**, 1417, 1969.

Pitt, D. and Coombes, C., The disruption of lysosome-like particles of *Solanum* tuber tissue during infection by *Phytophthora erythroseptica* Pethybr, *J. Gen. Microbiol.*, **53**, 197, 1968.

Pitt, D. and Galpin, M., Isolation and properties of lysosomes from dark-grown potato shoots, *Planta*, **109**, 233, 1973.

Politis, J., Cytological observations on the production of anthocyanins in certain Solanaceae. *Bull. Torrey. Bot. Club*, **86**, 387, 1959.

Ponstein, A.S., Bres-Vloemans, S.A., Sela-Buurlage, M.B., Van den Elzen, P.J.M., Melchers, L.S. and Cornelissen, B.J.C., A novel pathogen- and wound-inducible tobacco (*Nicotiana tabaccum*) protein with antifungal activity, *Plant Physiol.*, **104**, 109, 1994.

Poovaiah, B.W. and Reddy, A.S.N., Calcium and signal transduction in plants, *CRC Crit. Rev. Plant Sci.*, **21**, 185, 1993.

Pope, A.J., and Leigh, R.A., Some characteristics of anion transport at the tonoplast of oat roots, determined from the effects of anions on pyrophosphate-dependent proton transport, *Planta*, **172**, 91, 1987.

Pope, A.J. and Leigh, R.A., Dissipation of pH gradients in tonoplast vesicles and liposomes by mixture of acridine orange and anions, *Plant Physiol.*, **86**, 1315, 1988.

Pope, A.J. and Leigh, R.A., Characterization of chloride transport at the tonoplast of higher plants using a chloride-sensitive fluoresent probe, *Planta*, **181**, 406, 1990.

Poux, N., Sur la presence d'enclaves cytoplasmiques en voiede degenerescence dans les vacuoles des cellules vegetales. *Compt. Rend. Acad. Sci.*, Paris, **257**, 736, 1963.

Prasad, D.N. and De., D.N., Ultrastructure of release of *Rhizobium* and formation of membrane envelope in root nodule, *Microbios*, **4**, 13, 1971.

Preisser, J. and Komor, E., Analysis of the reaction products from incubation of sugarcane vacuoles with uridine-diphosphate-glucose: No evidence for the group translocator, *Plant-Physiol.*, **88**, 259, 1988.

Preisser, J. and Komor, E., Sucrose uptake into vacuoles of sugarcane suspension cells, *Planta*, **186**, 109, 1991.

Preisser J., Sprügel, H. and Komor, E., Solute distribution between vacuole and cytosol of sugarcane suspension cells: Sucrose is not accumulated in the vacuole, *Planta*, **186**, 203, 1991.

Pueyo, J.J., Chrispeels, M.J. and Herman, E.M., Degradation of transport-competent destabilized phaseolin with a signal for retention in the endoplasmic reticulum occurs in the vacuole, *Planta*, **196**, 586, 1995.

Raskin, I., Kumar, P.B.A.N., Dushenkov, S. and Salt, D.E., Bioconcentration of heavy metals by plants, *Curr. Opin. Biotechnol.*, **5**, 285, 1994.

Rausch, T., Butcher, D.N. and Taiz, L., Active glucose transport and proton pumping in tonoplast membrane of *Zea mays* L. coleoptiles are inhibited by anti-H$^+$-ATPase antibodies, *Plant Physiol.*, **85**, 996, 1987.

Rauser, W.E., Phytochelatins, *Annu. Rev. Biochem.* **59**, 61, 1990.

Rea, P.A., and Poole, R.J., Proton-translocating inorganic pyrophosphatase in red beet (*Beta vulgaris* L.) tonoplast vesicles, *Plant Physiol.*, **77**, 46, 1985.

Rea, P.A. and Poole, R.J., Vacuolar H$^+$-translocating pyrophosphatase, *Annu. Rev. Plant Physiol.*, **44**, 157, 1993.

Rea, P.A. and Sanders, D., Tonoplast energization: Two H$^+$-pumps, one membrane. *Physiol. Plant*, **71**, 131, 1987.

Rea, P.A., Li, Z-S., Lu, Y-P., Drozdowicz, Y.M., and Martinoia, E., From vacuolar GS-X pumps to multispecific ABC transporters, *Annu. Rev. Plant Physiol. Plant Mol. Biol.*, **49**, 727, 1998.

Record, R.D. and Griffing, L.R., Convergence of endocytic and lysosomal pathways in soybean protoplasts, *Planta*, **176**, 425, 1988.

Reid, R.J., Smith, F.A., and Whittington, J., Control of intracellular pH in *Chara corollina* during uptake of weak acid, *J. Exp. Bot.*, **40**, 883, 1989.

Reinhold, L. and Kaplan, A., Membrane transport of sugars and amino acids, *Ann. Rev. Plant Physiol.*, **35**, 45, 1984.

Renaudin, J.P., Compartmentation of ajmalicine in *Catharanthus* vacuoles, *Plant Physiol. Biochem.*, **27**, 613, 1989.

Rengel, Z., The role of calcium in salt toxicity, *Plant Cell Environ.*, **15**, 625, 1992.

Rennenberg, H., The fate of excess sulfur in higher plants, *Ann. Rev. Plant Physiol.*, **35**, 121, 1984.

Rentsch, D., Goerlach, J., Vogt, E., Amrhein, N. and Martinoia, E. The tonoplast-asscociated citrate binding protein of *Hevea brasiliensis*: Photoaffinity labeling, purification, and cloning of the corresponding gene, *J. Biol. Chem.*, **270**, 30525, 1995.

Rieder, S.E., Banta, L.M., Kohrer, K., McCaffery, J.M. and Emr, S.D. Multilamellar endosome-like compartment accumulates in the yeast vps28 vacuolar protein sorting mutant, *Mol. Biol. Cell.*, **7**, 985, 1996.

Robards, A.W. and Kidwai, P., Vesicular involvement in differentiating plant vascular cells, *New Phytol.*, **68**, 343, 1969.

Robinson, D.G. and Hillmer, S., Endocytosis in plants. *Physiol. Plant.*, **79**, 96, 1990.

Rocha Faenha, A. and de Meis, L., Reversibility of H^+-ATPase and H^+-pyrophasphatase in tonoplast vesicles from maize coleoptiles and seeds, *Plant Physiol.* **116**, 4487, 1998.

Rodrigo, I., Vera, P., Tornero, P., Hernandez-Yago, J. and Conejero,V., cDNA cloning of viroid-induced tomato pathogenesis-related protein P 23, *Plant Physiol.*, **102**, 939, 1993.

Rommero, A., Alamillo, J.M. and Garcia-Olmedo, F., Processing of thionin precursors in barley leaves by a vacuolar proteinase, *Eur, J. Biochem.*, **243**, 202, 1997.

Roos, W., Schulze, R. and Steighart, J., Dynamic compartmentation of vacuolar amino acids in *Penicillium cyclopium*: Cytosolic adenylates act as a control signal for efflux into the cytosol, *J. Biol. Chem.* **272**, 15849, 1997.

Rosen, B.P., and Silver, S., Ion transport in Prokaryotes, Acad Press, San Diego, 1987, 47.

Roy, A.T. and De, D.N., Studies on differentiation of laticifers through light and electron microscopy in *Calotropis gigantea* (Linn.) R.Br., *Ann. Bot.*, **70**, 443, 1992

Ruiz-Medrano, R., Jimenez-Moraila, B., Herrera-Estrella, L. and Rivera-Bustamante, R., Nucleotide sequence of an osmotin-like cDNA induced in tomato during viroid infection, *Plant Mol. Biol.*, **20**, 1199, 1992.

Runeberg-Roos, P., Kervinen, J., Kovaleva, V., Raikhel, N.V. and Gal, S., The aspartic proteinase of barley is a vacuolar enzyme that processes probarley lectin in vitro, *Plant Physiol.*, **105**, 321, 1994.

Saftner, R.A., Daie, J. and Wyse, R.E., Sucrose uptake and compartmentation in sugar beet tap root tissue, *Plant Physiol.*, **72**, 1, 1983.

Sakano, K. and Tazawa, M., Metabolic conversion of amino acids loaded in the vacuole of *Chara australis* internodal cells, *Plant Physiol.*, **78**, 673, 1985.

Salt, D.E. and Rauser, W.E., MgATP-dependent transport of phytochelatins across the tonoplast of oat roots, *Plant Physiol.*, **107**, 1293, 1995.

Salt, D.E. and Wagner, G.J., Cadmium transport across tonoplast of vesicles from oat roots: Evidence for a Cd^{2+}/H^+ antiport activity, *J. Biol. Chem.*, **268**, 12297, 1993.

Sandermann, H., Plant metabolism of xenobiotics, *Trends Biochem. Sci.*, **17**, 82, 1992.

Sangwan, R.S., Mathivet, V., and Vasseur, G., Ultratructural localization of acid phosphatase during male meiosis and sporogenesis in *Datura*: evidence for digestion of cytoplasmic structures in the vacuoles, *Protoplasma*, **149**, 38, 1989.

Sato, T., Ohsumi, Y., and Anraku, Y., Substrate specificities of active transport systems for amino acids in vacuolar membrane vesicles *of Saccharomyces cerevisiae*. Evidence of seven independent proton amino acid antiport systems, *J. Biol. Chem.*, **259**, 11505, 1984.

Satter, R.L. and Galston, A.W., Mechanisms of control of leaf movement, *Annu. Rev. Plant Physiol.*, **32**, 83, 1981.

Sauer, N. and Stadler, R., A sink-specific H^+/monosaccharide co-transporter from *Nicotiana tabacum*: cloning and heterologous expression in baker's yeast, *Plant J.*, **4**, 601, 1993.

Sauer, N. and Tanner, W., The hexose carrier from *Chlorella* cDNA cloning of a eucaryotic H^+-cotransporter, *FEBS Lett.* **259**, 43, 1989.

Sauer, N., Friedlander, K. and Graml-Wicke, U., Primary structure, genomic organization and heterologous expression of a glucose transporter from *Arabidopsis thaliana*, *EMBO J.* **9**, 3045, 1990.

Scannerini, S. and Bonfante-Fasola, P., Comparative ultrastructural analysis of mycorrhizal associations, *Can. J. Bot.*, **61**, 917, 1983.

Schimmoeler, F. and Riezman, H., Involvement of Ypt7p, a small GTPase, in traffic from late endosome to the vacuole in yeast, *J. Cell Biol.*, **106**, 823, 1993.

Schnabl, H., and Kottmeier, C., Determination of malate levels during the swelling of vacuoles isolated from guard cell protoplasts, *Planta*, **161**, 27, 1984.

Schrempf, M., Satter, R.L. and Galston, A.W., Potassium linked chloride-fluxes during rhythmic leaf movement of *Albizzia julibrissin*, *Plant Physiol.* **58**, 190, 1976.

Schroeder, J.I., and Hedrich, R., Involvement of ion channels and active transport in osmoregulation and signalling of higher plant cells, *Trends Biochem. Sci.*, **14**, 187, 1989.

Schroeder, J.I., Word, J.M. and Gassmann, W., Perspectives on the physioloigy and structure of inward-rectifying K^+ channels in higher plants: biophysical implications for K^+ uptake, *Annu. Rev. Biophys. Biomol. Struct.*, **23**, 441, 1994.

Schröppel Meier, G. and Kaiser, W.M., Ion homoeostasis in chloroplasts under salinity and mineral deficiency. 2. Solute distribution between chloroplasts and extrachloroplastic space under excess or deficiency of sulfate, phosphate or magnesium, *Plant Physiol.*, **87**, 828, 1988.

Schulz-Lessdorf, B. and Hedrich, R., Protons and calcium modulate SV-type channels in the vacuolar-lysosomal compartment - channel interaction with calmodulin inhibitors, *Planta*, **197**, 655, 1995.

Schumaker, K., and Sze, H., Calcium transport into the vacuole of oat roots: characterization of H^+/Ca^{2+} exchange activity, *J. Biol. Chem.*, **261**, 12172, 1986.

Schumaker, K.S., and Sze, H., Decrease of pH gradients in tonoplast vesicles by NO_3^- and Cl^-: Evidence for H^+-coupled anion transport, *Plant Physiol*, **83**, 490, 1987.

Schwencke, J., and de Robichon-Szulmajster, H., The transport of S-adenosyl-L-methionine in isolated yeast vacuoles and spheroplasts, *Eur. J. Biochem.*, **65**, 49, 1976.

Scott, M.P., Jurg, R., Müntz, K. and Nielson, N., A protease responsible for post-translational cleavage of a conserved Asn-Gly linkage in glycinin, the major seed storage protein of soybean, *Proc. Natl. Acad. Sci. USA*, **89**, 658, 1992.

Scott, S.V., Hefner-Gravink, A., Morano, K.A., Noda, T., Ohsumi, Y. and Klionsky, D.J., Cytoplasm-to-vacuole targeting and autophagy employ the same machinery to deliver proteins to the yeast vacuole, *Proc. Natl. Acad. Sci., USA*, **93**, 12304, 1996.

Seals, D.F. and Randall, S.K., A vacuole-associated annexin protein, VCaB42, correlates with the expansion of tobacco cells, *Plant Physiol.*, **115**, 753, 1997.

Seals, D.F., Parrish, M.L. and Randall, S.K., A 42-Kilodalton annexin-like protein is associated with plant vacuoles, *Plant Physiol.*, **106**, 1403, 1994.

Sharma, V. and Strack, D., Vacuolar localization of 1-sinapoyl glucose: L-malate sinapoyl transferase in protoplasts from cotyledons of *Raphanus sativus*, *Planta*, **164**, 507, 1985.

Shaul, O., Hilgemann, D.W., de-Almeida-Engler, J., Van Montagu, M., Inzé, D., Galili, G., First identification of a higher plant's Mg^{2+}/H^+ exchanger gene: A structural homolog of animal's

Na$^+$/Ca^{2+} exchangers, *Abstract, IX. Inst. Cong. Plant. Tiss. Cell Cult.*, IAPTC, Jerusalem, 1998, 48.

Shimmen, T. and MacRobbie, E.A.C., Characterization of two proton transport systems in the tonoplast of plasmalemma-permeabilized *Nitella* cells, *Plant Cell Physiol.*, **28**, 1023, 1987.

Shimony, C. and Friend, J., Ultrastructure of the interaction between *Phytophthora infestans* and leaves of two cultivars of potato (*Solanum tuberosum*), Orion and Majestic, *New Phytol*, **74**, 59, 1975.

Shiratake, K., Kanayama, Y. and Yamaki, S., Characterization of hexose transporter for facilitated diffusion of the tonoplast vesicles from pear fruit, *Plant Cell Physiol.*, **38**, 910, 1997a.

Shiratake, K., Kanayama, Y., Maeshima, M. and Yamaki, S., Changes in H$^+$-pumps and a tonoplast intrinsic protein of vacuolar membranes during the development of pear fruit, *Plant Cell Physiol.*, **38**, 1039, 1997b.

Sidel, V.W. and Solomon, A.K. Entrance of water into human red cells under an osmotic gradient, *J. Gen.Physiol.*, **41**, 243, 1957.

Sievers, A., Lysosomen-ähnliche Kompartimente in Pflanzenzellen, *Naturwissenschaften*, **13**, 334, 1966.

Sievers, A., and Schmitz, M., X-ray microanalysis of barium, sulfur and strontium in statolith compartments of *Chara* rhizoids, *Ber. Dtsch. Bot. Ges.*, **95**, 353, 1982.

Singh, N.K., Bracker, C.A., Hasegawa, P.M., Handa, A.K., Buckel,S., Hermondson, M.A., Pfankoch, E., Regnier, F.E. and Bressan, R.A. Characterization of osmotin: A thumatin-like protein associated with osmotic adaptation in plant cells, *Plant Physiol.*, **85**, 529, 1987.

Skou, J.C., The (Na$^+$, K$^+$)-activated enzyme system, in 'Perspectives in Modern Biology', Estrada, S. and Gilter, C., Eds, Academic Press, New York, 1974, 74.

Smart, C.M., Gene expression during leaf senescence, *New Phytol.*, **126**, 419, 1994.

Smith, J.I., Amouzou, E., Yamaguchi, A., Mclean, S. and Dicosmo, F., Peroxidase from bioreactor-cultivated *Catharanthus roseus* cell cultures mediates biosynthesis of alpha-3',4'-anhydrovinblastine. *Biotechnol. Appl. Biochem.*, **10**, 568, 1988.

Smith, M.T., Saks, Y. and Van Staden, J., Ultrastructural changes in the petals of senescing flowers of *Dianthus caryophyllus* L., *Ann. Bot*, **69**, 277, 1992.

Spanu, P., Boller, T., Ludwig, A., Wiemken, A., Faccio, A. and Bonfante-Fasolo, P., Chitinase in roots of mycorrhizal *Allium porrum*: regulation and localisation, *Planta*, **177**, 447, 1989.

Staal, M., Maathuis, F.J.M., Elzenga, J.T.M., Overbeek, J.H.M. and Prins, H.B.A., Na$^+$/H$^+$ antiport activity in tonoplast vesicles from roots of the salt-tolerant *Plantago maritima* and the salt-sensitive *Plantago media*. *Physiol. Plant.*, **82**, 179, 1991.

Steffens, J.C., The heavy metal-binding peptides of plants, *Annu.Rev. Plant Physiol. Plant Mol. Biol.*, **41**, 553, 1990.

Steudle, E., The biophysics of plant water compartmentation, coupling with metabolic processes and water flow in plant roots, in 'Water and Life: A Comparative Analysis of Water Relationships at the Organismic, Cellular and Molecular Levels', Somero, G.N., Osmond, C.B. and Bolis, C.L. Eds., Springer-Verlag, Berlin, 1992, 173.

Stevens, L.H., Blom, J.M. and Verpoorte, R., Subcellular localisation of tryptophan decarboxylase, strictosidine synthase and strictosidine glucosidase in suspension cultured cells of *Catharanthus roseus and Tabernaemontana divaricata*, *Plant Cell Rep.* **12**, 573, 1993.

Sticher, D., Hinz, V., Meyer, A.D. and Meins Jr. F., Intracellular transport and processing of a tobacco vacuolar beta-1,3-glucanase, *Planta*, **188**, 559, 1992.

Strack, D. and Sharma, V., Vacuolar localisation of the enzymatic synthesis of hydroxycinnamic acid esters of malic acid in protoplasts from *Raphanus sativus* leaves, *Physiol. Plant.*, **65**, 45, 1985.

Sturm, A., Sebkova, V., Lorenz, K., Hardegger, M., Lienhard, S. and Unger, C., Development- and organ-specific expression of the genes for sucrose synthase and three isoenzymes of acid beta fructofuranosidase in carrot, *Planta*, **195**, 601, 1995.

Swain, D. and De, D.N., Active dye uptake: A new mechanism of vital staining of some plant protoplasts, *Ind. J. Exp. Biol.*, **35**, 180, 1997.

Swanson, S.J., and Jones, R.L., Gibberellic acid induces vacuolar acidification in barley aleurone, *Plant Cell*, **8**, 2211, 1996.

Swanson, S.J., Bethke, P.C. and Jones, R.L., Barley aleurone cells contain two types of vacuoles: characterization of lytic organelles by use of fluorescent probes, *Plant Cell*, **10**, 685, 1998.

Sze, H., H^+-translocating ATPases of the plasma membrane and tonoplast of plant cells, *Plant Physiol.*, **61**, 683, 1984.

Sze, H., H^+-translocatings ATPases: Advances using membrane vesicles, *Ann. Rev. Plant Physiol.*, **36**, 175, 1985.

Takahama, V., A role of hydrogen peroxide in the metabolism of phenolics in mesophyll cells of *Vicia faba* L., *Plant Cell Physiol.* **30**, 295, 1989.

Takeshige, K. and Hager, A., Ion effect on the H^+-translocating adenosine triphosphatase and pyrophosphatase associated with the tonoplast of *Chara corallina*, *Plant Cell Physiol.*, **29**, 649, 1988.

Takeshige, K., Baba, M., Tsuboi, S., Noda, T. and Ohsumi, Y., Autophagy in yeast demonstrated with proteinase-deficient mutants and conditions, for its induction, *J. Cell Biol.*, **119**, 301, 1992.

Tanchak, M.A., Griffing, L.R., Mersey, B.G. and Fowke, L.C., Endocytosis of cationized ferritin by coated vesicles of soybean protoplasts, *Planta*, **162**, 481, 1984.

Thoiron, A., Thoiron, B., Demarty, M., and Thellier, M., Compartmental analysis of sulfate transport in *Lemna minor* L., taking plant growth and sulfate metabolism into consideration, *Biochem. Biophys. Acta*, **644**, 24, 1981.

Thom, M. and Komor, E., H^+ sugar antiport as the mechanism of sugar uptake by sugarcane vacuoles, *FEBS Lett.*, **173**, 1, 1984a.

Thom, M. and Komor, E., Role of the ATPase of sugar cane vacuoles in energization of the tonoplast, *Eur. J. Biochem.*, **138**, 93, 1984b.

Thom, M. and Komor, E., Electrogenic proton translocation by the ATPase of sugar cane vacuoles, *Plant Physiol.*, **77**, 329, 1985.

Thom, M. and Maretzki, A., Group translocation as a mechanism for sucrose transfer into vacuoles from sugarcane cells, *Proc. Natl. Acad. Sci., USA.*, **82**, 4697, 1985.

Thom, M., Leigh, R.A. and Maretzki, A., Evidence for the involvement of a UDP-glucose dependent group translocator in sucrose uptake into vacuoles of storage roots of red beet (*Beta vulgaris* L.), *Planta*, **167**, 410, 1986.

Thomas, P.L. and Isaac, P.K., An electron microscopic sutdy of intravacuolar bodies in the uredia of wheat stem rust and in hyphae of other fungi, *Can. J. Bot.*, **45**, 1473, 1967.

Thornton, R.M., The fine structure of Phycomyces I. Autophagic vesicles, *J. Ultr. Res.*, **21**, 269, 1968.

Thurman, D.A. and Rankin, J.A., The role of organic acids in zinc tolerance in *Deschampsia caespitosa*, *New Phytol.*, **91**, 629, 1982.

Tice, K.R., Parker, D.R. and De Mason, D.A., Operationally defined apoplastic and symplastic aluminium fractions in root tips of aluminium- intoxicated wheat, *Plant Physiol.*, **100**, 309, 1992.

Toepfer, E.W., Mertz, W., Polansky, M.M., and Roginski, E.F., Preparation of chromium-containing material of glucose tolerance factor activity from brewer's yeast extract and by synthesis, *J. Agric. Food. Chem.*, **25**, 162, 1977.

Tommasini, R., Martinoia, E., Grill, E., Dietz, K.J. and Amrhein, N., Transport of oxidized glutathione into barley vacuoles: evidence for the involvement of the glutathione S-conjugate ATPase, *Z. Naturforsch.* Sect C Biosci, **48**, 867, 1993.

Tomsett, A.B. and Thurman, D.A., Molecular biology of metal tolerances in plants, *Plant Cell Environ.*, **11**, 383, 1988.

Torii, K., and Lateis, G.G., Dual mechanisms of ion uptake in relation to vacuolation in corn roots, *Plant Physiol.*, **41**, 863, 1966.

Toriyama, H. and Jaffe, M.J., Migration of calcium and its role in the regulation of seismonasty in the motor cell of *Mimosa pudica* L., *Plant Physiol.*, **49**, 72, 1972.

Truchet, G. and Coulomb, P., Mise en evidence et evolution du systeme phytolysosomal dans les cellules des differentes zones de nodules radiculaires de pois (*Pisum sativum* L.). Notion d'heterophagie, *J. Ultrastruct. Res.*, **43**, 36, 1973.

Tuttle, D.L. and Dunn, W.A., Divergent modes of autophagy in the methylotrophic yeast *Pichia pastoris*, *J. Cell Sci.*, **108**, 25, 1995.

Urech, K., Dürr, M., Boller, T., Wiemken, A. and Schwenke, J., Localization of polyphosphate in vacuoles of *Saccharomyces cerevisiae*, *Arch. Microbiol.*, **116**, 275, 1978.

Van der Leij, M., Smith, S.J. and Miller, A.J., Remobilization of vacuolar stored nitrate in barley root cells, *Planta*, **205**, 64, 1998.

Van der Wilden, W., Herman, E.M. and Chrispeels, M.J., Protein bodies of mung bean cotyledons as autophagic organelles, *Proc. Nat. Acad. Sci. USA.*, **77**, 428, 1980.

Van Steveninck, R.F.M., Van Steveninck, M.E., Fernando, D.R., Edwards, L.B. and Wells, A.J., Electron probe microanalytical evidence for two distinct mechanisms of Zn- and Cd-binding in a Zn-tolerant clone of *Lemna minor* L., *Compt. Rend. .Acad. Sci.*, **310**, Paris, Ser. III, 671, 1990.

Van Steveninck, R.F.M., Van Steveninck, M.E., Fernando, D.R., Horst, W.J. and Marschner, H., Deposition of zinc phytate in globular bodies in roots of *Deschampsia casespitosa* ecotypes: a detoxification mechanism? *J. Plant Physiol.*, **131**, 247, 1987.

Vázquez, M.D., Poschenrieder, Ch., Barceló, J., Chromium VI induced structural and ultrastructural changes in bush bean plants (*Phaseolus vulgaris* L.), *Ann. Bot.*, **59**, 427, 1987.

Vázquez, M.D., Poschenrieder, Ch., Corrales, I. and Barceló, J., Change in apoplastic aluminium during the initial growth respose to to aluminium by roots of a tolerant maize variety, *Plant Physiol.*, **119**, 435, 1999.

Vázquez, M.D., Poschenrieder, Ch., Barceló, J., Baker, A.J.M., Hatton, P., and Cope, G.H., Compartmentation of zinc in roots and leaves of the zinc hyperaccumulator *Thlaspi caerulescens* J&C Presl., *Bot. Acta*, **107**, 187, 1994.

Verma, D.P.S., Kazazian, V., Zogbi, V. and Bal, A.K., Isolation and characterization of membrane envelope enclosing the bacteroids in soybean root nodules, *J. Cell. Biol.*, **78**, 919, 1978.

Villiers, T.A., Cytolysosomes in long-dormant plant embryo cells, *Nature*, **214**, 1356, 1967.

Villiers, T.A., Lysosomal activities of the vacuole in damaged and recovering plant cells, *Nature New Biol.*, **233**, 57, 1971.

Vögeli-Lange, R. and Wagner, G.J., Subcellular localization of cadmium and cadmium-binding peptides in tobacco leaves, *Plant Physiol.*, **92**, 1086, 1990.

Voss, M. and Weidner, M., Uridine-5'-diphospho-D-glucose-dependent vectorial sucrose synthesis in tonoplast vesicles of the storage hypocotyl of red beet (*Beta vulgaris* L. ssp. Conditiva), *Planta*, **173**, 96, 1988.

Wagner, G.J., Vacuolar deposition of ascorbate-derived oxalic acid in barley, *Plant Physiol.*, **67**, 591, 1981.

Wagner, G.J., Vacuoles, in Modern Methods of Plant Analysis, Vol. 1, Linskens, H.F. and Jackson, I.F., Eds, Springer Verlag, Berlin, Heidelberg, 1985, 105.

Wagner, G.J., Accumulation of cadmium in crop plants and its consequences to human health, *Adv. Agron.* **51**, 173. 1994.

Wagner, V.T. and Dumas, C., Morphometric analysis of isolated *Zea mays* sperm, *J. Cell. Sci.*, **93**, 179, 1989.

Wagner, G.J. and Lin, W., An active proton pump of intact vacuoles isolated from *Tulipa* petals, *Biochem. Biophys. Acta*, **689**, 26l, 1982.

Wagner, G.J., and Mulready, P., Characterization and solubilization of nucleated specific Mg^{2+}-ATPase and Mg^{2+}-pyrophosphatase of tonoplast, *Biochem. Biophys. Acta*, **728**, 267, 1983.

Wagner, W., Keller, F. and Wiemken, A., Fructan metabolism in cereals: induction in leaves and compartmentation in protoplasts and vacuoles, *Z. Pflanzenphysiol.*, **112**, 359, 1983.

Walker, D.J., Leigh, R.A. and Miller, A.J., Potassium homeostasis in vacuolate plant cells, *Proc. Natl. Acad. Sci. USA*, **93**, 10510, 1996.

Wang, J., Evangelou, B.P., Nielsen, M.T. and Wagner, G.J., Computer simulated evaluation of possible mechanisms for sequestering metal ion activity in plant vacuoles. I. Cadmium, *Plant Physiol.*, **97**, 1154, 1991.

Wang, J., Evangelou, B.P., Nielsen, M.T. and Wagner, G.J., Computer simulated evaluation of possible mechanism for sequestering metal ion activity in plant vacuoles, II Zinc., *Plant Physiol.* **99**, 621, 1992.

Wang, Y., Leigh, R.A., Kaestner, K.H. and Sze, H., Electrogenic H^+-pumping pyrophosphatase in tonoplast vesicles of oat roots, *Plant Physiol.*, **81**, 497, 1986.

Ward, J.M. and Schroeder, J.I., Calcium-activated K^+ channels and calcium-induced calcium release by slow vacuolar ion channels in guard cell vacuoles implicated in the control of stomatal closure, *Plant Cell*, **6**, 669, 1994.

Ward, Z.M., Pei, Z-M., Schroeder, J.I., Roles of ion channels in initiation of signal transduction in higher plants, *Plant Cell*, **7**, 833, 1995.

Wardlaw, I.F., Phloem transport: physical, chemical or impossible, *Ann. Rev. Pl. Physiol.*, **25**, 515, 1974.

Wardrop, A.B., Occurrence of structures with lysosome-like function in plant cells, *Nature*, **218**, 978, 1968.

Weiser, T., Blum, W. and Bentrup, F.W., Calmodulin regulates the Ca^{2+}-dependent slow-vacuolar ion channel in the tonoplast *of Chenopodium rubrum* suspension cells, *Planta*, **185**, 440, 1991.

Welch, R.M., Micronutrient nutrition of plants, *Crit. Rev. Plant Sci.*, **14**, 49, 1995.

Wiebe, H.H., The significance of plant vacuoles, *Bioscience*, **28**, 327, 1968.

Wiemken, A., and Dürr, M., Characterization of amino acid pool in vacuolar compartment of *Saccharomyces cerevisiae*, *Arch. Microbiol.*, **101**, 45, 1974.

Wilkins, M.B., The diurnal rhythm of carbon dioxide metabolism in *Bryophyllum*: the mechanism of phase-shift induction by thermal stimuli, *Planta*, **157**, 471, 1983.

Willenbrink, J., Die pflanzliche Vakuole as Speicher, *Naturwissenschaften*, **74**, 22, 1987.

Williamson, B., Acid phosphatase and esterase activity in orchid mycorrhiza, *Planta*, **112**, 149, 1973.

Williamson, R.E., and Ashley, C.C., Free Ca^{2+} and cytoplasmic streaming in the alga *Chara*, *Nature*, **296**, 647, 1982.

Wilson, C. and Lucas, W.J., Influence of internal sugar levels on apoplasmic retrieval of exogenous sucrose in source leaf tissue, *Plant Physiol.*, **84**, 1088, 1987.

Wilson, C.L., Stiers, D.L. and Smith, G.C., Fungal lysosomes or spherosomes, *Phytopathology*, **60**, 216, 1970.

Wilson, P.D. and Wheeler, K., Permeability of phospholipid vesicles to amino acids, *Biochem. Soc. Transact.*, **1**, 369, 1973.

Winter, H., Lohans, G. and Heldt, H.W., Phloem transport of amino acids and sucrose depending on the corresponding metabolite levels in the leaves of barley, *Plant Physiol.*, **99**, 996, 1992.

Winter, H., Robinson, D.G. and Heldt, H.W., Subcellular volumes and metabolite concentrations in barley leaves, *Planta*, **191**, 180, 1993.

Winter. H., Robinson, D.G. and Heldt, H.W., Subcellular volumes and metabolite concentrations in spinach leaves, *Planta*, **193**, 530, 1994.

Wodzicki, T.J. and Brown, C.L., Organization and breakdown of the protoplast during maturation of pine tracheids, *Amer. J. Bot.*, **60**, 631, 1973.

Wolf, A.E., Dietz, K-J, and Schroeder, P., Degradation of glutathione S-conjugates by a carboxypeptidase in the plant vacuole, *FEBS Lett.*, **384**, 31, 1996.

Wolff, A.M., Din, N. and Petersen, J.G.L., Vacuolar and extracellular maturation of *Saccharomyces cerevisiae* Proteinase A, *Yeast*, **12**, 823, 1996.

Wood, R.K.S., Physiological Plant Pathology, Blackwell Sci. Pub., Oxford and Edingburgh, UK, 1967.

Wright, K.M., Davies,T.G.E., Steele, S.H., Leigh, R.A. and Oparka, K.J., Development of a probenecid-sensitive Lucifer Yellow transport system in vacuolating oat aleurone protoplasts, *J. Cell Sci.* **102**, 133, 1992.

Wyn Jones, R.G., Salt tolerance, in 'Physiological Process Limiting Plant Productivity', Johnson, C.B., Ed., Butterworths, London, 1981, 271.

Yamada, S., Katsuhara, M., Kelly, W.B., Michalowski, C.B. and Bohnert, H.J., A family of transcripts encoding water channel proteins: Tissue-specific expression in the common ice plant, *Plant Cell*, **7**, 1129, 1995.

Yamaguchi-Shinozaki, K., Koizumi, M., Urao, S. and Shinozaki, K., Molecular cloning and characterization of 9 cDNAs for genes that are responsive to desiccation in *Arabidopsis thaliana* : Sequence analysis of one cDNA clone that encodes a putative transmembrane channel protein, *Plant Cell Physiol.*,**33**, 217, 1992.

Yamamoto, H., Suzuki, M., Kitamura T., Fukui, H. and Tabata, M., Energy requiring uptake of protoberberine alkaloids by cultured cells of *Thalictrum flavum*, *Plant Cell Rep.* **8**, 361, 1989.

Yamasaki, H., Sakihama, Y. and Ikehara, N., Flavonoid-peroxidase reaction as a detoxification mechanism of plant cells against H_2O, *Plant Physiol.*, **115**, 1405, 1997.

Zeiger, E., The biology of stomatal guard cells, *Ann. Rev. Plant Physiol.*, **34**, 441, 1983.

VIII
Retrospect and prospect

8.1 Definitions

An **organelle** can be defined as a membrane-bound subcellular entity with distinctive structure and functions, which can be isolated by centrifugation. Vacuoles meet these criteria. **Vacuoles** are plant cell organelles delimited by a single membrane, the **tonoplast**, with characteristic integral proteins, water channels (aquaporin) and proton pumps, as well as vacuolar-type tonoplast-bound ATPase and PPase. The lumen of the vacuole contains an aqueous acidic sap with different ions, molecules and enzymes, chiefly hydrolytic, capable of undertaking diverse functions as dictated by the physiology of the cell.

8.2 Significance of vacuoles

In terms of evolution, the primitive uni- or multi-cellular plants were aquatic. The evolution of land plants started with adaptation for mesophytic conditions and the development of a mode of conservation of water and maintenance of body structure with **rigidity** by turgor pressure. The cell wall *per se* is not rigid enough, as is obvious from wilted leaves or plucked flowers. Whereas vacuoles may not be obligatory for an aquatic plant, for maintenance of body rigidity they are a necessity for land plants. Evolutionary success came to those plants that could store water and maintain turgor through diverse adaptations, with concomitant evolution of vacuoles for myriad functions.

The **turgor pressure** developed by the vacuole is involved directly in the expansion or elongation of cells, when localised or uniform weakening of the cell wall orients turgor-driven cell elongation or expansion. The aerial hyphae of fungi, or even the cracking of the asphalt road by mushrooms are manifestations of turgor pressure. The fast elongation of the germinating spore, the extension of the pollen tube, the growth of the root hair or the pushing of the basidial cytoplasm to the basidiospores are examples of rapid elongation of the cell wall as well as the vacuole and maintenance of adequate turgor pressure. Turgor pressure is also responsible for all increases in cell dimension which lead to growth. Thus, vacuolar activity is necessary for plant growth. Similarly, all plant movements, directly or indirectly, involve vacuoles. Except for the movement of locomotion by cilia, all tropic and nastic movements are due to differential cell elongation or expansion caused by changes in turgor pressure.

The ability to hold a large amount of water in the vacuole leads to an increase in cell dimension only with light atoms and/or substances low in energy. Thus, a decent structure and strength can be maintained by the cheapest space-filling substance. Moreover, a **high surface/dry weight ratio** is necessary for obtaining nutrients. The large vacuole moves the cytoplasm to the periphery of the cell, where the cytoplasm forms a thin layer in order for exchange of gases and nutrients to occur readily. This is true even for deep-seated cells in a multicellular structure. In flowering plants, after fertilisation, the embryo sac undergoes rapid enlargement and the endosperm nucleus undergoes free nuclear division. The embryo sac wall, which becomes the endosperm wall, continues to press into the nucellar tissue and gradually consumes it. The basal cell of the developing embryo enlarges, becomes a large sac and functions as a haustorium. In both cases, at least at the early stage, the vacuole is used as a device for greater contact with the surrounding tissue for enhanced absorption of nutrients. Moreover, an adequate **fluidity of the cytosol**, which is an important aspect of intracellular activity, is assured by the supply of water from the vacuole. The process of fertilisation, for example, requires a degree of fluidity for release of the two sperm nuclei and their subsequent movement for fertilisation and secondary fusion.

The role of the vacuole in **storage** of diverse substances, from ions to very large molecules, oligomers, polypeptides and complex molecules, cannot be overemphasised. Whatever is absorbed or synthesised in excess by the cytoplasm, or is not immediately needed by the cell, is stored in the vacuole. However, the most significant storage substances are the hydrolysing enzymes. The sequestration of the deadly enzymes is of remarkable advantage to the cell which uses the vacuole for degrading undesirable substances or for future release in the cytosol when needed. The latency of vacuoles in this respect makes them homologous to the lysosomes.

In addition to storage, **homeostasis** of ions and small molecules is of great importance to the cell for operation under normal as well as under stress conditions and for differentiation. A wide range of cytosolic activities including pH regulation, salt stress, osmoregulation, etc., are aided and/or controlled by homeostasis of diverse ions, organic acids, amino acids, sugars, etc. Ca^{2+} homeostasis is esential for Ca^{2+} to act as the second messenger in controlling diverse physiological processes.

It is remarkable that some of the degrading enzymes are used **against pathogens**, and vacuoles also constitute the store-house for non-enzyme pathogenesis-related proteins. It is still more surprising that very similar proteins are induced by salt stress and pathogens. Osmotin produced by osmotic stress is antifungal. The vacuolar enzyme Ap-24 is homologous to P-23 which is virus-induced as well as ethylene-induced. In other words, closely similar genes or DNA-sequences are involved for diverse reactions to be carried out by the vacuole. Very often the vacuole is a **waste disposal** system, as shown by its activity in autophagy, heterophagy, detoxification and phytoremediation. It is also the default destination, that is, if any nutrient is absorbed or produced by the cytoplasm with any defect, it is dispatched to the vacuole. In many cases non-specific proteins, with no targeting information or with wrong information, find entry into the lumen, only to be broken down. It is also remarkable that vacuoles of leaf mesophyll cells are preferred as receptacles for garbage disposal. For example, if excess sulfate is taken up by the mesophyll or root cells it is sequestered in the mesophyll vacuoles. It is a very sensible way to dispose of un-utilised excess or metabolic waste product from the inventory by simple shedding of leaves!

A number of **biosynthetic** pathways operate inside the lumen of the vacuole. It is an enigma that a structure which is a station for the breakdown of various substances can also carry out biosynthesis. In fact, the various biochemical steps that occur at the tonoplast or in the lumen may be considered as another method of garbage disposal, in that the substances are immobilised or trapped inside. It is not the typical biosynthesis which connotes synthesis of substances to be utilised by the cell for some anabolic purpose. In other words, what cannot be synthesised or processed in the common milieu of the cytosol is done selectively in the vacuole. For example, glycosylation, which is necessary for biosynthesis of secondary metabolites like flavonoids, cyanogenic glucosides, glucoalkaloids, glucosinolates, etc., is also needed for conjugation of glutathione with xenobiotic electrophiles and possibly metal conjugates of mustard oil glucosides. The glutathione-conjugate is transported into the vacuole by a glutathione-conjugate pump, a member of the superfamily of ATP-binding cassette (ABC) transporters known for a multitude of transport functions.

The tonoplast is a unique membrane capable of housing diverse transporters and permitting numerous enzymatic reactions. A number of experiments suggest that it may be as important to the cell as the plasma membrane (Tazawa *et al.* 1987). Studies on the viability of giant cells of *Nitella* are illuminating. When the plasma membrane is permeabilised or disintegrated by an ice-cold hypertonic medium containing EGTA, the tonoplast holds the cytoplasm intact and living for

7 days. But when the tonoplast is destroyed, keeping the plasma membrane intact, the cells cannot survive for more than 24 hours.

Finally, it has been suggested that the vacuole provides direct protection to the nucleus. Evacuolated protoplasts are more sensitive to mutagens (UV) and other toxic substances than vacuolated ones (Griesbach and Lawson 1985). Many investigators suggest that the vacuole's ability to sequester toxic substances precludes the exposure of genetic material to potential mutagens. In a series of publications, Raven (1984, 1985, 1991, 1997) recorded the cost–benefit analysis of the vacuole with respect to nutrient uptake, energetics and transport of solutes, storage of water, photon absorption during photosynthesis, accumulation of defence materials, etc., in diverse plants. Raven (1997) concluded that the evolutionary costs of vacuoles are outweighed by their evolutionary benefits.

It may be an aphorism that the cell uses the vacuole as an additional utility room in its household for a diverse range of purposes.

Although vacuoles are involved in numerous activities of the cell, they are not involved in certain functions, especially secretion. They do not participate in mucilage secreting cells and nectaries, nor in the glands of carnivorous plants or in the glands for terpenoid, flavonoid or phenolic secretion. They do not have direct participation in salt glands and hydathodes. They are not involved in secretion of proteins, polysaccharides or root cap slime. The secretion from the stigma of *Solanum tuberosum* originates as osmiophilic droplets in the cytoplasm, accumulates in intracellular space and fills the base of the papillae, after rapturing and lifting the cuticular layer covering the stigma surface (Mackenzie *et al.* 1990). The essential oil of Lamiaceae are secreted outside the peltate trichome cells and the secretory product seems to remain accumulated in a subcuticular space and not in the large vacuoles. In the glandular trichomes the oil is released through the micropores (Ascenso *et al.* 1995).

8.3 Uniqueness of vacuoles

The plant cell vacuole is an exceptional organelle in many ways. It is virtually **ubiquitous** and is prominent in most cells. The only type of cells where vacuoles are rarely detectable are those where intense programmed metabolic activity takes place, e.g. meristematic cells of root tips or microsporogenous cells. But as soon as differentiation starts, vacuolation becomes prominent. Ubiquity, in such cases, almost simultaneously connotes **indispensibility**. Whereas a plant cell can live, at least for some time, without an organelle like the mitochondrion, the Golgi body, the plastid or even nucleus, no plant cell can live without a vacuole. There are examples of cells which generally lack one of the above organelles, but no cell exists without a vacuole for long. That it is indispensible has been shown by evacuolation studies by a number of workers. Vacuoles reappear right after evacuolation, even in suspension culture cells where turgor pressure and rigidity are maintained and waste disposal or homoeostatic requirements are provided by the culture medium. More remarkable is the observation that the vacuole-less daughter of a yeast *conditional vacuolar segregation* mutant, which failed to inherit proteolytic activities from the mother cell, formed vacuolar vesicles that fused into a new normal vacuole within 30 min (De Mesquita *et al.* 1997). Despite the fragility of its single outlining membrane the vacuole maintains its **integrity**. It is not known to undergo fusion with other organelles or membranes *in planta* and it is not normally attached to any membrane, microtubule, microfilament, intermediate filament or ribosome. When more than one similar or dissimilar vacuoles co-reside in the same cell they are not connected by any membrane or structure and maintain their individuality. On the other hand, the formation of a vacuole is an **irreversible** process. With exception, once the vacuole is formed, it exists for the rest of the life of the cell. It can undergo fragmentation or division, and its total obliteration or disintegration occurs prior to or during cell death. No natural process exists for evacuolation, except for contractile vacuoles in lower plants. Even after all the stored protein in protein storage vacuoles are utilised, the vacuoles are maintained.

Whereas other cellular organelles conduct only one major function, e.g. the nucleus for transcription of the genome, the mitochondria for aerobic respiration, the chloroplast for photosynthesis, the Golgi body for biochemical processing, the vacuole is a **multifaceted** organelle. It does not restrict itself to a single metabolic function like other single membrane-bound organelles, peroxisomes and spherosomes. To describe a vacuole as solely a water reservoir, an osmoregulator, a pH controller, a storage structure, a digestion chamber or a waste basket would not only be a misnomer, but would also be undermining its importance to the cell. It is a versatile organelle capable of undertaking different functions in different cells or at different times in the same cell. Moreover, there are cases in which two or more different types of vacuoles have different tonoplast composition and perform different functions. Closely similar proteins, which are induced by salt-stress or pathogens, are targeted to vacuoles. Similarly, homologous proteins responsible for anti-fungal activity and osmotic adaptation are vacuolar. Physical stress-induced proteins may be co-localised with pathogen-related proteins in the vacuole. However, the most remarkable instance of the multipurpose characteristic of the vacuole is that noted in yeast cells where cytoplasm-to-vacuole targeting and autophagy employ the same machinery to deliver proteins to the vacuole. Two related but distinct autophagic-like processes are involved in both biogenesis of vacuolar resident proteins and sequestration of substances to be degraded (Scott *et al.* 1996; Baba *et al.* 1997). An equally pre-eminent feature is that a multispecific ATP-binding cassette transporter of *Arabidopsis* tonoplast is competent in the transport of both glutathione-S conjugates of xenobiotics and chlorophyll catabolites for micro-autophagy (Lu *et al.* 1998). Besides, a tonoplast phytoremediation factor for cadmium is also a GS-conjugate transporter (Li *et al.* 1997). Active oxygen species formed by various conditions can be **potentially scavenged** by the vacuole. A confluence of such functions emphasises the role of vacuoles, not only in **house-keeping** but also in **fortification** against stress, disease and unwarranted oxidation.

Vacuoles also represent a unique **storage-retrieval** system. They are a sink-source structure for diverse nutrients and metabolites. Vacuoles may be used for: i) **diurnal** accumulation and release, e.g. organic acids in CAM plants; ii) **seasonal** deposition and retrieval, e.g. oligosaccharides in storage cells; or iii) **long-term** or permanent storage and retrieval, e.g. reserve food and secondary metabolites. The retrieval process does not involve blebbing or formation of vesicles and **no exocytosis** or evagination takes place, although the process of invagination or endocytosis involving fusion of vesicles and provacuoles is a regular feature for development of the vacuole. The most important contribution of the vacuole in the functioning of the cell is **homeostasis** of osmotic pressure, ions, pH, Ca^{2+}, amino acids, etc. Moreover, the vacuole may be considered as equivalent to an **extracellular compartment**, to put aside what is not immediately needed for cell metabolism. The usual extracellular substances secreted in the environment or apoplast are not retrievable, whereas substances set aside in the vacuole can be supplied to the cytosol on demand. On the basis of its ubiquity and multipotency, a vacuole should be regarded as an essential compartment which is modified according to the physical, physiological and metabolic necessities of the cell.

8.4 Homology

An appreciation of the homology of an organelle with other organelles in terms of its structure, chemical composition, origin, function, etc., is essential for comprehending its evolution, as well as the scope for its manipulation. The homology of the vacuole can be conveniently analysed with reference to the tonoplast vis-a-vis other membranes, and the vacuolar contents and functions vis-a-vis those of other organelles.

8.4.1 Homology of tonoplast with plasma membrane

One way to assess the relationship between two different membranes is to study their fusion. The first indication of similarity between the tonoplast and the plasma membrane came from the micromanipulation studies by Plowe (1931). He thrust a needle into a naked protoplast previously divested of its cell wall and pushed the vacuole towards the plasma membrane. At a place where the cytoplasmic surface of the tonoplast and the plasma membrane came into contact the membranes fused without releasing the cytoplasm, creating an aperture through which the vacuolar contents came out in the medium. The resulting cup-like protoplast regained a spherical form covered by a single membrane of dual origin. The routine fusion of plasma membrane with tonoplast during zoospore formation in a number of aquatic phycomycetes is another pointer to their similarity. The process of zoospore development starts with the formation of cleft-vesicles (large vacuoles) which, by their position, delimit the uninucleate spore initials at the periphery of the sporangium (Bessey 1950). Ultimately the cleft-vesicles fuse to form an extensive and continuous tonoplast deeply invaginated between the lobes of the zoospore initials. During this process, the plasma membrane has remained closely associated with the sporangium wall with no evidence of invagination or other changes. The newly developed tonoplast extends to contact and fuse with the plasma membrane, thereby separating the spores, and the plasma membrane of each spore is derived from parts of both tonoplast and sporangium plasma membrane.

Ganser *et al.* (1982) induced fusion of vacuoles with protoplasts in *Kalanchoe daigremontiana*, under the influence of an electrical field. The fusion of protoplasts with vacuoles occurs nearly as quickly as the fusion of isolated protoplasts or vacuoles with each other. They concluded that the plasma membrane and the tonoplast are biologically compatible in spite of their certain differences in physiological properties and chemical composition. Similarly, De and Swain (1983) showed that in the locular sap of ripening tomato fruits, free vacuoles may spontaneously fuse with each other, as well as undergoing heterotypic fusion with sub-protoplasts, cytoplasts and vacuoplasts. That the tonoplast could be fused with the plasma membrane and the resulting cell kept alive has been demonstrated by Lackney *et al.* (1990). Protoplasts from grape calli were fused with small vacuoles containing anthocyanins in an electro-fusion buffer containing 1.0% PEG after application of DC pulses. The novel grape vine cells with tonoplast segments in the plasma membrane remained viable for several hours.

As far as the chemical composition of the two membranes is concerned, Marty and Branton (1980) reported that the ratio of phosphotidyl choline and phosphotidyl-ethanolamine of tonoplast resembles those of the plasma membrane and ER. Both plasma membrane and tonoplast, which are the most sterol-rich membranes of the cell, are thought to be terminal membranes produced through sequential and antiparallel differentiation (Marty 1978). As has been described earlier (Chapter IV), filipin induces lesions in sterol-rich membranes. Whereas filipin treatment showed little effect on ER or nuclear membrane, both tonoplast and plasma membranes exhibited lesions, albeit in different frequencies and patterns (Marty 1985). That cytochemically the tonoplast may be similar to the plasma membrane has been indicated by osmium-impregnation techniques together with high-voltage transmission-electron microscopy of thick sections. Chaffey (1995) has shown that zinc iodide-osmium tetroxide or osmium ferricyanide stained the intercisternal spaces of the dictyosomes, ER lumen, central region of plasmodesmata and the perinuclear and perimitochondrial spaces. However, the plasma membrane, tonoplast and smooth coated vesicles within the cell remained unstained.

Both the tonoplast and the plasma membrane permit virtually free water flow driven by hydraulic or osmotic forces but not by any membrane pumps. Aquaporins form these water-selective channels, allowing water to pass while excluding ions and metabolites. Daniels *et al.* (1994) identified a specific plasma membrane-intrinsic constitutively expressed aquaporin RD-28 which is a homologue of the tonoplast aquaporin gamma-TIP. So far, no aquaporin has been specifically identified as a permanent constituent of any other plant cell membranes. The most important and ubiquitous of the tonoplast enzymes, ATPase, is also present in the plasma membrane. Although tonoplast ATPase is different from plasma membrane ATPase in structure and sensitivity to inhibitors, Magnin *et al.* (1995) have pointed out some fine biochemical similarities between the two.

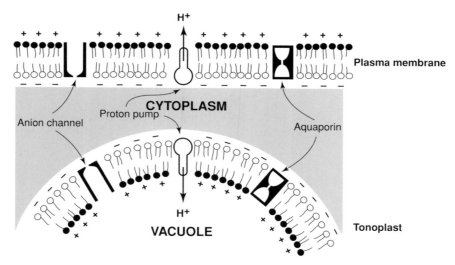

Figure 8.1 Schematic diagram depicting reverse orientation of the plasma membrane and tonoplast with respect to the cytoplasm.

It is also significant that both tonoplast and plasma membrane contain mobile charges which form a part of the transport system for ions. Benz et al. (1988) suggested that in *Halicystis parvula*, a marine alga, both tonoplast and plasma membrane have very similar electrical properties, both contain mobile charges with a total surface concentration of about 30 nmol m^{-2} and a translocation rate constant of about 500 s^{-1}. This is possible due to the sum total of various proton pumps and ion channels present in both the membranes in the same or opposite direction with reference to the cytosol. The inner surface of the plasma membrane and the cytosolic surface of the tonoplast are both negatively charged. In the same way the anion channels permit a passive influx of a high concentration of anions across the plasma membrane from the outside of the cell, so does the anion channel of the tonoplast permit passive efflux of anions from the vacuole to the cytoplasm. The energy-dependent proton pump ATPases are equally operative, but only in opposite directions. According to the sign conventions for endomembranes, the vacuolar lumen is considered electrically equivalent to the extracellular space (Bertl et al. 1992). In essence, the tonoplast and plasma membranes are placed inside the cell in opposite orientation with respect to the cytoplasm (Fig. 8.1). Thus it is likely that what is transported inside the lumen of the vacuole can be secreted out by the plasma membrane in the apoplast or extracellur space. Wink (1994) observed this in cell suspension culture of *Lupinus polyphyllus*. The cultured cells secreted a number of organic acids, many amino acids, as well as quinolizidine alkaloids in the medium. In addition, the cells secreted the typical vacuolar enzymes, viz. acid phosphatase, phosphodiesterase, DNase, alpha-mannosidase, alpha-galactosidase, beta-glucosidase, lipase and protease, in the extracellular medium, but not the cytosolic enzymes, such as glutamate dehydrogenase and malate dehydrogenase. In other words, vacuoles are equivalent to an extracellular compartment. Lately, Kunze et al. (1998) observed that the tonoplast and the plasma membrane behave similarly under certain physiological conditions. When treated with 2,4-D, tobacco cell cultures secreted newly synthesised vacuolar class I isoforms of chitinase, vacuolar α-mannosidase, as well as β-1,3-glucanase. Subsequently, results from the same laboratory (Hensel et al. 1998) indicated that a chitin-binding PR-4 protein is constantly secreted in the medium during subculturing in the presence of different combinations of plant growth regulators. The similarity of the two membranes is also obvious from the report by Frigerio et al. (1998) that when a form of phaseolin targeted for vacuoles is

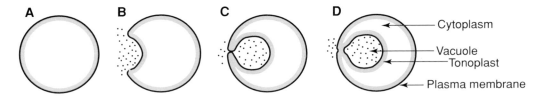

Figure 8.2 Suggested stages of a model for the evolution of the vacuole, by which the tonoplast has reverse orientation of the plasma membrane with respect to the cytosol. **a** Primaeval cell sac outlined by a single membrane (plasma membrane). **b** Invagination of the membrane needed for ingestion of exoplasmic solution with particulate matter (small circles) and /or bulk flow of solute. **c** Fusion of the membrane at the point of pinching of the invaginated mass. Note that the engulfment of the exoplasmic solution leads to fusion of the identical faces of the membrane. **d** Detachment of the inner sac (vacuole) leads naturally to reversal of the membrane with respect to the sap (cytosol) and sequestration of the exoplasmic solution.

overexpressed, the excess phaseolin is released in the culture medium. They concluded that targeting to the vacuole is less adaptable and secretion through the plasma membrane seems to be the safety valve for saturation of vacuolar sorting receptors.

8.4.1.1 Evolution of the vacuole

It is generally agreed that the primeval cell was a membrane-bound sac (liposome), formed spontaneously by assembly of amphipathic, possibly phospholipid, molecules from the prebiotic soup. The sac enclosed a self-replicating mixture of RNA, various ions, amino acids, sugars, fatty acids, nucleosides and nucleotides.

Two possible paths are suggested for the origin of vacuoles. The first possibility is that a large cell ingests a small liposome as heterophagy. In that case the membrane of the ingested liposome is not expected to turn inside-out. The second possibility is that the vacuole originated as an invagination of the cell membrane (plasma membrane), or by **fluid-phase endocytosis**. Such invagination is a common feature of many biomembranes. Invagination of the primeval cell membrane could have taken place i) when the cell is forced to shrink in surface area; ii) when bulk ingestion of external fluid is necessitated; or iii) when ingestion of particulate matter, incapable of diffusing across the membrane, is required.

It is proposed here that the vacuole originated by invagination of the plasma membrane leading to the inverted (**inside-out**) lipid bilayer of the tonoplast. This accounts for the reverse orientation of the proton pumps, anion channels and charge distribution of the tonoplast (Fig. 8.2). The reverse orientation of the tonoplast structure also suggests that it must have originated after the primeval plasma membrane acquired proton pumps, ion channels and some other transporters. In other words, the metabolic exigencies of the primeval cell brought about the origin of the vacuole.

8.4.2 Homology with other membranes

As far as the homology of the tonoplast with other membranes of the cell is concerned, there is no indication of similarity of the tonoplast with the outer and inner membranes of the nuclear envelope, mitochondria and chloroplast. ER and Golgi cisternae are also dissimilar to the tonoplast with reference to fusion properties, chemical composition, filipin-induced lesions, electron-staining, etc. However, biogenesis of vacuoles from ER, Golgi bodies, etc., as discussed in Chapter VI, may be considered. In numerous cases the ER directly enlarges or swells to form typical vacuoles. This development is evident from electron microscopic studies by various workers on germinating pollens, root tip cells of *Lupinus*, differentiation of *Chara* cells, leaf mesophyll cells, trichomes etc. Similar development of tannin vacuoles has been noticed in spruce cells in culture. It is likely that such direct development of a vacuole as a short-term or long-term storage structure would necessitate insertion of typical tonoplast proteins and enzymes in the ER membranes before it actually functions as a vacuole.

8.4.3 Homology with Golgi apparatus

The earliest indication of homology of vacuoles with Golgi apparatus was obtained by Bensley (1910), who studied root tips of onion with diverse fixatives. In young cells the vacuoles appeared as fine canals or a system of 'canaliculi' which enlarged and coalesced to form large vacuoles. On the basis of striking similarity of these canaliculi with the Golgi canals of animal cells, he suggested that the two are morphologically and physiologically equivalent. Guilliermond and Mangenot (1922) attempted silver impregnation technique in the root tip cells of barley and observed a dark network of canaliculi which led them to support the homology of vacuoles with Golgi bodies. In the following year, utilising the techniques of the animal cytologists, Dangeard (1923) concluded that the plant vacuolar substance and the animal Golgi material were identical.

Since then, numerous workers studying diverse cellular systems in plants using electron microscopic, electron cytochemical, organelle fractionation, biochemical and immunochemical methodologies, have established that vacuoles may originate from the Golgi apparatus. The steps in Golgi–vesicle–vacuole and ER–Golgi–vacuole systems have been discussed in Chapter VI. Moreover, fine biochemical aspects, as well as molecular details on the biogenesis of vacuoles, leave no doubt on the predominant role of the Golgi system in the formation of a vacuole and its subsequent development in many systems.

8.4.4 Homology with single-membraned organelles

Since the tonoplast is a single unit membrane, the homology of the vacuole may be sought with similarly outlined organelles, viz. spherosome, peroxisome and lysosome. Early workers identified the spherosomes as highly refractile bodies, lipid bodies or oil droplets. Kuehn (1985) defined the spherosome as a lysosome-like compartment in plants, which originates from ER and serves as a principal site of lipid storage. There have been occasional reports of association of acid phosphatase, esterase and other hydrolases with spherosomes. Presently the particle is not recognised as a distinct organelle but as a lipid body with the 'single-line' membrane being in fact a half unit-membrane (Yatsu and Jacks 1972).

Peroxisomes are single-membrane-bound organelles which usually contain one or more enzymes that use molecular oxygen to remove hydrogen atoms from specific organic substrates and produce hydrogen peroxide, which in turn is used by catalase to oxidise a variety of substances. A type of peroxisome present in certain seeds, where conversion of fats to sugars is accomplished by the enzymes of the glyoxylate bypass, is known as glyoxysome. The structure of the enveloping membrane which originates from ER, the contents and the function of peroxisome have no similarity with those of the vacuoles. Recently, Banjoko and Trelease (1995) have shown that peroxisomes have specific targeting signals which are not meant for vacuoles. Hence the two organelles are not considered to be homologous.

8.4.5 Homology with lysosomes

The first person to visualise vacuoles as homologous to lysosomes was Brachet (1957), who stated that lysosomes are identical to vacuoles as far as *in vivo* staining with basic dyes, e.g. neutral red and toluidine, is concerned. The origin of the vacuole through lytic processes aided by hydrolytic enzymes in membrane-enclosed cytosolic regions has been discussed in Chapter VI. In addition, the various lytic functions of the vacuoles have been discussed in detail in Chapter VII. The search for typical lysosomes in plants started after the discovery of lysosomes in animal cells.

Lysosome concept

In the early 1950s confusion existed over the presence of oxidative and hydrolytic enzymes in the mitochondrial fraction from various animal tissue homogenates. The problem was resolved when de Duve et al. (1955) demonstrated the presence of a number of hydrolases with acid pH optima in a sedimentable particulate fraction different from the mitochondrial fraction of rat liver tissue. Originally the five hydrolytic enzymes, acid phosphatase, acid ribonuclease, acid deoxyribonuclease, beta-glucuronidase and cathepsin, appeared to remain intact in the particles and indicated no soluble enzyme activity, a feature called latency. The enzyme activity could be released, i.e. latency could be broken, if the particles were treated with physical or chemical agents affecting the lipoprotein membrane. Subsequently, a whole range of hydrolytic enzymes including lipase, β-galactosidase, β-N-acetyl-glucosaminidase were also found in these particles which were termed 'lysosomes' (de Duve 1969). This biochemical idea of the compartmentation of hydrolytic enzymes was firmly established by cytochemical and electron microscopical studies (Novikoff et al. 1956). The lysosome is defined as a cellular organelle containing a number of hydrolases with acid pH optimum and outlined by a single membrane accounting for enzymatic latency. In addition, the concept includes the functional aspect of lysosomes, in that this organelle unifies the various steps of the intracellular digestion process in which the hydrolases act on the target substances. In animal cells, the primary lysosomes, also known as storage granules containing the enzymes, fuse with the endocytic vesicles which house the foreign extracellular material and form the secondary lysosome. The actual breakdown of the foreign substances takes place in the secondary lysosomes. This phenomenon is known as heterophagy, in contrast to autophagy in which some cellular structure or substance is engulfed and digested by the lysosome of the same cell (de Duve and Wattiaux 1966). Depending on the type of cell and the ingested substance, the lysosomes at various stages of digestion have been called food vacuole, digestive vacuole, pinocytic vacuole, cytolysosome, autophagic vesicle, multivesicular body, etc. At the final stage of digestion the lysosome appears as a residual body with undigested material, which may be expelled from the cell by exocytosis.

Lysosomes in plant cells

The development of a technique of sub-cellular localisation of enzymes by Gomori (1952) indicated the presence of hydrolysing enzymes in plant cells. By his method, or its modification, an enzyme can be detected in a tissue section or cell by its action on a suitable substrate leading to the formation of a product which makes a black or coloured precipitate. The presence of acid phosphatase in a particulate component of the cytoplasm of root tip cells was first reported by Jensen (1956). The presence of acid phosphatase as well as sulfatases in the root meristem cells of grasses was confirmed by Avers (1961). Subsequently, Walek-Czernecka (1962, 1965) demonstrated the presence of acid phosphatases, non-specific esterases, aryl-sulfatase, lipase, β-glucuronidase and β-galactosidase in particles which she called spherosomes in the cells of onion scale epidermis. Similarly, a search in pollens and pollen tubes of a large number of monocotyledonous and dicolyledonous plant species by Gorska-Brylass (1965) led to the detection of acid phosphatase, β-glucuronidase, acid deoxyribonuclease and aryl sulfatases in minute particles in the cytoplasm. Keeping the properties of lysosomes in animal cells in mind, Gahan (1965) studied lysosome-like particles in plant tissues which were carefully prepared. He demonstrated the presence of acid phosphatase in a particulate structure of the cytoplasm. Moreover, on the basis of incubation parameters with the substrate, he could demonstrate latency of the particles. The demonstration of both esterase and acid phosphatase substantiated his claim that these particles may be called lysosomes.

Parallel to the cytochemical studies, plant biochemists attempted to fractionate lysosomes from plant cells. However, using isolation procedures identical to those used for animal tissue failed. Working with germinating onion seeds, Harrington and Altschul (1963) were the first to isolate a light mitochondrial fraction similar to those obtained from animal cells, which showed acid phosphatase activity. The latency of the enzyme could also be demonstrated. Attempts by Matile *et al.* (1965) to demonstrate typical lysosomes in corn and tobacco seedlings met with partial success, since the hydrolases could not be definitely attributed to a specific fraction containing particulate structures. Semadani (1967) was more successful in isolating a light fraction which contained a number of acid hydrolases.

It was clear at this stage that the isolation of typical lysosome from plant cells was more to do with developing techniques for sedimentability and maintenance of latency, than of definition. Gahan (1967) proposed a definition that required a lysosome i) to possess a single limiting membrane; ii) to contain one or more acid hydrolases; iii) to possess the property of latency that could be affected by a series of processes known to affect membrane structure; and iv) to react cytologically with the acid haematin test, PARS reaction, or vital dyes. Although most of the reports do not precisely satisfy all the above criteria, it is now generally accepted that lysosomes are present in diverse plant cells.

At the ultrastructural level, dense particles exhibiting acid phosphatase activity have been demonstrated by Berjak (1968) in the root cap cells of maize. In a study of the differentiation of lignified collenchyma of *Eryngium*, Wardrop (1968) demonstrated the presence of distinct bodies containing acid phosphatase and beta-glucosidase with the aid of Gomori reaction at the electron microscope level. He emphasised the lysosomal function of these structures which are distinctly different from the large central vacuole. The existence of subcellular particles rich in lysosomal hydrolases in potato tuber tissue has been demonstrated by histochemical methods. Moreover, when the potato tubers were infected by *Phytophthora erythroseptica*, the lysosomes underwent swelling and disruption (Pitt and Coombes 1968). On the basis of morphological characteristics and enzymatic activity, Coulomb (1969) identified lysosome-like particles in the cells of gourd roots. A number of workers (Hall 1969; Malik *et al.* 1969; Ashford, 1970; Ashford and McCully 1970) asserted the presence of particulate structures or lysosomes in various meristematic cells of a number of flowering plants.

Like higher plants, lower plants also contain lysosomes. The presence of lysosomes in *Euglena* and other phytoflagellates has been established by Brandes *et al.* (1965), Ueda (1966) and Aaronson (1973). In the fungus *Botrytis cinerea*, Pitt and Walker (1967) and Pitt (1968) demonstrated the presence of acid phosphatase, several esterases and β-galactosidase in certain particulate structures with the aid of histochemical methods. Similarly, on the basis of biochemical, histochemical and ultrastructural observations, Ashworth and Weiner (1973) showed that the slime-mould *Dictyostellium discoidium* contained typical lysosomes. Hislop *et al.* (1974) located α-l-arabinofuranosidase at the particulate sites, similar to those containing acid phosphatase, in the cells of *Sclerotinia fructigena*. Thus, it appears that typical lysosomes are present in many types of plant cells. Summarising the available data on the presence and latency of hydrolases, Gupta (1985) contended that plant vacuoles are functionally analogous to animal lysosomes. As well as numerous similarities between vacuoles and lysosomes, certain aspects of gene expression and protein targeting experiments point to their homology. For example, when a typical lysosomal protein is synthesised in the cell of a transgenic plant, it may be transported to the vacuole. Neuhaus (1995) has shown that the human lysosomal enzyme (beta)-glucuronidase is targeted to the vacuole of plant cells, whereas a deletion mutant lacking the 15 C-terminal amino acids is secreted efficiently.

In his list of 'Some one-hundred 'somes'' , Kuehn (1985) defined the semantically related organelles attributed with lytic functions as follows:
- *Autolysosome* — digestive vacuoles formed by fusion of an autophagosome with cytoplasmic particles (teleolysosomes) containing acid hydrolase enzymes.
- *Autophagosome* — cytoplasmic vacuoles in rat liver cells into which have been sequestered cytoplasmic constituents of uncertain identity.
- *Cytosome* — a vacuole of unkown function, frequently found distributed in the cytoplasm of eukaryotes.

- *Endosome* — an endocytotic vesicle derived from the cell membrane that can fuse with lysosomes; probably the same as a receptosome.
- *Neolysosome* — a vesicle derived from Golgi-associated tubule buds which contains acid phosphatase and which grows into a free, mature lysosome.
- *Oleosome* — plant spherosomes that are rich in lipids but devoid of acid phosphatases and other lytic enzymes; specifically formed in fat-storing cells of developing seeds and fruits.
- *Heterolysosome* — a lysosome produced by fusion of primary lysosomes, originating from the Golgi, with other vesicles of cytoplasmic bodies.
- *Heterophagosome* — an endocytotic vacuole which is fused with other cellular vesicles or cytoplasmic bodies containing particulate material, but which is not fused with a lysosome to form a digestive vacuole.
- *Phagolysosome* — an endocytic vacuole which has taken up lysosomal hydrolases upon fusion with a primary lysosome or other digestive vacuole.
- *Phagosome* — an endosome which contains particulate material destined for hydrolytic digestion.
- *Phytolysosome* — a lysosome of plant origin.
- *Protophytolysosome* — an acid phosphatase-rich vesicle budded off from the Golgi saccule which is believed to be a conveyer of acid phosphatase to various sites in the cell.
- *Spherosome* — a lysosome-like compartment in plants which originates from the endoplasmic reticulum and serves as a principal site of lipid storage.
- *Vacuolysosome* — a lysosome that has formed by fusion of a lysosome with an 'empty' vacuole.

8.4.5.1 Vacuoles are not lysosomes

Assessment of all the studies on different tissues with diverse methodologies suggests that although many plant cells contain typical lysosomes, like animal cells, virtually all plant cells contain vacuoles with specific properties and functions. Plant lysosomes and vacuoles should be considered as two different organelles. It is not advantageous to consider vacuoles as enlarged or diluted lysosomes or lysosomes as minute or specialised vacuoles.

A comparative assessment of animal lysosomes, plant lysosomes and plant vacuoles points to their similarities and differences. Whereas lysosomes are chiefly concerned with **storage** of lytic enzymes, vacuoles are concerned with storage of various substances in addition to hydrolases. Whereas animal lysosomes demonstrate internal **catabolic functions**, plant lysosomes do not. Vacuoles demonstrate in-house catabolism as well as anabolism. Whereas animal lysosomes are capable of **exocytosis**, plant lysosomes and vacuoles exhibit no such process. All three organelles are surrounded by a single membrane. **Latency** is a characteristic of all typical lysosomes but vacuolar latency is demonstrated only in special situations of pathogenesis and senescence. The plant and animal lysosomal membrane is the *cordon sanitaire*, protecting the cytoplasm from the dangerous hydrolytic enzymes, and permitting only limited transport. Macromolecules, oligopeptides, disaccharides, and mononucleotides cannot normally cross the lysosomal membrane, whereas specific transport systems for a wide range of substances exist in tonoplast. Although proton pumps are operative in all three organelles to maintain an acidic lumen, secondary active transport and diverse ion channels operate widely only in vacuoles. Although aquaporins are not present in lysosomes but widely distributed in vacuoles, certain features of membrane peptides are found to be common.

8.5 PROSPECTS FOR GENETIC ENGINEERING OF TONOPLAST

The techniques of molecular genetic engineering have obviously opened a wide vista of improvement of agricultural and horticultural crops, and the revolution in the enhancement of quality and quantity of plant products due to genetic engineering has just began. In the last two decades only a few of the plants which have been improved by DNA-based strategies have actually been commercialised. This is due to the fact that DNA manoeuvring does not by itself permit direct metabolic engineering. In the milieu of the cell, no biochemical pathway operates in isolation, and for any metabolic engineering to succeed it has to be in consonance with the general metabolism of the cell, as well as of the tissue, the organ and the whole plant body. It is known that any perturbation in the primary metabolic pathway leads to pleiotropic effects which may produce some unacceptable results. Even if manipulation is made in the distal or terminal part of a pathway, its effect on other processes may have unknown consequences. Besides, a genetically engineered or transgenic plant may need special cultural practices or environmental requirements and may cause unforeseen agrotechnological and post-harvest problems. In other words, for successful genetic engineering, numerous possibilities have to be explored from the viewpoint of both the biochemical pathway and structure–function correlation. The following discourse is limited to the possibilities of engineering of tonoplast structure and function.

8.5.1 Engineering against water stress

One of the desiderata of genetic engineering is to develop crops which could be cultivated in very hostile environments, e.g. excessive heat, cold climate, physiologically or physically dry soil. Plant responses to water limitation are complex and involve changes in metabolism, hormone levels, growth and development. Since the tonoplast in collaboration with the plasma membrane is responsible for water relations of the plant from root-hair to xylem to mesophyll cells, the target of genetic engineering may be the tonoplast. The discovery of aquaporin and comprehension of its structure and function have great potential. By modifying the regulation and expression of specific aquaporins it would be possible to control and modulate water uptake, cell-to-cell water flow and hydraulic conductivity. The biochemistry and molecular genetics of aquaporins have already advanced to a level from where it will soon be possible to control production and expression of such water channels (Chrispeels and Maurel 1994). Some of these 27-kD proteins may be expressed in a tissue-specific manner and others may be induced by specific physiological conditions. For example, the aquaporin gene tobRB7 is expressed in roots of tobacco, trg-31, and rb-28 genes are induced by desiccation of pea and *Arabidopsis* respectively, and tmp-A and tmp-B are highly expressed by etiolation in *Arabidopsis*. On the other hand, the observation by Maurel *et al.* (1995) that water channel activity of the vacuolar membrane aquaporin alpha-TIP is regulated by phosphorylation, suggests that water loss may be controlled at the membrane *in situ*.

8.5.2 Engineering of proton pumps

In order to introduce crop plants to saline soils and develop salt-tolerant varieties, it is necessary to control the driving forces for sodium ion transport and the properties of the membrane carriers responsible for coupling this driving force to the movement of sodium ions. It has been established that tonoplast ATPase creates a proton gradient that serves as an energy source for the vacuolar accumulation of sodium from the cytoplasm via Na^+/H^+ antiport in the roots of glycophytes and halophytes. Reuveni *et al.* (1990) indicated that salt adaptation results in an alteration of the V-ATPase. Compared to the unadapted cells, the capacity for ATP hydrolysis and H^+ transport is four-fold greater for the V-ATPase of NaCl-adapted cells. Again, an increase in V-ATPase activity is expected to facilitate an increased influx of osmoregulatory solutes via secondary transport into the vacuole during turgor-driven cell expansion. In addition to osmoregulatory solutes, there would be a corresponding water-uptake in the vacuole.

Thus a major opportunity for tonoplast engineering is available through genetic modification of the proton pumps. As has been discussed in detail in Chapter IV, V-ATPase is composed of a hydrophilic catalytic V_1 complex and a hydrophobic membrane-spanning V_0 complex. Already the genes of a number of subunits have been cloned and characterised. For example, Wilkins (1993) isolated and identified the 69kD catalytic subunit from cotton. The genes encoding the 54kD B subunit of the major non-catalytic ATP-binding subunit have been characterised and sequenced by Berkelman et al. (1994). Wan and Wilkins (1994) indicated that the subunit-B is organised as a multigene family consisting of at least three genes. At this point it is important to note that a specific phospholipid environment is required for optimal ATPase activity (Kasamo and Nouchi 1987) and there are reports that changes in phospholipids and free sterol composition may contribute to salt tolerance (Mansour et al. 1994; Burgos and Donaire 1996). Therefore, any major engineering in V-ATPase may need concordant phospholipid engineering —a complex task indeed!

That fine techniques of molecular genetics are extremely potent for precise tonoplast engineering is shown by the use of antisense mRNA to inhibit the tonoplast H^+-ATPase in carrot by Gogarten et al. (1992). They have also presented direct evidence that V-ATPase drives osmotic uptake of water into the central vacuole facilitating cell expansion. A dramatic demonstration of the economic potential of V-ATPase remodelling may be expected in cotton fibres which are single-celled trichomes that differentiate from the outer epidermis of the ovule. The extraordinarily fast elongation of the cell is driven by the turgor pressure generated by the influx and accumulation of potassium and malate in the enlarging central vacuole (Basra and Malik 1983). The differential accumulation of V-ATPase activity of the tonoplast of developing cotton seed trichomes parallels the remarkable increase in the rate of cell elongation (Joshi et al. 1988), indicating the involvement of the proton pump in the transport and compartmentalisation of these osmoregulatory solutes during cotton fibre elongation. This is an example for cases where fast expansion of cells is desired. Deep-water rice varieties, where the panicle axis elongates as fast as the rising water level of the field in the rainy season, may also provide information on cell elongation and proton pumps.

The transport of malate to and from the vacuole is governed by V-ATPase in the case of CAM plants. This is another target for tonoplast engineering, since genes AtpvA, Atpv B, Atpv C and Atpv E have already been identified for subunits A, B, C and E of V-ATPase in *Mesembryanthemum crystallinum*. In this facultative CAM, V-ATPase is thought to represent up to 30% of tonoplast proteins (King and Lüttge 1991). In other words, the density of the proton pumps provides another parameter for improvement.

In conjunction with V-ATPase, V-PPase plays an important role in proton pumping and it has been characterised more as a back-up system for H^+-ATPase. Marked induction of V-PPase takes place by anoxia or chilling in rice and is maintained at high levels in the hypoxia-tolerant corn, together with the theoretical advantages of PPi-based metabolism under energy stress and the apparent need to maintain tonoplast-energisation. These findings led Carystinos et al. (1995) to hypothesise that V-PPase is an important element in the survival strategies of plants under hypoxic or chilling stress. This suggestion opens up a new avenue of breeding for stress resistance.

Along with ATPase and PPase, Ca^{2+}-ATPase in the tonoplast is also involved in regulation of ionic balance of the vacuole. Ca^{2+} is generally considered as an intracellular messenger, conveying information about the nature of a particular stimulus or stress impinging on the cell to target proteins that guide the cellular response. Although the tonoplast Ca^{2+}/H^+ antiport plays an important role in high capacity intracellular storage, high affinity Ca^{2+}-ATPase-driven Ca^{2+} transport has been proposed to play the most significant role in modulating cytosolic Ca^{2+} levels in the physiological range (Briskin, 1990). Recently, Ferrol and Bennett (1996) have identified a tomato gene (LCA) encoding two calmodulin-insensitive Ca^{2+}-ATPases which are differentially localised in the tonoplast and the plasma membrane. Malinström et al. (1997) cloned the cDNA (BCA1) corresponding to calmodulin-stimulated Ca^{2+}-ATPase from cauliflower tonoplast. It is not known whether the LCA (116 kD protein) and BCA1 (111 kD protein) are encoded by the same or different genes. Genes encoding two high- and low-affinity Ca^{2+}/H^+ antiporters *CAX1* and *CAX2* have been cloned by Hirschi et al. (1996). In addition to Ca^{2+}, CAX1 mediates Na^+/H^+ antiport as well as Cd transport. Again, CAX2 may function as a high affinity H^+/heavy metal cation antiporter with a role in metal homeostasis rather than Ca^{2+} signalling. SV channels are considered to be ubiquitous due to their detection in diverse plants and cells, and their proposed function in Ca^{2+}-induced Ca^{2+}-release could be important for the transduction of a variety of signals. All these data open up intriguing possibilities.

8.5.3 Engineering for storage and processing

Plant cells often utilise vacuoles for storage. Substances are directly transported across the tonoplast and stored, or precursor substances are transported which are processed, polymerised and/or conjugated. The whole system may be considered under: i) transport of the metabolite; ii) processing of the metabolites, if needed, by lumen enzymes; iii) storage; and iv) mobilisation and retrieval for use. There are numerous examples of each of these steps and, as expected, each one may be controlled by gene manipulation. Only a few examples will be presented here.

Sucrose transport and metabolism in the vacuole is understood to a certain extent. In sugar cane cells a series of tonoplast-bound enzymes, or enzymes closely associated with the tonoplast, provide a vectorial group translocation system for compartmentalisation of sucrose in excess of metabolic requirements (Thom and Maretzki 1992). Moreover, the specific phosphatase, sucrose phosphate phosphohydrolase, which is the last enzyme in the pathway of sucrose synthesis in plants, has been detected in vacuole preparations from red beet, sugar beet and immature sugar cane stems. During grape berry ripening, sucrose transported from the leaves is accumulated in berry vacuoles where it is converted to fructose and glucose by the enzyme β-fructosidase, the genes of which have been cloned (Davies and Robinson 1996). This process occurs in tomato and some other fruits as well. Again, sucrose synthase breaks sucrose to a reversible form so that it is used for onward biosynthetic reactions. The isozyme sII transcript of β-fructo-furanosidase is located in the vacuole sap where it controls sucrose storage and sugar composition in grape cells (Sturm *et al.* 1995). Knowledge of all the genes of sucrose metabolism is not far off and their manipulation is highly possible to improve fruit quality and ripening. Similar biosynthetic processes taking place inside the vacuole are seen in the case of inulin, a water-soluble non-structural carbohydrate. All four enzymes necessary for conversion of sucrose to inulin and all the low polymers of inulin saccharides are vacuolar in Jerusalem artichoke as well as in barley leaves.

Many of the commercially precious alkaloids are transported, synthesised and/or stored in vacuoles of a number of medicinal plants. The most widely studied plant is *Catharanthus roseus* which grows well even in neglected cultivation or harsh conditions, and synthesises more than 100 alkaloids (vide review by Moreno *et al.* 1995 and Chapter VII). The most important of these are the antileukemic alkaloids, vinblastine and vincristine, the antihypertensive alkaloid ajmalicine, and serpentine. Information on the gross metabolic pathways is almost complete. In essence, tryptamine synthesised in the cytosol is transported to the lumen of the vacuole, which finally forms strictosidine which is transported out to the cytosol and is converted to ajmalicine. Ajmalicine freely diffuses across the tonoplast and accumulates in the vacuoles by an ion-trap mechanism. Inside the vacuoles, ajmalicine is converted into serpentine by basic peroxides. The oxidized product remains trapped inside the vacuole, since the strongly basic serpentine cannot pass the tonoplast (Meijer *et al.* 1993). At least three important enzymes, viz. geraniol-10-hydroxylase, NADPH-cytochrome P-450 and strictosimidine β-glucosidase have been localised in the tonoplast and a number of enzymes including strictosidine synthase and peroxidase have been detected in the vacuole sap. Some of the genes of the enzymes have already been cloned and characterised. Alkaloid metabolism seems to be restricted to certain tissues and is modulated by different developmental and environmental mechanisms. Further, all cells of a given tissue may not be active in alkaloid biosynthesis. Similarly, in cell suspension cultures, alkaloid accumulation seems to be restricted to certain cells. The alkaloid storage cells have a much lower vacuolar pH than other cells in the population. These observations indicate that alkaloid synthesis and storage might take place in different cells. Thus, there exists an enormous scope for improvement of specific alkaloid production by manipulating many cellular parameters and biosynthetic pathways. Senescence can also be manipulated. Senescence involves activation of a number of senescence-associated genes (SAG). SAGs encode enzymes that are thought to be involved in cell degeneration and nutrient mobilisation. In the process, vacuoles are induced to store and release hydrolases. The regulation of SAG is multifactorial and one of them is vacuolar.

8.5.4 Engineering for phytoremediation and detoxification

Another prospect for genetic engineering is economic plants capable of withstanding various pollutants in soil, water and air. Amongst the different types of pollutants, heavy metals and xenobiotics are the major substances which largely affect most plants. Through absorption and/or adsorption, metals are transported to various tissues and cells where they are usually sequestered in vacuoles. Various mechanisms have been suggested for sequestering Zn ions in vacuoles of Zn-tolerant plants. According to a number of workers Zn may be conjugated with organic acids such as malate, citrate or oxalate, mustard oil glucosides, or phytate within vacuoles (Wang *et al.* 1992). Like Zn, cadmium and other heavy metals are also sequestered in the vacuole. Various mechanisms have been proposed for the transport of these toxic ions from the cytoplasm to the vacuole. These mechanisms include formation of soluble metal-organic acid complexes, metal-phytate, metal peptide and metal-peptide-sulfide complexes. In addition, Cd^{2+}/H^+ antiports may also provide one way for Cd accumulations (Salt and Wagner 1993). Phytochelatins, the cysteine-rich metal binding proteins, have an important role in cytoplasmic tolerance to toxic metals and in turn, both Cd and phytochelatin are accumulated in vacuoles. Salt and Rauser (1995) suggested that the MgATP-dependent vanadate-sensitive specific transporter, possibly belonging to the super-family of ATP-binding cassette transport protein (ABC transporter), is responsible for transport of Cd across the tonoplast of oat roots.

The metabolism and detoxification of xenobiotics by plant cells have been divided into three phases by Ishikawa (1992). In phase I, the compound is oxidised, reduced or hydrolysed to expose or introduce a functional group of the appropriate reactivity for phase II enzymes. In phase II, the activated derivative is conjugated with hydrophilic substances, such as glutathione, glucuronic acid or glucose. In phase III, the conjugate is excreted to the extracellular medium or sequestered in an intracellular compartment, such as a vacuole, possibly with the aid of tonoplast glutathione-S-(GS) conjugate transporter (Li *et al.* 1995). A number of studies on herbicide detoxification have produced a good deal of information on the process in general. The detoxification process involves the metabolism of herbicides to glucosides or GS-conjugates. The uptake of GS-conjugates is strongly ATP-dependent, but not by the tonoplast proton pumps. Kreuz *et al.* have shown that GS-conjugates can be efficiently accumulated in the vacuole due to direct energisation by MgATP. It is now known that the gene from yeast *YCF1* which confers resistance to cadmium and arsenic salts encodes a vacuolar GS-X pump resembling *HmMRP1*. Full-length cDNA of four classes of MRP1 genes has been cloned from *Arabidopis*. What is remarkable is that the GS-X pumps AtMRP1 and AtMRP2 are capable of vacuolar delivery of DNP-GS, GSSG, as well as chlorophyll catabolites and a number of herbicides (Szczypka *et al.* 1994; Lu *et al.* 1997; Lu *et al.* 1998). The overall transport capacity of AtMRP2 exceeds that of AtMRP1 by three to six times. The high capacity of AtMRP2 and moderate capacity with broader range of AtMRP1 for various herbicides, e.g. alachlor, atrazine, and symetryn, from the cytosol, are the targets for manipulation. There is a direct connection between oxidative stress caused by active oxygen species and GS-X pumps. In yeast, overexpression of a bZIP transcription factor, *YAP1* activates *YCF1* and *GSH1* and a multitude of oxidoreductases, leading to increased resistance to hydrogen peroxide, O-phenanthroline and heavy metals (Wemmie *et al.* 1994, Li *et al.* 1997). Phytochelatins (PC) which are elaborated in response to metal stress form PC-metal conjugates, are sequestered by a vacuolar ABC transporter. In yeast the *HMT1* gene has been identified as responsible for vacuolar delivery of the PC-Cd complex (Oritiz *et al.* 1992; Oritz *et al.* 1995). Further research on the molecular genetics of the MRP-subclass of ABC transporters in plants will pave the way for manipulation for xenobiotic resistance, pathogen resistance, protection from oxidative stress, control of senescence and related benefits.

8.5.5 Engineering for protein targeting

A large number of economically important proteins and protein conjugates are stored in vacuoles. The precursor forms of these proteins, from the site of synthesis, enter the lumen of the endoplasmic reticulum and, in many instances, appear to follow the bulk flow to the Golgi apparatus where they are sorted into vesicles for further transport to the vacuole. It has now been established that positive sorting information resides in the polypeptide structure itself and is located in N- or C-terminal propeptides or within the mature structure of the proteins. Thus sporamin, the storage protein of sweet potato and aleurain, a proteolytic enzyme of the barley aleurone cell are targeted by N-terminal propeptide. In contrast, C-terminal propeptides are responsible for targeting lectin in barley and chitinase in tobacco. A comparison of N-terminal propeptides of sweet potato sporamin A, barley alurain, potato 22kD protein and potato cathepsin D inhibitor show that a common motif of NPIR resides in all the four proteins. That the C-terminal propeptide may contain a sufficient vacuolar signal, has been noted in the case of lectins of barley and rice, and chitinase and β-1,3-glucanase of tobacco. Although a comparison of the C-terminal extensions of the lectins and the vacuolar hydrolases shows no amino acid identities, they are rich in hydrophobic amino acids and the characteristics may be recognised by the sorting machinery (Chrispeels and Raikhel 1992; Kirsch et al. 1996). Thus, recent work on targeting of vacuolar storage proteins provides a basis for improving the nutritional and processing properties of crops using genetic engineering. It is paradoxical that in some cases storage proteins may partly undergo proteolysis in the vacuole. The expression of the gene for vicilin, a vacuolar storage protein of pea, modified to encode an ER-retention signal at the carboxy-terminus of the protein, results in a dramatic (100-fold) increase in vicilin accumulation in the leaves of transgenic tobacco and alfalfa, compared with its unmodified counterpart (Wandelt 1992). So in certain cases, it may be desirable not to permit the protein to move into the vacuole and to harvest it from the cytosol, or to modify the protein against intravacuolar proteolysis.

On the other hand, deletion or modification of the signal motifs may bring about secretion of the protein in the apoplast or to some intracellular compartment. This is a handy tool for production of proteins by suspension culture cells where it would be desirable to have the metabolite protein secreted out in the medium of a bioreactor for facilitating downstream processing.

8.5.6 Engineering for disease resistance

A number of physiological and biochemical mechanisms have been developed by most plants as a method of defence against various pathogens. These include cell wall lignification, synthesis of anti-microbial low-mass compounds and the *de novo* expression of a number of pathogenesis-related (PR) proteins. Many of the PR proteins are enzymes. For example, five families of PR proteins have been identified in tobacco. Of these, β-1,3-glucanases and chitinases represent antifungal hydrolases that act synergistically to inhibit fungal growth *in vitro* and are stored in vacuoles. A 23kD PR protein (P-23) is induced in tomato when infected with citrus exocortis viroid (Rodrigo et al. 1993). P-23 accumulates in the vacuole and inhibits the growth of several phytopathogenic fungi. A novel pathogen- and wound-inducible antifungal 20kD protein was isolated from plants by inoculating them with tobacco mosaic virus (Ponstein et al. 1994). All these PR proteins are targeted to vacuoles and are responsible for necrosis in response to pathogenic infections. A number of PR proteins have already been manipulated and engineered to enhance disease resistance in different plants (vide review by Dixon et al. 1996).

8.5.7 Other possibilities

The production of osmolytes or proteins may be one way of scavenging undesirable radicals, as has been attempted in some transgenic plants. For example, Tarczynski et al. (1993) reported that the transfer of genes leading to increased amounts of mannitol in transgeneic tobacco make the plants more resistant to salinity stress. An increased drought tolerance has been reported in transgenic tobacco overexpressing fructans (Pilon-Smits *et al.* 1995). Proteinase inhibitors are known to be powerful anti-nutrients suppressing appetite and causing enzyme over-secretion. Transgenic tobacco plants expressing serine protease inhibitors in leaves show increased resistance against lepidopteran insect larvae (Ryan 1989). These inhibitors are present in vacuoles of fruit cells of wild tomato (*Lycopersicon peruvianum*) (Wingate *et al.* 1991). Their presence in vacuoles of different tissues may be engineered in crop plants as a defence mechanism against herbivores and insects. Another vacuolar enzyme, carboxypeptidase, which is widely expressed constitutively in certain tissues, can also be induced by physical stress (wounding) in tomato fruit pericarp (Mehta *et al.* 1996). Like the proteinase inhibitors, carboxypeptidase may have an additional function as a defence against pathogen attack.

- If tonoplast enzymes can be modified to tolerate high water stress and salt stress, crops can be grown in hostile environments.
- If vacuoles can store abundant amount of reserve food in the form of starch and proteins in different types of tissues, productivity of plants can be dramatically increased.
- If the tonoplast enzymes, along with the vacuolar lumen processing enzymes, can be properly engineered in consonance with the overall cellular metabolism, enormous amounts of very precious secondary metabolites and drugs can be conveniently produced by cells *in planta* or in bioreactors.
- If the proteinase inhibitors as defence arsenals can be judiciously produced and distributed in vacuoles of proper tissues or organs, predators and pathogens may be warded off.

8.6 Summary

Definition
An organelle is a membrane-bound and centrifugally separable subcellular entity with a distinct structure and function. Vacuoles are organelles which are delimited by a single membrane called the tonoplast with characteristic proton pumps, transporters and water channels, and which contain an aqueous acidic sap with dissolved substances including hydrolytic enzymes.

Significance
- Vacuoles are vital organelles of the plant cell. They are essential for maintaining **turgidity** of the individual cell and they support rigidity of plant organs. The vacuole provides the high surface/dry weight of the cell by holding a large amount of water which also provides intracellular fluidity for many essential metabolic functions.
- It is a ***storage structure*** of various substances from ions to large molecules. Whatever is absorbed or synthesised in excess of the cytoplasm's immediate need is sequestered in the vacuole. It may house hydrolysing enzymes and a number of proteins, lectins, etc. for diverse purposes.
- ***Homeostasis*** of ions and molecules permits regulation of pH, osmosis, salt stress, and makes it responsive to supply and demand of Ca^{2+}, organic and amino acids, sugars, etc.
- Vacuolar **hydrolases** are responsible for ***disposal of wastes*** of various nature through autophagy or heterophagy. Vacuoles may also conduct ***detoxification*** of xenobotics, sequestration of heavy metals and containment of pathogens.
- Certain **biosynthetic activities**, including conversion and final storage of secondary metabolites, are carried out within certain vacuoles.
- The tonoplast plays a crucial role in all these activities and in addition protects the cell from incoming external hazards and mutagens.

Uniqueness of vacuoles
- Vacuoles are *omnipresent* in all types of cells and are **indispensible** because of the multiplicity of their functions.
- Vacuoles maintain their **integrity** and do not attach themselves to any other structure or organelle. Even if similar and dissimilar vacuoles co-reside in the same cell, they maintain their individuality, unless induced.
- It is a **multifaceted** organelle and, unlike other organelles, it does not restrict itself to one function. It participates in house-keeping, maintenance of cell shape and turgidity, storage of various substances, homeostasis, biosythesis and processing, as well as in fortification against pathogens, chemical and physical stress, xenobiotics, unwarranted oxidation, etc.
- Vacuoles are unique in operating a diurnal, seasonal and long-term *storage–retrieval system*. Their capacity for homeostasis is unparalleled by any other organelle. In view of its homology with the plasma membrane and its capability to set aside what is not needed by the cytoplasm, the vacuole is equivalent to an **extracellular compartment within the cell**.

Homology of vacuoles
An appreciation of the homology of the tonoplast with other membranes and of the vacuole with similar structures is necessary for its comprehension and manipulation.
- Homology with **plasma membrane**: That the tonoplast is homologous to plasma membrane is evident from:
  i) the capacity of the two membranes to fuse with each other both *in planta* and *in vitro*,
  ii) the similarity in chemical composition, and electron cytochemical staining,
  iii) the similarity in distribution of electrical charge and other electrical properties,
  iv) the presence of aquaporin,
  v) the presence of H^+-ATPase and anion channels, albeit in opposite direction with respect to the cytosol,
  vi) the similarity in transport of certain amino acids and enzymes,
  vii) the capacity of both the membranes to secrete similar substances,

Evolution of vacuoles
On the basis of chemical composition, presence of diverse transporters and reverse orientation of the tonoplast with respect to the cytosol, it is proposed that the **vacuole has evolved through the invagination of the plasma membrane of the primeval cell**.
- Homology with **other membranes**: The tonoplast is possibly not homologous with other membranes, excepting the endoplasmic reticulum which may be directly converted to storage vacuoles.
- Homology with **Golgi apparatus**: In view of the derivation of vacuoles from Golgi and the direct contribution of Golgi vesicles to the vacuole, it is considered to be homologous with the Golgi.
- Homology with **single-membraned organelles**: Vacuoles are not considered to be homologous with any single-membraned organelles like spherosomes, peroxisomes or glyoxysomes.
- Homology with **lysosomes**: In many cells typical animal-like lysosomes co-exist with vacuoles and for various other reasons vacuoles are not considered as lysosomes. Animal lysosomes, plant lysosomes and vacuoles are homologous only up to the point of housing hydrolytic enzymes. Otherwise, plant lysosomes are only storage particles of hydrolases, whereas vacuoles are capable of undertaking multiple functions due to the presence of various transporters in the tonoplast.

Prospect of genetic engineering of tonoplast
- Engineering against **water stress** is possible by modification and regulation of expression of the tonoplast aquaporin. The aquaporin gene has alsready been isolated and cloned.

- Engineering of **proton pumps** may be a major target for manipulation against salt stress. The structure and molecular biology of V-ATPase has been studied in considerable detail and a number of genes of its various subunits have been cloned. In addition, V-PPase and Ca^{2+}-ATPase genes have also been identified and cloned.
- Engineering for **storage and processing** of diverse vacuolar substances has yet to gain momentum. Gene cloning and manipulation for storage and processing of sugars, and biosynthesis and sequestration of health-care alkaloids need adequate attention.
- Engineering for **phytoremediation and detoxification** has become an urgent necessity. An MRP subclass of ATP-binding cassette transporters on the tonoplast performs a very important function of vacuolar delivery of diverse xenobiotics. A number of these genes have been identified and cloned.
- Engineering for **protein targeting** is an important objective for improving the quality and quantity of storage proteins in economically important seeds. Gene cloning and genetic manipulation towards that end are being actively pursued.
- Engineering for **disease resistance** is currently attempted through manipulation of pathogenesis-related proteins.

8.7 REFERENCES

Aaronson, S., Digestion in phytoflagellates, in 'Lysosomes in Biology and Pathology', Vol. 3., Dingle, J. T., Ed., North Holland. Pub Co., Amsterdam and London, 1973, Chap 2.

Ascenso, L., Marques, N. and Pais, M.S., Glandular trichomes and vegetative and reproductive organs of *Leonotis* and *Leonurus* (Lamiaceae), *Ann. Bot.*, **75**, 619, 1995.

Ashford, A.E., Histochemical localisation of beta-glucosidase in roots of *Zea mays*. I. A simultaneous coupling azo-dye technique for the localisation of beta-glucosidase and beta-galactosidase, *Protoplasma*, **71**, 281, 1970.

Ashford. A.E. and McCully, M.E., Localization of naphtho (AS-B1) phosphatase activity in lateral and main root meristems of pea and corn, *Protoplasma*, **70**, 441, 1970.

Ashworth, J.M., and Weiner, E., The lysosomes of the cellular slime mould *Dictyostellium discoideum*, in 'Lysosomes in Biology and Pathology', Vol. 3, Dingle, J.T, Ed., North Holland Pub. Co. Amsterdam and London, 1973, Chap 3.

Avers, C. J., Histochemical localization of enzyme activities in root meristem cells, *Amer. J. Bot.*, **48**, 137, 1961.

Baba, M., Ohsumi, M., Scott, S.V., Klionsky, D.J. and Ohsumi, Y., Two distinct pathways for targeting proteins from the cytoplasm to the vacuole/lysosome, *J. Cell. Biol.*, **139**, 1687, 1997.

Banjoko, A. and Trelease, R.N., Development and application of an *in vivo* plant peroxisome import system, *Plant Physiol.*, **107**, 1201, 1995.

Basra, A.S. and Malik, C.P., Dark metabolism of CO_2 during fibre elongation of two cottons differing in fibre lengths, *J. Exp. Bot.*, **26**, 4, 1983.

Bensley, R.R., On the nature of canalicular apparatus of animal cells, *Biol. Bull.*, **19**, 179, 1910.

Benz, R., Büchner, K. H., and Zimmerman, U., Mobile charges in the cell membranes of *Halicystis parvula*, *Planta*, **174**, 479, 1988.

Berjak, P., A lysosome-like organelle in the root cap of *Zea mays*, *J. Ultrastruct. Res.*, **23**, 233, 1968.

Berkelman, T., Houtschens, K.A. and DuPont, F.M., Two cDNA clones encoding isoforms of the B subunit of the vacuolar ATPase from barley roots, *Plant Physiol.*, **104**, 287, 1994.

Bertl, A., Blumwald, E., Coronado, R., Eisenberg, R., Findlay, G., Gradmann, D., Hille, B., Köhler, K., Kolb, H.A., MacRobbie, E., Meissner, G., Miller, C. Neher, E., Palade, P., Pantoja, O., Sanders, D., Schroeder, J., Slayman, C., Spanswick, R., Walker, A. and Williams, A., Electrical measurements on endomembranes, *Science*, **258**, 873, 1992.

Bessey, E.A., Morphology and Taxonomy of Fungi, Blakiston Co. Philadelphia, 1950.

Brachet, J., Biochemical Cytology, Academic Press, New York, 1957, 52.

Brandes, D., Beutow, D.E., Bertini, F.J. and Malkoff, D.B., Role of lysosomes in cellular lytic processes. I. Effect of carbon starvation in *Euglena gracilis*, *Exp. Mol. Pathol.*, **3**, 583, 1965.

Briskin, D.P., Ca^{2+}-transporting ATPase of the plant plasma membrane, *Plant Physiol.*, **94**, 397, 1990.

Burgos, P.A. and Donaire, J.P., H^+-ATPase activities of tonoplast-enriched vesicles from non-treated and NaCl-treated jojoba roots, *Plant Sci.*, **118**, 167, 1996.

Carystinos, G.D., MacDonald, H.R., Monroy, A.F., Dhindsa, R.S. and Poole, R.J., Vacuolar H^+-translocating pyrophosphatase is induced by anoxia or chilling in seedlings of rice, *Plant Physiol.*, **108**, 641, 1995.

Chaffey, N.J., Structure and function in grass ligule: the endomembrane system of the adaxial epidermis of the membranous ligule of *Lolium temulentum* L. (Poaceae), *Ann. Bot.*, **76**, 103, 1995.

Chrispeels, M.J. and Maurel, C., Aquaporins: The molecular basis of facilitated water movement through living plant cells, *Plant Physiol.*, **105**, 9, 1994.

Chrispeels, M.J. and Raikhel, N., Short peptide domains target proteins to plant vacuoles, *Cell*, **68**, 613, 1992.

Coulomb, P., Mise en évidence de structures analogues aux lysosomes dans le mèristéme radiculaire de la Courge (*Cucurbita pepo* L. Cucurbitacée), *J. Microscop.*, **8**, 123, 1969.

Dangeard, P., Recherche de biologie cellulaire. Evolution du systeme vacuolaire chez les vegetaux, *Le Botaniste* **15**, 1, 1923.

Daniels, M.J., Mirkov, T.E. and Chrispeels, M.J., The plasma membrane of *Arabidopsis thaliana* contains a mercury-insensitive aquaporin that is a homolog of the tonoplast water channel protein TIP, *Plant Physiol.*, **106**, 1325, 1994.

Davies, C. and Robinson, S.P., Sugar accumulation in grape berries : Cloning of two putative vacuolar invertase cDNA and their expression in grape vine tissues, *Plant Physiol.*, **111**, 275, 1996.

de Duve, C., The lysosome in retrospect, in 'Lysosomes in Biology and Pathology', Vol. 1., Dingle, J.T. and Fell, H.B., North Holland Pub. Co., Amsterdam and London, 1969, 3.

de Duve, C. and Wattiaux, R., Functions of lysosomes, *Ann. Rev. Physiol.*, **28**, 435, 1966.

de Duve, C., Pressman, B.C., Gianetto, R., Wattiaux, R. and Appelmans, E., Tissue fractionation studies. 6. Intracellular distribution patterns of enzymes in rat liver tissue, *Biochem. J.* **60**, 604, 1955.

De, D.N. and Swain, D., Protoplast, cytoplast and sub-protoplast from ripening tomato fruits: their nature and fusion properties, in 'Plant Cell Culture in Crop Improvement', Sen, S.K., and Giles, K.L., Eds. Plenum Press, New York, London, 1983, 201.

De Mesquita, D.S.G., Shaw, J., Grimbergen, J.A., Buys, M.A., Dewi, L. and Woldringh, C.L., Vacuole segregation in the *Saccharomyces cerevisiae* Vac 2-1 mutant; Structural and biochemical quantification of the segregation defect and formation of new vacuoles, *Yeast*, **13**, 999, 1997.

Dixon, R.A., Lamb, C.J., Paiva, C.L. and Masoud, S., Improvement of natural defense responses, in 'Engineering Plants for Commercial Products and Application, Collins, G.B., Shepherd, R.J., Eds, N.Y. Acad. Sci. New York, 1996, 126.

Ferrol, N. and Bennet, A.D., A single gene may encode differentially localised Ca^{2+}-ATPases in tomato, *Plant Cell*, **8**, 1159, 1996.

Frigerio, L., de Vergilio, M., Prada, A., Fraro, F., and Vitale, A., Comparative analysis of regulatory mechanisms in the secretory pathway, Abstract, IX. Int. Cong. Plant Tiss. Cell Cult., IAPTC, Jerusalem, 1998, 141.

Gahan, P.B., Histochemical evidence for the presence of lysosome-like particles in root meristem cells of *Vicia faba*, *J. Expt. Bot.*, **16**, 350, 1965.

Gahan, P.B., Histochemistry of lysosomes, *Int. Rev. Cytol.*, **21**, 1, 1967.

Ganser, R., Vienken, J., Hampp, R., and Zimmerman, V., Electric field-induced fusion of isolated vacuoles and protoplast of different developmental and metabolic provenience, in

'Plasmalemma and Tonoplast : Their Functions in the Plant Cell, Marmé, D., Marré, E. and Hertel, E., Eds, Elsevier Biomedical Press, 1982, 225.

Gogarten, J.P., Fichmann, J., Braun, Y., Morgan, L., Styces, P., Taiz, S.L., Delapp, K. and Taiz, L. The use of antisense mRNA to inhibit the tonoplast H^+-ATPase in carrot, *Plant Cell*, **4**, 851, 1992.

Gomori, G., Microscopic Histochemistry, Principles and Practice, Univ. of Chicago Press, Chicago, 1952.

Gorska-Brylass, A., Hydrolases in pollen grains and pollen tubes, *Acta. Soc. Bot. Polon.*, **34**, 589, 1965.

Griesbach, R.J. and Lawson, R.H., Protoplast evacuolation, in 'The Physiological Properties of Plant Protoplasts', Pilet, P.E. Ed., Springer Verlag, Berlin, 1985, 99.

Guilliermond, A. and Mangenot, G. (1922) cited in 'The Cytoplasm of the Plant Cell', Chronica Botanica, Waltham, Mass, 1941. Chap XVII.

Gupta, H.S., Plant lysosomes: aspects and prospects, *Curr. Sci.*, **58**, 554, 1985.

Hall, J.L., Histochemical localization of beta-glycerophosphatase activity in young root tips, *Ann. Bot.*, **33**, 399, 1969.

Harrington, J.F. and Altschul, A.M., Lysosome-like behaviour in germinating onion seeds, *Proc. Fedn. Am. Socs. Exp. Biol.*, **22**, 475, 1963.

Hensel, G., Horstmann, C., Adler, K., Manteuffel, R., Kunze, G., and Kunze, I., Evidence for secretion of the vacuolar PR-proteins: Chitinase, beta-1,-3-glucanase and CBP-20 in tobacco, *Abstract, IX. Int. Cong. Plant Cell Tiss. Cult.* IAPTC, Jerusalem, 1998, 151.

Hirschi, K.D., Zhen, R.G., Cunningham, K.W., Rea, P.A. and Fink, G.R., CAX1, an H^+/Ca^{2+} antiporter from *Arabidopsis*, *Proc. Natl. Acad. Sci. USA*, **93**, 8782, 1996.

Hislop, E.C., Barnaby, V.M., Shellis, C. and Laborda F., Localization of (alpha)-L-arabinofuranosidase and acid phosphatase in mycelium of *Sclerotinia fructigena*, *J. Gen. Microbiol.*, **81**, 79, 1974.

Ishikawa, T., The ATP-dependent glutathione S-conjugate export pump, *Trends Biochem. Sci.*, **17**, 463, 1992.

Jensen, W.A., The cytochemical localization of acid phosphatase in root tip cells, *Amer. J. Bot.*, **43**, 50, 1956.

Joshi, P.A., Stewart, J. Mc. D. and Graham, E.T., Ultrastructural localization of ATPase activity in cotton fibre development during elongation, *Protoplasma*, **143**, 1, 1988.

Kasamo, K. and Nouchi, I., The role of phospholipid in plasma membrane ATPase activity in *Vigna radiata* L. (Mung bean) roots and hypocotyls, *Plant Physiol.*, **83**, 823, 1987.

King, R. and Lüttge, U., Electron microscopic demonstration of a head and stalk structure of the leaf vacuolar ATPase in *Mesembryanthemum crystallinum* L., *Bot. Acta* **104**, 122, 1991.

Kirsch, T., Saalbach, G., Raikhel, N.V. and Beevers, L., Interaction of potential vacuolar targeting receptor with amino-and carboxy-terminal targeting determinants, *Plant Physiol.*, **111**, 466, 1996.

Kreuz, K., Tommasini, R. and Martinoia, E., Old enzymes for a new job : Herbicide detoxification in plants, *Plant Physiol.*, **111**, 249, 1996.

Kuehn, G.D., Some one hundred 'Somes', *Trends. Biochem. Sci.*, **10**, 227, 1985.

Kunze, I., Kunze, G., Bröker, M., Manteuffel, R., Meins, Jr., F. and Müntz, K., Evidence for secretion of vacuolar <alpha>-mannosidase, class I chitinase, and class I <beta>-1, 3-glucanase in suspension cultures of tobacco cells, *Planta*, **205**, 92, 1998.

Lackney, V.K., Spanswick, R.M., Hirasuna, T.J. and Shuler, M.L., PEG- enhanced, electric field-induced fusion of tonoplast and plasmalemma of grape protoplasts, *Plant Cell Tis. Org. Cult.*, **23**, 107, 1990.

Li, Z-S., Lu, Y-P., Zhen, R.G., Szczypka, M., Thiele, D.J., and Rea, P.A., A new pathway for vacuolar cadmium sequestration in *Saccharomyces cerevisiae* : YCF1-catalyzed transport of bis (glutathionato)-cadmium, *Proc. Natl. Acad. Sci. USA*, **94**, 42, 1997.

Li, Z-S., Zhen, R-G.and Rea, P.A., 1-Chloro-2,4-dinitrobenzene-elicited increase in vacuolar glutathione-S-conjugate transport activity, *Plant Physiol.*, **109**, 177, 1995.

Lu, Y-P, Li, Z-Sand, Rea, P.A., AtMRP1 gene of *Arabidopsis* encodes a glutathione S-conjugate pump: isolation and functional definition of a plant ATP-binding cassette transporter gene, *Proc. Natl. Acad. Sci. USA*, **94**, 8243, 1997.

Lu, Y-P., Li, Z-S., Drozdowicz, Y.M., Hortensteiner, S., Martinoia, E., and Rea, P.A., AtMRP2, an *Arabidopsis* ATP-binding cassette transporter able to transport glutathione S-conjugates and chlorophyll catabolites: functional comparisons with AtMRP1, *Plant Cell*, **10**, 1, 1998.

MacKenzie, C.J., Yoo, B.Y. and Seabrook, J.E.A., Stigma of *Solanum tuberosum* cv. Shepody : Morphology, ultrastructure and secretion, *Amer. J. Bot.*, **77**, 1111, 1990.

Magnin, T., Fraichard, A., Trossat, C. and Pugin, A., The tonoplast H^+-ATPase of *Acer pseudoplatanus* is a vacuolar type ATPase that operates with a phosphoenzyme intermediate, *Plant Physiol.*, **109**, 285, 1995.

Malik, C.P., Sood, P.P. and Tewari, H.B., Occurrence of lysosome-like bodies in plant cells: Acid phosphatase reaction, *Z. Biol.*, **116**, 264, 1969.

Malinström, S., Askerlund, P. and Palmgren, M.G., A calmodulin-stimulated Ca^{2+}-ATPase from plant vacuolar membranes with a putative regulatory domain at its amino terminus, *FEBS Lett.*, **400**, 324, 1997.

Mansour, M.M., Van Hasselt, P.R. and Kuiper, P.J.C., Plasma membrane lipid alterations induced by NaCl in winter wheat roots, *Physiol. Plant*, **92**, 473, 1994.

Marty, F., Cytochemical studies on GERL, provacuoles and vacuoles in root meristematic cell of *Euphorbia*, *Proc. Nat. Acad. Sci. USA*, **75**, 852, 1978.

Marty, F., Analytical characterization of vacuolar membranes from higher plants, in 'Biochemistry and Function of Vacuolar Adenosine triphosphatase in Fungi and Plants', Marin, B.P., Ed., Springer Verlag, Berlin, 1985, 14.

Marty, F. and Branton, D., Analytical characterization of beet root vacuole membrane, *J. Cell Biol.*, **87**, 72, 1980.

Matile, P., Balz, Z.R., Semadani, E. and Jost, M., Isolation of spherosomes with lysosome characteristics from seedlings., *Z. Naturforsch. Sect B.* **20**, 693, 1965.

Maurel, C., Kado, R.T., Guern, J. and Chrispeels, M.J., Phosphorylation regulates the water channel activity of the seed-specific aquaporin alpha-TIP, *EMBO J.*, **14**, 3028, 1995.

Mehta, R.A., Warmbardt, R.D., and Mattoo, A.K., Tomato fruit carboxypeptidase: Properties, induction upon wounding and immunocytochemical localisation, *Plant Physiol.*, **110**, 883, 1996.

Meijer, A.H., Verpoorte, R.and Hoge, J.H.C., Regulation of enzymes and genes involved in terpenoid indole alkaloid biosynthesis in *Catharanthus roseus*, *J. Plant Res.*, **3**, 145, 1993.

Moreno, P.R.H., Van der Heijden, R. and Verpoorte, R., Cell and tissue cultures of *Catharanthus roseus*: A literature survey, II. Updating from 1988 to 1993, *Plant Cell Tiss. Org. Cult.*, **42**, 1, 1995.

Neuhaus, J.M., Abstract, Keystone Symp. Plant Cell Biology: Mechanisms, Pathways and Signals. New Mexico, 1995.

Novikoff, A.B., Beaufay, H. and de Duve, C., Electron microscopy of lysosome-rich fractions from rat liver. *J. Biophys. Biochem. Cytol.* **2**, 179, 1956.

Oritz, D.F., Ruscitti, T., McKue, K.F. and Ow, D.W., Transport of metal-binding peptides by HMT1, a fission yeast ABC-type vacuolar membrane protein, *J. Biol. Chem.*, **270**, 4721, 1995.

Oritz, D.F., Kreppel, L., Speiser, D.M., Scheel, G., McDonald, G., and Ow, D.W., Heavy metal tolerance in the fission yeast requires an ATP-bindning cassette-type membrane transporter, *EMBO J.*, **11**, 3491, 1992.

Pilon-Smits, E.A.H., Ebskamp, M.J.M., Paul, M.J., Jeuken, M.J.W., Weisbeek, P.J. and Smeekens, S.C.M., Improved performance of transgeneic fructan-accumulating tobacco under drought stress, *Plant Physiol.*, **107**, 125, 1995.

Pitt, D., Histochemical demonstration of certain hydrolytic enzymes within cytoplasmic particles of *Botrytis cinerea*, Fr. *J. Gen. Microbiol.*, **52**, 67, 1968.

Pitt, D., and Coombes, C., The disruption of lysosome-like particles of *Solanum tuberosum* cells during infection by *Phytophthora erythroseptica* Pethybr., *J. Gen. Microbiol.*, **53**, 197, 1968.

Pitt, D., and Walker, P.J., Particulate localization of acid phosphatase in fungi, *Nature*, **215**, 783, 1967.

Plowe, J.Q., Membrane in the plant cell I. Morphological membranes at protoplasmic surfaces, *Protoplasma*, **12**, 196, 1931.

Ponstein, A.S., Bres-Vloemans, S.A., Sela-Buurlage, M.B., Van der Elzen, P.J.M., Melchers, L.S. and Cornelissen, B.J.C., A novel pathogen- and wound- inducible tobacco (*Nicotiana tabaccum*) protein with antifungal activity, *Plant Physiol.*, **104**, 109, 1994.

Raven, J.A., A cost-benefit analysis of photon absorption by photosynthetic unicells, *New Phytol.*, **98**, 593, 1984.

Raven, J.A., Regulation of pH and generation of osmolarity in vacuolar land plants: cost benefits in relation to efficiency of use of water, energy and nitrogen, *New Phytol.*, **101**, 25, 1985.

Raven, J.A., Responses of aquatic photosynthetic organisms to increased solar UV-B, *J. Phytochem. Phytobiol*, B: Biology **9**, 239, 1991.

Raven, J.A., The vacuole: a cost-benefit analysis, in 'The Plant Cell Vacuole', Leigh, R.A. and Sanders, D., Eds, Acad Press, London, 1997, 59.

Reuveni, M., Bennett, A.B., Bresson, R.A. and Hasegawa, P.M., Enhanced H^+ transport capacity and ATP hydrolysis activity of the tonoplast H^+-ATPase after NaCl adaptation, *Plant Physiol.*, **94**, 524, 1990.

Rodrigo, I., Vera, P., Tornero, P., Hernandez-Yago, J. and Conejero, V., cDNA cloning of viroid-induced tomato pathogenesis-related protein P23, *Plant Physiol.*, **102**, 939, 1993.

Ryan, C.A., Proteinase inhibitor gene families: Strategies for transformation to improve plant defenses against herbivores, *Bioessays*, **10**, 20, 1989.

Salt, D.E. and Rauser, W.E., MgATP-dependent transport of phytochelatins across the tonoplast of oat roots, *Plant Physiol.*, **107**, 1293, 1995.

Salt, D.E. and Wagner, G.J., Cadmium transport across tonoplast of vesicles from oat roots: Evidence for a Cd^{2+}/H^+ antiport activity, *J. Biol. Chem.*, **268**, 12297, 1993.

Scott, S.V., Hefner-Gravink, A., Morano, K.A., Noda, T., Ohsumi, Y. and Klionsky, D.J., Cytoplasm-to-vacuole targeting and autophagy employ the same machinery to deliver proteins to the yeast vacuole, *Proc. Nat, Acad. Sci, USA*, **93**, 12304, 1996.

Semadani, E.G., Enzymatische characterisierung der Lysosomen äquivalente (Sphärosomen) von Maiskeimlingen., *Planta*, **72**, 91, 1967.

Sturm, A., Sebkova, V., Lorenz, K., Hardegger, M., Lienhard, S., and Unger, C., Development-and organ-specific expresssion of the genes for sucrose synthase and three isoenzymes of acid (beta)-fructofuranosidase in carrot, *Planta*, **195**, 601, 1995.

Szczypka, M.S., Wemmie, J.A., Moye-Rowley, W.S., and Thiele, D.S., A yeast metal resistance protein similar to human cystis fibrosis transmembrane conductance requlator (CFTR) and multidrug resistance-associated protein, *J. Biol. Chem.*, **269**, 22853, 1994.

Tarczynski, M.C., Jensen, R.G. and Bohnert, H.J., Stress protection of transgenic tobacco by production of the osmolyte mannitol, *Science*, **259**, 508, 1993

Tazawa, M., Shimmen, T. and Mimura, T., Membrane control in the Characeae, *Ann. Rev. Plant Physiol.* **35**, 95, 1987.

Thom, M. and Maretzki, A.J., Evidence of direct uptake of sucrose by sugarcane stalk tissue, *J. Plant Physiol.*, **139**, 555, 1992.

Ueda, K., Fine structure of *Chlorogonium elongatum* with special reference to vacuole development, *Cytologia*, **31**, 461, 1966.

Walek-Czernecka, A., Histochemical demonstration of some hydrolytic enzymes in the spherosomes of plant cells, *Acta. Soc. Bot. Polon.*, **34**, 573, 1965.

Walek-Czernecka, A., Mise en évidance de la phosphatase acide (monophosphoesterase II) dans les sphérosomes des cellules épidermiques des ecailles bulbaires d *Allium cepa.*, *Acta. Soc. Bot. Polon.*, **31**, 539, 1962.

Wan, C-I., and Wilkins, T.A., Isolation of multiple cDNAs encoding the vacuolar H^+-ATPase subunit B from developing cotton (*Gossypium hirsutum* L.) ovules, *Plant Physiol.*, **106**, 393, 1994.

Wandelt, C.L., Khan, M.R.I., Craig, S., Schroeder, H.E., Spencer, D. and Higgins, T.J.V., Vicilin with carboxy terminal KDEL is retained in the endoplasmic reticulum and accumulates to high levels in the leaves of transgenic plants, *Plant J.*, **2**, 181, 1992.

Wang, J., Evangelou, B.P., Nielsen, M.T., and Wagner, G.J., Computer stimulated evaluation of possible mechanisms for sequestering metal ion activity in plant vacuoles. II. Zinc., *Plant Physiol.*, **99**. 621, 1992.

Wardrop, A.B., Occurrence of structures with lysosome-like functions in plant cells, *Nature*, **218**, 978, 1968.

Wemmie, J.A., Szczypka, M.S., Thiele, D.J., and Moye-Rowley, W.S., Cadmium resistance mediated by the yeast AP-1 protein requires the presence of ATP-binding cassette transporter-encoding gene, YCF, *J. Biol, Chem.*, **269**, 32592, 1994.

Wilkins, T.A., Vacuolar H^+-ATPase 69-kilodalton catalytic subunit cDNA from developing cotton (*Gossypium hirsutum*) ovules, *Plant Physiol.*, **102**, 679, 1993.

Wingate, V.P.M., Franceschi, V.R. and Ryan, C.A., Tissue and cellular localization of proteinase inhibitors I and II in the fruit of wild tomato, *Lycopersicon peruvianum* (L.) Mill., *Plant Physiol.*, **97**, 490, 1991.

Wink, M., The cell culture medium - a functional extracellular compartment of suspension-cultured cells, *Plant Cell Tiss. Org. Cult.*, **38**, 307, 1994.

Yatsu, Y., and Jacks, T.J., Spherosome membranes, Half - unit membranes, *Plant Physiol.*, **49**, 937, 1972.

Index

*indicates an illustration or table

ABC-transporter 68, 72, 163, 213, 215*, 263
   Cd-complex removal 217
abscisic acid, hormones 90
absence of vacuole 50*
accumulation
   abiological bodies 202
   chlorophyll catabolites 202
   diurnal 252
   in infected cells
      beta-1-3 glucanase 206
      chitinase 206
   ions 168
   long-term 252
   of breakdown products 202
   seasonal 252
acetylesterases 103
acid phosphatase
   cytochemical localisation 13
   electron microscopy 15
   light microscopy 15
acridine orange 26
active oxygen species, see AOS 208
adaptor complexes, see AP 129
ADP-ribosylation factor see ARF 129
agglutinin 83
alachlor 216
albumin 103
   storage proteins 81
aleurone grain 1, 3, 80
   acid phosphatase 121
   conversion to vacuole 122
   hydrolysis 122
   lysis 122
   metamorphosis 122
   protease 121
   RNase 121
   vacuolar nature of 122
aleurone vacuole 149
   origin 121*
alfa-mannosidase 68
alfa-TIP 69*
algae 39, 59
alkaline phosphatase, see ALP 139
alkaloid accumulation by ion trap
      mechanism 195
alkaloids 88
   of *Catharanthus* 194
allocryptopine 88
ALP 139, 140, 142
   signals for 147
ALP/alternative pathway 141*, 142
alpha factor 138
   degradation 197
   internalisation 138*, 197
Al-Phytoremediation 219
Al-toxicity 219
aluminium 219
alurain 134
amine accumulation 175*
amino acid homeostasis
   in *Chara* 187
   in yeast 187
amino acids 104
   alanine 89
   arginine 89
   glutamine 89
   histidine 89
   leucine 89
   pools of 90
   serine 89
   tryptophan 89
   valine 89
aminopeptidase I, see API 141
aminopeptidases in cytoplasm 196

amorphous protein 103
angiosperm 47, 60
anion
    channel 172
    exchange systems 174
    uptake 172
annexin 68, 137
    as a signal transducer 186
anterograde
    traffic of vesicles 130*, 131
    transport: ER to Golgi 129, 130*
anthocyanin 103
    by glutathione-S-transferase 192
    dissolved in sap 85*
    metabolism in vacuole 191
    uptake 192
    vacuolar contents 87*
anthocyanin pigment
    vacuolar contents 85*
anthoxanthin 103
    dissolved in sap 85*
anthraquinone 93
antiport 171, 221
antisense technique for ATPase 261
AOS 208
    scavenging by flavonoids 219
    types of 219
AP in clathrin assembly 129
AP-1 129
    in plants 134
AP-2 129
    in plants 134
AP-3 129
    in plants 134
API 141
apoplastic route 190
aquaporin 69*, 164, 165*, 178
    genetic engineering of 266
    osmoregulation 178
ARF 129
arginine homeostasis 187
arsenite 217
arylsulfatase
    electron microscopy 16
    light microscopy 16
aspartic endopeptidase 193
aspartic proteinase 193
    asymmetry, tonoplast 64
ATPase
    in active transport 168

characteristics 170
electron microscopy 16, 15
H^+ ATPase 68
particle 65
V_0 domain 67
V_1 domain 67
vacuolar type 169
ATP-binding cassette transporter (see ABC-transporter)
autofluorescence 10
autolysis
    by tonoplast breakdown 212*
    for secondary wall formation 209
    in cellular differentiation 224
autolysosome 201
autophagic
    bodies in vacuole 200*
    vacuole 45, 211
autophagosome 199
autophagy 124, 149, 198, 223, 250
    and Cvt pathway 143
    by lysosome 257
    by starvation 200
    due to metabolic disorder 199
    due to starvation 199
    during senescence 212*
    in development of cotyledon 211
    in development of gametogenesis 211
    in development of laticifer 211
    in laticifer development 210
    in vascular differentiation 208
    in yeast 200
    sperm release by 211
    under starvation 197
autophagy, EM
    of PB 199
auxin hormones 90
azo dye technique 13

barium in *Chara rhizoids* 172
$BaSO_4$ in *Chara* 213
berberine 88
beta glycerophosphatase 208, 103*
bidirectional transport of vesicles 150
    biogenesis, biochemical aspects of 128*
biosynthesis
    inside vacuole 191, 222
    of alkaloids 194, 222
    of glycosides 191
    brick red pigment, vacuolar contents 87*

bryophyta 59
bulliform cells 48

Ca-binding protein 185
Ca^{2+}
   antiporter 72
   as cellular fire 184
   as second messenger 185
   ATPase 68, 72
   channels 137, 181, 184, 261
   homeostasis 183, 222
      role of vacuole 184
   regulates
      FV channel 184
      SV channel 184
   release
      ABA-voltage gated 183
      Auxin-ligand gated 183
   transporters 184
Ca^{2+}-ATPase 184, 261
Ca^{2+}/H^{+} antiport 184, 261
cadmium (Cd) 217
calcium carbonate crystals
   vacuolar solids 81*
calcium oxalate crystals
   vacuolar solids 80*
calmodulin 136, 185, 186
CAM 186, 89
CAM plant 186, 68
cambial cells 48*, 49
canalicular vacuoles 48
canaliculi
   to form vacuole 3*
   coalescence of 115
carboxypeplidase Y 137
   exclusively in vacuole 197
cardenolides 88
cardiac glucoside 93
cargo protein by VTC to Golgi 132
carrier proteins 163
CCV 129
CCVs in plants 133, 134
Cd
   complex 217
   detoxification 217
   factor, MRP1 homologue 217
   peptides in vacuoles 217
   phytochelatin sulfide complex 217
   pollution 217
   toxicity 217

Cd/glutathione stoichiometry 217
CD-MPR 134
cell
   basal 54
   cortical 79*
   cultured 56
   extensor 57
   flexor 57
   oil-secreting 56
   sap
      empty 9
      full 9
   secretory 57
   specialised 60
   suspensor 54, 55*
CGN 132
change in
   vacuolar volume 58*
changes in motor cell shape 182*
channel proteins 163
channels
   ligand gated 164
   mechanically gated 164
   voltage-gated 164
Chara 187
   vacuolation in 124
chelerythrine 88
chemical
   analysis, tonoplast 67
   composition 73
      tonoplast 64
   gradient 167
   potential difference 167
chemi-osmotic hypothesis 172
CI-MPR 134
cis-Golgi network, *see* CGN 132
clathrin AP complexes 130*
clathrin coats 129
clathrin-coated vesicles (CCV) *see* CCV 129
CLSM 117
coated vesicles 128, 129, 137
   CCV 150
   COP I 150
   COP II 150
   formation 129
coatomer
   complex, composition 129
   coated I (COP I) 129
   coated II (COP II) 129
coleorhiza 54

colloidal
   bodies, vacuolar contents 87*
   content, vacuolar contents 85*
   substances 103
   compact tissues, intact vacuole from 18
comparison
   of cytoplasm and sap
      ionic concentration 91*
   of salt water, fresh water and cell sap
      ionic concentration 91*
   of sap and external solution
      ionic concentration 91*
   of V-ATPases 71*
compartmentation of ions
   Cl^- 168*
   K^+ 168*
   Na^+ 168*
compartmentation of Na^+, Cl^- ions 177
Con A (concanavalin A) 68
confocal laser scanning microscopy (CLSM) 117
conglycinin 83
continuous gradient
   direct extraction of tonoplast by 24
contractile vacuole 39–41, 59
COP mediated traffic 129
COP I
   vesicles for bidirectional transport 129
   composition of 129
COP II 131
   structure 129
copy number of vacuoles 57
cortex 48
CPS 141, 142
   signals for 147
CPY 137, 140, 141, 142, 144
   pathway 141*, 142
Cr
   diffusion 218*
   Phytoremediation 218
   retention 218*
   storage 218
   toxicity 218
   uptake 218*
Crassulacean acid metabolism (see CAM) 89, 186
crystal sand
   vacuolar solids 80*
crystalline
   protein 103
   needles

vacuolar contents 87*
crystalloids 5
crystals, solid inclusions 79
CTPP(C-terminal propeptide) 145*, 146
   in hydrolases 146
   in lectin 146
   of invertase 134, 146
culture, suspension 56
cultured cell 56
Cvt
   complex 138
   pathway 141*, 142
cyanogenic diglucoside 122
cysteine protease 193
cystolith 81
   vacuolar solids 80*
cytochemical localisation 30
   of enzymes 13
   of esterases 13
   of hydrolases 13
   Gomori's method 13
cytochemistry of enzyme 15
   cytokinin hormones 90
cytolysosome 122, 123
cytoplasmic pH see pHc 175
cytoplasm-to-vacuole complex, see Cvt complex 138
cytotoxic 214
CytP450 dependent mono-oxygenase 214

dark violet pigment
   vacuolar contents 87*
DCCD
   inhibition of proton pump 174
   sensitivity 175
de novo
   Golgi bodies 116
   liposome 115
   origin 115
default mechanism 143, 147
defence against herbivores 264
defence response against pathogens 204
degradation
   by membrane enclosure 198
   cytoplasmic protrusion for 199
   of damaged cytoplasm 203*
   of extraneous protein in vacuoles 197
   of peroxisome 199
   of *Rhizobium* in root nodule vacuoles 205*
dense body vesicles 119

densitometry
    tonoplast 64
determination of osmotic value
    by freezing point depression 27
    by plasmolysis 27
    by pressure bomb technique 28
    by pressure probe technique 28
    by resonance frequency technique 28
    by vapour pressure equilibration 28
determination of pH
    by fluorescence probes 26
    by microelectrodes 26
    by P-NMR 27
    by weak acid/base distribution 25
detoxification, genetic manipulation for 263
detoxification
    metals 217
    herbicides 224
    SO_2 213, 224
    xenobiotics 214, 215*, 224
dhurrin 88, 93
dicyclohexylcarbodiimide, see DCCD 174
diesterase
    in fungi 119
    in vacuolation in fungi 119
difference in lumen and membrane protein
        sorting 140
diffusion
    facilitated 163
    potential 167
    simple 163
dihydrocopticine 88
dihydrozeatin
    hormones 90
dipeptidyl aminopeptidase A, see DPAP 141
    direct extraction of tonoplast 30
    by continuous gradient 24
    by discontinuous gradient 24
dissolved in sap
    anthocyanin 85*, 103
    anthoxanthin 85*, 103
    phenolics 85*
    polyphenols 86
    tannin 85*
dissolved substances 103
division of vacuole in yeast 127*
DNP-GS 216
docking 132, 136
    specificity 132
Donnan
    equilibrium 167, 176

phase 167
DPAP 141, 142
DPAP-A, signals for 147
DPAP-B, signals for 147
dual uptake -two mechanisms 165

electrical potential difference 167
electrochemical gradient 167
electron cytochemistry 13, 127
electron microscopy 30
    acid phosphatase 15
    arylsulfatase 16
    ATPase 16, 15
    IDPase 16
    ITPase 16
    of non-specific esterase 17
    plasma membrane 12
    tonoplast 12
    TTPase 16
electron probe analysis 28
elicitor-enzyme-vacuole-discharge model
        206, 207*
EM of ATPase in vesicles 126
embryo 54, 55*
    sac 51*, 52*, 53*
endocytosis 138*
endodermis 48
endomembrane
    enzyme localisation on 127*
    electron cytochemistry 126, 127*
endopeptidases exclusively in vacuole 196
endoplasmic reticulum (ER) see ER 124
endoproteinases 149
endosome 137, 138*, 140, 150, 151
endosperm 52*, 53*
engineering
    for disease resistance 264
    for inulin metabolism 262
    for sucrose metabolism 262
    of alkaloid genes 262
    of detoxification 267
    of disease resistance 267
    of phytoremediation 267
    of processing of products 267
    of proton pumps 260, 267
    of storage products 267
    of targeting economic proteins 267
    of tonoplast structure 260
    of V-ATPase 260
enzymatic steps of indole alkaloid
        biosynthesis 195*

enzyme cytochemistry
   acid phosphatase 14
   fixatives 13
   incubation for enzyme 13
   lead salt procedure by electron
      microscopy 14
   lead salt procedure by light microscopy 14
   post-incubation processing 14
   staining 13
enzymes
   arylsulfatases 103*
   endonucleases 103*
   exonucleases 103*
   lipases 103*
   lyases 103*
   nuclease 103*
   phosphodiesterases 103*
   phosphomonoesterases 103*
   phytase 103*
   tonoplast 67, 70, 171
   types
      aminopeptidase 94*
      endopeptidase 94*
      esterase 94*
      exopeptidase 94*
      glycosidase 94*
      peptidase 94*
epidermal flavonoid scavenging mesophyll
      peroxide 220*
epidermis 48
ER 124
   vacuolation from 124
   –Golgi–vacuole, EM 126
essential oil 103
esterase 208
ethylene metabolism 192
ethylene precursor
   hormones 90
evacuolation 208
extracellular compartment 252

feedback
   control, negative 213
   inhibition
      ion uptake 166
FFEM 64, 65
flavones 88
flavonoid metabolism in vacuole 192
flavonoids, protective function 220*
flexor Cell 57
flippase 117

fluid-phase endocytosis 201
fluorescein 26
fluorescence
   microscopy 30, 9
   Primary 9
   Secondary 10, 9
   probes
      determination of pH by 26
fluorochrome 10
fluorochroming by
   6-carboxy fluorescein 11
   acridine orange 10, 11
   BCECF 11
   BCECF-AM 11
   CDCFDA 12
   CF 11
   CF diacetate 11
   DASPMI 12
   FM4-64 12
   Lucifer Yellow 12
   monochlobimane (MCB) 11, 12
   rhodamin B 10
flux sizes 168*
formation of vacuole EM
   Endocytotic 125
   Pinocytotic 125
freeze-fracture electron microscopy (see
      FFEM) 64, 65
freezing point depression
   determination of osmotic value by 27
freezing–thawing, isolation of tonoplast by
   23
fructan
   metabolism 191
   storage 191
fructose 88
   full cell sap 85
fungi 42, 59
fusion 136*
   docking for 135
   mechanism of 135
   of membranes 150
   of vacuole 44, 50, 57
   of vesicles 150
   priming for 135
   stages
      docking 150
      priming 150
      tethering 150
      triggering 150
   tethering for 135

vacuole 3

gadolinium sensitive Ca^{2+} channel 185
garbage pathway 147
gas vacuoles 38, 59
genes
    for alkaloid
        biosynthesis 262
        transport 262
    for Ca-antiporter 261
    for Calmodulin Ca^{2+}-ATPase 261
    for Cd-resistance 263
    for GS-X pumps 263
    for V-ATPase 261
genetic engineering
    aquaporin 266
    tonoplast 259, 266
genetic manipulation
    detoxification 263
    phytoremediation 263
gentianose 88
GERL 123*, 149
    concept 124
gerontoplast 213
gibberelline
    hormones 90
globulin
    storage proteins 81, 103
glossopodium 46
gluconastartin 93
glucose 88
    transport 190
glucosinolates 93
glucosyl transferase 214
glutathione, *see* GSH 214
    conjugation 215
    -S-transferase, *see* GST 214
glutelin 103
    storage proteins 81
glycolipid, tonoplast 67
glycoproteins, tonoplast 67
glycosides 88
glycosylation in vacuole 191
Golgi
    bodies, origin from, EM 125*
    stain 2
    –vesicle–vacuole, EM 126
Gomori's method, cytochemical localisation
    13, 30
gross functions of vacuoles 220
group translocation

    of hexose 189
    of sucrose 189
GS
    conjugate pumps 215
    conjugate transporter 217
    -transferase 219
    -X pumps 215, 216, 219
GSH 214
GST 214
GTP binding protein 132, 136
GTPase 129, 132, 133
gymnosperm 47, 59
gypsum crystals, vacuole 1

H^+
    ATPase 68, 70
    -F-ATPase 70
    -P-ATPase 70, 71*
    -V-ATPase 70
halophyte 176
haustorium 54
herbicide 215
    detoxification 216
heterophagosome 204
heterophagy 201, 223, 250
    by lysosomes 257
    types of 201
hexose 88
homeostasis
    of amino acids 222
    of molecules 250
homogenisation, isolation of tonoplast by 23
homology of tonoplast with
    chloroplast envelope 255
    ER 255, 266
    glyoxisome 256
    Golgi body 256
    Golgi membrane 266
    Golgi-cisternae 255
    lysosome 256
    mitochondrial envelope 255
    nuclear membrane 255
    peroxisome 256
    plasmamembrane 266
    single membraned organelles 256
homology with other organelles 252
hordein 103
    storage proteins 81
hormones 104
    abscisic acid 90
    auxin 90

cytokinin 90
    dihydrozeatin 90
    ethylene precursor 90
    gibberelline 90
host-pathogen interactions 202, 223
HR 206, 223
hydraulic waterflow 165
hydrolases
    in differentiation 224
    in senescence 212
    in vacuoles 196
    released as HR 207*
hydrolysing enzymes 104
hydrolytic activity 196
    in aleurone grains 122
    in protein bodies 122
hypersensitive reaction, see HR 206

immunohistochemical EM of VPE 194
IMP 64, 65
IMP-tetrad 65, 66*
in bulk, intact vacuole 17, 18
    increase vacuolar volume 58*
independent sorting of membrane and lumen protein 139
indole alkaloid biosynthesis in vacuole 195*
infection thread debris in vacuole 205*
inheritance of vacuole in yeast 128*
    inhibitor, chymotrypsin 103
inorganic ions 104
    in vacuoles 90
inorganic phosphate, see Pi 172
inositol lipid in vesicle transport 134
intact vacuole
    by sand grinding 17
    from compact tissues 18
    from leaf tissues 19, 20
    from patch-clamp recording 20
    from stomatal guard cells 20
    from suspension cells 19
    in bulk 17, 18
    isolation 17
    yeast cells 22
integral proteins, see also TIP
    membrane (see MIP)
    tonoplast 73
internalisation of pathogens 202
intramembranous particles (see IMP)
    tonoplast 64, 65
intravacuolar abiological mass 203*

inulin 83
    metabolism in vacuoles 191
invagination
    of damaged cytoplasm in vacuole 203*
    of tonoplast 198
ion
    channels 163
        role in physiological process 173
    compartmentation 177
    fluxes 168
        active 169*
        in freshwater alga 169*
        in higher plants 169*
        in sea water alga 169*
    transport, general pattern 166
    transporters 163
    uptake
        feedback inhibition, 166
        general considerations 167
        two systems 166
ionic concentration
    comparison of cytoplasm and sap 91*
    comparison of salt water, fresh water and cell sap 91*
    comparison of sap and external solution 91*
IP_3-sensitive Ca^{2+} channel 185
isolation
    intact vacuole 17
    of tonoplast 22
        by direct method 23
        by freezing-thawing 23
        by homogenisation 23
        by osmotic shock 23
    of macromolecules 30
isoprenoids 84
    ITPase, electron microscopy 16

K^+, Ca^{2+} release from guard cell 181
K^+/H^+ antiport 213

latex vessels 84
laticifer 84
    development 210*
leaf tissues
    intact vacuole from 19, 20
lectin 68, 93, 147
lectin-aspartic proteinase localisation 194*
light microscopy
    Arylsulfatase 16

non-specific esterase 16
acid phosphatase 15
linamarin 88
lipid
   bodies 83, 120*
   transfer proteins (LTPs) see LTPs 117
lipids 103
   insertion of 117
   insertion in tonoplast 117
   tonoplast 67
   transfer to tonoplast 117
lipoxygenase 83, 103
   localisation of proteinase inhibitor 82*
LTPs 117
   non-specific (nsLTP) 117
lumen protein
   sorting 139, 140
   trafficking of 130*
LYCH uptake 201
lysis 209
lysosome
   concept 257
   definition 258
   in lower plants 258
   in plants 257
   isolation in plants 257
   latency 257
   vacuolar homology with 266
lytic
   activity 210, 211
      in differentialtion 208
      in enclosed cytoplasm 123
   functions 196, 223
   origin of vacuole 122
   process 149, 149

macroautophagy 199
major facilitator super family of transporter 190
malate
   exchange 174
   movement through ion channel 187
   transporter 186
malic acid
   efflux 186
   influx 186
malic-acid permease in CAM plant 187
mALP 141
manipulation
   for herbicide resistance 263

   for protein targeting 264
   of PR proteins 264
   of senescence genes 262
mannitol 88
mannose-6-phosphate receptor, see MPRs 134
mAPI 141
   markers, tonoplast 25, 30
mating pheromone 138
MCB 216
mCPS 141
mCPY 141
mechanism of *Mimosa* pulvinar movements
   role of aquaporin 183
   role of V-ATPase 183
membrane
   and luminal proteins to Golgi 141*
   integral proteins (see MIP)
   fusion 136*
   phospholipid constitution 177
   potentials 168*
   protein 140
   protein sorting 139
   trafficking 139, 140*
meristem 47
metabolic functions 222
metabolites
   in isolated vacuole 92*
   relative amount of in sap 92*
metal salt technique 13
Mg-ATPase 170
   microelectrodes, determination of pH by 26
micromanipulation 4
microscopy
   electron 30
   fluorescence 30, 9
   proteinase inhibitor, electron 82*
   proteinase inhibitor, light 82*
microspore 50
   mother cell 50*
   vacuolation of 50
Mimosa 183
MIP 68
model
   Cr-Phytoremediation 218*
   protein delivery in yeast 138*
   tonoplast orientation 254*
molecular
   chaperon 148

features of transport 133
  mechanism of vesicular traffic 133
morphine 88
MPR 139, 134
MRP 215
MRP genes for stress regulation 263
multidrug resistance associated proteins 215
multifaceted organelle 252
multiple delivery
  mechanism 143
  single targeting 143

Na$^+$/H$^+$
  antiport 177
  exchange 177
  in glycophyte 177
  in halophyte 177
nectary 54
N-ethyl maleimide-sensitive factor, see NSF 131
neurotoxin 89
  neutral lipid tonoplast 67
neutral red staining 5
Ni$^+$/H$^+$ antiport 218
nickel detoxification 217
non-specific entry for degradation 149
non-specific LTPs (nsLTP) 117
  in vacuole 118*
NSF 131, 150
  attachment proteins (SNAPs), see SNAP 131
N-terminal propeptide see NTPP 145*
NTPP 145*, 146, 134
nucleolytic activity 197

oil-secreting cell 56
oleosomes 121
one hundred 'somes' 258
organelle breakdown 198
  definition 249
  degeneration of 210*
organic acids 104
  ascorbic 89
  citric 89
  isocitric 89
  malic 89
  oxalic 89
  tartaric 89
  origin

from trans-Golgi network (TGN), see TGN127
  of tannin vacuole 124
osmoregulation 177, 221
  aquaporin 178
  by conversion
    malate starch 179
    sugar-starch 179
  by ion transport 179
  by sugars 178
  in halophytes 180
  mechanism 178
  role of Cl$^-$ 180
  role of K$^+$ 180
  role of Pi 180
osmosis 2
osmotic
  lysis 19
  pressure 27, 177
  shock 22
    isolation of tonoplast by 23
oxalate crystals 5
oxidative burst 207*
oxide-reductases, role in pH regulation 176

particle, ATPase 65
particular inclusions 103
partitioning
  vacuole/extravacuole 92*, 94
passive diffusion 167
patch-clamp
  recording from intact vacuole 20
  technique 29
pathogenesis related (PR) proteins 204
pathways
  ALP 151
  CPY 151
  Cvt 151
P-ATPase
  H$^+$ ATPase 70, 71*
PBM 139
  of *Rhizobium* 139
PC-Cd complex gene for vacuolar delivery 263
perfusion technique 28, 174
peribacteroid membrane see PBM 139
peripheral probe
  tonoplast 73
permease 68, 176, 221
peroxisome 199

pH homeostasis, role of phosphate 175*
pH regulation 173, 176, 221
phases of detoxification
    activation 214
    conjugation 214
    degradation 214
    elimination 214
    Substitution 214
pHc 175
phenolics 103
    dissolved in sap 85*
pheromone alpha factor degradation 138*, 197
phloem 48
phosphatidylinositol-3-phosphate see PtdIns(3)P 134
    phospholipid
    tonoplast 67
    translocator 117
phospholipidase A, B 119
pHv 173, 174
pHv-pHc correlation 175
physiological methods 25, 30
phytate 103
phytoalexins 223
    as defense arsenal 208
    release of 208
phytochelatin 217
phytoremediation
    genetic manipulation for 263
    of heavy metal 216, 224
    terminal point
        by glutathiane peptide 219
        by phytate 219
        by phytochelatins 219
phytosiderophore 218
Pi 172
Pi homeostasis 172
    pinocytotic vacuole 45
plasma membrane, see also PM
    tonoplast homology 253
    origin from 125*
plasmalemmasomes 201
plasmolysis 2
    determination of osmotic value by 27
PM 66, 253
PMF 170, 172
PM-tonoplast reverse orientation 254*
PM-tonoplast similarity
    aquaporin 253, 254*

chemical composition 253
electron staining 253
in electrical properties
    presence of anion channel 254*
    presence of proton pump 254*
    presence of transporters 254*
P-NMR
    determination of pH by 27
pollen
    grain vacuolation of 50
    tube vacuolation of 50
polyphenols
    dissolved in sap 86
polyphosphate 84, 103
polyterpenes 84
    pools of amino acids 90, 104
porins 163
positive signals 143
PPase 169, 261
    characteristics 170
    tonoplast 25
    tonoplast localisation 170
    vacuolar type 170*
PR proteins, types of 204
PrA 140, 142
prespore vacuole 45
    pressure bomb technique, determination of osmotic value by 28
    pressure probe technique, determination of osmotic value by 28
prevacuole 50, 134, 137, 140, 142
    as intermediate structures 137*
primary active transport 169, 221
primary lysosomes 123
priming 136*
    complex 129
primordial utricle 48
processing
    proconcanavalin A 193
    proglycin 193
    proproteins 193
    proricin 193
    storage proteins 193
    sugars 191
pro-CPY 144
production of hydrolases as defense response 204
proembryo 54, 55*
Prolamin 81, 103
prospects of tonoplast engineering 265

proteolysis of mutant proteins 197
proteolytic processing 144
protection of nucleus 251
protein
    crystalline 103
    insertion in tonoplast 117
    solid 103
    storage vacuole (PSV) see PSV 116, 117
    transfer to tonoplast 119
proteinase 68
    inhibitor 83
        localisation of 82*
proteinase A, see PrA 140
proteolytic activity 196
proton motive force, see PMF 170
proton pumps
    H^+-ATPase 169
    H^+-PPase 169
protopine 88
provacuolar synthesis
    alkaloid 125
    protoberberine alkaloids 125
    role of 138
    tannin 125
provacuole 124
    acid phosphatase 122
    enzyme accumulation in 122
    fusion of 122
    fusion, origin of vacuole 126
    role of 138
PR-proteins 223, 250
PSV origin 121
PtdIns(3)P 134, 135
pteridophyta 59
    pulsule, types of 41, 42
pulvinar movement 180
pulvini in leaves of Leguminosae 182
pulvinus 57
pusule 41
pyrophosphatase 68

quality control 147
quantitative estimation
    of vacuolar enzymes 94*

Rab protein 132, 133
    in yeast 133
    raphides, vacuolar solids 80*
regeneration of tonoplast 117
regulation of

ATPase 174
pHv 174
PPase 174
relative amount of metabolites in sap 92*
release of Ca^{2+} 186
resin 84, 103
    resonance frequency technique,
        determination of osmotic value by 28
retrograde traffic of vesicle 131
reverse orientation of tonoplast 255
    reversible change of vacuolar volume 57
reversible transformation
    ER to vacuole 125
    vacuole to ER 125
Rhizobium in vacuole 203
Ricin 103
RIP (Ribosome-inactivating protein) 83, 103
root
    cap 54
    nodule 203
    nodule vacuoles 205*
rubber 84, 103
ruthenium red 12

S-adenosyl methionine 93
salt gland 56
salt tolerance 221
    mechanism 176
sanguinarine 88
    sand grinding, intact vacuole by 17
    sap, full cell 85
saponin 93
saporin 103
scavenging
    of AOS 219, 224
    undesirable radicals 264
scutellum 54, 55*
secondary active transport 164, 170*, 221
secretion 138
    secretory cell 57
seismonastic movements in
    in Mimosa pudica 182
    in Robinia pseudoacacia 182
senescence of mesophyll 212*, 213
SER 117
    shape of vacuole 56
shoot meristem 54
sieve element 48
signal

for alpha-mannosidase 147
for ALP 147
for CPS 147
for DPAP-A 147
for DPAP-B 147
for chitinase A 146
peptide, absence of 147
lectin 147
signal peptides
for lectin 145, 146
for legumin 145, 146
for vacuolar targeting 145
signal transduction
by annexin 186
by Ca^{2+} 186
by CaM 186
sinigrin 93
smooth endoplasmic reticulum (SER) see SER 117
SNAP-25 131, 143, 150
SNARE
complex 132, 133
hypothesis 131, 132
hypothesis for vesicular protein transport 150
hypothesis in yeast 133
mechanism and pathways 143
mechanism in plants 133
pairing 135
pin 137
proteins 131
SO_2 pollution 213
solid inclusions, crystals 79
sorting signals 144, 151
absence of 147
for proteins 143
specialised cell 60
specificity for targeting vesicles 131
sphaeraphides, vacuolar solids 80*
spherical crystals, vacuolar contents 87*
spherical shape of vacuole 57
spheroplast 22
spherosome 83
SPS 189
stachyose 88
uptake 189
stages
of autophagy 199
of fusion 136*

of vacuole formation by fluid phase endocytosis 255
staining of vacuole 8, 9
steroids 88
stomatal guard cells, intact vacuole from 20
stomatal opening 180
storage in vacuole 250
storage proteins
albumin 81
globulin 81, 103
glutelin 81
hordein 81
vicilin 81
zein 81
strictosidine-beta-glucosidase (SG) on tonoplast 195
subcellular lysosome-like particles 258
subunit composition, V-ATPase 70, 71*
Suc/H^+ antiport
in tonoplast 189
for sucrose transport 189
sucrose 88
accumulation 188, 222
transport 188
synthesis 189
sucrose-phosphate-synthase, see SPS 189
sugars 88, 103
sulfate accumulation 213
sulfur 213
sulfuric acid, in *Desmarestia* 213
superfast cascade for motor cells 183
surface area of tonoplast 57
surface/dry weight ratio 249
suspension culture 56
suspension cells, intact vacuole from 19
suspensor cell 54, 55*
SV channels 181
symplastic route for sucrose 190
symport 171, 221
synergid 52*

tannin 46, 103
vacuole 56
dissolved in sap 85*
in rapid pulvinus 183
in slow pulvinus 183
in vacuole sap
-dye accumulation 3
targeting mechanism 143

targeting proteins for
   exoplasmic space 264
   secretion 264
   vacuole 264
TE 209
terminal propeptides 145*
tethering 136*
TGN 127, 129, 139
   to PVC 142
TIP 69*
tissues
   reproductive 50
   vascular 48
tonoplast 3, 66*, 115
   asymmetry 64
   ATPase, insensitive to vanadate 169
   channels
      fast-vacuolar (FV) 173
      K^+-vacuolar (VK) 173
      slow-vacuolar (SV) 173
   chemical analysis 67
   chemical composition 73, 64
   definition 249
   densitometry 64
   direct extraction of 30
      by continuous gradient 24
      by discontinuous gradient 24
   engineering
      against water stress 260
      for mobilisation and retrieval 262
      for processing 262
      for storage 262
      for water 260
      limitations 260
   enzymes 67, 70, 171
   fast growth of 149
   genes for herbicide detoxification 263
   genetic engineering of 259, 266
   glycolipid 67
   glycoproteins 67
   growth 116
   homology with plasma membrane 252
   integral proteins 73
   intramembranous particles (see IMP) 64, 65
   isolation of 22
      by direct method 23
      by freezing-thawing 23
      by homogenisation 23
      by osmotic shock 23

   lipids 67
   markers 25, 30
   neutral lipid 67
   peripheral probe 73
   permeability 178
   phospholipid 67
   PPase 25
   proteins 67
   regeneration 117
   surface area of 57
   tensile strength of 178
   transport processes through 171*
   ultrastructure 64
   V-ATPase 25
Torri-Laties hypothesis 166
trabeculae 46
tracheids 49
tracheory elements (see TE) 209*
trafficking of vesicles
   anterograde 131
   between ER, Golgi, vacuole 130*
   modes of 131
   of proteins 130*
   retrograde 131
transdifferentialtion
   of mesophyll to TE 209*
   of TE, stages of 209
trans-Golgi network (TGN) see TGN
   origin from 127
transport
   active 163
   apoplastic 165, 166, 167
   of molecules across membranes 163
   of molecules, types of 220
   passive 163
   pathways for proteins 151
   pathways for proteins from Golgi to vacuole 141*
   proteins 163
   symplastic 165, 166, 167
   transcellular 165
triggering 136*
t-SNARE 132, 150
   in plants 134
TTPase
   electron microscopy 16
tubular vacuole 117
tunica 48
turgor
   for growth 178

maintenance 249
movement 180, 222
    changes in central vacuole 182*
    changes in tannin vacuole 182*
    in guard cell 181*
    in pulvinal motor cell 182*
pressure 177, 178
role of 249

UDPGlc 214
    uptake 190
    in sucrose synthesis 189
ultrastructural interrelationship 126
unidirectional transport
    anterograde 150
    retrograde 150
uniport 171, 221
uniqueness of vacuole 266
uptake
    of anions 172, 221
    of ions 221
    of sugars 190
    of water 164
uridine diphosphate glucose, see UDPGlc 189

V_0 domain, ATPase 67
V_1 domain, ATPase 67
V-ATPase, tonoplast 25
v-, t SNARES
    anchorage of 135
    binding of 135
vac mutants 137
vacuolar
    contents
        anthocyanin 103 87*
        anthocyanin pigment 85*
        brick red pigment 87*
        colloidal bodies 87*
        colloidal content 85*
        crystalline needles 87*
        dark violet pigment 87*
        significance of 265
        spherical crystals 87*
    enzymes
        in algae 95*
        in dicots 99*
        in ferns 96*
        in fungi 95*, 96*
        in gymnosperm 96*, 97*
        in monocots 97*, 98*, 99*

        in myxamoeba 96*
        in phytoflagellate 95*
        in yeast 95*
        quantitative estimation of 94*
    homology with
        glyoxisome 266
        lysosome 266
        peroxisome 266
        spherosome 266
    hydrolysis in TE formation 209*
    origin from membranous components 124
    perfusion 197
    peroxidase 220
    pH see pHv 173
    processing 222
    processing enzyme, see VPE 193
    solids
        calcium carbonate crystals 81*
        calcium oxalate crystals 80*
        crystal sand 80*
        cystolith 81, 80*
        rahpides 80*
        sphaeraphides 80*
    sorting receptor 134
    volume
        change in 58*
        reversible change of 57
        variation of 57
vacuolar/cytoplasmic concentration of amino acids 188
vacuolation
    by accumulation of biosynthesis-150
    by lipid utilisation 120*
    by SER synthesis of alkaloids 150
    by SER synthesis of tannin 150
    ER in 123
    from Golgi 150
    from plasma membrane 150
    lipid utilisation for 119
    of microspore 50
    of pollen grain 50
    of pollen tube 50
    protein bodies in 121
vacuole
    as amino acid reservoir 187
    as extracellular compartment within the cell 266
    as fortification 252
    as indispensible 266

as multifaceted 266
biosynthesis inside 250
breakdown and formation 2
buffering capacity of 175
contents 1
    nitrate 1
    phosphate 1
    salts of Na, K, Ca, Mg 1
    Sulfate 1
    tannin 1
definition 249
development 115, 116
discovery 1
division 128*
evolution of 255
flavonoids in 219
for house-keeping 252
for storage retrieval 252
formation of 123*
gypsum crystals 1
indispensibility of 251
integrity of 251
irreversible formation of 251
number of 56
origin of 255
protein targeting, see vpt 137
protoplast fusion 253
rigidity of 249
segregation 128
significance of 249
terminal of phytoremediation 219
uniqueness of 251
vacuole/extravacuole partitioning 92*, 94
vacuome 3
vapor pressure equilibration
    determination of osmotic value by 28
    variation of vacuolar volume 57
vascular tissues 48
V-ATPase 49, 73*, 142
    H^+ ATPase 70
    H^+ ATPase 221
    in vacuolar malate transport 186
    proteins 70
    subunit composition 70, 71*
vesicle
    budding, model for mechanism 129

formation
    at Golgi 131
    generalised concept 135
    stages of 150
fusion 132, 136*
transport 128*
transport in yeast 132
tubular clusters, see VTC 131
vessels 49
    vicilin, storage proteins 81
VK channels 181
volume of vacuole 56, 57
volutin 84, 103
VPE, activity of 193
vps mutants 137
VPS-dependent pathway 142
vpt mutant 137
v-SNARE 132, 150
v-SNARE in plants 134
VTC 131, 132, 133
VTC in plants 132

waste disposal in vacuole 250
water expulsion vacuole (see WEV)
water relations 2
weak acid/base distribution
    determination of pH by 25
WEV 42, 43*, 59
whole body transport, of ions 166

xenobiotic 214, 215
    removal from cytosol 214
xylem 48

yeast 187
    cells, intact vacuole 22
Ypt protein 132, 133

zein 81, 103
zinc 216
    accumulation 216
    hyperaccumulator 216
    in vacuole 217
    phytoremediation of 216
    storage 217
    tolerance 216